TREATISE ON ANALYTICAL CHEMISTRY

A comprehensive account in three parts

PART I

THEORY AND PRACTICE

PART II

ANALYTICAL CHEMISTRY OF INORGANIC
AND ORGANIC COMPOUNDS

PART III

ANALYTICAL CHEMISTRY IN INDUSTRY

TREATISE ON
ANALYTICAL CHEMISTRY

PART I

THEORY AND PRACTICE

VOLUME 10

Edited by I. M. KOLTHOFF
School of Chemistry, University of Minnesota

and PHILIP J. ELVING
Department of Chemistry, University of Michigan

with the assistance of ERNEST B. SANDELL
School of Chemistry, University of Minnesota

WILEY–INTERSCIENCE

a division of John Wiley & Sons, Inc., New York–London–Sydney–Toronto

TREATISE ON ANALYTICAL CHEMISTRY

PART I
THEORY AND PRACTICE

VOLUME 10:

SECTION E (*Continued*)

Application of Measurements

Chapters 102–103

SECTION F

Principles of Instrumentation

Chapters 104–105

SECTION G
Preparation for Analytical Research
and Utilization

Chapters 106–107

AUTHORS OF VOLUME 10

CHESTER C. CARSON
WILLIAM S. HORTON
RICHARD KIESELBACH
C. D. LEWIS
M. G. MELLON

GORDON D.
    PATTERSON, JR.
JOHN A. SCHMIT
FRED H. TINGEY

Authors of Volume 10

Chester C. Carson

General Electric Company, Schenectady, New York, Chapter 103

William S. Horton

National Bureau of Standards, Washington, D.C., Chapter 103

C. D. Lewis

Plastics Department, E. I. du Pont de Nemours & Company, Wilmington, Delaware, Chapter 105

M. G. Mellon

Department of Chemistry, Purdue University, Lafayette, Indiana, Chapter 107

Gordon D. Patterson, Jr.

*E. I. du Pont de Nemours
& Company, Wilmington,
Delaware,* Chapter 104

John A. Schmit

*Instrument Products Division, E. I. du Pont de
Nemours & Company,
Wilmington, Delaware,*
Chapter 102

Fred H. Tingey

Idaho Nuclear Corporation, Idaho Falls, Idaho,
Chapter 106

PART I. THEORY AND PRACTICE

CONTENTS—VOLUME 10

SECTION E: Application of Measurements (*continued*)

103. Gas Analysis: Determination of Gases in Metals. By *William S. Horton* and *Chester C. Carson*...................... 6017

SECTION F: Principles of Instrumentation

104. General Concepts of Instrumentation for Chemical Analysis.

105. Continuous Automatic Instrumentation for Process Applications. By *C. D. Lewis*

SECTION G: Preparation for Analytical Research and Utilization

106. Design of Experiments in Analytical Chemistry Investigation.
By Fred H. Tingey................................... 6405

TREATISE ON ANALYTICAL CHEMISTRY

A comprehensive account in three parts

PART I

THEORY AND PRACTICE

PART II

ANALYTICAL CHEMISTRY OF INORGANIC
AND ORGANIC COMPOUNDS

PART III

ANALYTICAL CHEMISTRY IN INDUSTRY

SECTION E: Application of Measurements
(*Continued*)

Part I
Section E

Chapter 102

GAS ANALYSIS:
THERMAL CONDUCTIVITY

By Richard Kieselbach,* *Engineering Physics Laboratory*, and John A. Schmit, *Instrument Products Division, E. I. du Pont de Nemours & Company, Wilmington, Delaware*

Contents

* Deceased.

I. INTRODUCTION

A versatile instrumental technique applicable to many gas analysis problems consists of the measurement of thermal conductivity. The technique is based on the fact that the thermal conductivity of a mixture of two gases assumes a value intermediate between the conductivities of the pure components of the mixture, in a manner related to their relative concentrations. In most cases the relationship is approximately linear over moderate ranges of concentration. The measurement is solely quantitative, requiring advance qualitative knowledge of the gas composition. The method is particularly attractive because of the relative simplicity of the required apparatus, which produces an electrical output signal that can be directly calibrated and recorded in terms of sample concentration. Concentrations from the percentage range to, in some cases, parts per million can be measured with good precision with conventional equipment; refinements can extend the sensitivity of the equipment into the parts-per-billion range. The apparatus is particularly well suited to the continuous analysis of flowing gas streams.

Although thermal conductivity analysis is applied most unequivocally to binary gas mixtures, it can be, and frequently is, used in applications where only one component of a multicomponent mixture varies in concentration, or where the thermal conductivity of one component is markedly different from that of the other components of the mixture.

Perhaps the greatest utility of thermal conductivity analysis lies in the differential detection of a component or components of a gas mixture by means of an auxiliary chemical or physical separation process. The concentration of acidic gases, for example, can be measured by comparing the thermal conductivity of a sample with that of the same sample after treatment with an alkaline absorbent. This approach permits the ingenuity of the analytical chemist to extend the applicability of the technique almost without limit.

Historically, thermal conductivity analysis has been in use for about 50 years, primarily for the analysis of flue and automobile exhaust gases and certain industrial process streams. The method was first demonstrated as a practical analytical technique in 1908 by Koepsel (62–65). Shakespear (96,111) developed the first commercial thermal conductivity cell outside of Germany, calling it the "katharometer." The technique was first reviewed in the United States by Palmer and Weaver (87). A comprehensive textbook on the subject was prepared by Daynes (22) in 1933, and the subject was again reviewed by Weaver (129) in 1951. The last two publications are especially recommended to the serious student of this subject. With the recent advent of gas chromatography, use of thermal conductivity analysis has expanded considerably, because of its nearly ideal qualifications as a detector for the binary gas mixtures produced by that separation technique. Considerable effort has been devoted to the refinement of apparatus for this application and to the quantitative study of its performance characteristics. There has resulted a significant advance in the utility of the technique over the past 5–10 years.

II. PRINCIPLES

A. BASIS OF THE TECHNIQUE

1. Heat Transfer through Gases

Heat can be transferred through a gas by three fundamental mechanisms: conduction, convection, and radiation. In conduction, the kinetic energy of the molecules of the higher-temperature gas is transferred by collision to molecules of lower kinetic energy, that is, those of the cooler portion of the gas. Convection involves the bulk transfer of masses of heated or cooled gas, as, for example, in a chimney. Radiation involves the conversion of heat to electromagnetic energy, which can travel through a gas from one body to another not in direct contact with the gas.

In general, an instrument intended for analytical applications of thermal conductivity measurement is designed and used so as to minimize or eliminate, the effects of convection and radiation. (There are cases, however, in which the effects of convection can be used to advantage. This subject is discussed in Section II.F.1.a.)

2. Measurement of Thermal Conductivity

a. Direct Measurement

If two parallel plates of equal area are separated by a stagnant layer of gas and are held at different temperatures, T_1 and T_2, the Fourier expression for heat flow through the gas under steady-state conditions can be written as

$$q = \frac{kA(T_1 - T_2)}{d} \tag{1}$$

where q is the quantity of heat transferred in unit time, k is the thermal conductivity of the gas, A is the area of the plates, and d is the distance between the plates.

In practice, it is more convenient to measure the thermal conductivity of gases with apparatus of a concentric cylindrical form than with parallel plates. In one form, this apparatus consists of an electrically heated fine wire (e.g., platinum), stretched axially in a hole bored in a metal block. The heat flow equation for this configuration can be derived from equation (1) (51) and is

$$q = \frac{2\pi kL(T_r - T_w)}{\ln (r_w/r_r)} \tag{2}$$

where L is the length of the wire, r_r and r_w are the radii of the wire and the cell wall, respectively, and T_r and T_w are the temperatures of wire and wall, respectively.

The heat flow is directly related to the electrical power dissipated in the wire, which can be readily measured as a function of the voltage and current, and the temperature of the wire is readily measured as a function of its electrical resistance. The temperature of the metal block can easily be controlled and measured. In principle, therefore, the measurement of the absolute thermal conductivity of a gas would appear to be a relatively simple matter, given a knowledge of apparatus dimensions.

However, the precision with which such an absolute measurement can be made is far from adequate for most analytical purposes. Useful analytical data can be obtained from temperature changes of less than 10^{-5}°C in thermal conductivity measurements, but it is apparent that absolute measurements of such small quantities would be extremely difficult. Therefore analytical thermal conductivity almost invariably involves a differential measurement, in which the thermal conductivity of one gas is compared with that of another in a single apparatus. The reader concerned with the absolute measurement of thermal conductivity is referred to the critical treatment of this subject by Daynes (22), and to an extensive body of literature which is beyond the scope and intent of this chapter.

b. Differential Measurement

The principle of differential measurement is illustrated schematically in Fig. 102.1. Two filaments of matched dimensions and resistance are mounted in similar cavities in the cell block. The filaments are connected as adjacent arms in a Wheatstone bridge, the other two arms of which comprise fixed resistors. The bridge contains a potentiometer for balancing any slight inequalities in the bridge arms. Sufficient potential is applied to the bridge to heat the filaments, and bridge unbalance is detected with the meter. Initial "zeroing" of the instrument is accomplished by filling both cavities of the cell with the same gas and adjusting the potentiometer for zero bridge output. Then, when different gases are placed in the cavities, the bridge output becomes a measure of the difference in their thermal conductivities. The effect of cell block temperature variations is reduced by the symmetry of the apparatus to a second-order consideration.

For analytical purposes, there is no need to compute the actual difference in thermal conductivities measured by this procedure. Indeed, the

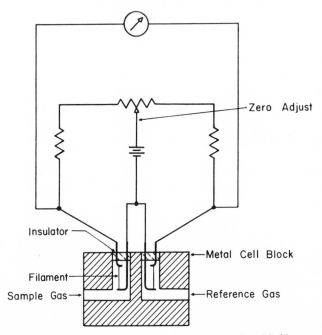

Fig. 102.1. Simple thermal conductivity cell and bridge.

value of such a computation is highly suspect for a variety of reasons that will become apparent in the following discussion. Instead, the usual practice involves the direct empirical calibration of the bridge output voltage in terms of concentration. This is accomplished by maintaining a gas of fixed composition (called the reference gas) in one cavity of the cell and noting the bridge output produced by various known mixtures of sample and reference gas placed in the other cavity of the cell.

For many simple applications of thermal conductivity analysis, no further technical information beyond that already presented, except in regard to the selection of apparatus is necessary to a successful application. Commercially available apparatus is well adapted to the try-it-and-see approach, and several instrument manufacturers will furnish complete apparatus, calibrated for a specific analysis. For other than the simplest cases, however, a more detailed consideration of the physical–chemical and instrumental factors involved is desirable, both to appreciate the scope and limitations of the technique and to avoid a number of pitfalls not immediately obvious. These factors are discussed in the following sections.

B. THERMAL CONDUCTIVITY OF GASES

1. Thermal Conductivity Table

There are a great many published data on the absolute values of thermal conductivity of various gases and vapors. It would appear at first thought that a tabulation of such data would be of great practical value in the prediction of instrumental response for analytical purposes. Unfortunately this is not the case. One reason for this apparent anomaly lies in the fact that a variety of errors which must be corrected in absolute measurements are automatically eliminated in an empirical analytical calibration. Consequently, the precision of the absolute data is far less than that of the analytical measurement in which they might be used.

In a well-designed thermal conductivity cell, a small amount of heat, 1–4% of the total (90,129), is lost by convection, radiation, and conduction through the terminals of the heated wire. This error, which must be evaluated in each absolute measurement, is of little practical significance for analysis. The thermal conductivity of a gas varies with its temperature, so that an absolute measurement must involve the evaluation of an appropriate mean temperature between wire and cell wall. The error from this source is minimized by use of a small temperature difference between wire and wall. For analytical purposes, however, a relatively large temperature difference is ordinarily desirable, so that data obtained under the two conditions are not directly comparable. Thermal diffusion results in a concentration of the lighter component of a gas mixture near the wire, where its presence produces a disproportionate effect on the conductivity measured (22). Thus the quantity measured in the analysis of a mixture is not representative of that measured on a single gas. In addition to these discrepancies between absolute and analytical measurements, a far more serious effect can lie in the nonadditivity of thermal conductivities of gas mixtures. This subject is dealt with in Section II.2.C.

A table of thermal conductivities has been compiled by Weaver (129) and is reproduced in Table 102.I. It includes "best values" selected by Laby and Nelson (69) and Daynes (22) and values interpolated or extrapolated by Weaver to a common reference temperature. As Weaver emphasizes, the data should be considered only as guideposts, because of a variety of assumptions made in their selection. The student concerned with absolute values should consult the original sources.

In view of the above discussion, it is clear that the table is of value to the analyst only in a qualitative sense. The data indicate, for example,

TABLE 102.I
Thermal Conductivity of Gases[a]

$$k_0 \times 10^5$$

Gas	Thermal conductivity, cal cm^{-1} sec^{-1} °C^{-1}	$R_0{}^b$	$R_{100}{}^b$	References
Inorganic gases				
Air	5.8	1.00	1.00	22,23,39,40,57,69,75, 79,97,110,130
Hydrogen	41.6	7.15	7.10	3,22,23,48,56,57,69, 83,84,110,130
Deuterium	34.0	5.85	. . .	3,56,83,84,105
Helium	34.8	5.97	5.53	18,22,23,57,69
Nitrogen	5.8	1.00	1.00	22,23,41,48,69,110
Oxygen	5.9	1.01	1.01	22,23,41,69,83,84, 110
Neon	11.1	1.90	1.84	18,22,57,69,130
Chlorine	1.9	0.32	. . .	22
Argon	4.0	0.68	0.70	18,22,23,48,57,69, 104
Krypton	2.1	0.36	. . .	14,18,19
Xenon	1.2	0.21	. . .	18
Carbon monoxide	5.6	0.96	0.96	22,23,48,57,69,104
Carbon disulfide	3.7	0.29	. . .	22
Carbon dioxide	3.5	0.61	0.70	22,23,48,57,66,69,97, 110
Water	. . .	. . .	0.78	22
Nitric oxide	5.7	0.98	. . .	22
Hydrogen sulfide	3.1	0.54	. . .	22
Ammonia	5.2	0.90	1.04	22,23,66,69
Sulfur dioxide	2.0	0.35	. . .	22,23
Hydrocarbons				
Methane	7.2	1.25	1.45	22,69,75,104
Deuteromethane	7.1	1.22	. . .	105
Ethane	4.4	0.75	0.97	22,69,75
Deuteroethane	3.7	0.64	. . .	105
Propane	3.6	0.62	0.83	75, 104
Deuteropropane	3.1	0.52	. . .	105
n-Butane	3.2	0.55	0.74	75
Isobutane	3.3	0.57	0.78	75
n-Pentane	3.1	0.54	0.70	22,75
Isopentane	3.0	0.52	. . .	22
n-Hexane	3.0	0.51	0.66	22,75
n-Heptane	. . .	. . .	0.58	22
Cyclohexane	. . .	. . .	0.58[c]	22
n-Hexylene	2.5	0.43	0.62	22
Ethylene	4.2	0.72	0.98	22,69
Acetylene	4.5	0.78	0.90	22
Benzene	. . .	0.37	0.58	22

TABLE 102.I (*continued*)

Gas	Thermal conductivity, cal cm^{-1} sec^{-1} °C^{-1}	$R_0{}^b$	$R_{100}{}^b$	References
Halogen compounds				
Carbon tetrachloride	. . .	. . .	0.29	22
Chloroform	1.6	0.27	0.33	22
Dichloromethane	1.6	0.28	0.36	22
Methyl chloride	2.2	0.38	0.53	22
Methyl bromide	1.5	0.26	0.35	22
Methyl iodide	1.1	0.19	0.25	22
Difluoromethane	2.0	0.34	. . .	97
Ethyl chloride	2.3	0.39	0.54	22
Ethyl bromide	. . .	0.30^b	. . .	22
Ethyl iodide	. . .	0.24^b	. . .	22
Alcohols				
Methyl alcohol	3.5	0.59	0.73	22
Ethyl alcohol	. . .	. . .	0.70	22
Ketones				
Acetone	2.4	0.41	0.56	22
Ethers				
Methyl ethyl	. . .	. . .	0.77	42
Methyl propyl	. . .	. . .	0.72	42
Methyl butyl	. . .	. . .	0.66	42
Ethyl	. . .	. . .	0.75	42
Ethyl propyl	. . .	. . .	0.67	42
Ethyl butyl	. . .	. . .	0.63	42
Propyl	. . .	. . .	0.62	42
Propropyl	. . .	. . .	0.64	42
Propyl butyl	. . .	. . .	0.57	42
Butyl	. . .	. . .	0.53	42
Esters				
Methyl acetate	1.6	0.42	. . .	22
Ethyl acetate	. . .	. . .	0.54	22
Amines				
Methyl	. . .	0.66^c	. . .	69
Dimethyl	. . .	0.61^c	. . .	69
Ethyl	. . .	0.58^c	. . .	69
Trimethyl	. . .	0.57^c	. . .	69
Propyl	. . .	0.52^c	. . .	69
Diethyl	. . .	0.52^c	. . .	69
Isobutyl	. . .	0.51^c	. . .	69
n-Amyl	. . .	0.48^c	. . .	69
Di-*n*-propyl	. . .	0.44^c	. . .	69
Triethyl	. . .	0.46^c	. . .	69

[a] After Weaver (129).[1] Courtesy of Academic Press, Inc.

[b] Conductivity of gas relative to that of air at 0° and 100°C.

[c] Values reduced by Weaver from data recorded at different temperature.

that the measurement of hydrogen in air is relatively easy. The measurement of ammonia in air, however, can present serious problems, because the table shows that the conductivity of ammonia is higher than that of air at 100°C and lower at 0°C. At some intermediate temperature, no measurement may be possible.

2. Effect of Temperature

No definite relationship has been established between the thermal conductivity of a gas and its temperature. Empirical equations of varying degrees of complexity have been formulated to describe this relationship for various specific pure gases (2,37,79,101,102,108,109,127), but are not of sufficient generality to be of interest for analytical purposes. A reasonable approximation is given by Weaver (129):

$$k_T = k_0(1 + aT) \tag{3}$$

where k_T is the thermal conductivity at temperature $T°C$, k_0 is the conductivity at 0°C, and a is a constant temperature coefficient. The value of a can vary from about 0.002 for helium to 0.01 for heavy vapors, such as benzene (129). Equation (3) is by no means rigorous, but it can be used for interpolation between small temperature differences.

In general, the thermal conductivities of gases tend to converge with increasing temperature. However, because of differing temperature coefficients, the difference between the conductivities of certain gases can change in sign as well as in magnitude with changing temperature. An outstanding example is that of ammonia–air mixtures, as noted above. A summary review of the subject (45) shows plots of thermal conductivity versus temperature of twelve common gases. The curves reveal not only different degrees but also different directions of curvature. Empirical data on methanol–nitrogen mixtures (44) show a maximum in the curve of thermal conductivity versus concentration, at temperatures over 100°C.

The nonuniform effect of temperature on the thermal conductivities of various gases is not ordinarily of serious consequence in the more common analytical applications of thermal conductivity measurement. However, it serves to emphasize the need for control of apparatus temperature. The possibility of anomalous instrument readings must be assessed by careful calibration, particularly when dealing with mixtures of gases not greatly different in thermal conductivity.

The difference in the thermal coefficients of conductivity of various gases can be used to advantage in the analysis of ternary gas mixtures. This subject is discussed further in Section II.F.1.b.

3. Effect of Pressure

To a first approximation, the thermal conductivity of a gas is independent of pressure, except at pressures so low that the mean free path approaches cell dimensions, typically 5 mm Hg (38). This is so nearly true that the effect can safely be ignored in most analytical applications at or near atmospheric pressure. The slight existing effect of pressure on thermal conductivity is most apparent in the case of polar molecules, and apparently results from association (31,32,49,50,107). Pressure control is warranted in such cases, where the utmost possible accuracy is required.

Pressure variation can alter instrument response, even though the thermal conductivity of the gas is not directly affected. In most practical analytical cells, a small amount of heat is transferred by convection. Since convection effects are proportional to gas density, a small change in heat transfer can result from pressure changes (38). Transient or pulsating pressure changes can enhance convective heat transfer, resulting in spurious signals or "noise." A perhaps obvious effect of subatmospheric pressure is the possibility of in-leakage of air (15). These instrumental factors are further discussed in Section II.E.

C. THERMAL CONDUCTIVITY OF GAS MIXTURES

A natural first intuitive assumption is that the electrical signal from a thermal conductivity cell will be linearly related to the relative concentrations and thermal conductivities of two mixed gases. This simple relationship can indeed be considered adequate over small concentration ranges, particularly where the conductivities of the two gases differ widely. The approximation, however, is justified only as a nonlinear relationship by small linear interpolations. The actual relation can be divided into two separate factors: (1) the effect of the relative concentrations of two gases on the thermal conductivity of the mixture, and (2) the electrical response of the detector to thermal conductivity changes. The significance of the two factors can be seen in Fig. 102.2. The lower curve, indicating the change in conductivity with composition of a gas mixture, suggests that maximum sensitivity to concentration change would be observed near 100% hydrogen. The upper

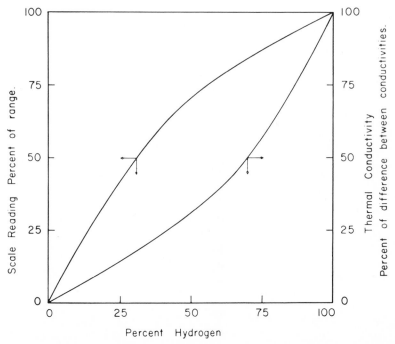

Fig. 102.2. Thermal conductivity and cell output for hydrogen–air mixtures. After Weaver (125). (Courtesy of Academic Press, Inc.)

curve, showing electrical output as a function of concentration, shows that, in fact, the opposite is true.

The two relationships between gas composition and electrical signal are so complex as to be of little practical value to the analyst. They will be discussed briefly in this and the following section, in order to explain the mechanisms involved and to provide a proper perspective for the large volume of published empirical data on the subject.

Wassiljewa (128) developed the expression for the thermal conductivity of a mixture of gases 1 and 2:

$$k_{12} = \frac{k_1}{1 + A_{12}(p_2/p_1)} + \frac{k_2}{1 + A_{21}(p_1/p_2)} \tag{4}$$

where k = thermal conductivity, p = partial pressure, and

$$A_{12} = \frac{1}{4} \left\{ 1 + \left[\frac{\mu_1}{\mu_2} \left(\frac{M_2}{M_1} \right)^{1/2} \left(\frac{1 + S_1/T}{1 + S_2/T} \right) \right]^{1/2} \right\}^2 \times \left(\frac{M_1 + M_2}{2M_2} \right)^{1/2} \left(\frac{1 + S_{12}/T}{1 + S_1/T} \right) \tag{5}$$

where μ = viscosity, M = molecular weight, S = Sutherland's constant $\simeq 1.5 \times$ boiling point, T = temperature. (A_{21} is obtained from A_{12} by interchanging subscripts.)

Daynes (22) summarized the fit of experimental data from various sources to the Wassiljewa equation, finding deviations as great as 50%, even in the case of simple molecules.

Lindsay and Bromley (71) subsequently modified equation (5) by empirical adjustment of exponents for the best fit to their data. They then succeeded in fitting equation (4) to data on 16 binary mixtures of simple gases in a total of 85 different compositions. Their average deviation was 1.5%, although individual errors as high as 10% occurred. Bennett and Vines (6) confirmed this work for nonpolar gases of similar viscosity and molecular weight, but observed large deviations and thermal conductivity maxima in the case of mixtures of polar and nonpolar gases. Schmauch and Dinerstein (94) observed deviations as great as 27%, even with nonpolar gas mixtures, such as n-hexane in helium.

Empirical (12) and semiempirical (7,47) equations have been developed but are of limited utility.

The very desirable characteristics of a thermal conductivity cell as a detector for gas chromatography have stimulated a considerable amount of effort directed toward the empirical determination of the relative molar response of the detector. Relative response data on a large number of gases and vapors in hydrogen, helium, and nitrogen have been reported (4,13,43,52–55,61,78,94,125). Linearity of response is reported for mixtures in hydrogen and helium, but severe nonlinearity has been observed in nitrogen, including reversal of sign (8). Some workers report a response proportional to the square root of molecular weight in homologous series (43,52–55), while others have observed a proportionality with weight per cent (13).

Hoffmann (46) has considered the thermal conductivity of gas mixtures in detail. In the light of the influence of molecular collision diameter upon viscosity and, hence, A_{12} [equation (5)], he shows that chromatographic detector response proportional to weight per cent is to be expected in the case of straight-chain homologous series. More compact structures and the inclusion of heavy atoms in the molecule reduce the influence of molecular weight on response. Mixtures of gases which produce a like response with a given carrier gas can be expected to show a response proportional to mole fraction. Using gas chromatographic data, Hoffmann succeeded in obtaining a good fit of the original Wassiljewa equation to binary mixtures of organic vapors in hydrogen and in helium, by calculating values relative to that of a standard mixture rather than absolute values, thus minimizing the influence of uncertainties in physical constants and experimental data. However, large errors were observed in some cases with mixtures of organic vapors in nitrogen, where the differences in thermal conductivity to be measured are smaller.

The empirical gas chromatographic data indicating a linear variation of thermal conductivity with concentration are valid for the concentration ranges ordinarily encountered in gas chromatographic effluents, usually well under 10 mol %. In general, such linearity over narrow concentration ranges is typical in cases where the difference in the thermal conductivities of the two gases is large. However, these data cannot be extrapolated to wider ranges or to mixtures of gases having similar thermal conductivity, where nonlinearity will usually become apparent.

D. CELL RESPONSE

The electrical signal produced by a thermal conductivity cell in response to a change in thermal conductivity depends on the design of the particular cell and on the associated electrical circuit and current. Two different basic types of thermal conductivity cells are in common use today: the hot-wire and the thermistor cell. In the latter type, the sensitive element is a small (typically 0.015-in. diameter) bead of fused metallic oxides, suspended from fine platinum electrical lead wires. The bead usually is protected with a thin glass coating. The quite different electrical responses of these two cell types to thermal conductivity changes have been derived by Schmauch and Dinerstein (94). Their derivation is based on the assumption that only one element of a differential thermal conductivity cell is exposed to gas of varying composition. The bridge circuit thus can be drawn, in simplified form, as in Fig. 102.3, where R_m represents the variable resistance of the hot wire or thermistor, and the other resistances are fixed.

1. Hot-Wire Cell

The equation relating the output voltage, E_o, of a hot-wire cell to the various geometrical, electrical, and thermal parameters is

$$E_o = C_w \left(1 - \frac{E_o}{IR_L}\right) \left[\frac{k_1(1 + E_o/IR_m)(1 - E_o/IR_L) - k_2}{k_2} k_1^{1/2}\right] \quad (6)$$

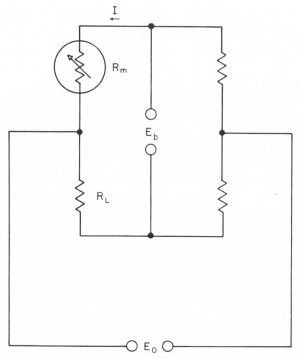

Fig. 102.3. Simplified bridge circuit.

where k_1 and k_2 are the thermal conductivities of the gas before and after a composition change, and C_w, the cell factor, is

$$C_w = \left(\frac{aJ}{R_m}\right)^{\frac{1}{2}} \alpha(T_r - T_w)^{\frac{3}{2}} \frac{R_0 R_L}{(R_L + R_m)} \tag{7}$$

The cell constant, a, for a coaxial wire in a cylindrical cell is

$$a = \frac{2\pi L}{\ln{(r_w/r_r)}}$$

In equation (7), J is Joule's mechanical equivalent of heat, α is the temperature coefficient of resistance of the wire, and R_0 is the resistance of wire at 0°C.

Ordinarily E_o is much smaller than IR_m or IR_L. Consequently, the rather unwieldy equation (6) can be approximated by

$$E_o \approx C_w \left(\frac{k_1 - k_2}{k_2} k_1^{\frac{1}{2}}\right) \tag{8}$$

It is evident from equation (8) that response is approximately linear with conductivity for small increments, but highly nonlinear for large increments. Fortunately, the former is the more usual case in gas analysis applications.

[Ray (90) has reported that the sensitivity of a chromatographic thermal conductivity detector increases when a carrier gas of low thermal conductivity is employed, in apparent contradiction to equation (8). His treatment, however, assumes a constant bridge supply voltage. Under these conditions, a low-conductivity carrier gas results in a high value of T_r, so that the response observed reflects a change in the cell factor, C_w, rather than in the thermal conductivity relationship.]

The cell factor, C_w, defined in equation (7) includes the various adjustable parameters determining the sensitivity of the cell to thermal conductivity changes. Of most practical significance are the temperature difference, $T_r - T_w$; the temperature coefficient, α; and the wire resistance, R_0—all of which can be maximized within limits for maximum sensitivity. A small additional increase results from increasing the ratio of load resistance, R_L, to hot-wire resistance, R_m, thus maximizing the last term of the equation. These factors are considered in more detail in Section II.-2.-E.

The various parameters in equations (7) and (8) generally are somewhat uncertain; in addition, the usual thermal conductivity cell deviates from the ideal geometry assumed for the purpose of calculation. Consequently, there is some merit in an empirical approach to the analysis of cell response, in which the unknown factors are lumped into a single experimentally or determined constant. The relation between the current through the hot wire and the voltage drop across the wire can be derived as follows:

$$I^2 R_m = aJk(T_R - T_w) \tag{9}$$

$$R_m = R_{T_w}[1 + \alpha(T_R - T_w)] \tag{10}$$

where R_{Tw} is the resistance of the wire at cell wall temperature. Hence

$$R_m = R_{T_w}1 + \frac{\alpha I^2 R_m}{aJk} = \frac{R_{Tw}}{1 - (\alpha I^2 R_{Tw}/aJk)} \tag{11}$$

$$E = I R_m = \frac{I R_{T_w}}{1 - (\alpha I^2 R_{T_w}/aJk)} \tag{12}$$

The quantity aJk in equation (12) can be evaluated readily for a given cell and gas from a series of simultaneous measurements of current (with a milliammeter in series with the wire) and voltage (with a volt-meter or potentiometer in parallel with the wire). Plots of such measure-ments with different gases are shown in Fig. 102.4. The quantity aJk was evaluated for the air and helium curves from data obtained at relatively low wire temperatures, and points predicted by equation (12) were plotted as the circles in Fig. 102.4a. This figure shows that the data fit equation (12) up to a wire temperature of about 200°C. (Tem-perature is indicated by the dashed lines of constant resistance.) Above this temperature, however, the curves tend toward parallel straight lines. This deviation results from radiant heat transfer, neglected in the ele-mentary mathematical analysis of cell response, and puts an effective upper limit to the benefit of increased wire temperature.

The curves can be considered as roughly analogous to the plate charac-teristic curves of a vacuum tube, in which gas composition substitutes for grid bias voltage. The cell output voltage on changing gas composi-tion can be obtained from the curves by plotting a "load line," represent-ing the resistor R_L in series with the filament, where the slope of the line represents the resistance of R_L, and the abscissa intercept indicates the bridge supply voltage. The output E_o is read off the voltage axis be-tween the intersections of the filament curves and load line. Thus (Figure 102.4a), for a load resistor $R_L = 21$ Ω (equal to filament resistance at 100 mA) and a supply voltage of 4.2 V, the output at 23°C wall tem-perature is about 0.45 V on changing from helium to air. The filament current is 100 mA in helium and 80 mA in air. The filament tempera-ture is 36°C in helium and 157°C in air.

For the same value of $R_L = 21$ ohms, the cell output on changing from helium to air could be doubled by raising the bridge supply voltage to 6.5 V. In this case, however, the filament temperature will vary be-tween 77° and 270°C.

It can be shown (20) that, for a given supply voltage, maximum sen-sitivity is obtained when $R_L = R_m$. Thus, in the above example, a slightly greater increase in E_o would be obtained by raising the value of R_L as well as that of the supply voltage. This matching of R_L to R_m is obtained automatically in one type of cell, in which another identical filament is used as a load resistance in place of a fixed resistor. The characteristics of this type of cell are discussed in Section III.A.2.a.

The increased sensitivity obtained by increased supply voltage in the above example involves a higher filament temperature. The gain derived from this increase becomes limited by the effect of radiant heat transfer, as well as by such practical considerations as thermal decomposition of the sample and oxidation or evaporation of the filament.

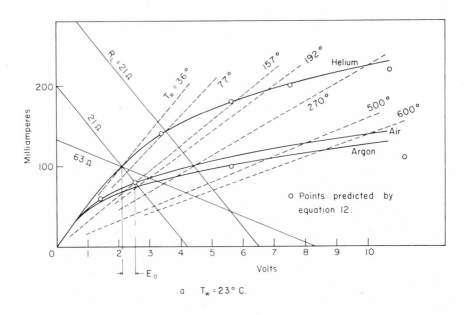

a. $T_W = 23° C.$

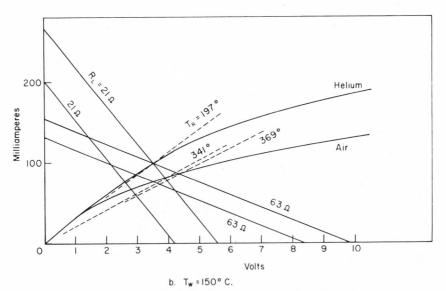

b. $T_W = 150° C.$

Fig. 102.4. Characteristic voltage–current curves for a hot-wire cell.

Another approach to increasing cell output is to maximize the signal for a given filament temperature, rather than for a given supply voltage. This can be accomplished, as shown by equation (7), by increasing the value of R_L (while simultaneously increasing the supply voltage to maintain the same filament temperature). Thus, in the example above, Fig. 102.4a shows that increasing R_L from 21 to 63 ohms, and increasing the supply from 4.2 to 8.4 V, doubles the cell output without increasing the initial filament temperature and with a smaller temperature rise. (This approach is limited in practice by the power dissipation limitations of precision resistors.)

As noted above, the divergence of the curves ceases to be appreciable when the temperature difference between filament and wall exceeds about 200°C. Thus the curves show that the benefit of increased current realized in the case of helium–air becomes negligible for argon–air. The increased load resistance, however, doubles the signal in both cases.

The effect of increased wall temperature is shown in Fig. 102.4b, where $T_w = 150°C$. The divergence of the helium and air curves is smaller than at lower wall temperature. The curvature and slope at low currents are less pronounced, because of the higher initial filament resistance. Consequently, if the same load resistance is used with the supply voltage producing 100 mA filament current at $T_w = 23°C$, the current will drop and the output voltage on changing from helium to air will be approximately halved. If, however, the current is restored to the initial 100 mA by increasing the supply voltage, the output will be restored almost to the initial value.

The curves shown in Fig. 102.4 are presented for illustrative purposes and should not be construed as being representative of all cases. It is possible, for example, for individual curves to cross, when marked changes in thermal conductivity occur with changing temperature. Similar curves can readily be produced for specific systems with very simple equipment and can be of much value in establishing optimum operating conditions.

2. Thermistor Cell

Whereas the resistance of a wire is linearly related to its temperature, that of a thermistor is an exponential function of temperature:

$$R_m = R_0 \exp B \left(\frac{1}{T} - \frac{1}{273} \right)$$

where B is a constant, and T is the thermistor temperature in degrees Kelvin. The value of B depends on the materials of which the thermistor is made. (Although B is here assumed constant, its value actually drops gradually with increasing temperature.) The equation for the output of a thermistor cell is (94)

$$E_o = C_t \left(1 - \frac{E_o}{IR_L}\right) \left[\frac{k_1(1 + E_o/IR_m)(1 - E_o/IR_L) - k_m}{k_m} k_1^{1/2}\right] \quad (13)$$

that is, the same as for a hot-wire cell, except that the cell factor C_t is

$$C_t = \left(\frac{aJ}{R_m}\right)^{1/2} \left(-\frac{BR_m}{R_0 T^2}\right) (T_R - T_w)^{3/2} \frac{R_0 R_L}{R_L + R_m} \quad (14)$$

For a thermistor, a spherical cell geometry would be ideal from the standpoint of mathematical analysis. In that case, the cell constant, a, would be (127)

$$a = \frac{4\pi r_r r_w}{r_w - r_r}$$

Such ideal geometry, however, is not approached in the usual cell design, and the value of a must be evaluated empirically.

As in the case of the hot-wire cell, the response of a thermistor cell can be approximated by an equation analogous to equation (8):

$$E_o \approx C_t \left(\frac{k_1 - k_2}{k_2} k_1^{1/2}\right) \quad (15)$$

The nonlinear and negative temperature coefficient of resistance of a thermistor complicates the interpretation of the response equation. Furthermore, an extremely important interdependence of I, R_L, R_m, and T_R is not explicit in the equations. Consequently, the graphical analysis outlined above for the hot-wire cell becomes even more desirable for the thermistor cell. Current–voltage curves for a typical thermistor cell are shown in Fig. 102.5. At low current levels, the thermistor, like the hot wire, shows an ohmic increase in voltage with current. As the temperature of the thermistor rises, however, the negative temperature coefficient of resistance causes a reversal of slope of the curve, whereby increasing current results in decreasing voltage. (The curves for $T_w = 150°$ were not carried to a high enough current level to display the slope reversal.)

It is immediately apparent from the curves for $T_w = 23°C$ that a thermistor, unlike a hot-wire cell, has an optimum load resistance and supply voltage for maximum output voltage. This optimum is closely

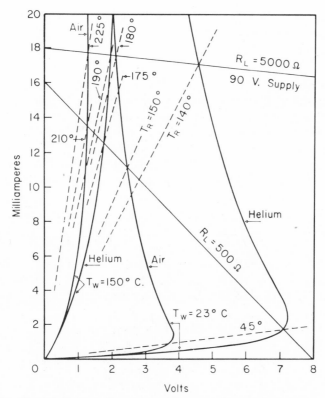

Fig. 102.5. Characteristic voltage–current curves for a thermistor cell. After Kieselbach (59). (Courtesy of *Analytical Chemistry*.)

approached by the 500-Ω load line shown in the figure. In this case, the output on changing from helium to air is 5.5 V, with a corresponding variation of T_R of 105°. From the curves of Fig. 102.4a, it can be seen that, even with an infinite load resistance, the maximum output of the hot-wire cell for a corresponding temperature change would be less than 1 V. At $T_w = 150°$, however, the output of the thermistor cell drops drastically, and, unlike that of the hot-wire cell, cannot be restored by increased current to the level obtained at $T_w = 23°$. Empirical data displaying the optimum sensitivity at low ambient temperature and a maximum (but low) sensitivity at high temperature have been reported (5).

Thus the thermistor cell offers the advantage of high output level at low values of T_w, obtained with low values of T_R, whereas the hot-wire cell is operable to high values of T_w with relatively small loss of output. Other important distinctions will be discussed in the following sections.

3. Effect of Convective Heat Transfer

Although the analysis of cell response has been treated on the assumption of purely conductive heat transfer, ordinary operating conditions of some types of thermal conductivity cell can lead to a significant amount of convective heat transfer. This factor warrants separate consideration, because of the marked anomalies in response which can result.

Where fast response is desired, as in the case of a gas chromatographic detector, it is common practice to cause the gas to flow directly over the detecting element. Under these conditions, a significant fraction of the heat from the hot wire or thermistor can be carried out of the cell by the gas, rather than be conducted directly to the cell wall. This effect is, of course, aggravated if the cell is constructed of material of low thermal conductivity, such as stainless steel or glass. For gas flowing axially through a concentric cylindrical cell, Bohemen and Purnell (8) have calculated that cell output voltage on changing gas composition can be approximated by the following expression:

$$E_o = K(T_R - T_w)\left[2\pi(k_1 - k_2) + \frac{m(C_{p1} - C_{p2})}{2L}\right] \qquad (16)$$

where K is a proportionality constant, m is the molar gas flow rate, and C_{p1} and C_{p2} are the heat capacities of the gases at constant pressure. Bohemen and Purnell observed that thermal conductivity changes are usually accompanied by heat capacity changes of opposite sign. Furthermore, the thermal conductivities of gases tend to converge with increasing temperature, while heat capacities are relatively less dependent on temperature. Consequently, with gases of low thermal conductivity, the effect of convection increases with rising temperature, being most pronounced when the thermal conductivity difference between the gases is small.

Figure 102.6 shows the effect of convection as a function of flow rate and filament temperature (8). The data were obtained with a gas chromatograph operating at constant carrier-gas flow rate, a variable fraction of the column effluent being passed through the thermal conductivity detector. The cell consisted of a coiled tungsten filament disposed axially in a 5-mm-bore glass tube, the latter held at 0°C. The carrier gas was nitrogen; the sample, acetone. The change of sign of the signal with

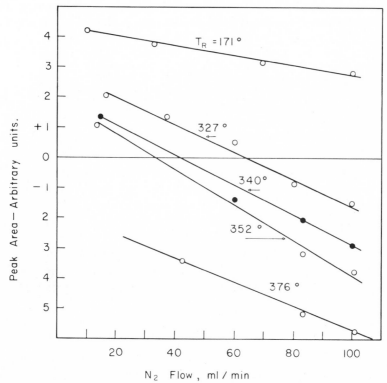

Fig. 102.6. Effect of flow rate on the response of a through-flow cell. After Bohemen and Purnell (8). (Courtesy of *Journal of Applied Chemistry*.)

increasing flow rate at high filament temperature obviously can lead to severe errors in measurement.

It seems probable that the observed effect would have been less severe had the cell wall been constructed of metal rather than glass. Nevertheless, the observation points to the need for caution in the use of high filament or thermistor temperatures in direct-flow cells.

Fortunately, cell designs are now available in which fast response is obtained without direct flow over the filament, so that convective heat transfer is eliminated. These designs are discussed in Section III.A.1.c.

E. SENSITIVITY

The useful sensitivity of a thermal conductivity cell is determined, as in any measurement, by its signal-noise ratio, where the signal is controlled by the cell response factors discussed above, and the noise represents the response of the cell to variables other than thermal conductivity changes. Young (131) has proposed a useful measure of sensitivity in terms of pQ_0, where

$$pQ_0 = \log (1/Q_0)$$

and Q_0 is the gas concentration change producing a signal equal to twice the observed noise level, usually expressed in moles per liter.

For the favorable case of organic vapors in helium, values of pQ_0 for typical thermal conductivity cells operated near room temperature are in the range of 6–7. However, it is possible to increase sensitivity by two orders of magnitude ($pQ_0 = 9$) by careful attention to the various systematic sources of noise (60), which are discussed in Sections II.E.1–7.

1. Johnson Noise

The ultimate sensitivity of a thermal conductivity cell is, in principle, limited by the thermal agitation of the electrons in the hot wire or thermistor, which produces a random noise voltage:

$$E_{rms} = \sqrt{5.496 \times 10^{-23} \times R_m \times T_R \times \Delta f}$$

where T_R is in degrees Kelvin, and Δf is the bandwidth in hertz.

The peak-to-peak noise level is generally of more interest than the rms value and, for random noise, is observed to be about five times greater. Peak-to-peak Johnson noise for a typical hot-wire cell is in the order of 10^{-8} V. For a thermistor cell, a typical value is about 10^{-7} V. The author has measured noise levels as low as about 2–3 times Johnson noise in both types of cell, under the most favorable operating conditions. Thus it is safe to say that noise sources other than Johnson noise will be limiting in the practical case. These other noise sources, discussed below, are all systematic in nature, leading to resistance or voltage changes through various direct mechanisms.

2. Flow Variations

As discussed in Section II.D.3, a cell through which the gas passes directly responds not only to thermal conductivity but to heat capacity as well. Such a cell, in fact, can be used as an extremely sensitive flowmeter, and cells designed for this purpose are commercially available (Gow–Mac Instrument Co., Madison, N.J.). Although it is possible in principle to eliminate this source of noise by maintaining a constant flow (95), it is extremely difficult to maintain a constancy of flow compatible with the ultimate sensitivity of the cell. In most practical cell designs, flow sensitivity is eliminated by permitting the gas to reach the sensitive element only by diffusion. As discussed in Section III.A.1.c, it is possible by appropriate choice of geometry to make such diffusion cells extremely rapid in response, while of negligible flow sensitivity. It is important to note, however, that gross flow through a diffusion path to the cell is possible when the gas pressure undergoes transient or oscillatory changes. Such changes ordinarily can be eliminated by appropriate surge tanks or pulsation dampers in the gas flow system.

3. Shock and Vibration

The heated wire or thermistor can produce a rising column of heated gas in the cell by thermal convection. The path of this column can be altered by shock or vibration of the cell, producing large noise signals. It has been shown, in the case of thermistor cells, that this effect can be eliminated by surrounding the thermistor with a metal baffle or screen spaced to bring the boundary layers of stagnant gas on thermistor and baffle into contact, thus eliminating thermal convection (60) (see Section III.A.1.c). The spacing for this purpose is approximately 0.015 in. Presumably a similar technique could be employed with hot-wire cells, although no data are available. Hot-wire cells ordinarily are quite shock sensitive and must be suitably isolated from vibration for high-sensitivity applications.

4. Ambient Temperature and Bridge Current Variations

Any departure from symmetry between the sample and reference elements of a thermal conductivity cell can result in noise produced by the differential effects of ambient temperature and bridge current variations. Consequently, a cardinal principle in the design of a cell is the attainment of identical electrical and thermal characteristics in the two elements. To this end, the best cells are made with the two cavities closely spaced in a block of metal of high thermal conductivity (e.g., brass), preferably massive, to minimize the effect of rapid ambient changes. Stainless steel and glass should be avoided for high-sensitivity applications, if possible, because of their relatively poor conductivity. (The conductivity of the cell block is of much less consequence in the case of thermistors than with hot wires. This results from the fact that the lower heat dissipation of the thermistor produces a smaller temperature gradient in the block.)

The geometry, electrical resistance, and temperature coefficient of resistance of the detecting elements also must be matched for best performance. In addition, their thermal time constants must match, to avoid response to transient variations.

In general, these ideals of symmetry are never attained in practice. Consequently, ambient temperature and bridge current must be controlled for high-sensitivity applications. Most of the noise observed in commercial thermal conductivity apparatus arises from this source, even with controlled temperature and current. It is possible, however, to compensate for residual asymmetry by appropriate design of the bridge circuit, as discussed in Section III.A.2.c.

5. Cell Contamination and Chemical Attack

A common, but not always obvious, source of noise lies in the nature of the sample, rather than of the cell itself. A contaminant, such as chromatographic column liquid phase, depositing on the electrical insulator supporting the filament or thermistor may produce a conductive path in parallel with the detecting element. This effect can be more severe with thermistors than with hot wires, because of the relatively high resistance of the former.

Thermal decomposition products, depositing on the element, may alter its thermal or electrical properties. Hot-wire elements, commonly made of tungsten or Kovar (nickel–cobalt–iron alloy), can be oxidized by excessive operating temperature in an oxidizing gas. The oxides of which thermistors are composed can be reduced by exposure to hydrogen. It is not possible to generalize about these chemical sources of noise, other than to say that they usually result in a gradually increasing noise level, often culminating in an abrupt short or open circuit. Means for their control must be devised for each special case. (See Section III.A.1.a.)

6. Leakage

A common, perhaps obvious, source of noise is leakage into or out of the cell or its associated gas system. Such noise usually is characterized by random signal changes over relatively long periods, for example, 1–10 min. It is commonly but erroneously assumed that contamination of a gas stream through a small leak is impossible if the gas is under pressure. Actually, the random energy distribution among gas molecules permits diffusion against a flowing stream through a leak. Consequently, a leaktight system is essential to high-sensitivity applications.

7. Electrical Noise

Electrical noise can arise from a variety of sources external to the cell proper. It is impossible to discuss this subject in detail in this chapter; therefore it must be assumed that one choosing to design his own circuits is familiar with the subject. Here it suffices to say that noise from external electrical sources can be made entirely negligible, as compared with noise from other sources discussed above, by proper circuit design. An easy test of the existence of electrical noise is the substitution of fixed-precision resistors for the sensitive elements of the cell.

F. SELECTIVITY

Although thermal conductivity analysis is inherently a technique best suited to binary mixtures, a variety of methods can be employed to extend its usefulness to the analysis of multicomponent mixtures. These methods can be conveniently subdivided into physical or instrumental and chemical methods.

1. Instrumental Selectivity

a. CONVECTION COMPENSATION

Although convective heat transfer ordinarily is thought of as a source of error in thermal conductivity measurements, its characteristics can be turned to advantage in some analyses of multicomponent mixtures. Convective heat transfer through a gas is a function of the molar heat capacity of the gas, a property not directly related to thermal conductivity. Consequently, a means for the measurement of both convective and conductive heat transfer through a gas can provide additional data whereby the composition of a ternary mixture can be determined. This can be accomplished with a cell in which the cavities are enlarged in diameter to enhance thermal convection. Since the heat capacity of a gas is inversely proportional to pressure, while its thermal conductivity is nearly independent of pressure, Minter (81) was able to calibrate such a cell for a three-component mixture in terms of readings taken at two different gas pressures.

Minter and Burdy (82) subsequently simplified the procedure by designing a cell and bridge circuit in which convective and conductive effects are compensated automatically, permitting direct calibration for a single component of a mixture in the presence of varying concentrations of two other components. A schematic diagram of the Minter and Burdy circuit is shown in Fig. 102.7. The cell contains four identical filaments. Two filaments, in cavities of small bore (ca. 5 mm), respond only to thermal conductivity. The other two, in cavities of relatively large bore (ca. 20 mm), respond to both conductivity and convection. One pair of large and small cavities receives sample gas; the other pair, the reference gas. The filaments are wired so as to place the sample cavities on one side of the bridge, and the reference on the other. Two fixed resistors, R, of equal value are connected in series with the filaments of the small conductivity cavities.

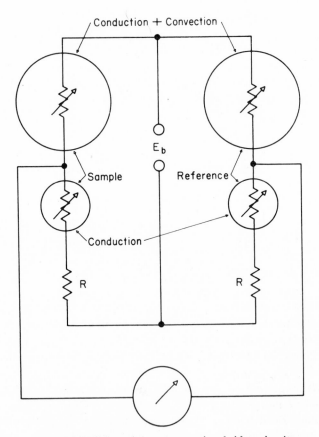

Fig. 102.7. Convection compensation bridge circuit.

Assume, for the moment, that the resistors R are omitted from the circuit, and that all four cavities contain the same gas. The bridge will be balanced, but the filaments in the large cavities will be cooler than those in the small cavities, because of the added effect of convective heat transfer. If the thermal conductivity of the sample gas is now increased, the bridge will unbalance because of the higher sensitivity of the hotter filament to conductivity changes. On the other hand, an increase in sample heat capacity will unbalance the bridge in the opposite direction, because of the insensitivity of the small cavity to convection.

The relative response of the bridge to thermal conductivity and heat capacity changes can be altered by selecting "annulling resistors," R, of such value as to reduce the fractional resistance change in both arms of the bridge with a change in thermal conductivity. In practice, the annulling resistors are selected by trial and error to give the desired response to one component of a ternary mixture, using known samples. Table 102.II shows the degree of selectivity obtained in the analysis of hydrogen in a mixture with carbon dioxide and air (82).

TABLE 102.II
Selectivity to Hydrogen by Convection Compensation[a]

H₂, %	E_0, mV				
	With annulling resistors			Without annulling resistors	
0.0	+0.0	+0.20	+0.11	0.0	+3.90
2.5	25.77		25.74		
5.0	51.67		51.65	−10.55	
CO₂, %	0.0	2.5	5.0	0.0	5.0

[a] Courtesy of *Analytical Chemistry*.

Figure 102.8 shows a sectional view of a commercially available convection compensation cell (80). The apparatus has been successfully applied to the analysis of mixtures of more than three components, although such applications necessitate a fortuitous combination of thermal properties in the components of the mixture. In general, the application of the technique requires a large difference between the thermal properties of the compound of interest and those of the other components of the mixture. For example, the method has been applied successfully to the measurement of either hydrogen (very high thermal conductivity)

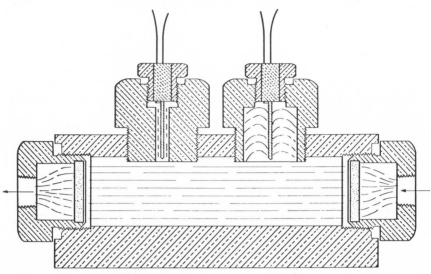

Fig. 102.8. Cross-section of convection compensation cell. (Courtesy of Mine Safety Appliance Co.)

or carbon dioxide (very low thermal conductivity) in a mixture with variable concentrations of carbon monoxide, nitrogen, oxygen, and hydrocarbons. The method fails, however, in the determination of oxygen in a mixture with nitrogen and carbon dioxide, where the thermal properties of oxygen and nitrogen are very similar.

b. THERMAL COMPENSATION

The differing temperature coefficients of thermal conductivity of various gases provide another means for enhancing selectivity. Bolland and Melville (9,10) observed that the thermal conductivity of hydrogen falls rapidly below 200°K, and that the conductivity of a ternary mixture was almost independent of hydrogen concentration at about 110°K. Cooling their cell with liquid air, they were able to calibrate for hydrogen concentration in terms of the power required to heat the filament to 110° and 300°K.

A more convenient means for accomplishing the same result was developed by Fleming (119). His apparatus, designed for use at room temperature, employs two filaments exposed to the sample gas and connected in adjacent arms of the bridge, so that their effects on bridge output are in opposition. One filament, of low resistance, is operated at about 200° above ambient temperature, while the other, of high resistance, rises only about 60° above ambient. The relative lengths of the filaments are adjusted to approximately equalize their responses to a change in hydrogen concentration. Fleming observed that the response of a filament to hydrogen rose linearly with increasing filament temperature, whereas that to carbon dioxide went through a maximum of opposite sign. Consequently, he was able to adjust his instrument to obtain preferential response to either gas in a mixture with air by connection of suitable fixed resistors in series with the two filaments. His apparatus was designed especially for the analysis of combustion products in engine exhausts, essentially air, carbon dioxide, and hydrogen. It can also be used, however, to measure the Btu content of a fuel gas by adjusting the relative response to hydrogen and the hydrocarbons present.

2. Chemical Selectivity

By far the most powerful means of obtaining selectivity in thermal conductivity gas analysis involves a chemical treatment of the sample, whereby the reference and sample elements of the thermal conductivity cell are exposed to the sample before and after treatment, respectively. In principle, any chemical reaction may be employed which produces a change in the thermal conductivity of the sample as a function of the concentration of the desired component. The most straightforward example is the absorption of the component, for example, carbon dioxide by Ascarite. Combustible gases can be measured with enhanced sensitivity in air by catalytic combustion. A special case is the separation of the components by gas chromatography, although the classification of that technique as an adjunct to thermal conductivity analysis is somewhat bizarre (see Part I, Vol. 3, Chapter 37 of this Treatise). The chemical reactions that can be employed for selectivity are limited only by the ingenuity of the analytical chemist and cannot be dealt with exhaustively in this chapter. Examples of practical applications are given in Section III.C.

III. PRACTICE

A. APPARATUS AND ACCESSORIES

A number of instrument companies manufacture packaged thermal conductivity apparatus, which can be obtained precalibrated for a particular analysis. The purchase of such equipment obviously is the simplest approach to any routine analysis problem. However, the wide variety of existing apparatus designs warrants consideration of their special characteristics, as discussed below, in relation to a specific application.

1. Cell Designs

Thermal conductivity cells can be classified according to the type of sensitive element (i.e., hot wire or thermistor), the materials of construction, and the geometry of the cell.

a. SENSITIVE ELEMENTS

The choice between hot-wire and thermistor cells for a given application depends on a number of parameters. The hot-wire cell offers a definite advantage in providing good sensitivity over a wide range of temperature; such cells have been used from —250° to over 500°C. Thermistors, because of their negative temperature coefficient of resistance, lose sensitivity rapidly with increased temperature and tend to instability above 300°C.

On the other hand, thermistors provide a relatively large signal when the cell is near room temperature, and therefore offer a sensitivity advantage under those conditions. Furthermore, the larger signal is obtained with a smaller differential temperature, $T_r - T_w$, than in the case of the hot wire. A typical thermistor bead measures only about 0.015 in. in diameter, a factor for consideration when small volume and fast response are desirable.

The relatively high temperature of the hot wire can be undesirable with thermally unstable samples, and the wire can be subject to burnout in an oxidizing atmosphere. Thermistors, on the other hand, even though normally glass enclosed, can be destroyed by reduction if operated in a strongly reducing atmosphere, such as hydrogen.

The resistance of a thermistor at operating temperature ordinarily is about 10 times that of the usual hot wire. Consequently, the bridge power requirements are less, a factor for consideration in the case of portable equipment. On the other hand, the higher resistance of thermistors makes them more subject to noise arising from the deposition of conductive contamination on the glass-to-metal seals.

Generally, the matching of hot-wire pairs is easier than that of thermistor pairs. Consequently, a hot-wire cell is usually less subject to drifts resulting from ambient temperature and bridge current variations. It is possible, however, to compensate for mismatching, as discussed in Section III.A.2.c.

In view of the variety of factors involved, the choice between hot-wire and thermistor cells must be made in the light of the particular application. Within each type, there are, in addition, several choices, as discussed below.

(1) Hot-Wire Elements

Hot-wire elements can be made of several different metals and alloys, depending on the application. The wire must be strong enough for durability in the small diameter required, typically 0.001–0.005 in. It should be inert to the sample. For sensitivity, it should have a high resistance and a high temperature coefficient of resistance. The metals most commonly employed are platinum, tungsten, Kovar (a nickel–cobalt–iron alloy), and nickel. The electrical characteristics of these metals are given in Table 102.III.

TABLE 102.III
Temperature Coefficients of Resistance and Resistivities
of Some Common Metals (20°C)

Metal	α, ohm/ohm °C $\times 10^3$	Resistivity, ohm cm $\times 10^6$
Kovar	3.8	49
Nickel	6.0	6.8
Platinum	3.0	10
Platinum—10% iridium	1.2	24.6
Tungsten	4.6	5.5

Where the oxide coating of the base-metal wires gives inadequate corrosion resistance, they can be protected. Gold-plated tungsten and polytetrafluoroethylene-clad tungsten filaments are available (36). Platinum filaments fused in glass are also used for extremely corrosive samples (26,116).

Commercially available hot-wire cells employ filaments in a variety of configurations. The majority use a helically coiled filament for compactness, consistent with reasonable resistance. Figure 102.9 shows a typical bare, coiled filament of tungsten, supported from the leads of a glass-to-metal hermetic seal. The 25°C resistance of this filament is 18 Ω. Ruggedness is improved by winding the filament on a thin, hollow glass mandrel (97), as shown in Fig. 102.10. One manufacturer (116) winds a platinum filament between two hard-glass tubes, which are then fused together to produce a rugged, corrosion-resistant element. Less common in commercial equipment is the bare, straight wire of classical design, such as that in Fig. 102.11.

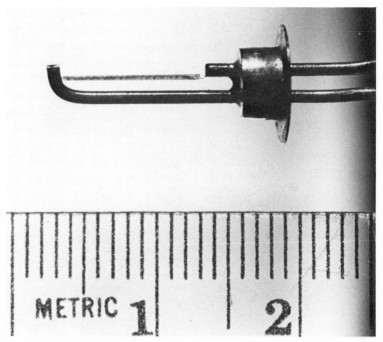

Fig. 102.9. Coiled tungsten filament. (Courtesy of Gow–Mac Instrument Co.)

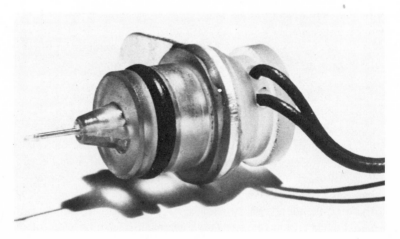

Fig. 102.10. Platinum filament wound on glass mandrel. (Courtesy of Leeds and Northrup Co.)

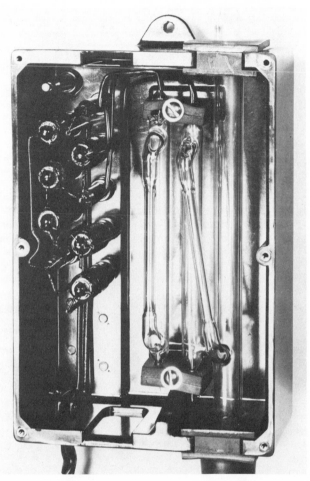

Fig. 102.11. Glass cell with straight platinum filament. (Courtesy of Leeds and Northrup Co.)

The rigid support furnished by the mandrel or glass enclosure offers the advantage of improved durability and resistance to shock and vibration. It is undesirable, however, when rapid concentration changes are to be measured, since the added heat capacity introduces a large thermal time constant, *ca.* 60 sec. (The thermal time constant of a bare filament is a small fraction of 1 sec.)

A convenient and inexpensive source of tungsten filaments consists of incandescent lamp bulbs, which are remarkably well matched because of their mass production. Lamp filaments have been successfully employed in very simple cell designs by Hebler and Hamilton (118) and by Stuve (103).

Filaments are also readily available in the form of model airplane glow plugs. These plugs, which employ high-temperature ceramic insulation, can conveniently be screwed into a cell and are readily replaced. Felton (28,29) has reported their use at ambient temperatures as high as 500°C.

(2) Thermistors

The thermistors almost invariably employed for thermal conductivity applications are 0.015-in.-diameter glass-coated beads, supported by 0.001-in. platinum–iridium leads. They are normally mounted on glass-to-metal hermetic seals (Fig. 102.12). They can be obtained in pairs matched as to resistance and temperature coefficient of resistance over the current range normally employed.

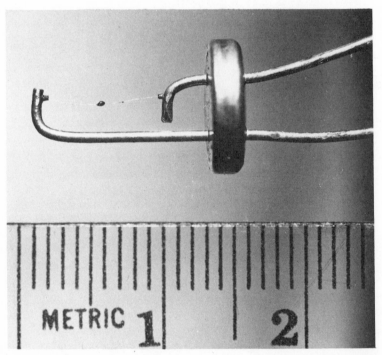

Fig. 102.12. Glass-coated thermistor. (Courtesy of Fenwell Electronics, Inc.)

Bead thermistors are available in 25°C resistances from 500 Ω to 18 MΩ. However, unavoidable manufacturing tolerances require the selection of matched pairs from a large number of units. Consequently, the range of resistance values available in pairs for thermal conductivity applications is more limited.

Table 102.IV shows the electrical and thermal characteristics of matched thermistor pairs available from one manufacturer (30).

TABLE 102.IV
Electrical and Thermal Characteristics of Thermistors[a]

Resistance at 25°C, Ω	Approximate temperature coefficient, %/°C	Dissipation constant, mW/°C	Time constant, sec
2,000 ± 25%	−3.9	0.16	1
8,000 ± 20%	−3.9	0.1	1
100,000 ± 15%	−4.6	0.1	1

[a] Courtesy of Fenwal Electronics, Inc.

The thermal time constants presented in the table are the values measured with the bead suspended in still air. The time constants are very much shorter in a gas of high thermal conductivity, such as helium.

The 8000-Ω thermistors are useful at ambient temperatures from room temperature up to about 150°C; above this point their resistance and sensitivity fall off drastically. The 100,000-Ω thermistors can be used up to 300°C. However, it is generally impractical to use the 100,000-Ω thermistors over the entire range of temperature, because the extreme variation in thermistor resistance over this range presents problems in impedance matching, as discussed in Section III.A.4.

The susceptibility of thermistors to damage from reducing atmospheres, mentioned above, apparently arises from imperfections in their glass coating. Cowan and Stirling (17) have found that hydrogen-resistant thermistors can be obtained by a destructive acceptance test, involving an extended exposure of the candidates to hydrogen at an elevated temperature.

b. CELL MATERIALS

From the standpoint of temperature stability and the avoidance of convective effects, the ideal materials for cell construction are metals of high thermal conductivity (e.g., copper or brass), and such materials should be employed where possible. When corrosion problems necessitate the use of low-conductivity materials such as stainless steel or glass, consideration should be given to the possibility of instability and anomalies in cell response.

One such anomaly was observed by the author in the response of an experimental detector for gas chromatography, in which a thermistor was mounted in a cell constructed of polytetrafluoroethylene (PTFE). With helium as a carrier gas, the tails of the chromatographic peaks went below the base line, and then gradually rose to the normal zero level. The explanation of this effect is as follows: In the absence of a sample, the PTFE cell wall was heated appreciably above the temperature of the block by conduction from the thermistor through the high-conductivity helium. During passage of the low-conductivity sample, the cell wall cooled. Then, when the sample had passed, the thermistor was exposed to a wall temperature below the equilibrium value, and consequently was cooled to a temperature below its equilibrium level, resulting in a negative cell output signal.

Effects of this type are more severe with hot wires than with thermistors because of the higher power levels and consequent higher temperature gradients involved. When low-conductivity cell materials must be employed, the best compromise is to construct the cell of thin-walled tubing that can be immersed in a water or oil bath or, better, cast in copper or aluminum.

c. CELL GEOMETRY

(1) Flow Sensitivity and Speed of Response

Most applications of thermal conductivity analysis involve measurements on flowing gas streams. Consequently, a primary consideration in the design of the cell is the avoidance of flow sensitivity. This aim is achieved in most designs by providing only a single connection between the cell cavity and the flowing stream, so that gas can travel to the cavity only by diffusion. Shakespear's original "katharometer" employed this principle (111). Typical modern cell designs employing diffusional flow are shown in Figs. 102.8 and 102.13. The cell in Fig. 102.8 also employs screens at inlet and outlet, to minimize turbulence in the flow channel. The speed of response of a diffusion cell of this type to a composition change is limited by the volume of the cavity and the length of the diffusion path, ordinarily resulting in a time constant of the order of 20 sec (24).

A cell is not seriously influenced by flow, provided that the flow is constant and small (Reynolds number < 100) (88). Peters (112) em-

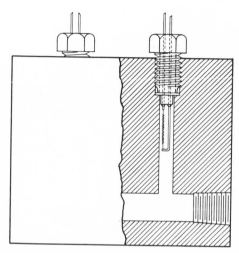

Fig. 102.13. Partial section of diffusion cell. (Courtesy of Gow-Mac Instrument Co.)

ployed this fact in the design of a cell of somewhat faster response, shown in Fig. 102.11. In this type of cell, a small, constant flow through the cavity is maintained by thermal convection, utilizing the chimney effect produced by the hot wire. Flow variations in the main gas stream do not influence this flow because both cell connections enter the main stream at the same point. Small cells (0.25 cc) employing this principle display a time constant of about 10 sec (24).

With the advent of gas chromatography, the need arose for much faster thermal conductivity cells in order to detect the rapidly varying gas concentrations produced by chromatographic columns. Severe distortions of chromatographic peaks can result from an excessive detector time constant (93). Since the carrier-gas flow rate is normally held constant in a chromatograph, it is possible to pass the gas directly through the cell cavity without undue disturbance to the cell response; in fact, most commercial chromatographs employ this through-flow detector geometry.

The response time of a through-flow cell is limited primarily by the volume of the cavity and the gas flow rate. Since cells of extremely low volume are possible (10,92), very rapid response can be obtained. Using a thermistor mounted in the center of a 2-mm-bore tube, the author has achieved a time constant of about 0.05 sec.

When sensitivity is an important consideration, however, minor uncontrolled flow variations can be a severe source of noise in through-flow

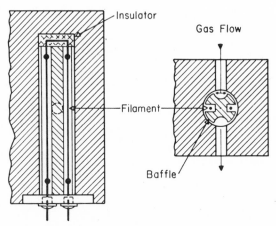

Fig. 102.14. Hot-wire cell with recessed filaments. After Schmauch (123).

cells. Consequently, several cell designs have evolved to combine the benefits of fast response and low flow sensitivity path with a geometry designed to prevent convection over the sensitive element.

Schmauch (123) has designed a hot-wire cell of negligible flow sensitivity, having a time constant of about 2 sec. His cell, shown in Fig. 102.14, provides narrow ($\frac{1}{16}$-in.-wide) recesses in a metal rod to enclose the straight filaments. The annular space between the rod and its enclosure serves as a low-resistance flow path for the gas. A similar scheme was employed by Kieselbach (60) in the design of a thermistor cell. His thermistor is contained in a double-walled enclosure of perforated nickel, as shown in Fig. 102.15. The assembly is mounted in a $\frac{1}{4}$-in.-diameter through-flow cavity. The response of this cell is approximately equivalent to that of the Schmauch cell.

In a somewhat different approach to flow stability, Kolloff (67) has designed a modification of the flow-through cell, shown in Fig. 102.16. The flow path through this cell is a spiral, reducing flow variations and, in consequence, flow noise. A number of other novel improvements, all of which contribute to a reduction in noise and drift, are also included in the construction of the cell. The filament illustrated here is one cartridge of a normal four-cartridge configuration of the detectors. With this design, noise levels in the order of less than a μ, and drift of a few microvolts per hour, have been obtained, making possible at least fiftyfold amplification of the detector signal and thereby greatly increasing the usable sensitivity of the detector.

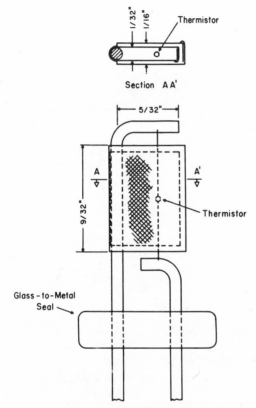

Fig. 102.15. Screened thermistor (60).

An extremely rapid thermistor cell designed by the Phillips Petroleum Company is commercially available from several instrument companies. As shown in Fig. 102.17, the Phillips cell employs the same principles as do the Schmauch and Kieselbach cells, but with greatly reduced dimensions. The time constant of this cell is only 0.07 sec. The cell employs a PTFE liner, leading to the anomalous response described in Section III.A.1.b at low signal levels. However, the author has constucted an all-metal cell of similar configuration, which was suitable for operation at signal levels as low as 50 μV full scale.

(2) High Temperature

Gas chromatography has also stimulated the design of cells suitable for use at high temperature (21,27,28,29,85). The major tempera-

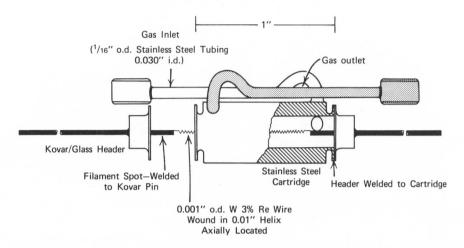

Gas Inlet
('/₁₆" o.d. Stainless Steel Tubing
0.030" i.d.)

Gas outlet

Kovar/Glass Header

Filament Spot—Welded
to Kovar Pin

Stainless Steel
Cartridge

Header Welded to Cartridge

0.001" o.d. W 3% Re Wire
Wound in 0.01" Helix
Axially Located

Thermal Conductivity Cartridge

Fig. 102.16. Modified flow-through detector cell. (Courtesy of Hewlett-Packard, Avondale Division.)

ture limitation—the increased electrical conductivity of the usual glass insulators at high temperature—is overcome by the use of ceramic high-temperature insulators, as in the glow plugs mentioned in Section III.A.1.a.(1). Lead wires can also be swaged in magnesia insulation in a metal tube, which then can be mounted conveniently with conventional tubing fittings (27).

Cells suitable for operation to 500°C are now available from most manufacturers of gas chromatographs.

(3) Miscellaneous

Cells have been designed in varying degrees of elaboration to meet needs in chromatography now served by commercial equipment (21,58,68,91). Since these designs offer no special advantages, they will not be discussed here.

A glass cell of only 0.05-cc volume has been designed for microanalysis (10); a metal cell of only 0.003-cc volume, for use in gas chromatography (92).

Means for improving the thermal symmetry of a cell have been described (115,117). This objective is accomplished by placing a movable metal rod or sleeve in each cell cavity, whereby the effective heat transfer path length between filament and wall can be adjusted. This device,

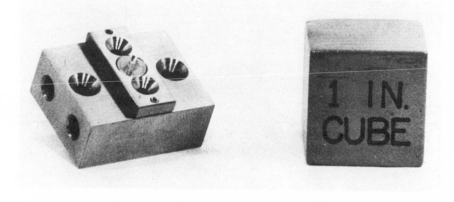

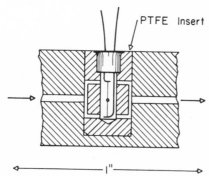

Fig. 102.17. High-speed diffusion cell. (Courtesy of Greenbrier Instruments, Inc.)

according to the patent disclosures, makes it possible to achieve such symmetry that the effects of ambient temperature on cell output are eliminated. A similar result can be achieved somewhat more simply by electrical means, as described in Section III.A.2.

2. Bridge Circuits

A number of bridge circuits of varying degrees of complexity can be employed with thermal conductivity cells, the choice being determined by considerations of sensitivity, stability, ease of manipulation, and cost.

a. SIMPLE BRIDGE CIRCUITS

In the simplest and most commonly employed bridge, depicted in Fig. 102.18a, the resistance of the fixed resistors is equal to that of the fila-

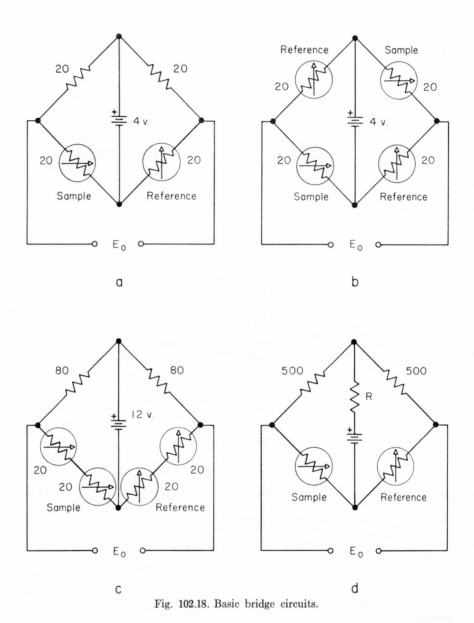

Fig. 102.18. Basic bridge circuits.

5982

ments or thermistors. In addition to its simplicity, this bridge has the advantages of operation with a low-voltage power supply and of providing maximum output power. As discussed in Section II.D.1, the sensitivity of such a bridge can be increased, in the case of hot-wire cells, by increasing the resistance of the resistors, R, and simultaneously raising the supply voltage. Thus, by increasing the fixed resistance values from 20 to 100 Ω and the supply voltage from 4 to 12 V, sensitivity is raised by a factor of 1.7 (24).

The bridge of Fig. 102.18b offers double the sensitivity of that of Fig. 102.18a, while requiring no increase in voltage. It has the further advantage of maintaining equal resistance in all bridge arms, irrespective of operating temperature. Cells containing four filaments intended for use in this bridge configuration are commercially available (35). The same manufacturer also furnishes eight-filament cells, in which two filaments are wired in series in each bridge arm, for further doubling of sensitivity. The bridge of Fig. 102.18b is not suitable for use with thermistors because their negative temperature coefficient of resistance would lead to a rapid temperature rise and destruction of the thermistors.

Still greater sensitivity can be obtained from the four- or eight-filament cell by use of the circuit of Fig. 102.18c. This produces a signal 3.4 times as large as that of Fig. 102.18a (24).

The negative temperature coefficient of resistance of thermistors can be turned to advantage in a regenerative bridge circuit, shown in Fig. 102.18d. The functioning of this circuit is as follows. Assume that the thermal conductivity of the sample is lower than that of the reference gas. The temperature of the sample thermistor will rise, and its resistance will fall. Consequently, more current will flow in the left side of the bridge. Because of the presence of resistor R, however, the bridge supply voltage will be reduced. This reduced voltage will result in a cooling of the reference thermistor, so that the unequal division of current between the two sides of the bridge will be augmented, and the output signal will be increased.

It will be seen that, if the resistance of resistor R is sufficiently high, an unstable condition can arise, whereby the bridge, once unbalanced, will regeneratively proceed to a permanent unbalance, in which virtually all of the bridge current flows through one side of the bridge. Consequently, care must be employed in the selection of R, the optimum value of which will vary with operating temperature. An approximately five- to ten-fold increase in sensitivity can be attained with this circuit without instability.

b. Zero Adjustments

Except in applications involving low sensitivity and large signals, it is usually necessary in practice to provide both coarse and fine zero-adjusting controls in the bridge. A typical arrangement is shown in Fig. 102.19a. The fine adjustment employs a relatively high resistance potentiometer shunted by a 0.5-Ω fixed resistor, because 0.5-Ω potentiometers are not readily available.

Another method of providing zero adjustments, shown in Fig. 102.19b, is particularly advantageous when the bridge and controls must be separated by a long cable, since the resistance of the cable is not inserted in the bridge arms. It is also possible to provide additional degrees of fine control with additional potentiometers and resistors across the output.

c. Compensating Bridge Circuits

As discussed in Section II.E.4, perfect electrical and thermal symmetry is never attained in a practical thermal conductivity cell. Consequently, changes in ambient temperature and bridge current are reflected in the

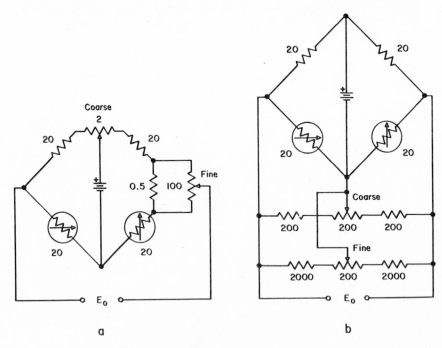

Fig. 102.19. Zero adjustment circuits.

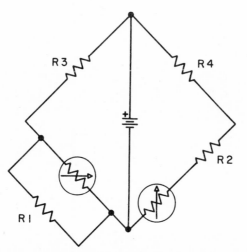

Fig. 102.20. Bridge compensated with series and shunt resistors. After Cherry and Foley (121).

output signal. However, it is greatly possible to reduce the effect of asymmetry by electrical compensation in the bridge.

One method of compensation was disclosed by Cherry and Foley (121) and, independently, by Littlewood (72). The circuit is shown in Fig. 102.20. The shunt, $R1$, and/or the series, $R2$, resistors are selected under operating conditions to produce equal response in both bridge arms to changes in bridge current. The bridge unbalance produced by the added resistors is corrected by appropriate adjustment of the values of resistors $R3$ and $R4$. This compensation scheme suffers the disadvantage of requiring the use of odd-value, fixed resistors, which, for stability, should be wire wound. The scheme is not very suitable for use with thermistor cells, because the large variation of thermistor resistance with temperature requires different values of compensating resistance for different operating temperatures.

Jones (124) developed a somewhat more flexible compensating bridge circuit, shown in Fig. 102.21. This circuit, applicable to both thermistor and hot-wire cells, permits the use of standard components and potentiometers instead of selected fixed resistances and, in addition, permits the adjustable components of the bridge to be remote from the thermal conductivity cell. The Jones bridge operates on the principle that, if the slopes of the voltage–current curves of both thermistors (or both hot wires) are identical at the operating current, supply voltage

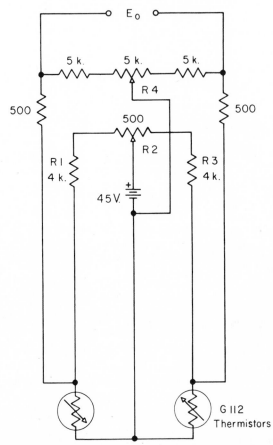

Fig. 102.21. Bridge compensated by equalization of dynamic resistances. After Kieselbach (59). (Courtesy of *Analytical Chemistry.*)

changes will produce no output signal. Accordingly, potentiometer $R2$ is adjusted to vary the division of current between the two thermistors until both are at points on their curves of equal slope, as determined by observing the bridge output for a fixed change in supply voltage. This procedure results in a bridge unbalance, since the voltage across the two thermistors are different. This unbalance is corrected by adjustment of $R4$ in the output voltage-dividing network, to restore an output null after each change in the setting of potentiometer $R2$.

Perfect current compensation is achieved by the Jones circuit over a 10% range of supply voltage. The circuit also results in a reduction of ambient temperature effects by a factor of 20.

The author has tested the circuit, using a four-filament, 20-Ω hot-wire cell in the configuration of Fig. 102.18b, with filaments substituted for resistors $R1$ and $R3$ of Fig. 102.21. The $R2$ was 2 Ω, and a 12-V supply was used, with helium in the cell. Compensation equal to that attained with the thermistor cell was achieved.

An important consideration in the application of compensating bridge circuits is the fact that compensation is achieved only for steady-state conditions. Any difference in the thermal time constants of the sensitive elements will result in a transient response to sudden power supply variations, even though gradual changes have no effect. Consequently, optimum performance of these circuits requires that the power supply be well filtered to damp line voltage transients, and that the cell be well insulated to prevent sudden temperature changes.

d. Isothermal Bridges

When large changes in thermal conductivity are to be measured, there is danger of damage to the hot wire or thermistor, because of the corresponding large temperature changes. This danger is obviated, however, in a circuit in which the temperature of the sensitive element is held constant by automatic adjustment of the bridge current, the output signal then becoming the change in bridge current.

An isothermal bridge circuit is shown schematically in Fig. 102.22. The sample and reference elements of the cell are connected in two independent bridges in which the fixed resistances have values such that the bridges will be balanced when the sensitive elements are at the desired operating temperature. The bridges are powered by two cathode followers, the voltage supply to which need not be regulated. Separate amplifiers amplify the output of each bridge to drive the cathode followers in a direction to minimize bridge output. Thus the filaments or thermistors are maintained at nearly constant temperature, to a degree determined by amplifier gain. The differential signal between the two cathodes represents the output.

In addition to providing burnout protection, the isothermal circuit has the advantage of reducing the effective thermal time constant of the sensitive element by a factor proportional to amplifier gain. The circuit is thus desirable when extremely rapid changes are to be measured, or where relatively sluggish glass-enclosed filaments are used. The reason for this time-constant reduction lies in the fact that the departure from thermal equilibrium is limited to the small degree necessary for corrective amplifier action.

The circuit, aside from its complexity, suffers one fundamental disadvantage from the standpoint of sensitivity. In most closed-loop circuits,

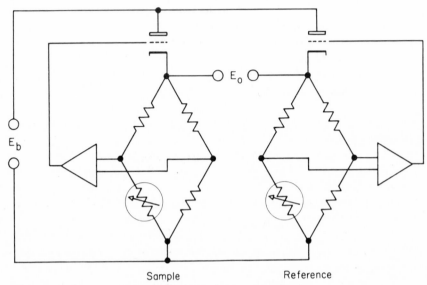

E_b

E_o

Sample   Reference

Fig. 102.22.   Isothermal bridge circuit.

noise originating in the amplifier is reduced by the negative feedback. In this circuit, however, the frequency-response-limiting element is the filament or thermistor. Consequently, amplifier noise is not fed back except at frequencies to which the sensitive element can respond. For the large-signal applications for which the circuit is intended, this factor is, of course, of little consequence.

Several variations of the isothermal bridge circuit, involving electronic and electromechanical rebalance systems, have been patented (122).

e. Multiarm Bridges

A number of special-purpose bridges have been devised, incorporating six or more arms, that is, three or more detecting elements (11,114,120). The purpose of these bridges is to permit intercomparison of the conductivities of more than two gases, which is accomplished by switching the meter or recorder appropriately.

3. Power Supplies

Thermal conductivity bridges can be powered by either direct or alternating current. The latter offers the advantage of permitting the use

of a simple step-down or constant-voltage transformer as the supply
(89). However, since it requires a nonstandard ac potentiometric recorder
or amplifier and demodulator, ac bridges are ordinarily restricted to
custom assemblies which include a recorder designed for the application.
Although ac operation eliminates the effects of stray thermal emf's in
the circuit, it suffers from susceptibility to stray capacitance, inductance,
and ac pickup from power lines. Since the advantages of ac operation
are marginal, the remainder of this discussion will be devoted to dc
power supplies.

a. Batteries

The primary consideration in the selection of a power supply is the
stability of the output current. From this consideration alone, ordinary
automobile storage batteries are well suited to the application. Figure
102.22a shows a simple supply, comprising battery, on–off switch, rheostat, and milliammeter. For a typical hot-wire cell, a 50-Ω rheostat
and a 500-mA meter are suitable; for a typical thermistor cell, a 500-Ω
rheostat and 50-mA meter.

Dry cells can be used with thermistor cells, because of the lower power
requirements of the latter. They are not recommended for continuous
service, however, since the cost of replacement will soon justify the purchase of a line-operated supply.

b. Line-Operated Supplies

The maintenance requirements of batteries usually warrant consideration of line-operated supplies. A simple circuit adequate for most purposes is shown in Fig. 102.23b.

When high sensitivity is required, electronically regulated supplies
are desirable because of the greater degree of stability obtainable. The
elaborate vacuum-tube-regulated supplies of the past are now largely
superseded by solid-state supplies, which are available in profusion in
compact packages from many manufacturers. Very high stability is
achieved with a simple and inexpensive circuit (Fig. 102.24) designed
by the author for use with the Jones thermistor bridge (Fig. 102.21).
The regulation of this circuit reduces the effect of a 20% line voltage
change to a 0.002% output voltage change and, with the Jones bridge,
to a cell output change of $<0.3\ \mu V$.

The circuit of Fig. 102.24 incorporates two distinctive features which
contribute to the stability of the overall circuit. The zener diode is incorporated in a bridge circuit with the 1500-Ω, 15-KΩ, and 68-Ω resistors,

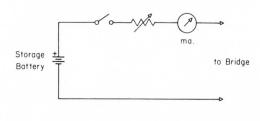

a.

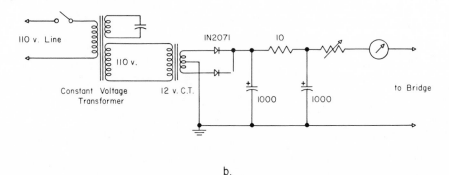

b.

Fig. 102.23. Simple bridge power supply circuits.

which has the effect of a first-order compensation for the dynamic impedance of the diode. Consequently, the stability of the bridge output is very much greater than that of the voltage across the diode itself.

The grounded potentiometer permits adjusting the thermal conductivity bridge output voltage relative to ground to approximately zero. This

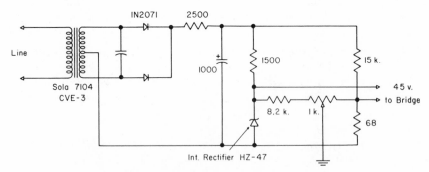

Fig. 102.24. Highly regulated bridge power supply.

adjustment minimizes the effects of insulation leakage on the signal, an important consideration at low signal levels.

c. Constant-Current Supplies

Constant-voltage operation of a bridge presupposes that the voltage at the bridge *terminals* is constant. In many industrial applications, however, the bridge may be separated from the power supply by a long cable of appreciable resistance, as well as by electrical connectors of appreciable contact resistance. Under these conditions, it is desirable to employ a constant-current rather than a constant-voltage source to eliminate the effects of these external resistances.

A constant-current source can be obtained by use of a constant-current ballast tube between power supply and bridge. More refined control is possible by electronic regulation of the voltage drop across a resistor in series with the bridge (59). Constant-current supplies are also readily available commercially.

Constant-current supplies are not suitable for use with thermistor bridges because of the regenerative bridge action produced (see Section III.A.2.a).

4. Readout Devices

The bridge output is ordinarily measured with either a D'Arsonval meter or a potentiometric recorder. The factor involved in the selection of the appropriate meter or recorder is that of impedance matching for maximum sensitivity, as discussed below.

a. Meters

Meters are used primarily for portable instruments. Since meters draw power from the bridge, the circuitry ideally should be designed for maximum bridge power output rather than voltage output. Such a bridge is one in which all arms are equal in resistance, and in which the meter resistance equals the bridge resistance.

In general, the resistance of standard milli- and microammeters of reasonable cost is considerably higher than that of the usual hot-wire cell, so that the second criterion for sensitivity is not readily met. For example, the resistance of a typical 500-μA meter is 200 Ω. Hence a 100-mV output from a typical 20-Ω cell would bring the meter to midscale. The resistance of a typical 50-μA meter is 2000 Ω, so that 100 mV would still bring the meter only to midscale. It is clear, therefore, that portable meter instruments are limited in sensitivity. In practice, the applications for which such instruments ordinarily are employed require little sensitivity, and the necessary compromise is of negligible importance.

A means of reducing meter sensitivity is generally desirable if a wide range of gas concentration is to be measured. Although this objective can be accomplished by reducing bridge current, such a procedure requires calibration of the instrument at several current levels. A more satisfactory approach is to provide an adjustable resistance in parallel with the meter (or in series with it), so that a single calibration can be employed for all sensitivity levels.

b. RECORDERS

(1) Potentiometer

Most recorders used with thermal conductivity cells are of the potentiometric type, drawing no current from the bridge when the recorder is at balance. However, a current is drawn when the recorder-balancing motor is moving, so that it is still necessary to ensure that the recorder amplifier input impedance is higher than the resistance of the bridge. This is a matter for little concern in the use of hot-wire cells, since the input impedance of all modern electronic recorders is of the order of several hundred ohms or more, well above the resistance of the hot-wire bridge.

In the case of thermistor bridges, however, bridge resistance can be in the order of 1000 Ω, so that it becomes important to select a high-impedance recorder. Recorders are available with input impedances of 30,000 Ω and higher.

The sensitivity of modern recorders ranges from a minimum of 100 μV full scale up to any desired value. Consequently, an appropriate selection is determined only by the requirements of the application.

Sensitivity adjustment in conjunction with a recording potentiometer is accomplished most readily with a voltage-dividing network across the bridge output, which can be either stepwise or continuous. The resistance of the voltage divider, or attenuator, must be selected with care to compromise between loading the bridge and supplying adequate power to the recorder amplifier. As a rule of thumb, the attenuator resistance should be at least 10 times the bridge resistance, but the recorder input impedance should never be exceeded. Figure 102.25 shows a simple attenuator circuit suitable for use with a 1000-Ω thermistor bridge. The slider is shown at the point presenting maximum resistance to the recorder, slightly above the midpoint. (The resistance presented to the recorder is the parallel combination of the resistances in the two

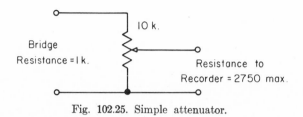

Fig. 102.25. Simple attenuator.

halves of the divider circuit.) Thus this circuit requires that the recorder input impedance be at least 2750 Ω.

(2) Self-Balancing Bridge

It is possible to dispense with the potentiometer circuit of the recorder, using its slide-wire as an active arm of the bridge, as shown in Fig. 102.26. In this case, the recorder amplifier and motor drive the slide-wire in a direction to reduce bridge output to zero, so that recorder pen position is a measure of bridge resistance change, rather than voltage change. The range of the instrument can be adjusted by varying the value of resistor R.

The major advantage of this circuit is the elimination of the separate voltage reference required in the potentiometric recorder. This advantage is all but negated, however, by the requirement for a nonstandard recorder and by the current availability of maintenance-free, zener-regulated potentiometer reference voltages.

5. Temperature Control

The degree of ambient temperature control required for a thermal conductivity cell will, of course, depend on the sensitivity requirements and on the symmetry of the cell (see Section II.E.4). In many applications involving large thermal conductivity changes, no thermostatting is required, particularly in manual operation, where frequent zero adjustments are not objectionable. If the cell is to be operated at ambient temperature, the effects of drafts and other interfering factors can be reduced to a negligible degree simply by placing the cell in a well-insulated box. One or two inches of Fiberglas batting is adequate.

For continuous operation at high sensitivity, a thermostat becomes mandatory to avoid frequent zero adjustments. An oven controlled to within ±1°C is adequate for measurements at the highest possible sensitivity, provided that a temperature-compensated bridge circuit is used.

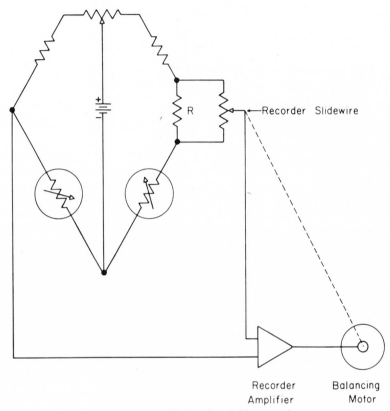

Recorder Slidewire

Recorder   Balancing
Amplifier   Motor

Fig. 102.26. Self-balancing bridge circuit.

(See Section III.A.2.c.) The temperature cycle introduced by the common
on–off type of controller can be isolated from the cell by providing adequate insulation between the oven walls and the thermal conductivity
cell. If proportional oven temperature control is employed, this expedient
becomes unnecessary.

When the thermal conductivity detector is used in connection with
the gas chromatograph, temperature control becomes far more critical.
This is especially true when temperature programming is employed. The
carrier-gas temperature may be different by 200–300°C from the cell
temperature, and in the case of subambient separations the differences
may be even greater. If the temperature is not adequate to compensate
for these differences, the detector drift makes high-sensitivity analysis
impossible. A heated, insulated transfer zone between the column oven

and the detector with a proportional temperature controller will, in large measure, minimize this problem. When a heated transfer zone is employed, it is advisable to operate this zone at a temperature a few degrees higher than that of the detector.

In these high-sensitivity gas chromatographic detectors, additional precautions must also be employed. Usually the detector is enclosed in a relatively large thermal mass heat sink, and the entire assembly insulated. Temperature control is then accomplished by placing heaters in both the detector block and the heat sink and using a proportional controller to regulate the power to the heater. The type and the placement of the temperature sensors used for power control to the detector are other important factors in achieving adequate temperature control. If control is required over a wide range of temperatures, a platinum-tape sensor is preferred. For temperature control over a relatively narrow range, a thermocouple will suffice.

By using a large thermal mass heat sink and a heated transfer zone, the detector drift can be reduced to the order of microvolts per hour. Such detectors, although extremely stable, require relatively long times to achieve thermal stability and are often operated at nonoptimum temperatures because of these long equilibration times.

6. Flow Control

In batch gas analysis, the most common sampling method involves filling the cell by purging with the sample, stopping the flow, and reading the indicator while the cell contains a static sample. A rubber-bulb aspirator is commonly employed when the sample is not under pressure. The time required to ensure complete cell purging depends on the flow rate, the cell and piping volume, and the extent of adsorption of the sample by the surfaces of the equipment. The required sampling time is readily determined by purging until duplicate readings are obtained.

An alternate batch sampling method involves evacuation of the cell and sample introduction by suitable valving. This method is to be preferred when the supply of sample is limited.

In the case of flowing streams, the degree of flow control required depends on the flow sensitivity of the cell. If the appropriate diffusion-type cell is employed, and the sample is at constant pressure, a simple needle valve control and a rotameter indicator are adequate for most purposes. The flow rate will be determined by the requisite response time and ordinarily will be in the order of 10–100 cc/min. For high-sensitivity measurements, a pneumatic flow controller should be employed.

A suitable instrument is the Moore Products Company (Philadelphia) differential flow controller, Type BDL or BUL.

An important consideration in flowing stream analysis is the possibility of back diffusion in the cell exhaust line, where the exhaust discharges to the atmosphere. Contrary to the common misconception, it is possible for air to diffuse against the direction of flow of a flowing stream. To prevent reading errors from this source, a suitably long, small-bore discharge line should be provided. For most purposes, 1–2 ft of $\frac{1}{4}$-in. tubing is adequate.

If the sample may contain suspended solids or liquids, a filter is mandatory to prevent cell contamination. Sad experience has shown that such a filter is a desirable precaution even when a clean sample is expected.

B. CALIBRATION

The technique employed for calibration of a thermal conductivity cell is dependent on the specific mixture involved and the auxiliary chemical or physical separations performed before the resulting mixture is introduced into the cell. As a result, no single calibration method can be used in all cases. In the selection of a calibration method, the analytical requirements of the particular measurement must also be kept in mind. In cases in which extreme precision is not required and the measurements are carried out over a small range of concentration with large differences in thermal conductivity, a single-point calibration is often sufficient. For example, the measurement of 0–10% carbon dioxide in air with a ±10% precision requirement can easily be accomplished with a single-point calibration even though the detector may be in a nonlinear portion of its response curve. On the other hand, for the determination of hydrogen in helium, admittedly a special and extremely difficult measurement, a large number of calibration points are needed. Regardless of the precision required, a linearity curve should be constructed for each mixture being measured. The linearity curve establishes the range of concentrations for which the detector is linear as well as the degree of curvature in the nonlinear portion of the response curve. After the degree of curvature has been established, and the possible errors resulting from this nonlinear response have been calculated, the degree and frequency of calibration can be determined.

In precise work, especially that involving a preseparation technique, frequent (i.e., daily) calibration checks must be made, since both sensitivity and linearity can change with small changes in instrumental variables. If the measurements are being made in the linear portion of the response curve and no preseparation steps are involved in the technique, less frequent calibrations are required. Generally, a change in cell parameters has a greater effect on zero stability than on the sensitivity and linearity of the detector. Hence the zero stability is a good indication as to whether calibration checks are required.

These observations have assumed that the gas stream and the detector with its associated controls are the only devices involved in a particular measurement; however, this is generally not the case, as the gas chromatograph is the most common application for the detector. In some respects, the use of a chromatographic preseparation method changes the calibration requirements. With the addition of the chromatograph, other variables must be considered: the column, the sample-introduction system, the temperature control, and the flow control. Of these, the most important are the chromatographic column and the flow control of the carrier gas. In most modern gas chromatographs, the temperature and flow are sufficiently well controlled and maintained so that they need not be of concern. The sampling system and the column, then, are the remaining variables to be taken into consideration in calibration. The quantitative problems associated with these variables can usually be detected when a linearity curve is constructed as part of the calibration procedure. A typical calibration-procedure linearity curve is shown in Fig. 102.27. In the construction of a linearity curve, increasing concentrations of the sample are measured and detector response is plotted against concentration. The following points are marked on this curve: B, minimum detectable concentration; A, extrapolation from minimum detectable level to zero; C, maximum linear response; and D, maxi-

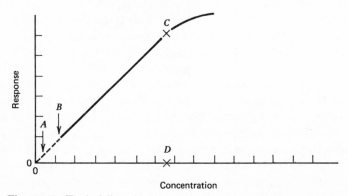

Fig. 102.27. Typical linearity curve for thermal conductivity detectors.

mum linear concentration. With the thermal conductivity detector, the plot should go through zero when extrapolated from the minimum detectable level and should have a linear dynamic range (linear increase in response as a function of an increase in concentration) of approximately 1×10^4.

Potential problems in calibration and analysis can usually be detected from linearity curves. As a rule, sampling difficulties produce somewhat smaller linear dynamic ranges and curves that do not go through zero. These sampling problems are usually associated with leaks in the sampling system or adsorption of the sample by the surfaces of the sampling system. Adsorption or reaction problems are generally reflected as S-shaped linearity curves with much reduced linear dynamic ranges. These nonlinear responses must be corrected if precise results are to be obtained.

1. Preparation of Standards

The preparation of accurate standards, the major problem in calibration, becomes more acute when low-level concentrations are required. Volumetric or monometric methods can be employed for the preparation of known mixtures, especially when relatively large concentrations are involved.

A very convenient way to prepare mixtures, where applicable, is to measure the components into an evacuated compressed-gas cylinder, computing concentrations from the ideal gas law. Mixtures so prepared to order are available from the Matheson Company, Rutherford. N.J. When using this method, it is important to use a cylinder valve with a large orifice, so that fractionation of the gas on withdrawal for calibration does not occur, and to avoid a partial pressure of any component in excess of its vapor pressure at room temperature. The method is not applicable to gases that might react with iron or iron rust. Alternatively, a gasometer can be employed to measure and store the mixture.

The disadvantages of the static method are stratification of the gases within the pressure vessel, leaks, contamination, and reaction or adsorption of the gases. The stratification problem can be minimized but not eliminated by agitation of the gases before sampling. A further disadvantage of the static method is the difficulty in changing the concentration of one of the constituents after the mixture has been prepared.

The adsorption problem is of particular concern with low concentrations of polar materials. Although it is possible to employ static mixtures containing small amounts of polar compounds (e.g., 100 ppm methanol in helium) without difficulty, this possibility must be checked before calibration. The adsorption problem can be conveniently solved by measuring the detector response with a full pressure vessel and then emptying the vessel over a period of a few hours and comparing the results.

When adsorption, reactivity, or instability prevents the use of stored standards, a dynamic calibration scheme becomes necessary, in which the components of the mixture are metered together as flowing streams. Valves and flowmeters can be used for this purpose, or the known vapor pressure of a liquid at a given temperature can be exploited by saturation of a diluent gas at a controlled temperature. In addition, several particularly ingenious techniques have been described.

a. DIFFUSION TUBES

Figure 102.28 shows a diffusion apparatus suitable for preparing low-concentration mixtures of a gas with the vapor of a liquid, originally

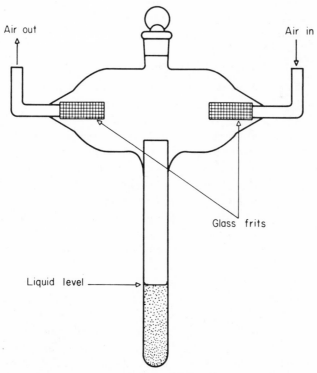

Fig. 102.28. Calibration apparatus using diffusion flow control (1). (Courtesy of *Analytical Chemistry*.)

described by McKelvey and Hoelscher (77) and refined by Altschuller and Cohen (1). The latter authors have analyzed and tested the performance of the apparatus quite thoroughly, finding that vapor mixtures up to about 1% by volume can readily be prepared.

A variation of this device is commercially available from Analytical Instruments Development, Inc., West Chester, Pa. This device is shown schematically in Fig. 102.29.

The diffusion rate of the vapor into the gas stream is described by the following equation:

$$r = \frac{DMPA}{RTL} \ln \frac{P}{P - p} \qquad (17)$$

where r = diffusion rate (g/sec),
　　　D = diffusion coefficient of vapor into diluent gas (cm²/sec),
　　　M = molecular weight of vapor,
　　　P = total pressure in cell (atm),
　　　A = cross-sectional area of diffusion tube (cm²),
　　　p = vapor pressure of liquid at temperature T (atm),
　　　R = gas constant (liter-atm/mole-°K),
　　　T = temperature (°K),
　　　L = length of diffusion path (cm).

The most difficult variables from equation (17) to control in a diffusion tube with a given gas are P, the total pressure in the cell; T, the temperature; and L, the length of the diffusion path.

The first two variables can be satisfactorily regulated by use of adequate gas pressure control and a thermostatted oven to contain the cell. The use of a bulb for a sample reservoir decreases the dependence of the diffusion rate as a function of change in path length due to loss of the sample. By the use of the method developed by Fuller et al. (33) for calculation of the diffusion coefficient of a compound, the molecular formula and the composition of a gas stream being known, the output of these devices can be predicted to about 5%. The output can also be measured by its weight loss over a period of time.

The diffusion tube is a useful device for compounds with widely varying vapor pressures. Martin et al. (76) have evaluated diffusion tubes for compounds ranging in vapor pressure from that of butane to that of squalene. The outputs obtained over this range of compounds varied from approximately 1000 µg/min for butane to 0.05 µg/min for squalene. Because the output of these devices can be both calculated on a theoretical basis and measured on an empirical basis, they should serve as a primary standard.

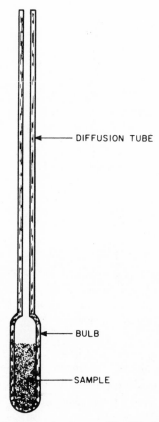

Fig. 102.29. Cross-sectional diagram of commercial diffusion tube. (Courtesy of Analytical Development Instrument, Inc.)

b. Permeation Tubes

O'Keeffe and Ortman (86) of the Taft Engineering Center have described the construction and showed the utility of the permeation tube for the generation of primary standard gas mixtures. A cross-sectional diagram of a typical permeation tube is shown in Fig. 102.30. A permeation tube is usually constructed by sealing liquid into a Teflon tube of known dimensions. The tube is allowed to equilibrate at a constant temperature for an induction period lasting several days. After the permeation has come to equilibrium, the rate is constant over a period of several months. The exact mechanism of diffusion through the plastic

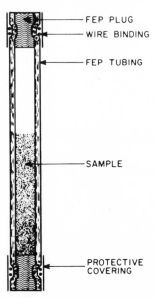

Fig. 102.30. Cross-sectional diagram of commercial permeation tube. (Courtesy of Analytical Development Instrument, Inc.)

is not well understood, and therefore the permeation rate cannot be predicted. Accurate calibration of the tubes, once they have been constructed, can be achieved by measuring the weight loss as a function of time at a constant temperature.

The variables which control the output of a permeation tube are the vapor pressure of the sample, the temperature, the length of the permeation path, and the solubility and diffusion coefficient of the sample in the polymer used for the tube. The wall thickness and cross-sectional area of the tube are probably also involved in the permeation rate. With a given tube and sample, the output is then dependent on temperature, as the other variables either are fixed by the polymer and the tube dimensions or are temperature dependent. In order to change the exit-gas concentration with a given tube, only the temperature need be varied. As the output of these tubes is directly proportional to the path length, the output can also be changed by constructing a new device of different path length.

If several gas concentrations are required within the range of the permeation tube, several tubes of varying lengths probably constitute the most convenient method of obtaining these mixtures.

c. Permeation Wafers

For very low gas concentrations the use of either diffusion tubes or permeation tubes is not satisfactory. The authors of reference 76 have designed a device to produce concentrations in the range of 0.0001–0.005 μg/min. This device, shown schematically in Fig. 102.31, is a modification of the permeation tube but employs a wafer rather than a tube, thereby reducing the gas output.

The sample is contained in a glass vial, sealed against the wafer by means of a heavy compression spring. By proper adjustment of the spring, a pressure of up to 30 atm can be contained in the vial. Aside from the materials of construction and the specific compound used, the

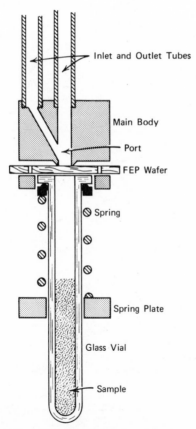

Fig. 102.31. Cross-sectional diagram of commercial permeation wafer. (Courtesy of Analytical Development Instrument, Inc.)

primary variable with the wafer device is the temperature. The output of any given device is proportional to $1/T°C$.

The major disadvantage of the wafer device is the difficulty in determining its output, as no theory is currently available for predicting the exit-gas concentration. The weight of the permeation wafer and its low output make weight loss versus time an impractical method of calibration. Thus far, this device can serve only as a secondary standard.

d. COMPARISON OF THE PERMEATION–DIFFUSION DEVICES

With the three devices just described, a range of gas concentrations can be generated over about 7 orders of magnitude, depending on the choice of device, temperature, dimensions, and other variables. A summary of the output rates for these three methods is shown in Fig. 102.32

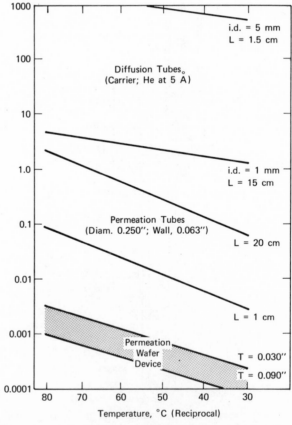

Fig. 102.32. Comparison of output rates of diffusion and permeation apparatus. (Courtesy of Analytical Development Instrument, Inc.)

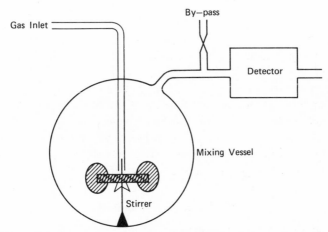

Fig. 102.33. Calibration apparatus using exponential dilution (73). (Courtesy of Butterworth and Co., Ltd.)

for butane. Similar relationships can be developed for other compounds by either calculation or measurement.

e. Exponential Dilution Flask

The exponential dilution flask described by Lovelock (73,74) is shown diagrammatically in Fig. 102.33. With this apparatus, a steady stream of diluent gas flows through the mixing vessel and then passes to the thermal conductivity cell or to the inlet of a gas-sampling system. The sample gas is introduced into the apparatus immediately upstream of the mixing vessel by means of a hypodermic syringe. The concentration of the exit gas decreases with time according to the following relationship:

$$C = C_0 \exp - \left(\frac{Vt}{u} \right) \qquad (18)$$

where C = gas concentration at time t, C_0 = initial gas concentration, u = gas flow rate, and t = time from injection of sample.

With the exponential dilution flask, it is theoretically possible to prepare known mixtures over an extremely wide range of concentrations down to the noise level of the instrument. The exponential method is suitable for preparing calibrated gas mixtures using a wide range of compounds, even though the vapor pressure of a particular compound may be quite low.

The variables associated with the exponential dilution technique, with a given vessel and sample, are temperature, flow control of the diluent gas, and concentration of the sample gas added to the flask. Temperature and flow rate can be adequately controlled by proper attention to the design of the system. The most difficult variable to control is the concentration of sample injected into the vessel. In order to increase the accuracy of the system, relatively large samples should be used for the initial concentration to decrease gas-measuring errors.

The exponential dilution flask is an especially valuable device when calibrations over a very wide range of concentrations are required. It is not well suited, however, to single-point or limited-point calibration. In addition, adsorption effects occurring when very polar samples are used can be significant at the lower concentrations.

2. Equivalent Standards

In some applications, the quantity of interest is not the actual sample composition, but rather some process variable which is reflected in sample concentration changes. An example is the analysis of automobile engine exhaust to facilitate carburetor adjustment. In such cases, a standard can be prepared containing, for example, a mixture of air and carbon dioxide equal in conductivity to the multicomponent exhaust gas at the optimum carburetor setting. This mixture can be sealed in the reference cavity of the cell, and the instrument used simply as a null indicator during carburetor adjustment. Because of the possibility of leakage, however, the usual procedure is to use air as the reference gas and to calibrate the indicator reading against a standard.

The preparation of some standards entails considerable care, using laboratory equipment, whereas a convenient, portable standard may be desirable for field-testing an instrument. In such cases, it is possible, after initial calibration, to prepare mixtures of other gases having thermal conductivities equal to those of known sample concentrations. For example, Cherry (16) found that mixtures of oxygen and air could be used for checking the calibration of a cell intended to measure low concentrations of water in air. As shown in Table 102.V, this equivalent

TABLE 102.V
Concentrations of Oxygen–Air and Water–Air Mixtures
of Equivalent Thermal Conductivity (16)[a]

Oxygen, %	Equivalent water, %
27.3	0.79
32.7	1.61
34.7	1.92

[a] Courtesy of *Analytical Chemistry*.

standard has the additional advantage that easily prepared high concentrations of oxygen correspond to more difficulty prepared low concentrations of water. The validity of an equivalent standard must, of course, be established empirically for any given cell and operating conditions.

C. APPLICATIONS

The early and well-established applications of thermal conductivity gas analysis have been thoroughly treated by Daynes (22) and Weaver (129). Weaver estimated that by 1951 more than 100,000 instruments had been produced, half of them designed for exhaust gas analysis. Another large fraction are intended for flue gas analysis, particularly for oil-burner adjustment. These rather simple and insensitive instruments usually employ either a desiccant to remove water vapor from the sample or a wick to saturate the gas with water. The latter method is preferred, since it eliminates the need for periodic replacement of the desiccant.

Other applications discussed by Daynes and Weaver include the analysis of metallurgical heat-treatment furnace atmospheres, as well as the measurement of oxygen in flue gas and in blast furnace feed, hydrogen and nitrogen in synthetic ammonia manufacture, oxygen and ammonia in nitric acid production, sulfur dioxide in the contact sulfuric acid process, and volatile solvents in air. Since the subject of applications has been treated most competently in both depth and breadth by these authors, the material will not be repeated here. Instead, a few examples will be discussed, to illustrate the approaches to various types of application.

In addition to the common applications mentioned above, the most general current use of thermal conductivity analyzers undoubtedly lies in detectors for gas chromatography. In this application, analysis of virtually any material displaying appreciable vapor pressure is possible by virtue of the separating power of the chromatographic column and the freedom of choice of carrier gas. Although this subject is discussed in detail in Part 1, Vol. 3, Chapter 37, of this Treatise, the increased sensitivity of thermal conductivity detectors since the publication of that treatment permits analyses not possible at that time. An example of the sensitivities currently obtainable is given in Fig. 102.34. In Fig. 102.34a the sensitivity normally expected with thermal conductivity detectors, that is, a minimum detectable level for neon of about 15–20 ppm, is shown. With new detector design, the detector output can be amplified 50–100 times and still produce acceptable base lines, as illustrated in Fig. 102.34b. The room air sample was diluted so that the neon concentration was 2.5 ppm. With this amplification, a minimum

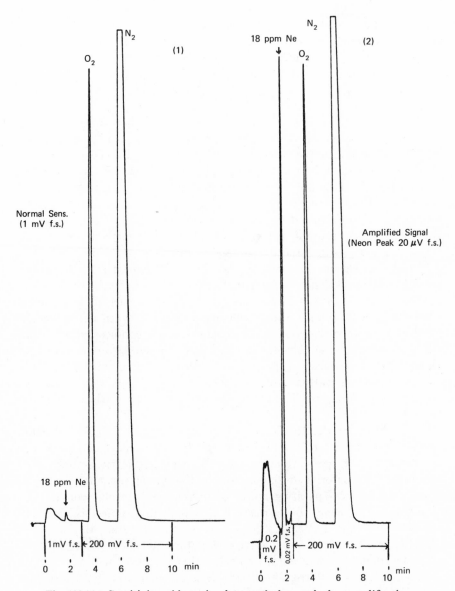

Fig. 102.34a. Sensitivity of hot-wire detector before and after amplification.

Fig. 102.34b. Minimum detectable signal for neon with amplified output. Operating conditions for 34A and 34B: instrument: gas chromatograph, Hewlett-Packard Model 7622A; columns: $12' \times \frac{1}{8}''$ o.d. packed with 60/80 mesh molecular sieve 5A; column temp.: 40°C; detector temp.: 125°C; Flow rate: helium at 24 ml/min; Bridge current: 300 mA. (Courtesy of Hewlett-Packard, Avondale Division.)

detectable level is in the range of 1 ppm. Similar improvements in minimum detectable levels can also be expected for other compounds.

Most applications are more specialized than in the case of gas chromatography. A good example of a special-purpose instrument is that devised by Stewart and Squires (100) for the measurement of parahydrogen in orthohydrogen. Their application required a small-volume cell capable of immersion in liquid nitrogen. This they constructed of glass tubing, which contained a single filament of tungsten supported from tungsten leads sealed in glass. The signal was sufficiently large to permit calibration directly in terms of filament voltage drop, rather than by the more conventional bridge circuitry. Measurement errors were about 0.1% over the range of 0–25% parahydrogen.

Twigg (106) used a similar single-filament cell to study the equilibration of hydrogen–deuterium mixtures. In this case the mixture was analyzed in the cell, exposed to a red hot nickel wire, and reanalyzed.

Combustible materials in flue gases can be determined by a thermal conductivity measurement before and after passage of the sample through a furnace (114). A simpler method involves the use of catalytically active platinum filaments in the cell (113). This latter method, commonly employed in so-called explosimeters, actually does not measure thermal conductivity, since the filament temperature rise is produced by the heat of combustion, rather than a change in conductivity. In construction and circuitry, however, the instruments are identical to thermal conductivity apparatus.

An ingenious and time-saving instrument for simultaneous carbon and hydrogen determinations in organic materials was developed by Sommer et al. (99). Their apparatus is shown in Fig. 102.35. A mixture of sample

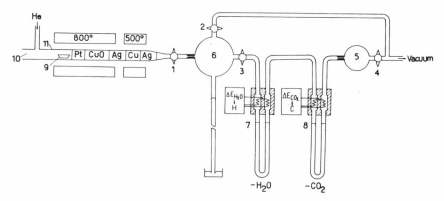

Fig. 102.35. Apparatus for simultaneous carbon–hydrogen determination (99).

and oxidizing agent is heated in a helium atmosphere, and the resulting combustion products are freed of interfering substances with silver and copper gauze. The remaining water vapor and carbon dioxide are swept with helium into an evacuated apparatus containing two differential thermal conductivity cells. A desiccant is placed between the two sections of the first cell, so that cell output represents water concentration. An alkaline carbon dioxide absorbent between sections of the second cell causes the output of that cell to represent carbon dioxide concentration. Table 102.VI shows the precision and accuracy of the method.

A rather elaborate application was described by Kieselbach (59), using an instrument designed for the automatic determination of total dissolved organic matter in chemical plant waste streams. This instrument, iself a miniature chemical plant, removed carbonates from the sample by precipitation, converted organic matter to carbon dioxide by wet combustion, and swept the carbon dioxide through a thermal conductivity cell with a stream of oxygen. The gas stream was freed of acid mist with granulated zinc, of halogens with antimony, and of water with Drierite. A differential thermal conductivity measurement was made, using Ascarite to absorb the carbon dioxide. In spite of its complexity, the instrument is capable of reliably detecting sample concentration changes as low as 1 ppm carbon, which corresponds to 0.005% carbon dioxide in the oxygen stream.

A specialized multicell instrument was described by D'Ouville and Howe (25) for the determination of methane, carbon monoxide, carbon dioxide, nitrogen, and hydrogen in Fischer–Tropsch synthesis gas. The separate components were determined by differential measurements in-

TABLE 102.VI
Precision and Accuracy of Carbon–Hydrogen Apparatus

Compound	Weight, g	Per cent carbon		Per cent hydrogen	
		Theoretical	Found	Theoretical	Found
Benzoic acid	0.5370	68.84	68.50	4.95	5.04
	0.6255		68.89		5.06
	1.1881		68.63		4.79
Acetanilide	0.6704	71.09	70.95	6.71	6.33
p-Nitrophenol	0.9033	51.80	52.21	3.62	3.71
Benzyl disulfide	0.9192	68.25	68.48	5.73	5.57
Chlorodinitrobenzene	1.9102	35.58	35.60	1.49	1.45
	1.5780		35.35		1.42

volving catalytic combustion and absorption. Although such elaborations are feasible, gas chromatography is now much to be preferred for the analysis of multicomponent mixtures.

An indication of the range of applicability of thermal conductivity gas analysis is given in Table 102.VII, which shows examples of binary

TABLE 102.VII
Examples of Gas Mixtures Analyzed with Commercial Thermal
Conductivity Instruments
I. Binary Mixtures

Minor component	Major component
Ammonia	Air
Argon	Helium
Carbon dioxide	Air
Carbon dioxide	Flue gas
Carbon dioxide	Helium
Freons[a]	Air
Helium	Air
Hydrocarbons	Helium
Hydrogen	Air
Hydrogen	Carbon dioxide
Hydrogen	Chlorine
Hydrogen	Hydrogen chloride
Hydrogen	Methane
Hydrogen	Nitrogen
Hydrogen	Oxygen
Methanol	Air
Methyl bromide	Air
Natural gas	Air
Nitrogen	Argon
Oxygen	Air
Oxygen	Hydrogen
Propane	Air
Sulfur dioxide	Air

II. Multicomponent Mixtures

Compound	Background
Carbon dioxide	Hydrogen and nitrogen
Carbon dioxide	Hydrogen, carbon monoxide, nitrogen oxygen, hydrocarbons
Ethylene oxide	Carbon dioxide and air
Hydrogen	Hydrocarbons, nitrogen, ammonia
Hydrogen	Carbon monoxide, hydrocarbons, nitrogen

[a] Du Pont's registered trademark for its fluorocarbons.

mixtures taken from the applications bulletins (26,70,36) of various manufacturers. Also listed are multicomponent mixtures capable of analysis with a convention-compensating cell (80).

REFERENCES

1. Altshuller, A. P., and I. R. Cohen, *Anal. Chem.,* **32,** 802 (1960).
2. Archer, C. T., *Phil. Mag.,* **19,** 901 (1935).
3. Archer, C. T., *Proc. Roy. Soc. (London),* **A165,** 474 (1938).
4. Arnett, E. M., M. Strem, N. Hepfinger, J. Lipowitz, and D. McGuire, *Science,* **131,** 1680 (1960).
5. Bennett, C. E., S. dal Nogare, L. W. Safranski, and C. D. Lewis, *Anal. Chem.,* **30,** 898 (1958).
6. Bennett, L. A., and R. G. Vines, *J. Chem. Phys.,* **23,** 1587 (1955).
7. Bock, H., *Ann. Physik,* **8,** 134 (1950), through *Chem. Zentr.,* **1951,** I, 3306, through *Chem. Abstr.,* **48,** 11860g (1954).
8. Bohemen, J., and J. H. Purnell, *J. Appl. Chem.,* **8,** 433 (1958).
9. Bolland, J. L., and H. W. Melville, *Nature,* **140,** 63 (1937).
10. Bolland, J. L., and H. W. Melville, *Trans. Faraday Soc.,* **33,** 1316 (1937).
11. Boreham, G. R., and F. A. Markoff, "Fuel Gas Analysis: An Apparatus Incorporating a Multi-cell Thermal Conductivity Detector," in *Gas Chromatography 1960,* R. P. W. Scott, ed., Butterworths, London, 1960, p. 412.
12. Brokaw, R. S., *Ind. Eng. Chem.,* **47,** 2398 (1955).
13. Browning, L. C., and J. O. Watts, *Anal. Chem.,* **29,** 24 (1957).
14. Brüche, E., and W. Littwin, *Z. Physik,* **67,** 362 (1931).
15. Bullowa, J. G. M., *Instruments,* **7,** 180 (1934).
16. Cherry, R. H., *Anal. Chem.,* **20,** 958 (1948).
17. Cowan, C. B., and P. H. Stirling, "The Selection and Operation of Thermistors for Katharometers," in *Gas Chromatography,* V. J. Coates, H. J. Noebels, and I. S. Fagerson, eds., Academic, New York, 1958, p. 165.
18. Curie. M., and A. Lepape, *Compt. Rend.,* **193,** 842 (1931).
19. Curie, M., and A. Lepape, *J. Phys. Radium* (7), **2,** 392 (1931) through Weaver, ref. 124.
20. Dal Nogare, S., and R. S. Juvet, Jr., *Gas–Liquid Chromatography,* Interscience, New York, 1962, p. 201.
21. Davies, A. J., and J. K. Johnson, "A Thermal Conductivity Detector for Use at High Temperatures in Vapour Phase Chromatography," in *Vapour Phase Chromatography,* D. H. Desty, ed., Butterworths, London, 1957, p. 185.
22. Daynes, H. A., *Gas Analysis by Measurement of Thermal Conductivity,* Cambridge University Press, London, 1933.
23. Dickens, B. G., *Proc. Roy. Soc. (London),* **A143,** 517 (1934).
24. Dimbat, M., P. E. Porter, and F. H. Stross, *Anal. Chem.,* **28,** 290 (1956).
25. D'Ouville, E. L., and K. E. Howe, *Petrol. Refiner,* **26,** No. 6, 526 (1947).
26. Engelhard, Charles, Inc., *Bull.* **3M-3-58.**
27. Faley, R. L., and J. F. Long, *Anal. Chem.,* **32,** 302 (1960).
28. Felton, H. R., "A Novel High-Temperature Gas Chromatography Unit," in *Gas Chromatography,* V. J. Coates, H. J. Noebels, and I. S. Fagerson, eds., Academic, New York, 1958, p. 131.

29. Felton, H. R., and A. A. Buehler, *Anal. Chem.,* **30,** 1163 (1958).
30. Fenwal Electronics, Inc., *Bull.* **EMC-4.**
31. Foz Gazulla, O. R., M. Colomina, and J. F. Garcia de la Banda, *Anales Real. Soc. Espan. Fis. Quim. (Madrid),* **44B,** 1055 (1948), through *Chem. Abstr.,* **43,** 4553b (1949).
32. Foz Gazulla, O. R., and S. S. Peréz, *Anales Fis. y Quim. (Madrid),* **39,** 399 (1943), through *Chem. Abstr.,* **42,** 6187i (1948).
33. Fuller, E. N., P. D. Schettler, and J. C. Giddings, *Ind. Eng. Chem.,* **58,** No. 5, 19 (1966).
34. Gow–Mac Instrument Co., *Bull.* **FIL 2-62-1M.**
35. Gow–Mac Instrument Co., *Bull.* **TCTH-1-62-3M.**
36. Gow–Mac Instrument Co., *Bull.* **GMH-3-61-3M.**
37. Gregory, H. S., *Proc. Roy. Soc. (London),* **A149,** 35 (1935).
38. Gregory, H. S., and C. T. Archer, *Phil. Mag.* (7), **1,** 593 (1926).
39. Gregory, H. S., and C. T. Archer, *Proc. Roy. Soc. (London),* **A121,** 285 (1928).
40. Gregory, H. S., and C. T. Archer, *Phil. Mag.* (7), **15,** 301 (1933).
41. Gregory, H. S., and S. Marshall, *Proc. Roy. Soc. (London),* **A118,** 594 (1928).
42. Gribkova, S. I., *J. Exptl. Theoret. Phys. (USSR),* **11,** 364 (1941), through Weaver, ref. 124.
43. Grob, R. L., D. Mercer, T. Gribben, and J. Wells, *J. Chromatog.,* **3,** 545 (1960).
44. Harvey, D., and G. O. Morgan, "Factors Affecting Thermal Conductivity Detectors in Vapour Phase Partition Chromatography," in *Vapour Phase Chromatography,* D. H. Desty, ed., Butterworths, London, 1957, p. 74.
45. Hawkins, G. A., *Trans. Am. Soc. Mech. Engrs.,* **70,** 655 (1948).
46. Hoffman, E. G., *Anal. Chem.,* **34,** 1216 (1962).
47. Horrocks, L. A., D. G. Cornwell, and J. B. Brown, *J. Lipid Res.,* **2,** 92 (1961).
48. Ibbs, T. L., and A. A. Hirst, *Proc. Roy. Soc. (London),* **A123,** 134 (1929).
49. Ishikawa, F., and K. Yagi, *Bull. Inst. Phys. Chem. Res. (Tokyo),* **22,** 12 (1943), through *Chem. Abstr.,* **41,** 6787b (1947).
50. Ishikawa, F., and I. Yamaryo, *Bull. Inst. Phys. Chem. Res. Chem. Ed. (Tokyo),* **23,** 311 (1944), through *Chem. Abstr.,* **42,** 5289d (1948).
51. Jakob, M., and G. A. Hawkins, *Elements of Heat Transfer,* 3rd ed., Wiley, New York, 1957, p. 30.
52. Jamieson, G. R., *Analyst,* **84,** 74 (1959).
53. Jamieson, G. R., *J. Chromatog.,* **3,** 464 (1960).
54. *Ibid.,* **3,** 494 (1960).
55. *Ibid.,* **4,** 420 (1960).
56. Kannuluik, W. G., *Nature,* **137,** 741 (1936).
57. Kannuluik, W. G., and L. H. Martin, *Proc. Roy. Soc. (London),* **A144,** 496 (1934).
58. Keppler, J. G., G. Dijkstra, and J. A. Schols, "Vapour Phase Chromatography at High Temperatures," in *Vapour Phase Chromatography,* D. H. Desty, ed., Butterworths, London, 1957, p. 222.
59. Kieselbach, R., *Anal. Chem.,* **26,** 1312 (1954).
60. *Ibid.,* **32,** 1749 (1960).
61. Killheffer, J. V., and E. Jungermann, *J. Am. Oil Chemists' Soc.,* **37,** 456 (1960).
62. Koepsel, A., *Verhandl. Deut. Physik. Ges.,* **10,** 814 (1908), through Weaver, . ref. 124.
63. *Ibid.,* **11,** 237 (1909), through Weaver, ref. 124.

64. Koepsel, A., *Chem. App.*, **3**, 377 (1908), through Weaver, ref. 124.
65. *Ibid.*, **3**, 401 (1908), through Weaver, ref. 124.
66. Kornfeld, G., and K. Hilferding, *Z. Physik. Chem. Bodenstein Band,* **792** (1931), through Weaver, ref. 124.
67. Kolloff, R., Pittsburgh Conference on Analytical Chemistry and Spectroscopy, Cleveland, Ohio, 1971.
68. Kreyenbuhl, A., *J. Chromatog.*, **4**, 130 (1960).
69. Laby, T. H., and E. A. Nelson, "Thermal Conductivity: Gases and Vapors," in *International Critical Tables*, E. W. Washburn, ed., McGraw–Hill, New York, 1929, Vol. 5, p. 213.
70. Leeds & Northrup Co., *Folder* **ND46-91**(6).
71. Lindsay, A. L., and L. A. Bromley, *Ind. Eng. Chem.*, **42**, 1508 (1950).
72. Littlewood, A. B., *J. Sci. Instr.*, **37**, 185 (1960).
73. Lovelock, J. E., "Argon Detectors," in *Gas Chromatography 1960*, R. P. W. Scott, ed., Butterworths, London, 1960, p. 16.
74. Lovelock, J. E., *Anal. Chem.*, **33**, 162 (1961).
75. Mann, W. B., and B. G. Dickens, *Proc. Roy. Soc. (London)*, **A134**, 77 (1931).
76. Martin, A. J., F. J. Debbrecht, and G. R. Umbreit, Technical Paper, Analytical Instrument Development, Inc., West Chester, Pa., 1970.
77. McKelvey, J. M., and H. E. Hoelscher, *Anal. Chem.*, **29**, 123 (1957).
78. Messner, A. E., D. M. Rosie, and P. A. Argabright, *Anal. Chem.*, **31**, 230 (1959).
79. Milverton, S. W., *Phil. Mag.*, **17**, 397 (1934).
80. Mine Safety Appliances Co., *Bull.* **0716-2.**
81. Minter, C. C., *Anal. Chem.*, **19**, 464 (1947).
82. Minter, C. C., and L. M. J. Burdy, *Anal. Chem.*, **23**, 143 (1951).
83. Nothdurft, W., *Ann. Physik*, **28**, 137 (1937), through Weaver, ref. 124.
84. *Ibid.*, **28**, 157 (1937), through Weaver, ref. 124.
85. Ogilvie, J. L., M. C. Simmons, and G. P. Hinds, Jr., *Anal Chem.*, **30**, 25 (1958).
86. O'Keeffe, A. E., and G. C. Ortman, *Anal. Chem.*, **38**, 760 (1966).
87. Palmer, P. E., and E. R. Weaver, *Natl. Bur. Std. (U.S.) Tech. Paper* **249,** 1924.
88. Peterson, A. C., A. J. Madden, Jr., and E. L. Piret, *Ind. Eng. Chem.*, **46,** 2038 (1954).
89. Purcell, J. R., and R. N. Keeler, *Rev. Sci. Instr.*, **31**, 304 (1960).
90. Ray, N. H., *Nature,* **182**, 1663 (1958).
91. Ryce, S. A., P. Kebarle, and W. A. Bryce, *Anal. Chem.*, **29**, 1386 (1957).
92. Sasaki, N., K. Tominaga, and M. Aoyagi, *Nature*, **186**, 309 (1960).
93. Schmauch, L. J., *Anal. Chem.*, **31**, 225 (1959).
94. Schmauch, L. J., and R. A. Dinerstein, *Anal. Chem.*, **32**, 343 (1960).
95. Scott, D. S., and A. Han, *Anal. Chem.*, **33**, 160 (1961).
96. Shakespear, G. A., *Proc. Phys. Soc. (London)*, **33**, 163 (1921).
97. Sherratt, G. G., and E. Griffiths, *Phil. Mag.* (7), **27**, 68 (1959).
98. Snowden, F. C., and R. D. Eanes, *Ann. N.Y. Acad. Sci.*, **72**, Art. 13: 764 (1959).
99. Sommer, P. F., W. Sauter, J. T. Clerc, and W. Simon, *Helv. Chim. Acta*, **45**, 595 (1962).
100. Stewart, A. T., and G. L. Squires, *J. Sci. Instr.*, **32**, 26 (1955).
101. Stolyarov, E. A., *Zh. Fiz. Khim.* **24**, 279 (1950), through *Chem. Abstr.*, **44**, 6694f (1950).

102. Stops, D. W., *Nature,* **164,** 966 (1949).
103. Stuve, W., "A Simple Katharometer for Use with the Combustion Method," in *Gas Chromatography 1958,* D. H. Desty, ed., Butterworths, London, 1958, p. 178.
104. Trautz, M., and A. Zündel, *Ann. Physik,* **17,** 345 (1933), through Weaver, ref. 124.
105. Trenner, N. R., *J. Chem. Phys.,* **5,** 382 (1937).
106. Twigg, G. H., *Trans. Faraday Soc.,* **33,** 132 (1937).
107. Ubbelohde, A. R., *J. Chem. Phys.,* **3,** 219 (1935).
108. Ubbink, J. B.. *Physica,* **13,** 629 (1947), through *Chem. Abstr.* **42,** 6188a (1948).
109. *Ibid.,* **13,** 659 (1947), through *Chem. Abstr.,* **42,** 6188c (1948).
110. Ulsamer, J., *Z. Ver. Deut. Ing.,* **80,** 537 (1936), through Weaver, ref. 124.
111. U.S. Pat. 1,304,208 (May 20, 1919), G. A. Shakespear.
112. U.S. Pat. 1,504,707 (Aug. 12, 1924), J. C. Peters, Jr. (to Leeds & Northrup Co., Inc.).
113. U.S. Pat. 1,562,243 (Nov. 17, 1925), M. Moeller (to Siemens and Halske Aktiengesellschaft).
114. U.S. Pat. 1,681,074 (Aug. 14, 1928), C. H. Porter.
115. U.S. Pat. 1,698,887 (Jan. 15, 1929), R. H. Krueger (to Charles Engelhard, Inc.).
116. U.S. Pat. 1,715,374 (June 4, 1929), R. H. Krueger (to Charles Engelhard, Inc.).
117. U.S. Pat. 1,802,713 (Apr. 28, 1931), W. O. Hebler (to Charles Engelhard, Inc.).
118. U.S. Pat. 1,918,702 (July 18, 1933), W. O. Hebler and W. F. Hamilton (to Charles Engelhard, Inc.).
119. U.S. Pat. 2,596,992 (May 20, 1952), J. G. Fleming (to Cambridge Instrument Co., Inc.).
120. U.S. Pat. 2,633,737 (Apr. 7, 1953), R. D. Richardson (to Cambridge Instrument Co., Inc.).
121. U.S. Pat. 2,734,376 (Feb. 14, 1956), R. H. Cherry and G. M. Foley (to Leeds & Northrup Co., Inc.).
122. U.S. Pat. 2,759,354 (Aug. 21, 1956), R. H. Cherry and G. M. Foley (to Leeds & Northrup Co., Inc.).
123. U.S. Pat. 2,926,520 (Mar. 1, 1960), L. J. Schmauch (to Standard Oil Co. of Indiana).
124. U.S. Pat. 3,080,745 (Mar. 12, 1963), W. L. Jones (to E. I. du Pont de Nemours & Co.).
125. Van de Craats, F., "Some Quantitative Aspects of the Chromatographic Analysis of Gas Mixtures, Using Thermal Conductivity as Detection Method," in *Gas Chromatography 1958,* D. H. Desty, ed., Butterworths, London, 1958, p. 248.
126. Varkaftik, N. B., and I. D. Parfenov, *J. Exptl. Theoret. Phys. (USSR),* **8,** 189 (1938), through Weaver, ref. 124.
127. Walker, R. E., and A. A. Westenberg, *Rev. Sci. Instr.,* **28,** 789 (1957).
128. Wassiljewa, A., *Physik. Z.,* **5,** 737 (1904).
129. Weaver, E. R., "Gas Analysis by Methods Depending on Thermal Conductivity," in *Physical Methods in Chemical Analysis,* W. G. Berl, ed., Academic, New York, 1951, Vol. 2, p. 387.
130. Weber, S., *Ann. Physik,* **82,** 478 (1927), through Weaver, ref. 124.
131. Young, I. G., "The Sensitivity of Detectors for Gas Chromatography," in *Gas Chromatography,* H. J. Noebels, R. F. Wall, and N. Brenner, eds., Academic, New York, 1961, p. 75.

Chapter 103

GAS ANALYSIS: DETERMINATION OF GASES IN METALS

By William S. Horton, *National Bureau of Standards, Washington, D.C.*, and Chester C. Carson, *General Electric Company, Schenectady, New York*

Contents

Contents (*Continued*)

I. INTRODUCTION

A. IMPORTANCE

It seems very unlikely that the experienced analytical chemist is unaware of the importance to be attached to knowledge of the gaseous impurity (or additive) content of metals. Too much has been written in the last 15–20 years (e.g., 126,190) to leave the chemist in such a state of ignorance. However, it may be of value to record here briefly some of the effects, both physical and chemical, of which gaseous elements are the cause. Not only the more widely recognized effects but also, perhaps, a few with which chemists may not be acquainted will be mentioned. Presumably it is unnecessary to dwell on the fact that these effects, and their possible cause by even minute quantities of gas, are what lend importance and draw attention to the determination of gaseous content.

In this chapter only the determination of oxygen, hydrogen, and nitrogen will be considered. This is not to deny the importance of other gases, such as chlorine and fluorine or carbon monoxide and carbon dioxide, existing as these molecules in the metal, but rather partially to limit the extent of the chapter and to recognize that these three are probably the gases most often determined. Even with this limitation the chapter is extensive because sufficient details for all methods are not available in any one other place. Although some general references to specific methods will be mentioned, it will often be true that more particulars are given in this chapter. The additional detail will not only aid the novice but also clarify significant factors in the analyses.

Obvious effects to be mentioned are the changes induced in the various types of strength (tensile, shear, etc.) and the concomitant properties of brittleness and hardness, specific elongation properties, and creep. Many such effects were known over 70 years ago. For example, nitrogen has been said to reduce the ductility of steel, and nascent hydrogen causes iron to become brittle (153). Iron and steel immersed in acid absorb hydrogen and become brittle (159). Some of this earlier work has been checked and not only confirmed but also supplemented (3). The elasticity of hydrogen-embrittled iron may be restored by removal of the gas. The presence of oxygen is attended by a large increase in strength. Iron in acidulated water, made cathodic, readily absorbs hydrogen at a voltage below the decomposition voltage of the electrolyte. The removal of gases from ferrous alloys markedly changes the microstructure and increases the density of the alloy.

Later evidence has accumulated on other metals (132). The tensile strength of zirconium is increased by hydrogen. Oxygen and nitrogen combine with this metal to form relatively high-strength alloys possessing good elongation values in the annealed condition. However, consider also that a surface oxide film produced by heating in air is sufficiently abrasive to reduce the life of a drawing die. The extreme hardness frequently attributed to beryllium is localized in a thin film of oxide on the surface. Often this is sufficiently hard to scratch glass.

Interestingly, it has also been pointed out that nitrogen can result in radioactivity when present in metals in nuclear radiation fields (42). For example, 76.5-sec O^{14} and 20.3-min C^{11} have been found as a result of (p, n) and (p, α) reactions, respectively, with N^{14}. Additive oxygen has produced a broad low-temperature anomaly in the heat capacity of gadolinium between 1.3° and 5°K (61). The anomaly is said to be the result of ordering of the spin $\frac{7}{2}$ Gd^{3+} ions of Gd_2O_3 molecules.

The presence of oxygen in metals can also influence the rates of reaction, such as absorption of hydrogen. Samples of zirconium having the room-temperature surface oxide present showed only a slow rate of reaction with hydrogen at 150°C (125). However, heating in vacuum to 700°C for 1 hr—which causes the surface oxide to diffuse into the bulk of the metal—gave specimens with a rate of reaction 7700 times as great. The aqueous corrosion resistance of tested Zircaloy-2 was found to decrease markedly after vacuum annealing to dissolve the oxide film (167).

Perhaps this is sufficient to convince the uninitiated that the gas content of metals is important enough to warrant efforts to determine this factor with reliable precision and accuracy. As will be seen in the next section, this need was recognized early by only a few scientists and only in approximately the last 15 years has been made an object of much study.

B. BRIEF HISTORY

Probably the earliest determination of gases in metals was made by Graham in 1863 (116). Placing a piece of meteoric iron in a porcelain tube which was heated to redness by a flame, he pumped off gases with a Sprengel mercury vacuum pump. By this method he found his sample to contain 2.85 volumes of gas—mostly hydrogen, although some carbon monoxide and nitrogen were also evolved—per volume of metal. Using horseshoe nails, he found 1 volume of metal to yield 2.66 volumes of gas: 52% CO, 30% H_2, the remainder CO_2 and N_2. About the same time Cailletet (41) observed that H_2, N_2, and CO were liberated from iron during fusion, a sort of qualitative analysis. Fifteen years later it was reported that carbon in steel reacted with porcelain tubes heated to around 1000°C and produced these same gases (215a). This was the first evidence of interference for this type of analysis. However, by heating iron to 1100°C in a quartz tube and keeping the iron out of contact with the tube, CO, CO_2, and H_2 were determined and nitrogen calculated by difference (109).

An interesting procedure was once used in Germany (206). Samples of iron and steel bored under water liberated gas that could be collected. This was found to be mostly hydrogen. Repetition under rape seed oil and under mercury demonstrated that chemical reaction with the water was not the source of the gas. Apparently the frictional heat of the drilling action was the cause.

An early dissolution method (114) was carried out by dissolving iron in a slightly acid solution of copper potassium chloride at 40°C. Carbon monoxide was determined. Of course the hydrogen had no analytical significance with this technique.

Austin (15) fused wires or bars of various metals other than iron and steel to attempt complete recovery of the gases. Kjeldahl procedures, used originally for organic nitrogen determinations, were modified in 1913 by Herwig for use with iron (138).

A serious effort was made in 1918 by Alleman and Darlington (3), to improve the high-temperature "degassing" methods, particularly in regard to the porosity of the containing vessels (tubes, etc.), the use of gas flames, and the lowness of the temperatures. These workers described two furnaces one of which was capable of attaining 1000° and the other 1500°C. Although reaction occurred with the silicon quartz tube used, the open-hearth steel specimen was completely fused and released about 18.6 cc of gas per gram of metal, of which 80% was CO. The amounts of various gases were, in decreasing order, $CO > H_2 > N_2 > CO_2$. No free oxygen was found. Alleman and Darlington noticed that in later, extended-time experiments both oxygen and nitrogen appeared, and in approximately the same ratio as in air. This led them to construct a third furnace in which an exterior surrounding chamber was evacuated to avoid diffusion of air. This furnace was used for "fractional vacuum-fusion" experiments in which the gases released at successively elevated temperatures were analyzed. The results showed the strong tendency to retain nitrogen.

Goerens and Paquet (110) believed, as did Austin, that it was necessary to fuse the metal in order to get all the gases out. Desiring to fuse completely within an evacuated system, they decided to try alloying iron in order to reduce the temperature. These authors settled on tin and on antimony in a magnesia crucible heated to 1150°C.

Probably no great strides were taken until a cooperative study (268) was made just before 1937, utilizing specially prepared heats of steel and iron. The study was under the joint sponsorship of the Iron and

Steel Division of the American Institute of Mining and Metallurgical Engineers and of the U. S. National Bureau of Standards. The main result of this work was an indication that the vacuum-fusion method, as then practiced, yielded accurate results for the steels employed for reference. A most useful by-product was a set of seven steels and one iron for which "best values" were chosen. For many years, these materials, made available to the public by NBS, were used as reference standards by which methods and apparatus could be checked. These aspects are discussed in Section VII of this chapter. Many versions of the vacuum-fusion apparatus have since appeared.

Another step forward was made with the advent of the Guldner–Beach furnace tube (128) for vacuum fusion, dispensing with the heavy brass furnace caps. Most furnace designs after 1950 followed the principles of this one.

The cooperative study reported by NBS in 1937 indicated that results for nitrogen and hydrogen obtained by vacuum fusion were reasonable. Since then, application to other metals has led to some reserve about nitrogen determinations, for which the Kjeldahl method is the outstanding procedure.

Other highlights in the history of these developing methods of analysis include the introduction in 1940 of the inert gas-fusion method (242), the introduction of the platinum bath technique in 1954 and 1955 (36,120), the use of isotopic oxygen in 1949 (122), and the introduction of practical spectrographic techniques in 1956 (93). First in 1954 (213a) and then in 1959 the beginnings of activation analysis were reported (26a,56a,180a). In 1965 (205a) the study of infrared absorption spectroscopy for determining oxygen in germanium was begun.

II. PHYSICAL METHODS

A. METALLOGRAPHIC EXAMINATION

The presence of gases in metals, aside from their presence in voids, presents at least three possibilities when viewed under the microscope with high magnification: there may be no visible evidence of the gas, there may be precipitation as inclusions, and the gas may be present as both precipitate and nonprecipitated impurity. In the latter two cases there is sometimes the possibility of qualitative or even quantitative determination of the impurity. The metallographer working with a given metal for an extended period of time learns to recognize typical grain shapes of inclusions such as oxides, nitrides, or hydrides of the metal.

Either by preparing standards or by reference to other methods it is possible to quantify the observations. Probably the chief advantage to be gained by this technique is that, when metallographic observation is to be made anyway, an estimation of the gas content using the same specimen may save much time and yet be adequate for the purpose.

An example of the metallographic method for determining gaseous constitutents in metals, in this case hydrogen in zirconium alloys (137), is available. A set of standard micrographs was prepared by the addition of measured amounts of hydrogen to Zircaloy-2 process tubes that had been in nuclear reactor service for $2\frac{1}{2}$ years. Part of a test loop, the tubes were not exposed to high-intensity radiation or to temperatures exceeding 300°C. The tube material is said to have been comparable to rolled sheet, although the crystallographic texture was different. Quantitative hydriding was performed above 550°C after bringing vapor-blasted specimens to temperature in vacuum. Presumably the vacuum-heating time was sufficient to remove all residual hydrogen previously present. After the addition, the specimens were homogenized by holding at temperature for 24 hrs. Furnace cooling permitted the hydrogen to precipitate as the hydride phase. The cooled specimens were mounted by a standardized metallographic procedure, and bright field micrographs prepared with standard photographic paper. Specimens containing 15, 60, 120, 220, 410, 580, and 810 ppm of hydrogen were prepared. The photographs reproduced appear quite different with respect to the number of hydride crystals visible per unit area. A minor improvement might have been made by preparing a 30-ppm specimen rather than the one containing 580 ppm. Since the eye generally responds in a logarithmic sense, a geometric series is advantageous. Also, the 410- and 580-ppm micrographs are not as easy to distinguish as the others. However, the purpose of the analyses may, of course, dictate more specimens in a region of hydrogen content critical to the user. If the suggestion were adopted and if only the numerical value corresponding to the most similar micrograph were reported for an answer, it appears that a relative standard deviation at least as good as 30% might be expected. This would be for unknowns *within* the range and represents the matching process only.

The metallographic method is said to be useful when massive pieces show inhomogeneous hydrogen distribution. Thus a significant concentration gradient occurring over a cross section the size of a hot-extraction specimen can be examined in detail not available to the hot-extraction technique. Hydrogen concentrations may also be estimated in relation to other microstructural details such as cracks or phase transformations. On the other hand, only hydride present as the hydride phase will be observed. Ideally the metallurgical history of unknowns and standards should be the same. There appears to be some confusion about the "Sampling error." Although there are cases in which the entire specimen cannot be handled by vacuum methods, such specimens are not necessarily analyzed better by the metallographic method. "Preparation of a standard micrograph requires a visual scan of the specimen, followed by selection of an area where the hydride density appears representative of the whole" (137). This is certainly also true if one is analyzing an incoming shipment of material.

Borrowing another metallurgical technique, Vasil'ev et al. (274c) related X-ray diffraction data to the oxygen concentration in the surface layers of titanium specimens.

B. MECHANICAL PROPERTIES

As indicated in Section I.A, the properties of metals are often drastically affected by the presence of small amounts of gaseous impurity. Consideration of these properties as indicative of impurity content is perhaps worthwhile, although it is not seriously suggested that in general this is likely to be useful. Consider some of the data first.

In some cases hardness may serve as an indicator of oxygen content. An advantage of using hardness measurements is that often a small sample may suffice, alternatively, a metallographic specimen prepared for microscopic examination may also be employed for a hardness measurement. However, hardness may be due to one or more of a number of causes other than the gaseous impurity content, such as the carbon content or the extent of hot or cold working. In other cases there may be no obvious relation. Zirconium specimens, for example (124c), heated for 1 hr in nitrogen did not show a significant positive correlation between weight gain for a given temperature and Rockwell or Knoop hardness. The same was true when the specimens were heated for 1 hr in air and for 24 hr in nitrogen. There was evidence that the hardness dropped rapidly for the initial weight gain and then more slowly rose again. On the other hand, niobium (175b) showed a very significant relation between oxygen content and Vickers harness, as long as the same type of material was examined. Bars from sintered niobium pentoxide–niobium carbide mixtures with oxygen content between 0.026 and 2.81 wt % had well-correlated hardness values between 52 and 451. "Pure" niobium specimens had from 0.079 to 0.55% oxygen and hardness between 116 and 274 in perfect rank correlation. However, the 0.60%

oxygen specimen had a 250 hardness value in this group, as compared with 451 for an 0.59% content specimen of the first-mentioned group. Also, the 274 hardness value of the pure niobium was associated with 0.55% oxygen, while 270 was obtained for a 2.26% specimen in the first group. For niobium oxygen alloys quite good rank correlation was obtained within the group also. Determination of hardness, therefore, may be useful when the metal is produced routinely with nearly constant impurity content and consistent treatment. Then the tests are made with minimal effort, time, and extra expense on specimens already prepared for another purpose.

In Table 103.I are shown results (51) of tensile strength studies for six specimens of zirconium. The data, taken from graphs, are indicative of the increase in strength at lower temperatures due to the presence of oxygen. There is a larger temperature coefficient for the specimens containing oxygen than for those without, so that above 600°F it would be difficult to distinguish between the two. For analytical purposes, of course, room-temperature tests would be preferable, and presumably an even larger difference than that shown would be exhibited. However, taking precision into account and the apparent change due to the amount of oxygen, it is not likely that differences in oxygen content of less than 150 ppm can be distinguished.

There are inherent difficulties in an approach of this kind. The main objection is to the precision with which the mechanical properties can generally be measured. Normally tests are extensively replicated in order to obtain an acceptable average value. Also militating against the concept is the usual specimen size for mechanical tests, except perhaps for hardness. Since other constitutents in an alloy produce mechanical changes, one needs calibration data for all pertinent alloys. It seems likely from these considerations that mechanical testing results may

TABLE 103.I

Tensile Strength of Zirconium with and without Oxygen in 10^3 psi

Temperature, °F	Clean	+0.08% O_2	Clean	+0.2% O_2	Clean	+0.08% O_2
100	33	57	37.5	63	35	64
300	21.5	37	28.5	41	23	42
500	18.5	25	23	27.5	18.5	28
600	18	22	20	21	18	21

be used only for indications of gross differences and even for this purpose only when the matrix material is well characterized with respect to the test involved.

III. CHEMICAL METHODS

A. CHLORINE, BROMINE, AND IODINE

This general method depends on heating the sample for some extended period of time in a stream of pure, dry chlorine or some other form of halogen. When the procedure is applicable, the metallic constituents are converted to volatile chlorides which are swept away from the reaction zone in the gas stream. Oxide constituents are not attacked and remain behind to be determined by appropriate means. Although the method has been criticized (261), it has been used with titanium (212) and titanium-base alloys (77). In the procedure described in reference 77 the solid sample is mixed with graphite and chlorinated in an atmosphere of argon; the products of reaction are subsequently isolated, excess chlorine being removed with antimony. The carbon monoxide produced, which is stoichiometrically related to the original oxygen present, is oxidized to the dioxide, absorbed, and weighed.

A very similar method uses bromine vapor (53). In this variation helium can serve as the carrier gas. Iodine has also been used, but as a solution—in aqueous ferrous iodide (63) or in anhydrous methyl alcohol (227). These methods have been used for iron and steel, which dissolve in the reagents, leaving behind carbides and oxides to be determined by suitable means.

B. OTHER HALOGENATION METHODS

The chlorination method using dry chlorine has been employed for zirconium (226) and has been modified, supposedly for greater convenience, by substituting dry hydrogen chloride (269). In general, of course, these methods suffer from a lack of specificity. Not only oxides but also carbides and other unremoved material remain behind. The success of the various methods depends strongly on how well the residue is characterized. In some cases a method applicable to a base metal is inapplicable to an alloy. This is true, for example, in the case of zirconium and uranium–zirconium alloy (237).

Hydrogen fluoride was used to determine oxygen present in zirconium in a form other than zirconium oxide (240). While dissolving the metal in the acid, the oxygen evolved was caught in a separate flask and ultimately determined as iodine. The range determined was between 50 and 170 μg on about 80-mg samples (0.06–0.22%).

Bromine trifluoride has also been used. This is said to liberate molecular oxygen at 75° from oxides of all metals which form volatile fluorides or fluorides soluble in BrF_3 (143).

Aqueous hydrogen chloride has been used for oxygen in chromium (1a,238). The procedure consisted of heating the sample to a high temperature in vacuum. Dissolution of the vacuum-"annealed" specimen in dilute hydrochloric acid was said to leave Cr_2O_3 behind. Furthermore all oxygen was supposedly converted to this oxide during the heating period. The residual oxide may be determined by a variety of methods. In particular, Horton and Brady (152a) introduced an alkaline persulfate fusion, converting the oxide to chromate, followed by colorimetric determination. The annealing–dissolution procedure has been criticized on the basis that black, oxide-bearing chromium forms a colloidal suspension which runs through the filter paper (179).

A chrome–manganese–nickel steel has been analyzed for nitrogen, using perchloric acid (260). The finely divided sample is dissolved in the acid with a bit of dilute sulfuric acid. After partial evaporation the residue is redissolved and treated essentially by a Kjeldahl procedure.

C. SULFUR AND HYDROGEN SULFIDE

Germanium has been analyzed for traces of oxygen by treating the metal with sulfur vapor between 700 and 800°C (16). Under these conditions the element is said to be converted to the sulfide, and oxygen to sulfur dioxide. The latter substance is determined colorimetrically with fuchsine–formaldehyde reagent. The sensitivity is reported to be 2 ppm.

Silver surfaces have been analyzed for oxygen by the use of hydrogen sulfide (259). One-millimeter-diameter wire, after being washed with solvent, is exposed to the dry gas at around 500° for about 30 min. Silver oxide is said to convert to the sulfide and water. Because hydrogen sulfide cannot diffuse through the metal, only the surface oxide is thereby determined. Total oxygen may be determined by the hydrogen reduction method.

A related method has been used in Japan (157), although for minerals rather than metals. A mixture of iron pyrites and an oxygen material is heated in vacuum. The evolved gases—SO_2, H_2S, and H_2O—are measured and analyzed. By a scheme of fractional vacuum fusion, whereby the reaction is carried out at successively higher temperatures, different oxygen-containing compounds may be determined, such as $FeSO_4$, $Fe_2(SO_4)_3$, and Fe_3O_4.

D. HYDROGEN REDUCTION

The obvious use of hydrogen to remove oxygen and be determined as water has been applied (19,20) to nonferrous metals such as lead, lead–tin alloys, and copper. Although the original method used pressure differentials to determine the water, other means are obvious to the chemist wishing to adapt this principle.

In an unusual method involving hydrogen reduction a metal sample is heated in a graphite boat under a few millimeters of mercury pressure of hydrogen at 1100° (164). The cooled gas is purged through a nickel–thorium catalyst and then through cinnamoyl chloride heated to about 65°. Apparently, oxygen has become water which evolves HCl quantitatively from the organic acid chloride. The HCl is absorbed in water and titrated. It would appear that care must be exercised to avoid any reduction by the graphite, as this would lead to low results.

Dulski and Raybeck (73a) found that they could get a good measure of the "interstitial" nitrogen (the nitrogen which is not in the form of one of a series of nitrides) in aluminum-killed steel by heating steel particles (about −80 mesh) in a hydrogen atmosphere at 550–600°C for about 10 min to convert such nitrogen into gaseous ammonia, which they then determined by Nesslerization. They also explained some discrepancies in ester–halogen results on the basis that the size of the aluminum nitride particles in the steel can affect the results significantly.

E. ALUMINUM REDUCTION

About 25 years ago the Iron and Steel Industrial Research Council of Great Britain encouraged work on a variety of methods for determining oxygen in metals. A method which had recently been published (117) and which depended on the greater affinity of oxygen for aluminum than for iron was included in the studies (118,257). In this procedure a 5- to 10-g specimen of unknown iron or plain carbon steel was inserted between two pieces of low-oxygen aluminum, and the sandwich placed in a graphite boat. In vacuum or in purified hydrogen (presumably one could use purified argon also) the combination was heated to 1150°C for 1 hr. After cooling, the iron–aluminum melt was easily separated from the boat and placed in a beaker for treatment with dilute HCl (1:1). The metal generally dissolved quite readily, leaving behind alumina in proportion to the original oxygen content of the iron. Treatment with nitric acid and, if necessary, other oxidants served to ensure the dissolution of carbides. The solution was filtered with ashless filter paper (sometimes with the addition of ashless pulp), washed, dried, ignited, and weighed.

The method is interesting and was said to be easier to learn and to operate than vacuum fusion. A claim that it was also less time-consuming was obviously based on that point in time (1938) and is certainly no longer valid. It appears likely that, although the concept and its application are interesting, they do not provide a more useful technique than others currently in use. However, for certain metals and alloys this method may provide an alternative checking procedure when the occasion warrants it. It must be remembered that aluminum scavenging of the oxygen will work with some metals and alloys but not with others. Simple thermodynamic calculations suggest, however, that difficulty would be encountered with only a few, such as calcium and magnesium. Since the procedure depends on dissolution of extraneous metals and carbides, this aspect may be serious for metals other than iron. The phase diagram for aluminum and the respective metal will also have a bearing on the application.

F. KJELDAHL DETERMINATION OF NITROGEN

This important technique for the determination of nitrogen in metals evolved from the determination of nitrogen in aqueous and organic media. It is generally agreed that, with the possible exception of that contained in relatively large, abnormal voids, the nitrogen in metals can in almost all instances by hydrolyzed to form ammonia. Essentially, therefore, the Kjeldahl technique consists of converting the nitrogen to ammonia under the influence of an acid attack (or, in some cases, under the influence of a base), distilling the ammonia after making the medium strongly basic, and then measuring the amount of ammonia either titrimetrically with a standard acid (9,205,129,156) or colorimetrically by forming the amber-colored, mercury–ammonia complex with Nessler's reagent (28). The colorimetric technique is the more sensitive one and has the advantage that it is not significantly affected either by entrainment of caustic solution during the distillation–separation of ammonia or by traces of acid vapors in the atmosphere (62).

The Kjeldahl (also called the Allen) technique is still the one most widely used for determining nitrogen in metals. It has the advantage of having proved itself for virtually all metals (one notable exception

being electrical silicon steel) and of being a relatively simple and economical technique. The time required for dissolution of the metal and
distillation of the ammonia is usually too long for quality-control work
during the production of a metal, although Kempf and Abresch (169)
report the surprising analysis time of 10 min for the readily soluble
nitrogen in steel. Fusion (either vacuum or inert gas) or spectrographic
techniques, discussed in Section IV.F, VI, and VII, respectively, can
be made rapid enough for quality-control work with respect to the nitrogen content during the production of a metal.

1. Determination of Nitrogen in Steel

The Kjeldahl technique will be described in terms of its application
to the determination of nitrogen in steel. A generally applicable procedure will be outlined.

The sample should preferably be in the form of millings or drillings,
obtained without excessive heating to avoid reaction with the nitrogen
in the atmosphere. With care, samples can be obtained which require
no further treatment in most cases; however, if a coolant is used or
if contamination is suspected, the samples should be washed with suitable
solvents such as acetone followed by benzene.

The amount of sample to be used for the determination varies with
the amount of nitrogen anticipated and the manner in which the final
measurement is to be made. If the measurement is made by titration,
the sample weight should be large enough to more than meet the detectability limit of the titration. In the case of a colorimetric measurement following Nesslerization, there is an upper limit as well. A sample
weight of 0.5–1.0 g is usually suitable.

The most critical step in the analysis is the solution of the sample
(129,156). Although the nitrogen in a carbon or some steels may be
found hydrolyzed to ammonia immediately upon the solution of the
sample in a dilute acid, such as hydrochloric or sulfuric, other steels,
for example, those containing even small amounts of titanium or niobium,
may require very rigorous treatment, such as long digestion in fuming
sulfuric acid to obtain all of the nitrogen in the form of ammonia. Consequently, it is advisable to use a rigorous treatment until tests prove
that a faster, more convenient one will suffice for a particular steel.

The sensitivity that can be attained is largely dependent on the size
of the blank; therefore, efforts should be made to minimize extraneous
nitrogen. The analyses should be done in a separate room to keep contamination from the atmosphere to a minimum. Also, acids and other

reagents should be kept in this room and reserved for Kjeldahl analyses. Water which is distilled and then deionized has been found satisfactory for this procedure; however, some workers prefer to use for most steps distilled water that has been treated with alkali and then boiled, sometimes with the addition of Devarda's alloy.

The weighed sample is transferred to flask A in Fig. 103.1, along with about 20 ml of water. With a pipet, 5 ml of concentrated sulfuric acid is measured out and placed in the flask. Most steels require no other acid for solution; however, some steels may need one or more acids such as hydrochloric in addition to the sulfuric. The flask is placed on a rack at low heat, and when solution is complete the equivalent of 1 ml of 30% hydrogen peroxide (129) and 5 drops of 48% hydrofluoric acid (172) is introduced. These reagents aid in the decomposition of

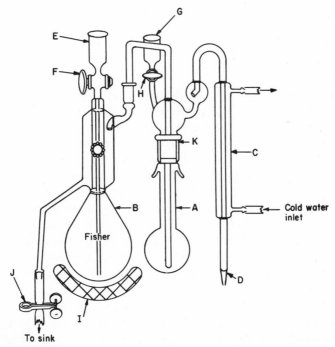

Fig. 103.1. Improved Kemmerer-Hallett type of ammonia distillation apparatus for the determination of nitrogen in metals. A, 100-ml flask containing the dissolved sample; B, boiler; C, condenser; D, condenser outlet. drawn to a tip $\sim \frac{1}{16}$ in.; E, thimble for introducing water; G, thimble for introducing alkali; I, heating mantle; J, pinch clamp; K, 29/42 standard taper ground-glass joint. (Courtesy Fisher Scientific Co).

some materials without leading to loss of nitrogen. The flask is then heated with swirling over an open flame until fuming starts, when the flask is placed on a rack again and fumed vigorously for 3–4 hr. The temperature should be high enough to produce a noticeable refluxing action of the sulfuric acid.

About 20 ml of water is introduced after the flask has cooled to room temperature, and the flask is then put back on a rack at low heat to loosen and dissolve the salts. Complete solution of all the salts is not essential; in fact, it has been reported (156) that, if chromium salts are found to go into solution, this is a sign that the fuming was not carried out at a sufficiently high temperature.

After the sides of the flask are washed down with a little water, the determination is ready for the next step: steam distillation of the ammonia from a caustic medium. The apparatus shown in Fig. 103.1 has proved to be suitable for the purpose. The use of a ground-glass joint to which the reaction flask A is connected is an improvement over the Parnas–Wagner type of apparatus (129), which requires that the reaction mixture be introduced into the distillation flask through a thimble such as G in the figure and then removed by suction in preparation for the next determination.

Steam is generated by boiling water, slightly acidified with sulfuric acid, in B, which is kept more than half full. Suitable solids such as silicon carbide (129) can be placed in the flask for even boiling. Before a flask with a sample is attached to the apparatus, about 50 ml of distillate is collected through I by bringing the water in B to a boil, passing cold water through the condenser C, placing an empty flask with a little caustic at A, and then closing off F, H, and J long enough to collect the distillate. Normally this suffices to clean out the system.

The empty flask is then removed, J is opened, water is added through F to restore the liquid in B to its original level, F is closed, and the flask with the sample is attached at K. After a receptacle is placed below D to catch the distillate, 20 ml of 40% sodium hydroxide is introduced through G, and H and J are closed.

About 25 ml of distillate is collected, the heating rate being adjusted to effect this collection in about 10 min. If the nitrogen is to be measured colorimetrically by Nesslerization, a suitable collector is a 50-ml graduated cylinder that can then be closed with a ground-glass cover. It was found that at least 25 mg of nitrogen in the form of ammonia could be distilled off with 25 ml of distillate. Also, with the amounts normally present in steel, <200 ppm, the distillate could be collected by simply positioning D in the neck of the receptacle. If more than about 200

μg of nitrogen is to be collected, D should probably be immersed in excess boric acid solution, or, if the nitrogen is to be measured by titration, either in a weak solution of boric acid (0.15%) or in an excess of a standard acid solution which is then back-titrated with a standard base.

A reagent blank must also be determined. Since the reagents may contain reducible nitrogen which may form ammonia upon solution of a steel sample, it is good practice to add a small, weighed amount of a low nitrogen standard such as National Bureau of Standards No. 10d to the blank run and then deduct the nitrogen contributed by the standard.

If the nitrogen is to be measured by Nesslerization, the distillate is diluted to 50 ml and placed in a water bath at 30°C (172) for at least 15 min. Then 1 ml of Nessler's reagent, prepared according to Beeghly (28), is added and the color is allowed to develop for 5 min at 30°C; at this temperature the color is stable up to about 30 min (172). Measurements at 410 mμ using a cell with a depth of 1 cm indicated linearity up to 180 μg of nitrogen (172). The colorimetric calibration is made with aliquots of a standard solution of ammonium chloride. If the amount of ammonia Nesslerized is found to be too high to be measured with sufficient precision or so high that the Nesslerized solution becomes clouded by a precipitate, it is possible to redistill the ammonia and then Nesslerize an aliquot (267). In such a case, the Nesslerized solution is transferred to the distillation flask, together with 25 ml of water and a sufficient excess of potassium iodide crystals to turn the color of the solution from a turbid orange to a clear, faint yellow, and the distillation is then carried out in the usual manner (267).

A series of determinations, using Nesslerization, on NBS steel No. 121B (0.012% nitrogen on certificate and 0.011% nitrogen from the Bureau's Ferrous Laboratory) yielded a mean value of 111 ppm with a standard deviation of 10 ppm (172). (The Ferrous Laboratory is no longer an official group of the NBS. Standard samples are handled by the NBS Office of Standard Reference Materials; R. E. Michaelis, Coordinator for Metals Standards. Steel 121B, a type 321 stainless, has been replaced by steel 121C, which is not yet certified for nitrogen.) Duplicate determinations with nitrogen concentrations ranging from 60 to 380 ppm showed agreement to better than 10 ppm between duplicates (172).

Although normally not as sensitive or as precise as Nesslerization for small amounts of nitrogen, titration has the advantage of greater rapidity and no upper limit for the range of nitrogen concentrations.

A popular indicator is the bromocresol green–methyl red mixture proposed by Ma and Zuazaga (186). A suitable preparation is 0.075 g of bromocresol green and 0.05 g of methyl red in 100 ml of methanol (6); however, other ratios have also been reported. Standard acids and bases used are usually 0.01N, and the titration is carried out with a microburet. Although this titration is not as sensitive or precise as the Nessler colorimetric technique, one titration system which apparently compares very well will be discussed later.

2. Discussion and Variations on Nitrogen in Steel

Starting with the solution step, it is, of course, desirable to use a minimum amount of acid to keep the blank low. A number of acids and acid combinations are used (9,28,169,57,156), and differences of opinion exist on the best way to dissolve a given steel. Nitric acid should be avoided because of the possibility of forming ammonia by reduction with nascent hydrogen. It is reported that perchloric (129,156) and phosphoric (156) acids can cause low results by oxidation if the samples are fumed with either of these acids. Apparently, this difficulty does not arise if phosphoric acid is used together with sulfuric acid (9,57). Some procedures call for the addition of a catalyst such as selenium (9,57), whereas other investigators (129,156) find this unnecessary.

The addition of potassium sulfate (9) or potassium bisulfate (9,62) is often advocated to raise the boiling point of the sulfuric acid if a fuming treatment is to be employed, but some investigators (129,156,180c) do not consider these salts useful. The usual addition of 2–3 g of the salt in 10 ml of sulfuric acid has been termed insufficient, and twice the amount of salt, 2 g in 5 ml of the acid, has been recommended (156). Because the rigorous fuming technique is applied for the benefit of the relatively small amount of residue left after solution of the sample, some procedures call for a fuming treatment on the residue alone after it is separated by filtration (9) or by centrifuging (156,180c). A small quantity of barium chloride can be added to the sulfuric acid solution of the steel, so that the precipitated barium sulfate can bring down the fine particles during centrifuging (156,168).

Because some metal nitrides are more difficult to break down than others, it is often possible to obtain interesting information about the distribution of the nitrogen in a steel by controlling the severity with which the steel is dissolved (170,177,183,285). The stability of the nitrides increases in the order Fe-Mn-Cr-V-Ti (177), although the situation is somewhat complicated by the existence of some carbonitrides, oxynitrides, and nitrides of varying composition (170,177). The separation and determination of nitrides by this means needs more study, but there is much agreement in significant areas. Thus electrolytic dissolution or dissolution in a weak acid such as oxalic or hydrochloric does not break down titanium nitride or carbonitride (177), whereas fuming sulfuric acid has this effect (129). Klyachko et al. (176b) dissolved a chromium steel sample electrolytically at 0.2 A/cm² in a solution containing 15% NaCl and 2.5% tartaric acid, and determined the nitrogen content of the electrolyte and of both the soluble and the insoluble portion of the anodic residue after treatment of the residue with 1:1 HCl. The three fractions were said to be, respectively, nitrogen present in the steel in solid solution, nitrides present in the steel as $(Cr, Fe)_2N$, and stable CrN. Lumley (183) was able to separate chromium nitrides for study by dissolution of the matrix in dilute hydrochloric acid.

Beeghly's work on the separation of aluminum nitride from steel is of considerable interest (29). In this procedure the sample is dissolved in an ester–halogen mixture such as bromine and methyl acetate or iodine and ethyl acetate. Beeghly's evidence indicates that one of the nitrides left undissolved by this treatment is aluminum nitride and that this nitride can then be filtered out on an asbestos mat and broken down with hydrochloric acid for a nitrogen determination. The method is not specific for aluminum nitride because other nitrides, such as those of zirconium and silicon, appear to contribute to this fraction. Caution must therefore be exercised when interpreting the data (262c). However, the technique has made possible some interesting conclusions on the behavior of nitrogen in steel (285,29). Bauer (24) found that with his steels he could use methyl alcohol instead of an ester and then determine the aluminum, rather than the nitrogen, spectrophotometrically. He found niobium nitride to interfere. The use of hydrogen reduction to determine interstitial or "dissolved" nitrogen in aluminum-killed steel was mentioned in Section III.D.

The Kjeldahl technique has been found inadequate for the determination of nitrogen in silicon steel, higher results being obtained by vacuum fusion when certain heat treatment and rolling operations have taken place (246,176,156). Satisfactory methods for this determination have been found to be vacuum fusion in a platinum bath (89), fusion in a caustic medium (166), and isotope dilution (216). The separation of some silicon–nitride forms can be accomplished by Beeghly's ester–halogen technique (29,272,239).

Newell (211) found he could Nesslerize the sample solution directly rather than the ammonia distillate. The colloidal mercury–ammonia complex was stabilized with gum arabic, and a blank had to be obtained with the same amount of salts and reagents. The hydroxide precipitates had to be filtered out or allowed to settle.

An excellent color reaction that can be employed for a photometric determination of the ammonia is the reaction of ammonia with phenol and hypochlorite to form a blue color (17a,137a,161b,208,226b,280a,282b). This reagent gives a greater sensitivity and range than Nesslerization, contributes a lower blank, and forms a colored product which is not colloidal in nature. The use of nitroprusside as a catalyst has been reported to be advantageous (17a,280a). Kallman et al. (161b) add iodide to the reagent. Upon automating the measurement, Willis (282b) decreased the time for color development from 40 to <5 min by heating to 95°C. Weatherburn (280a) reported maximum sensitivity with color development at 75°C.

Still another colorimetric approach is based on a colorimetric measurement of the pH of the distillate (223). Measurement of the conductivity of the distillate has also been employed (99a,167a).

Kawamura et al. (167a), working with steel, reported improvement over the sensitivity of Nesslerization when they developed a color by using chloramine T and bis(pyrazolinone) solution to obtain rebazoic acid. The sensitivity was further improved by solvent extraction of the colored product.

Shmelev and Ryutina (235a) used a polarographic technique for determining the ammonia in the Kjeldahl distillate, but reported a precision of only ±20 ppm. They also found it possible to mask the interferences and eliminate the distillation step, although at the cost of still poorer precision.

Milner and Zahner (205) reported a titration system which agreed with Nesslerization values to within 1 μg of nitrogen in the range 50–250 μg of nitrogen. One hundred milliliters of distillate was caught in 25 ml of 0.15% boric acid and then titrated with 0.01N acid, using a mixture of methyl red and alphazurin for an indicator. The end point was sharp, the color change being effected with 1 drop of the acid, and agreed well with an automatic potentiometric titration.

For the final measurement of evolved ammonia, Bradstreet (35) has described a number of titration systems and indicators, including Kjeldahl's iodometric titration, which is still preferred by some workers.

3. Application to Metals Other than Steel

For other metals, as for steel, the most critical step is usually the solution of the metal. The sample solution is then put through the usual procedure of steam distillation in a Kjeldahl apparatus, and the ammonia in the distillate is measured colorimetrically or titrimetrically.

The ASTM procedure for titanium (10) prescribes a mixture of hydrochloric acid with some fluoboric acid, HBF. The latter is a substitute for hydrofluoric acid, which is rough on glassware. Any insoluble residue is either (*1*) separated by decantation, treated with sulfuric acid and a crystal of selenium, and heated to fuming; or (*2*) filtered on a piece of filter paper, treated with a mixture of hydrofluoric acid, hydrogen peroxide, and sulfuric acid, and heated to fuming. The resulting solution in either case is then combined with the main solution for distillation in the Kjeldahl apparatus.

The ASTM procedure for zirconium (11) calls for solution in a mixture of hydrochloric acid with some hydrofluoric acid added dropwise. Heating on a water bath is carried out if necessary.

The solution step for molybdenum can be accomplished with a mixture of hydrochloric and hydrofluoric acids heated in a platinum dish. Hydrofluoric acid and hydrogen peroxide are added as needed to initiate and continue the reaction (12). In another procedure the sample is heated to fuming with sulfuric acid and a little phosphoric acid (190). Most laboratories have some hydrofluoric acid and hydrogen peroxide present for the solution step.

Niobium, tungsten, and tantalum can be suitably dissolved by digesting with sulfuric acid, some hydrofluoric acid, and a little hydrogen peroxide, and then heating sufficiently to drive off the hydrogen peroxide and the hydrofluoric acid (134,190). Chromium can be treated with a mixture of 1:1 sulfuric acid and fluoboric acid (7).

The determination of low amounts of nitrogen, especially in refractory metals, poses special problems that have been studied by Kallman et al. (161b,162), who made a particular effort to diminish the blank. By using capped polyethylene bottles to dissolve molybdenum and tungsten with HF and H_2O_2, they reduced the blank to about 2 or 3 μg of nitrogen (162). The capped bottles greatly reduced the necessity for replenishing the reagents, and the containers did not catalyze the decomposition of hydrogen peroxide, as do platinum containers (142). More recently, working with the same two metals and also niobium and tantalum, Kallman and his associates reduced the blank to <0.5 μg (161b). They found potassium dichromate in a hydrofluoric acid–phosphoric acid medium to be a better oxidant than hydrogen peroxide in this study, in which they used Teflon-capped bottles. They found it important to minimize the surface area of the sample and used whole chunks that had been abraded or etched. Their measurement of the ammonia finally distilled was made photometrically with a phenol-hypochlorite reagent.

The determination of nitrogen in alkali metals, of course, constitutes a special case. Solution of the sample must take place under a protective atmosphere such as argon. The sample can be dissolved in water in a modified Kjeldahl apparatus, and any nitrogen that is in the form of an alkali nitride or that is reduced to ammonia is distilled off as ammonia and measured spectrophotometrically (105,44). Kirtchik (175a), working with potassium, reacted the sample in an excess of phosphoric acid under a layer of hexane and an atmosphere of argon. The hexane was then distilled off and discarded, a known excess of sodium hydroxide solution was added, and the ammonia was distilled off. One laboratory found it advantageous to use n-butyl alcohol, with or without hexane, in Kirtchik's technique in order to reduce the surface tension between the aqueous and the organic layers (175a).

Anthanis (6a) determined the nitrogen and other gases in cesium by refluxing in a high-vacuum system (the cesium boiled at 250°C, and the transfer tube was kept at 450°C) and then analyzing the extracted gases with a mass spectrometer.

Mention should perhaps be made of the Dumas method and the alkali-fusion method for the determination of nitrogen in a variety of metals. In the Dumas method the sample is fused with copper oxide, and the nitrogen in the sample is evolved in elemental form and measured gasometrically. Kern and Brauer (171) found the method satisfactory for niobium and tantalum and preferred cuprous to cupric oxide. In the alkali-fusion method the sample is fused with an alkali at about 600°C in a stream of hydrogen, and the ammonia formed from the nitrogen is absorbed in water or a weak acid and measured. The method has been found satisfactory for such metals as silicon steel (165b), niobium (134,274b), boron, tungsten, and molybdenum (30), and some prefer it for steels which contain sparingly soluble nitrides and carbonitrides (93a).

In a combination of the Dumas and alkali-fusion methods, Lapteva et al. (180a), working with a mixture of BN and H_3BO_4, determined the nitrogen in the BN gasometrically by extracting it after fluxing with Na_2CO_3, K_2CO_3, and NiO in the weight ratio of 1.2:9.8:3. Addition of the NiO was found to diminish foam and spatter. Yamaguchi et al. (285a) employed oxidative fusion with Pb_3O_4 in a helium atmosphere for the convenient, rapid determination of nitrogen in steel down to 10 ppm. After passage of the evolved gases over hot copper oxide, hot copper, and molecular sieve 5a to remove oxygen, hydrogen, and carbon dioxide, the nitrogen was measured by thermal conductivity.

In Section IV.E on hot extraction, there is reported the substantial extraction of nitrogen from molybdenum at 1800°C (190) and 2000°C (82) and from tungsten and tantalum at 2000°C (82).

G. COMBUSTION

For some metals it has been possible to burn the sample in oxygen and to measure the product water as a reliable method for determining hydrogen. Molybdenum has been studied (100) in this way. The sample is burned in oxygen at 1100°C. Fused magnesium oxide sand is used to absorb the molybdic oxide produced and to prevent its sublimation in the combustion tube. The water produced by oxidation of the hydrogen present is absorbed in anhydrous magnesium perchlorate and weighed. Although an average recovery of 99% and a standard deviation of 8 ppm in the 13- to 457-ppm range was claimed, the data indicate a systematic trend. Low-range results are high, and high-range results are low. The details presented are insufficient to suggest the source of this very real discrepancy. The use of zirconium hydride to provide the standard "added" hydrogen may contribute to the difficulty. In spite of the bias, however, the method appears to be useful for hydrogen contents of 50 ppm and higher.

H. ELECTROCHEMICAL METHODS

Any technique which offers promise of instant and continuous measurement of significant gaseous constitutents in molten metals can be expected to excite much attention and investigation, for it would lead to improved quality control and to increased production in the plant. Therefore it is not surprising that many interesting papers have appeared in recent years on the determination of oxygen in molten steel and various other metals (182,264b) by means of an oxygen-ion, solid-electrolyte device. In this device it is sought to carry out the reaction

$$O_2 \text{ (gas, } T, P) \rightleftharpoons 2O \text{ (dissolved in metal, } T)$$

(in which gaseous molecular oxygen at pressure, P, and temperature, T, is isothermally dissolved in a molten metal) under reversible conditions in an electrolytic cell, such as

$$O_2 \text{ (gas, } T, P)/\text{electrolyte}/2O \text{ (dissolved in metal, } T)$$

If the electrolyte conducts only oxygen ions, then the Nernst equation applies in the form

$$E = \frac{RT}{nF} \ln \left[\frac{(a_o)^2}{P_{O_2}} \right]$$

where E = the open-circuit potential of the cell,

R = the gas constant in appropriate units,

n = the number of electrons involved in the electrochemical oxidation and reduction steps ($n = 4$ in this case),

F = the Faraday constant,

a_o = the chemical activity of the dissolved oxygen,

P_{O_2} = the partial pressure of the oxygen.

Under isothermal conditions, a_o can usually be considered to be proportional to (% O), the concentration of the dissolved oxygen, and it can then be shown that ln (% O) is directly proportional to E to provide the basis for an emf device.

Three materials are considered superior for the oxide-conducting electrolyte (264b): ZrO_2, HfO_2, and ThO_2 doped with other oxides. For molten steel, stabilized ZrO_2 appears to be the best choice at the present time and has been investigated by many workers since the pioneering research of E. Bauer and H. Preis in Germany in 1937. However, work by Nakagawa and Shiga (207a) indicates that Al_2O_3 may be a possible electrode material.

Fischer and Ackerman (94b,94c,94d,94e) succeeded in using the technique to measure accurately the oxygen content of molten steel. Their cell had platinum leads and air for the reference electrode, that is,

Pt, air/stabilized ZrO_2/O (molten steel, Pt)

and was applied under carefully controlled laboratory conditions. Development work is continuing at an accelerated pace to accomplish the successful transition of this technique from laboratory to shop conditions (94f,207a,271a). Pargeter (213c) used the Fischer–Ackerman cell with calcium-stabilized zirconia in a considerable study of the determination of oxygen in molten steel, and his comparisons with vacuum-fusion values indicated that the technique could detect changes in oxygen content of less than 10 ppm. Satisfactory experience with molten steel containing from 10 to 2000 ppm has been reported (104a).

Recently, Kulkarni et al. (179a) reported good results for the continuous measurement of oxygen in copper at 1200°C. They used stabilized zirconia as the electrolyte. Tests with air as well as oxides indicated that air provided the best oxygen reference potential, followed by

Ni/NiO, Co/CoO, and Fe/FeO in order of preference. The oxide Cr/Cr_2O_3 was found to be entirely unsatisfactory.

A commercial device, resulting from the further pursuit of Pargeter's work, has recently become available from Leigh Instruments, Ltd., Carleton Place, Ontario, Canada. Two types of probes are offered, one with air and the other with certain solid materials as a reference (271a).

IV. VACUUM METHODS

A. INTRODUCTION

The decade 1950–1960 probably saw the greatest surge in the application of vacuum methods for determining gases in metals. By this time technology had increased the availability of many metals heretofore infrequently used. Extensive use of zirconium in the atomic energy field led to serious investigation of the properties of this metal as a function of oxygen, nitrogen, and hydrogen content. For example, it was established that the presence of even 0.08% oxygen markedly increases the tensile strength of crystal-bar zirconium, more so at room temperature than at elevated temperatures (51). The most important effect of nitrogen appears to be on corrosion resistance. Above 40–50 ppm of this impurity seems to cause rapid corrosion, as compared to purer crystal-bar zirconium. The addition of tin, particularly if iron is also present, as in the Zircaloys, increases the tolerance toward nitrogen. Oxygen has only a slight effect up to 0.5%, whereas hydrogen at the 0.1% level causes appreciably faster corrosion than is observed with specimens containing 50 ppm or less (265). It should be remembered, in assessing data of this kind, that earlier methods of analysis were not always accurate. As a result, allowable limits of impurity became established empirically, and may not be satisfactory for use with today's more accurate methods.

The history of vacuum methods of analysis for oxygen, hydrogen, and nitrogen goes back a long way (see Section I.B), from simple thermal decomposition in vacuum to vacuum fusion. In an earlier volume of this series (150) a general introduction to vacuum methods was given. Here more details with respect to specific methods will be provided.

B. MERCURY AMALGAMATION

The general technique of solvent extraction has been used with mercury as solvent for determining the oxygen content of both sodium

(31a,217,188) and tin (241). These are classified as vacuum methods because at some point or at least intermittently a moderate vacuum is used. For both sodium and tin the mercury amalgates the body of metal and the latter is removed. With this gone, after sufficient washings what is left is generally assumed to be oxide. Because rather low oxygen contents are desirable, gravimetric estimation of the remainder is generally not feasible. In the case of sodium, the oxide residue is dissolved in water and titrated with the appropriate concentration of acid. The amalgam is treated with water, and the latter titrated with an appropriate acid in order to determine the sample weight. An alternative method for the oxide residue is flame photometric determination of the sodium equivalent. This is particularly advantageous when residues other than sodium oxide are present. The originators (217) of the amalgamation method for oxygen in sodium claim that hydroxide and peroxide do not cause error if the sodium is heated to 400°C before sampling. From the data available for sodium one concludes that the precision is a function of the amount of oxygen present. In the range from 0.004 to 0.11% the standard deviation is about 0.003%; from 0.07 to 1.14%, it appears to be about 0.024%. (The overlapping range is due to the grouping of samples.)

For tin the residue is determined by a blue phosphomolybdate colorimetric technique. A modification of the technique (49) has been alleged to yield results for oxygen in sodium below 40 ppm with a reproducibility of 1.5 ppm. The procedure has been simplified and the quantity of mercury employed in a determination reduced by Dorner and Kummerer (73). The simplifications involve both apparatus design and a modification of procedure permitting some amalgam to remain within the reaction bulb. The latter is accounted for, and the appropriate correction made. The standard deviation of this variation appears to be a function of the oxygen content. For five samples ranging from 33 to 187 ppm the standard deviations ranged from 5 to 21 ppm.

The amalgamation technique can also be used to determine the hydrogen in sodium. Meacham and Hill (202) reported the ability to determine first the hydride plus dissolved hydrogen gas and then the hydroxide. Advantage is taken of a difference in thermal stability between sodium hydride and hydroxide in the presence of sodium amalgam. Dissolved hydrogen is released as the sodium dissolves in mercury or amalgam. The hydride decomposes at 150–200°C, while hydrogen from the hydroxide is released upon reduction by sodium in the amalgam at temperatures above 320°C. Keough (170a) modified the Meacham–Hill procedure in several ways, for example, extruding the sample into a secondary container, reducing the number of heating–cooling–evacuating cycles, and improving the yield from hydroxide standards by adding calcium to the amalgam.

C. VACUUM HEATING

A combined chemical and vacuum method used for determining oxygen in chromium apparently originated from comments in an article dealing mainly with the preparation of pure chromium (1a). The method as used later (238) required one to heat the specimen of metal in vacuum at 800°C for 2 hr, cool, dissolve the metal in 1.2N hydrochloric acid, and filter, wash, ignite, and weigh the residue as chromium (III) oxide. For low oxygen contents the gravimetric aspect is a drawback, since 100 ppm oxygen in a 1-g sample would weigh only 0.0003 g as the metal oxide, although with micro-weighing techniques the problem may be overcome. In order to avoid this difficulty, however, a colorimetric method was developed (152a). This procedure is as described above, except that ignition occurs in a nickel crucible. To the ash (from a 1-g sample) 3 g of potassium persulfate and 5 g of potassium hydroxide are added and the mixture is fused. When the cooled fusion mixture is dissolved in water, any chromic oxide present appears as potassium dichromate in a basic solution. Measurement of the filtered and diluted solution on a spectrophotometer at 370 mμ gives the amount of chromium, and therefore of oxygen, by reference to a standard curve previously prepared. Recoveries from a few experiments performed directly on chromic oxide indicated at least 99%. Since interferences are apparently few and the method is straightforward, it would appear to warrant further development if only to serve as a comparison alternative. The initial heating, or vacuum annealing, is instituted in order to allow the oxygen to migrate sufficiently so that it will precipitate completely upon cooling. The choice of acid concentration for selectively dissolving the metal but not the oxide may be critical.

An interesting development for analysis of tantalum films has utilized a xenon flash discharge lamp (127). The light energy of the flash is absorbed by the film and is apparently capable of vaporizing tantalum from a glass substrate. The extreme temperature also dissociates the tantalum compounds within a vacuum system. Collection of these gases and analysis by means of a gas chromatograph are reported. Gases found when the temperature attained an estimated 5000°C included hydrogen, argon, nitrogen, and carbon monoxide. The work reported emphasized nitrogen because of the particular interests of the experimenter.

In a series of experiments tantalum films on glass were prepared by sputtering in argon, the partial pressure of nitrogen in the sputtering system being regulated from 5×10^{-6} to 1×10^{-3} torr. The data obtained by flash analysis of the films appear to be related to the pressure by $N = aP$, where N is the atomic per cent of nitrogen in the film, a is a constant, and P is the partial pressure of nitrogen during sputtering. The demonstrated precision of the method is not impressive; on the other hand, it is recognized that micro sample sizes are involved (<1 mg). The limited data indicate a standard deviation of (at best) 50 relative % for nitrogen concentrations up to about 20 at. %. Above this point the precision may level off to about 5 at. %. (The data show some evidence of being relatively more precise at the higher nitrogen levels.) This method appears to be most useful for experiments similar to those reported, wherein nitrogen content and electrical resistivity were studied as a function of the nitrogen partial pressure in the sputtering atmosphere.

In Section E.5 on the determination of hydrogen, determination in sodium by amalgam refluxing (170a,202) is discussed.

D. DISTILLATION

Similar to the other methods whereby the matrix metal is removed, leaving the oxide residue, is one reported for oxygen in magnesium and in calcium (189). The magnesium or calcium of the oxides can be determined in the residue after the metals have been distilled away in vacuum. Chemical or spectrographic methods are well known for the determination. For example, a filtration with the disodium salt of EDTA may be used. The usual problems associated with residue methods are present, of course, but for metals such as those mentioned the procedure may be quite useful. Bradley and Hand (33c) applied the technique to the determination of oxygen in lithium and reported good results. They introduced the sample by sectioning a cylindrical specimen in a glove box, rather than by extrusion, and they considered this method to be less subject to possible errors (incomplete transfer, radial segregation of oxides). Distillation of the lithium was carried out at 700°C from a molybdenum crucible, and the remaining lithium oxide was determined by either acid titration or flame spectrophotometry. Walker et al. (277a) reported good results for oxygen in sodium when they used a similar procedure.

E. HOT EXTRACTION AND DETERMINATION OF HYDROGEN

1. Principle

In hot extraction the gases are extracted from the metal without melting the specimen. The technique apparently dates back about 100 years to the time when Graham (116), then Master of the Mint in Britain, observed that hydrogen formed a substantial portion of the gases evolved when ferrous materials were heated in a vacuum. Today, the technique is the most common one for the determination of the hydrogen content of metals. The extracted gases are usually collected in a known volume and then analyzed in one of several ways.

The choice of the extraction temperature is influenced by the metal to be analyzed and the objective of the analysis. For determination of the hydrogen content, it normally suffices to be able to attain a temperature of about 1000°C. The hydrogen is also evolved during the vacuum-fusion analysis of a metal; however, hot extraction is normally preferable for this determination, not only because it is more convenient, but also because it yields a higher precision. The greater precision is due to the fact that, just as in Graham's observations, most of the extracted gas is usually hydrogen, whereas in vacuum fusion hydrogen is usually a minor constituent and the hydrogen blank is higher.

Hot-extraction temperatures of the order of 1000°C are sufficient to break down metal hydrides and to impart a relatively rapid rate of diffusion to the released hydrogen, as well as to any hydrogen that is present interstitially. The hydrogen is collected as H_2. In the case of some metals, such as titanium, the extracted gas is almost pure hydrogen (47), whereas for other metals there may be varying amounts of other gases. Thus steel releases some carbon monoxide and a little nitrogen. Also, Fagel et al. (82) report that, along with the hydrogen, they were consistently able to extract the oxygen and much of the nitrogen from molybdenum, tungsten, and tantalum by hot extraction at temperatures of 2000°C and higher in a vacuum-fusion apparatus. Mallett and Griffith (192) extracted the nitrogen from molybdenum at 1800°C. The extraction was carried out in a graphite crucible, and the oxygen was collected as carbon monoxide. Situations are encountered, of course, in which more than one gas is of interest; however, these are exceptions to the main application of hot extraction: the determination of hydrogen in metals.

In the case of metals that dissolve much hydrogen, such as titanium, zirconium, or vanadium (74b), some rather substantial amounts of this gas can be encountered; these can often be measured by macro-techniques (54). In the case of other metals, such as iron, aluminum, and steel, however, the hydrogen, usually present interstitially, may be well under a part per million, and a sufficiently sensitive technique, one of which is described below, is needed to measure the hydrogen content.

2. Apparatus

Figure 103.2 shows an apparatus for the hot-extraction technique (45) in which the evolved gases are analyzed by oxidation and fractional freezing. The specimen is placed in the hot zone of the quartz tube C, which is heated by a resistance furnace. The samples are introduced via the mercury lift A, which is suitable for solid specimens that can be handled with a magnet. A more general method of sample introduction utilizes the vacuum lock shown in Fig. 103.5. Except for the quartz tube C and the tubing containing the copper oxide oxidant K, the system is constructed of borosilicate glass with most of the tubing 0.5 in. in diameter. A mercury diffusion pump and a mechanical pump are connected to stopcock 5.

The Pirani gage (or an equivalent thermal conductivity gage) serves as a convenient means of following he course of the extraction and as one means of analyzing the extracted gases (46). The stopcock leading

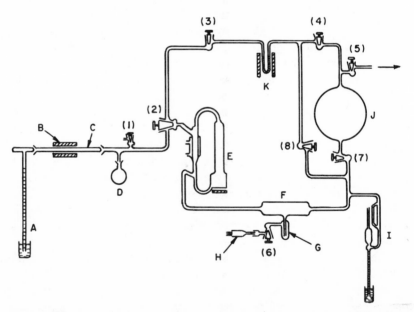

Fig. 103.2. Apparatus for hot-extraction technique (45). A, mercury lift; B, resistance furnace, hinged type; C, quartz tube; D, receptacle for analyzed samples; E, Naughton–Uhlig mercury diffusion pump; F, calibrated volume for gas collection; G, freeze-out trap; H, Pirani gage; I, McLeod gage; J, reserve volume for gas collection; K, copper oxide oxidant. (Courtesy of *Analytical Chemistry*.)

to it is opened only when the freeze-out trap G is immersed in a cold medium at about $-80°C$ to trap out condensables.

A comparison will show that this apparatus is essentially the same as the vacuum-fusion apparatus shown in Fig. 103.6 except that, since a significant gettering problem is not normally present in hot extraction, the diffusion pump used to transfer the gases rapidly from the vacuum-fusion furnace to the collecting–circulating diffusion pump has been eliminated. Otherwise, the same considerations of design and gas analysis apply. If only the hydrogen is of interest, the analysis of the gases is simplified.

3. Procedure for the Determination of Hydrogen in Steel

The amount of hydrogen present in steel varies with the constituents and the history of the steel (43a) but is normally far below the amount which would be found if the molten steel were allowed to come to equilibrium with the moisture in the atmosphere. Hydrogen contents of only a few parts per million or less are commonly measured; therefore, a fairly sensitive and precise method is required. For reasons that will be discussed later, the procedure to be described applies to steels with a carbon content of $<0.5\%$.

Sampling and sample preparation must take into account the fact that, except for some austenitic steels, the diffusivity of hydrogen in steels is significant even at room temperature (75). Also, the sampling of solid steel must consider the fact that hydrogen segregates toward the center, as the solubility of hydrogen in steel increases with temperature. Finally, the sample should be a single piece weighing about 3–7 g with a small surface/volume ratio, for example, a cylinder with a diameter of about $\frac{1}{4}$–$\frac{3}{8}$ in. and a length of about $\frac{1}{2}$–1 in.

A steel bath is sampled in one of several ways (14c,75,p. 210,200a,281) that call for rapid cooling followed by storage in Dry Ice or liquid nitrogen until analysis. One of the simplest and most widely used methods is to bring an evacuated length of borosilicate glass or Vycor tubing with a diameter of about $\frac{1}{4}$–$\frac{3}{8}$ in. into contact with a spoonful of aluminum-killed molten steel and quickly water-quench the sample as soon as it is drawn into the tube. Quenching in liquid nitrogen has been reported to yield very clean, oxide-free surfaces that need no further preparation (197). In an improved, more elaborate sampler (48), developed by the Acid Open Hearth Research Association, Inc., and available from Burrell Corporation, the evacuated glass tube, brought to a thin bubble at the contact point, is enclosed in a copper mold and then a graphite casing. Some aluminum wire is placed in the mold to "kill" the steel, and a detachable cap permits direct sampling of the molten bath by immersing the sampler below the slag layer. A sampling device sold by the Laboratory Equipment Corporation produces "banjo-shaped" samples, the stem of which is used for gas determinations and the circular portion for X-ray or spectrographic analyses. The molten steel is killed upon contact with an aluminum seal (202c).

Tests by the authors have indicated that, when a sample is cut reasonably rapidly from the specimen with water cooling, the loss of hydrogen is completely negligible as compared to that occurring when carbon dioxide snow is used for cooling. Similarly, a sample from a large body of steel should be taken as rapidly as possible, whether by trepanning or cutting, with the generous use of a coolant.

With the hot-extraction apparatus in readiness, the specimen is taken out of the Dry Ice or liquid nitrogen storage, brought to room temperature by immersing in acetone, and then dried in a warm air stream, filed lightly, and weighed. If a clean vise, file, and pair of forceps have been used and the filed surface is untouched by the fingers, the specimen can be introduced into the apparatus; otherwise, it should first be washed in suitable solvents, such as acetone and benzene, and then dried. In about 2 min the air introduced with the specimen is sufficiently exhausted. The total time of exposure of the specimen to room temperature should be no more than about 10 min, during which time the loss of hydrogen is found to be negligible and well below 2% of the total.

In preparation for the determination, the system is first outgassed with the copper oxide K (Fig. 103.2) at a temperature of about 325°C or somewhat higher, and the furnace B at about 1000°C. It has been observed that, if a pressure of 0.005 μ Hg is measured on the McLeod gage, the blank correction can normally be neglected.

After the system is closed with stopcock 7, the sample is moved into the hot zone with a magnet, a ferromagnetic "pusher" being used in the case of nonferromagnetic steel. The time of extraction depends on the size of the specimen and, to a minor extent, on the hydrogen content. Specimens less than $\frac{3}{8}$ in. in diameter require only about 15 min, after which time it has been observed that the rate of gas evolution is normally <0.2 μl/min (45). The trap G is kept at about −80°C, a temperature which is conveniently obtained by cooling acetone with small additions of liquid nitrogen (45), and a suitable correction is applied for the effect of the trap on the calibrated volume F, in which the gases are collected. If the pressure appears to be too high, the gases can be expanded into reserve volume J, which increases the volume from 1 to 7 liters; however, the analysis can be carried out only on the gases in volume F.

For the analysis, stopcock *2* is turned toward the copper oxide oxidant, and the gases are circulated until the pressure ceases to drop (usually 5–10 min). This indicates that all of the hydrogen has been converted to water and frozen out in trap G. (The water can also be removed by absorption with a column of magnesium perchlorate, but the absorbent should be checked daily.)

The CO is converted to CO_2, so that the number of molecules remains the same, while the nitrogen is unaffected. The gases are collected in volume F, and the total drop in pressure is equal to the partial pressure of the hydrogen in the original gas mixture. The concentration of hydrogen is calculated from the pressure, (room) temperature, and volume of the gas, using the ideal gas law, $PV = nRT$, as shown in Section IV.F on vacuum fusion, and is equal to KP/W, where K is a constant, P is the partial pressure of the hydrogen, and W is the sample weight. If the volume is 1 liter and the room temperature 27°C, the concentration of hydrogen, in parts per million, is 0.1078 P/W.

The vapor pressure of water at −80°C, which is 0.4 μ Hg, is no cause for concern if the freeze-out trap F is a part of the calibrated volume F. However, it would not be advisable to have trap F located in the section containing the copper oxide oxidant, for example, because significant amounts of water vapor would be pumped into the calibrated volume when the gases were collected after oxidation.

Analyzed specimens are collected in receptacle D, a piece of ferromagnetic material being used to push each specimen out of the hot zone. The system is then evacuated to prepare it for the next specimen.

4. Variations for the Determination of Hydrogen in Steel

Any system capable of measuring the hydrogen content of steel can also be applied to many other metals; therefore, it is useful to consider the variations that have been developed in the case of steel.

It has already been mentioned that, at least for low-carbon steels, hot extraction is preferable to fusion in a graphite crucible. Though the latter technique releases the hydrogen faster, it usually yields a gas mixture in which the hydrogen is a minor constituent, it ordinarily has a significant and somewhat variable blank, and it does not offer the convenience and simplicity of the hot-extraction technique. Cooperative programs (13,14a) and work by Martin et al. (199) show that hot extraction gets the same results with better precision for low-carbon steels, stainless steels, and nickel. Hydrogen determinations by vacuum fusion, however, are still of interest for steels with higher carbon contents, some workers reporting higher results by this technique. Klyachko and Izmanova (176) recommend vacuum fusion rather than hot extraction for steels with more than 0.5% carbon, and suggest that higher carbon contents lead to the formation of more stable hydrides, possibly of a Me–C–(H) type.

The tin-fusion technique (43b), in which the sample is dropped into a pool of molten tin at 1000–1150°C, was designed to combine the fast release of hydrogen achieved by vacuum fusion with the high fraction of hydrogen in the extracted gases, the low temperature of operation, and the low blank rate achieved by hot extraction. Also, some workers doubted that hot extraction was sufficient even for low-carbon steels and that melting in a vacuum was necessary for the extraction of all the hydrogen. However, Carson (45) and Martin et al. (199) found that hot extraction gave the same results more conveniently and provided better precision. Carson also found that for cylindrical specimens with a diameter of about 0.25 in. and a weight of about 3 g the extraction times for the two techniques were very similar, <15 min, when the hot-extraction temperature was 975°C.

When Newell (210) described the hot-extraction method in 1940, he had used an extraction temperature of 600°C and found the results to agree with those of vacuum fusion at 1700°C. Temperatures of about 650°C are also commonly employed. At such temperatures, the extracted gas can often be assumed, without serious error, to be pure hydrogen (210,281). Although some workers (176,196) report that they get more hydrogen as they increase the extraction temperature (usually for steels with >0.5% C), Carson (45) found that he obtained the same results, but in half the time, on specimens weighing ~3 g when he increased the extraction temperature from 650° to 975°C. However, Geller and Sun (101) and Hill and Johnson (141) have reported that, unless an extraction temperature of >800°C is used, a "residual" amount of hydrogen can remain in the specimen in the form of methane. More work needs to be done in this area, particularly with respect to the carbon and some other constituents of the steel, to resolve the matter. In the meantime, as a general rule for steels with <0.5% C, it appears advisable to use a temperature of >800°C. The authors prefer a temperature of about 1000°C because the extraction is then relatively rapid.

Though induction heating is feasible for hot extraction, Martin et al. (199) found a resistance heater preferable. Temperature control is difficult with induction heating. In addition, resistance heating gives lower blanks, because the entire hot zone can be outgassed at temperature before the analyses.

Figure 103.3 shows a simplified, economical apparatus for the hot-extraction technique (46). It will be noted that the furnace section is part of the calibrated volume, so that a transfer pump is eliminated; however, a calibration must be made for the effect of the furnace on the pressure of the gas as measured at room temperature. Since the evolved hydrogen remains in contact with the sample, the latter retains some hydrogen in equilibrium with the hydrogen atmosphere. According to Sievert's law, hydrogen solubility is proportional to the square root of the partial pressure of hydrogen. Therefore, if the solubility of hydrogen in steel is about 6 ppm at 1000°C in 1 atm of hydrogen (23), it can be calculated that, if the volume of the system is 1 liter and the sample weight is 3 g, the amount of hydrogen remaining in the steel is about 0.06 ppm when the initial amount is 3 ppm and about 0.04 ppm when the initial amount is 1 ppm. This equilibrium solubility can be added to the result calculated from the observed pressure, but such a small correction is normally negligible. This method has been called the "equilibrium pressure" method by McKinley in his application of it to titanium (201b).

Since steel does not yeild pure hydrogen, but can also yield significant amounts of CO and N_2, the gas should be analyzed for hydrogen. With

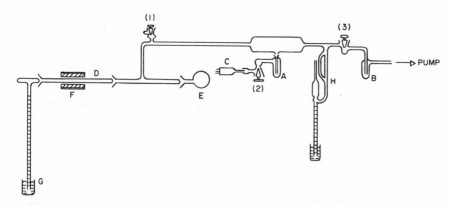

Fig. 103.3. Simplified apparatus for hot-extraction technique (46). A, trap; B, trap for mechanical pump; C, Pirani tube; D, quartz tube; E, receptacle for analyzed samples; F, resistance furnace; G, mercury lift; H, McLeod gage. (Courtesy of *Transactions of American Foundarymen's Society.*)

the apparatus shown in Fig. 103.3 the analysis is accomplished by measuring the thermal conductivity on the Pirani gage, as well as the pressure on the McLeod gage. The gases usually evolved are H_2, CO, and N_2; and, since the isosteres CO and N_2 have virtually identical thermal conductivities, the gases can be treated as a binary mixture. A calibration is made with pure hydrogen as well as with pure nitrogen to relate the thermal conductivity readings to pressure readings at 100% and 0% hydrogen. If a linear relationship is assumed, the fraction of hydrogen in a gas mixture is then calculated from the position of the measured thermal conductivity between these two extremes at the measured pressure (46). The high thermal conductivity of hydrogen makes possible a normally acceptable precision by this technique. In this apparatus, a small correction is made for the hydrogen which diffuses through the hot zone of the quartz tube (74a) when the extraction temperature is about 1000°C. The correction is determined experimentally for different pressures of hydrogen. If the gases are transferred from the hot zone into a fixed volume for the analysis, as in Fig. 103.2, the final transfer should be made with a Naughton–Uhlig (209) type of diffusion pump or a Toepler pump to maintain a constant volume for the collected gases.

A thermal conductivity cell (236), such as a Pirani gage, can also be used for the analysis of the extracted gases when they are transferred to a constant volume. Of course, this technique is not suitable for very low pressures, at which the calibration curves converge.

An elegant and popular method for the analysis of the gases for hydrogen involves the use of a palladium filter (75,199,67) or, better still, a silver–palladium filter (154) (to prevent embrittlement). Since hydrogen is the only gas that diffuses through the palladium or silver–palladium membrane heated to a suitable temperature, usually 400–600°C, the hydrogen can be separated from the other gases either by permitting it to escape into the atmosphere or by pumping the filtered hydrogen into a known volume for direct measurement. It is good practice to make sure that the gases are dry, for example, by the use of a freeze-out trap (75); otherwise an error can result from the following reaction at the hot palladium surface:

$$H_2O + CO \rightarrow H_2 + CO_2$$

The analysis can also be carried out by gas chromatography (232). Either the gases can be pumped into a small space and then swept out by a carrier gas (94a), or they can be first adsorbed on a column of molecular sieve pellets cooled to −196°C and then swept by a carrier gas into the gas chromatograph after being released by heating to 300°C

(199,181a). Separation into isotopes of hydrogen is also possible by gas chromatography (101a). The use of peak areas yields a better precision than does the use of peak heights (199).

Normally, nothing is gained by employing a mass spectrometer for the analysis (198). Its use is of value, however, when unusual gaseous constituents are to be checked. Martin et al. (199) obtained the same results with a mass spectrometer as they did by oxidation–separation, gas chromatography, thermal conductivity, and diffusion through a silver–palladium membrane.

Although in the past the gases were frequently transferred to an Orsat type of apparatus and analyzed at atmospheric pressure (256,17), this is rarely done today. For one thing, the amount of gas required for this technique is relatively large, whereas the trend in regard to the hydrogen content of steel has been downward.

Gulbransen and Andrew (125a), in working with zirconium and its alloys, collected the hydrogen evolved as 1200°C by freezing it in a liquid helium trap at 4°K. It was then released for pressure measurement in a taut-metal-membrane capacitance manometer.

It is generally agreed that the amount of water vapor given off during hot extraction can be neglected in the case of properly prepared, low-carbon steel samples. For example, mass spectrometric analyses by the authors showed no water. Measurement of the water vapor, when it is deemed advisable, can also be carried out with a modified McLeod gage so as not to cause condensation of the water vapor, or, preferably, by passing the water vapor over an asbestos preparation of low-carbon ferromanganese at 750°C in order to decompose the water to hydrogen (17,253). Though it is theoretically possible to convert all of the hydrogen to water vapor and to measure it as such with a modified McLeod gage (69), this procedure is not recommended because at low pressures large errors can result from the adsorption and release of water on the walls of the apparatus (224).

Working with pickled specimens of commercial mild steel wire, Sachs and Odgers (229) concluded that appreciable amounts of hydrocarbons can be emitted by such specimens under hot-extraction conditions at 600°C. In their tests with an apparatus that made use of a palladium filter for hydrogen separation, they got more hydrogen when they thawed out a fraction trapped out at −79°C and circulated it over hot ferromanganese. The increase sometimes was more than 50%. The formation of hydrocarbons is further supported by the observation of Sachs that there is decarburization at the surface of pickled steels subjected to vacuum extraction (228). Unfortunately, the Sachs–Odgers tests did not differentiate between water and hydrocarbons. It would be of interest to use a mass spectrometer for the analysis of the gas extracted from pickled or cathodically charged steels.

In connection with steel specimens that have been charged with hydrogen by pickling or cathodic charging, it is important to keep exposure to room temperature as brief as possible (284), because some of the hydrogen may be concentrated near the surface and escape at normal temperatures. Franklin and Potts (95a) report that a coulometric technique appears to give direct determinations, from the areas under the maxima of polarograms, of the hydrogen codeposited with metals through which it can diffuse, such as iron, nickel, and palladium.

Hydrogen determinations connected with welding present a number of special problems in sampling and analysis (32b,55). These are fully discussed by Coe (55). A sampling technique must achieve the aim of retaining as much as possible of the hydrogen present at the instant that the weld is made; therefore a rapid-quenching arrangement is advisable. If the sample is to be introduced into a vacuum system, an efficient sample-introduction system is necessary to keep exposure to room temperature at a minimum. An alternative procedure is to first collect the hydrogen given off at room temperature, preferably over mercury, and then, when the evolution at room temperature has virtually ceased, to determine the remainder of the hydrogen by vacuum extraction at about 1000°C and to add the two fractions.

The determination of hydrogen in cast irons also offers some additional problems. The molten metal should be sampled by a chill-cast technique and should have the surface machined lightly. Dawson and Smith (68) found an extraction temperature of 1070°C to be suitable. Gray cast irons and cast irons with more open structures are likely to have hydrogen as a minor constituent in the extracted gases (155). Small amounts of methane in the extracted gases have been reported by some workers (229).

5. Specific Metals

As in the case of steel, a temperature of about 1000°C can be used to extract hydrogen from metals with a sufficiently high melting point, such as cobalt (249a), titanium, vanadium, chromium, and niobium. The extracted hydrogen can then be analyzed and measured by any of the techniques described above. Since these metals do not possess the high hydrogen diffusivity of iron and ferritic steels (75), they can be handled more easily. Hence they can usually be stored at room temperature, although any cutting should still be done with the liberal use of a coolant. The samples are again prepared by filing lightly. Relatively high hydrogen contents are possible, so that a small sample often suffices. Titanium specimens should be well under 1 mm in thickness.

Carson and Rossettie (47a) found that they could determine the hydrogen content of titanium after only 5 min of extraction at ~1100°C by making the specimen sufficiently thin (~0.05 mm) and applying an average value for the residual gas other than hydrogen.

Powdered or granulated forms of these metals are conveniently prepared for analysis by wrapping in platinum foil. Since such samples cannot be prepared by filing the surface clean, the hot-extraction procedure is normally applicable only to oxidation-resistant metals which are prepared in a vacuum or an inert atmosphere.

Consolazio and McMahon (58) heated the specimen by making it a filament and passing a current through it. This technique is feasible for a number of metals, such as titanium.

Several techniques for the determination of hydrogen in aluminum have been reported. Dardel (66) melted the specimen and then lowered the pressure and related the hydrogen content to the pressure at which the first bubble appeared. Griffith and Mallett (121) used the tin-fusion technique at 525°C on aluminum alloys, reporting a sensitivity of about 0.05 ppm of hydrogen, and found that hot extraction at 525°C did not remove all of the hydrogen when it was present to greater than 0.5 ppm. Brandt and Cochran (38), however, found the results of hot extraction at 550–575°C to be in agreement with those of tin fusion at 650°C for aluminum and its common alloys. Gillespie (107), comparing a tritium-tracer technique with the hot-extraction technique of Brandt and Cochran, obtained the same results but found his tracer technique (applicable only when tritium can be introduced into the molten aluminum) to be more sensitive. It therefore appears that hot extraction at 550–600°C is a suitable technique for the determination of hydrogen in aluminum and its common alloys. However, since hot extraction takes more than an hour (38), tin fusion is more advantageous when rapidity of analysis, rather than convenience, is important.

The surface preparation of an aluminum sample is very important. The most suitable procedure (38,225) was found to be a light, dry machining of the surface of a cylindrical specimen, followed by a benzene wash. The blank for the surface contribution is determined by taking a specimen that has been analyzed by hot extraction and putting it through the steps for another analysis.

The same considerations and temperatures apply to the determination of hydrogen in magnesium. The determination has also been carried out for magnesium using a tin bath at 450°C (191) and a lead bath at 600°C (31).

Copper and its alloys appear to offer no difficulties in extraction or in surface effects (as long as the surface is filed clean), an extraction temperature of about 650°C being used (75). However, if the copper is not well reduced and low in oxygen, the hydrogen may be present almost entirely as water, as it would be in tough pitch copper. In this case a vacuum-fusion procedure at a temperature of 1100°C or higher is necessary to reduce the water to hydrogen and carbon monoxide. There are indications that, although the hydrogen in phosphorus-deoxidized copper appears to be present as water, it is recovered by hot extraction; however, there is need for more work in this area (75).

Because Peterson and Fattore (220) encountered excessive volatilization in the application of the hot-extraction technique to calcium, they preferred the tin-fusion technique at 670°C for hydrogen determination. Fusion in a copper bath at 1150°C has been reported to give good results for hydrogen in beryllium (230b).

Finally, the alkali metals pose additional problems because of their reactivity with water and other substances and because of their low melting point. Special precautions must be taken in handling them to prevent hydrogen pickup from the water vapor in the atmosphere (191). Pepkowitz and Proud (218) sealed a sample in an iron capsule and then heated the capsule in a vacuum at 700°C to extract the hydrogen through the iron container. Glass et al. (108) improved on this technique by using a stainless steel Swagelok reducer with a thin iron "window," thus providing a reusable capsule and eliminating the welding operation. Lithium and magnesium–lithium alloys have also been analyzed for hydrogen with a tin bath at 450°C (96). Malyshev et al. (193a) determined the hydrogen in lithium by simply heating 5–10 mg of the metal in a special molybdenum crucible in a vacuum up to a temperature of 650°C. Anthanis (6a) determined the hydrogen in cesium by refluxing 15-g samples in a high vacuum and analyzing the extracted gases with a mass spectrometer.

6. Commercial Equipment

Of the commercial equipment (226a) available, the LECO Hydrogen Determinator (Laboratory Equipment Corporation, St. Joseph, Mich., 49085) uses induction heating to heat a specimen in a graphite crucible up to 1250°C. The extracted gases in the glass system are transferred to a calibrated volume with a mercury diffusion pump; there their pressures are measured with a McLeod gage. The partial pressure of the hydrogen is then determined from the pressure drop as the hydrogen is converted to water over hot copper oxide and absorbed on magnesium perchlorate. A false bottom transports the analyzed sample into a container. The LECO Simplified Hydrogen Determinator, an inexpensive, less precise instrument for rapid, quality-control work, does not analyze the evolved gases and includes the furnace section as part of the collecting volume.

The Hytest Analyzer (De La Rue Frigistor, Ltd.) is not a vacuum instrument but employs argon as a carrier gas to determine hydrogen with the aid of a katharometer. It is mentioned, therefore, with more detail in Section VI.

Metals such as titanium and titanium alloys, in which high concentrations of hydrogen are possible, lend themselves to the interesting technique of neutron radiography (30a). Hydrogen has a much higher mass attenuation coefficient toward neutrons than most metals have. Therefore, by using a technique with thermal neutrons similar to the X-ray absorption technique, high total or local concentrations of hydrogen can be detected (128b,153a,282a). The precision of the technique is rather poor; for example, hydrogen can be detected in a 20-mil thickness of zirconium only when its concentration is about 200 ppm or more. However, the technique can serve to detect abnormal situations.

F. VACUUM-FUSION ANALYSIS

1. Principle of the Method

When most metals are melted in vacuum in the presence of an excess of carbon, gas which consists simply of carbon monoxide, hydrogen, and nitrogen is produced. The carbon monoxide normally results from the reduction of oxides in the metal and is, therefore, a measure of the oxygen content of the metal. The hydrogen and nitrogen are released by decomposition and diffusion, although additional processes, such as displacement by carbon, are possible in the case of nitrogen (247). The gas mixture can then be withdrawn rapidly and separated in one of several ways to determine the quantity of each gas, normally with the aid of the perfect gas law, $PV = NRT$, after measuring the relatively low pressure of the gas in a known volume at room temperature. Sensitive measurements of microgram quantities of each gas are possible with this technique.

Sloman et al. (247) have discussed in detail the thermodynamic relations involved when oxides or nitrides are analyzed by dropping them into a graphite crucible containing molten iron, called an iron "bath." They concluded that the reactions most likely to be responsible for evolution of gases were reaction of the solid oxide or nitride with solid carbon to give carbon monoxide or nitrogen and a solution of the specimen metal in molten iron (when an iron bath is used). Although the formation of the specimen metal carbides is recognized, this reaction is apparently secondary and does not influence the theoretical thermodynamic equilibrium involved in gas evolution. However, these carbides are generally solid at the temperatures employed for analysis. Their formation and precipitation, particularly in an iron bath wherein iron also contributes a solid carbide, leads sooner or later to high viscosity and solidification. The precipitate, called "kish," markedly interferes with the physical process of gas removal and with the dissolving of later metal specimens for analysis.

By combining the Gibbs energy of formation of a particular oxide with that of carbon monoxide, a value for the pressure of CO in equilibrium with the oxide at any temperature may be calculated according to

$$\log (P_{CO}) = \frac{\Delta G_1 - \Delta G_2}{4.58T}$$

Sloman et al. (247) used this concept to estimate required temperatures for analyzing metals not previously attempted. Experience had shown that Al_2O_3 evolved gas quite adequately in their apparatus at 1600°C, and the equation yielded the result of 2.4 torr for this temperature. With appropriate thermodynamic data, the temperatures required to yield the same pressure were computed for ZrO_2, UO_2, TiO, and ThO_2 as 1670, 1710, 1785, and 1950°C, respectively. Experimental and analytically satisfactory temperatures were found to be 1625, 1700, 1750, and 1900°C, respectively. Similar calculations were made for the nitrides, but the agreement was not particularly good. The difficulty was attributed in part to the poor thermodynamic data available for the nitrides.

Before making the measurement of temperatures for good analytical work, the authors studied the effects of hot metal film deposited on the inside of the apparatus. The aim was to see how long the film could be tolerated before the amount of CO that it reabsorbed seriously affected the analysis. This effect whereby evaporated metal recondenses and reabsorbs the gas to be measured, known as "gettering," constitutes a serious problem in some cases. The use of tin as an additive to the iron bath represented an early attempt to overcome the problem. When a diluent metal is added at the same time as a specimen, this "flux" is intended to keep the specimen-metal concentration low in the bath in order to reduce gettering. Tin is also supposed to decrease gettering by evaporating and overlying evaporated iron and/or specimen metal.

Although the theory presented has large elements of validity, there is one aspect not treated which has very important bearing on the analysis of metal samples. The thermodynamic treatment assumes that a definite compound oxide (or nitride) is being decomposed, whereas, in fact, for many metals the low oxygen contents are present as solid solutions. The Gibbs energy of formation of a solid solution is different from that of the definite oxide and depends on concentration as well as on temperature. It has been experienced (152b) that a vacuum-fusion analytical procedure adequate for ZrO_2 is not satisfactory for a much lower ratio of oxygen dissolved in zirconium. Modification of the theory to include this aspect is, of course, relatively simple, but the newly required data are more difficult to obtain.

The case for hydrogen is similar to that for oxygen and nitrogen except that for the equilibrium reaction carbon is not needed. Simple decomposition of the hydride (or solid solution of hydrogen in metal) to yield hydrogen gas and specimen metal dissolved in iron is adequate.

At present the technique is used chiefly for the determination of the oxygen contents of metals. Under the usual vacuum-fusion conditions the hydrogen is also extracted from the metal, and a useful value for the hydrogen content can be obtained, particularly when the hydrogen level is relatively high, as in the usual commercial titanium. However, if the hydrogen content is of prime concern, then the hot-extraction technique (i.e., heating to temperatures below melting) is normally more convenient and precise. The determination of nitrogen by vacuum fusion is a more complex subject. Until fairly recently, analysts in general viewed such nitrogen determinations as unreliable. However, Fassel et al. (89) and Gerhardt et al. (104), using procedures described in Section IV.F.6, report nitrogen results which agree well with those obtained by other techniques.

A satisfactory vacuum can be achieved with a mercury diffusion pump backed up by a mechanical pump, while the necessary reducing conditions are usually obtained by melting the metal in a graphite crucible heated with an induction heater. Often the metal must be fused with a suitable molten bath of another metal, usually iron or platinum.

The quantity of gas evolved is still sometimes measured gravimetrically, but is now usually determined either manometrically or with a gas chromatograph. This measurement has also been carried out by other means, which will be discussed in Section IV.F.6.

The subject of vacuum-fusion analysis will now be treated in more depth by presenting the development of the technique, the operation of a vacuum-fusion apparatus for the determination of gases in steel, some of the variations in apparatus and procedure for this determination, some problems encountered in vacuum-fusion analyses, the application of the technique to different metals, and, finally, a brief description of some commercial pieces of equipment.

2. History

It is generally considered that the vacuum-fusion method stems from work done by von Oberhoffer and Schenck (277) in Germany and published in 1927 and from similar work done by Jordan and Eckman (160) in the United States and published in 1925. Jordan and Eckman, using gravimetric measurements with an oxidation and absorption train having a vacuum pump at one end, established the superiority of an inductively heated graphite crucible (which also provided the carbon for the reduction of oxides) over ceramic crucibles or heated, "exploding" sample wires for the determination of oxygen in metals. Oberhoffer and Schenck developed equipment employing the vacuum concepts and gasometric measurements which give the method its scope and sensitivity. Later, Sloman and his associates did a considerable amount of work in England (37,243,244,95) on the application and theory of the method for ferrous and nonferrous metals. A comprehensive survey of the methods developed up to 1952 is given by Yeaton (286).

It soon became clear that the vacuum-fusion method offered a far more rapid, precise, and accurate way of determining the oxygen in metals than any existing wet methods, and the number of workers in the field steadily increased, as did improvements and variations of the technique. Recent reviews have been published by Mallett and Kallman (192a) and Dallman et al. (63a).

3. Furnace Assembly

Difficult problems were involved in the construction of the furnace. High temperatures had to be attained in the crucibles, and the amount of gas given off or absorbed by the furnace was required to be relatively small. Also, handling and cleaning had to be convenient.

Gradually, the furnace assembly housing the graphite crucible became less and less cumbersome, culminating in the Guldner–Beach furnace, shown in Fig. 103.4, which was described in a paper (128) in 1950. This type of furnace, usually with modifications, was quickly adopted

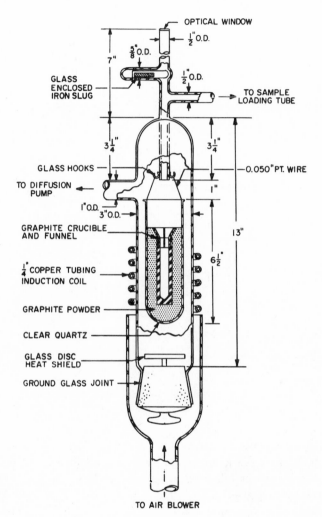

Fig. 103.4. Guldner–Beach type of furnace. (Courtesy of *Analytical Chemistry*.)

and is still in very wide usage. The graphite crucible, fitted with a slitted graphite funnel, is positioned in very loosely packed graphite powder, usually 200 mesh, in a quartz container. The graphite powder serves as insulation to retain the heat in the crucible and to prevent the walls of the quartz container and of the outside shell, made of borosilicate glass, from becoming too hot. This is desirable in order to reduce the amount of blank gases. The outside shell is cooled by an air blower,

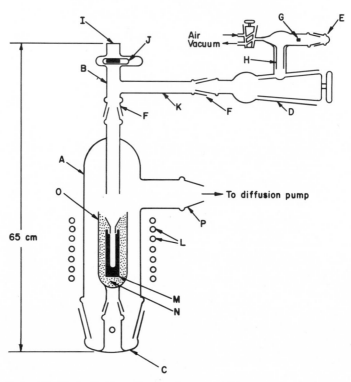

Fig. 103.5. Modified Guldner–Beach furnace with vacuum lock for sample introduction.

rather than by a water jacket as had previously been done. The container is suspended from the glass hooks by wires of platinum (or a suitable substitute such as molybdenum). A wax such as Apiezon W is used to seal the ground-glass joint, which Horton and Brady (152a) found much more convenient to replace with a butt seal. The glass-enclosed iron slug serves to protect the optical window from spattered metal during the melting of a sample and to cut down the deposition of vapors on the window. The temperature in the crucible is measured through the window with an optical pyrometer.

Figure 103.5 shows a modification (developed cooperatively by the laboratory of the Allegheny–Ludlum Steel plant and Wilt Laboratory Glass Blowing, Inc., both of Watervliet, N.Y.) of the Guldner–Beach furnace. A sample-introduction system is also shown. The crucible container rests on a pedestal in order to eliminate the inconvenience of positioning the container by platinum wires. The outside envelope D is made of quartz, which is much less subject to thermal strains, particularly at the ring seal. Parts B, C, and D are of borosilicate glass; cleaning the furnace is facilitated by having part B removable. All ground-glass joints are sealed with a sealing wax such as Apiezon W. Part D, which will be described later, is a sample-introduction device for inserting a sample into the sample chamber without breaking the vacuum. Hetherington (140a) designed a jacketed furnace to maintain the chamber wall at 100°C, both to keep the wall from overheating and yet to be hot enough to diminish adsorption of evolved gases.

Most often, the success or failure of an analysis is determined by the conditions that are brought about in the graphite crucible and its vicinity.

4. Apparatus and Functions

Numerous designs of the vacuum-fusion apparatus have been developed to carry out the functions of removing the gases out of and away from the fused metal and measuring them. The gases are usually transferred to a constant volume, where they are analyzed and their pressures are accurately measured. Figure 103.6 shows a vacuum-fusion apparatus in one of its simpler forms. The parts and functions, and later the opera-

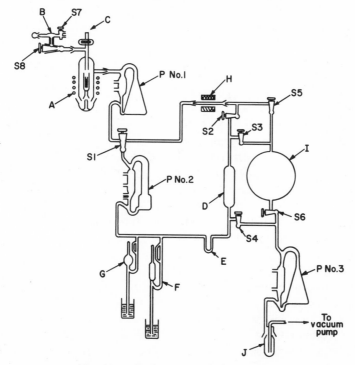

Fig. 103.6. Basic vacuum-fusion system.

tion, of the apparatus will be described. Except where indicated, the system is constructed of borosilicate glass.

Since the gases, at the low pressures involved, are either in slip flow or molecular flow (39), it is desirable to have large-bore tubing ($\sim$12 mm or more) of minimum length to achieve rapid flow of gases in the system. The two-way stopcock and all other stopcocks are precision vacuum stopcocks greased with Apiezon T or other suitable vacuum grease. The two-way stopcock should have a relatively large bore, such as 10 mm. However, it is possible to have stopcocks S2–S6 of a lesser bore (4–10 mm) so that they will be easier to operate and maintain without greatly affecting the time for an analysis. Subsequently, we will consider mercury cutoffs, solenoid-operated valves, and greaseless stopcocks as substitutes for greased stopcocks.

All of the ground-glass joints are sealed with a wax, such as Apiezon W, except for the cap E (Fig. 103.5) on the sample-introduction system, which is greased.

The furnace assembly and the sample-introduction system are as shown in Fig. 103.5. An induction heater with an output of about 2.5 kW or higher energizes the copper coil.

Rapid removal of the gas evolved from the melted sample is essential. This is the function of the two-stage mercury diffusion pump, P No. 1. It is also desirable, of course, to have a short connection with a relatively large diameter (20 mm or more) between the furnace assembly and the diffusion pump.

The gases must be pumped into a constant volume. This is accomplished by having the gases transferred into a known volume by P No. 2, a two-stage mercury diffusion pump with a Naughton–Uhlig modification (209), which assures a constant gas volume against the forepressure end of the pump. The other terminals of the constant volume are the stopcocks $S2$, $S3$, and $S4$ and the mercury level in the McLeod gage. A volume of about 1 liter is convenient. If a larger volume is found desirable or necessary, an expansion of about 5 liters can be obtained by opening $S3$. The small volume can be calibrated against a comparable known volume attached temporarily or against the known volume of the McLeod gage (207). The U-bend in the known volume serves as a cold trap in the subsequent analysis of the evolved gases. The two-stage mercury diffusion pump, P No. 3, and the mechanical vacuum pump are used to evacuate the system. Condensable vapors are frozen out with liquid nitrogen at trap J.

The function of the copper oxide–Misch metal mixture between $S1$ and $S2$ is to oxidize the gases. It is commonly put in a trap and heated by a resistance furnace to a suitable temperature, usually about 325°C, to oxidize hydrogen to water and CO to CO_2. Satisfactory results have also been reported with oxidized copper wire (183) and a kaolin–copper oxide mixture (59). In any case, care must be taken that the oxidant does not prevent the easy flow of gases. If oxidized copper wire is not used, the oxidizing mixture can be pressed or extruded into rods which will leave ample empty space when loosely packed into the trap.

The type and the number of McLeod gages can be varied considerably. The writers have found it convenient to use two McLeod gages, one with a range of about 0–300 μ Hg for measuring the usual pressures encountered with good precision and one of about 0–5000 μ Hg to extend the range of readable pressures.

5. Procedure for Steel

We will now follow the procedure for the vacuum-fusion analysis of a low-alloy steel for the determination of the oxygen content. Essentially, this involves dropping the cleaned specimens one after another into the heated graphite crucible and analyzing the evolved gas.

The sampling of molten steel was discussed in Section IV.E (14c,75, 200a,281).

Temperature control during the section of solid steel to obtain a specimen for analysis usually does not appear to be important. It is, nevertheless, good practice to keep the specimen surface cool with the aid of a coolant.

To prepare the system for analysis, the graphite crucible is positioned as shown in Fig. 103.5.

On the assumption that the rest of the system up to the two-way stopcock has been evacuated, stopcocks $S2$, $S5$, $S6$, and $S8$ and the two-way stopcock are closed while the rest remain open. If $S8$ is open (e.g., after freshly greasing), then $S7$ should be closed. With the vacuum pump operating, the two-way stopcock, $S1$, is cracked very slightly toward the furnace, because, if it is opened too much, graphite will be blown through the system, necessitating cleanup and reassembly. As a guide, the pressure on the McLeod gage can be kept between 5000 and 10,000 Torr. The two-way stopcock is cracked more and more until it is safe to have it wide open toward the furnace. The heat can then be turned on for P No. 1.

The temperature of the furnace is brought up gradually, because rapid heating can also disturb the graphite powder. A temperature of 2300–

2400°C is maintained for at least a couple of hours and then is allowed to drop down to about 1200°C or less.

, With the aid of a magnet, about 5 g of iron and about 0.5 g of tin are dropped into the crucible and outgassed at 1650°C until the pressure reading on the McLeod gage indicates that a suitable blank rate has been reached. This will vary with the apparatus and the level of oxygen expected in the sample (152a). For steel containing about 30 ppm of oxygen, a blank of about 2 liter-microns of CO is satisfactory. The outgassing temperature should be maintained for the minimum time, both at this point and during subsequent analyses, because the iron eventually diffuses through the crucible, forming one or more beads or lumps, and the blank rate will in time become intolerable. Thus a greater number of samples can be analyzed in the crucible if outgassing times are kept to a minimum. Probably the beneficial effect of the tin is derived from its deposition on surfaces in the vicinity of the crucible to prevent adsorption of evolved gases (26). This adsorption is not very important when the oxygen level in the steel is substantial (e.g., 60 ppm or more), but it becomes increasingly significant as the oxygen level becomes lower. This is evident from the data shown in Table 103.II. Although the magnitude of the tin effect will vary with the design of the apparatus, the addition of the tin is advisable at higher oxygen levels and is necessary at low levels to combat adsorption, or gettering, of the gases. In this respect, manganese is considered to be a bothersome metal which promotes gettering. Somiya et al. (252), however, found that manganese

TABLE 103.II

Effect of Tin Addition on the Determination of Oxygen in Steel

Steel sample	Oxygen content, ppm	
	Without tin addition	Tin added before each determination
1. Bur. Stds. #7	1075	1062
2. Bur. Stds. #5	85, 94	97, 85, 92
3. Bur. Stds. #4	15, 19	22, 23, 18
4. Vac. melted, V-287	24	39
5. Vac. melted, V-312	11, 8	25, 29
6. Vac. melted, V-280	0.3, 0.4, 0.5, 1.5, 5.8	6.3, 6.6, 7.3, 7.1, 7.0, 6.6, 7.6, 6.0, 6.7, 8.6, 7.3
7. Vac. melted, V-278	5, 1.3	12
8. Vac. melted, V-286	15, 2.6	21

gave no trouble in their apparatus if it was present to an extent of <1% or if the sample was dropped into a bath of tin or platinum.

A solid sample weighing about 1–3 g is usually suitable. The sample is degreased by washing first with benzene and then with acetone. With the aid of a clean file in good condition and a clean vise which is used for no other purpose, the entire sample is filed to remove all surface oxides. Handling of the sample is done with a pair of tweezers. After being weighed, the sample is introduced into the system through the vacuum lock.

When the initial charge of iron and tin has been sufficiently outgassed, the temperature is dropped to 1200°C or less, a piece of tin weighing 0.1–0.3 g is dropped into the crucible, and the temperature is raised again to 1650°C. Then, $S2$, $S3$, and $S4$ are closed; the time is noted; the glass-enclosed iron slug is moved to protect the optical glass at the top of the furnace; and the sample is dropped into the iron bath at 1650°C. Outgassing takes place until the blank rate is reached, usually within 15 min. When the outgassing characteristics of the steel are known, the outgassing time can be fixed. The power is shut off between runs to diminish diffusion of iron through the crucible.

The pressure of the gases collected (shown by many workers to be only hydrogen, CO, and nitrogen) in the "small" volume is measured with the U-bend trap immersed in a medium of about −80°C. Such a bath can be obtained in a number of ways, for example, by the use of a Dry Ice–acetone or a Dry Ice–methanol slush, or by the addition of small amounts of liquid nitrogen to acetone (45) or ethanol (112). If it is preferred to use a lower temperature, −98°C can be attained with small additions of liquid nitrogen to methanol in order to make a slush of the solid and liquid alcohol (152a). At this temperature virtually all of the water is frozen out; this merely constitutes an added precaution at this point because usually the water collected is negligible. From a previously prepared chart, a correction is made for the small drop in pressure due to the cold trap. This corrected pressure will be called P_i. If the pressure is above the forepressure against which P No. 2 can operate, $S5$ and $S6$ are closed, the gases are expanded into the expansion volume by opening $S3$, the pressure is measured after a few minutes, and $S3$ is again closed so that only the gases in the small volume will be analyzed.

The gases are now oxidized by circulating them over the copper oxide. For this purpose, $S5$ is closed, $S2$ is opened, and the two-way stopcock, $S1$, is turned to the copper oxide. The U-bend is now immersed in liquid nitrogen (−195°C). The gases are circulated until the pressure ceases

to drop, which normally should not take much more than 5 min. Then
S2 is closed, and the gases are again collected in the small volume. During the oxidation the hydrogen is converted to water and the carbon monoxide to carbon dioxide. Both the water and the CO_2 are frozen out in the trap. Therefore, only the nitrogen, which is unaffected by the copper oxide, contributes to the pressure, P_{N_2}, measured at this point.

Now S4 is opened to remove the nitrogen. When the pressure is <0.1 μ Hg, S4 is again closed and the liquid nitrogen around the trap is replaced by acetone at $\sim$ −80°C. This releases the CO_2, the pressure of which P_{CO_2}, can then be measured directly; the correction for the presence of the cold trap is made as usual. Of course, P_{CO_2}, equals the pressure of the CO originally evolved. If the hydrogen evolved is of interest, its pressure, P_{H_2}, can be obtained by difference, that is,

$$P_{H_2} = P_i - P_{CO_2} - P_{N_2}$$

To prepare for the next sample, S4 is opened, S1 is turned toward P No. 1, and the furnace and a piece of tin weighing 0.1–0.3 g is dropped into the crucible with the temperature of crucible <1200°C. The next sample is introduced into the system through the vacuum lock, and the temperature of the furnace is quickly raised to 1650°C. When the pressure reading indicates that gas is evolving at the blank rate, S4 is closed, the time is noted, and the next sample is dropped into the crucible.

The blank is obtained by proceeding as for the analysis of a sample, using the average time of outgassing.

The weight of each of the three gaseous constituents evolved from the sample is calculated from the ideal gas law, $P_G V = NRT$, where P_G and V are the pressure and the volume of the measured gas, N is the number of moles of the gas, R is the gas constant in the proper units, and T is the absolute temperature (normally the room temperature) of the gas. If W_G is the weight of the gas and M_G is its molecular weight, then

$$W_G = \frac{M_G P_G V}{RT} \tag{1}$$

or, in parts per million,

$$(\text{ppm gas}) = \frac{M_G P_G V \times 10^6}{RTW_S} \tag{2}$$

where W_S is the weight of the sample. Equation (2) could be put into the form (ppm gas) = $K_G P_G / W_S$ by combining all the constants into K_G. Of course, in the case of hydrogen and nitrogen M_G represents the molecular weights of these gases. However, in calculating the oxygen content of the steel, it will be recalled that the oxides in the steel are reduced by carbon to form CO, which subsequently is converted to CO_2, so that $M_G = 16$ in the calculation.

This procedure for vacuum-fusion analysis, using iron as a bath medium, is used chiefly for determining the oxygen content. It is generally agreed that the nitrogen results obtained by this procedure are to be questioned (166,273), although some workers have found these values to be satisfactory (13,94a,112). Masson and Pearce (200) got good results in some cases when the sample was outgassed at temperatures as high as 2240°C, and they suggested that the higher temperatures help to release the nitrogen absorbed by the graphite crucible. Other work (115), however, indicated that the unreleased nitrogen was in the iron–carbon bath. Somiya et al. (252) found the vacuum-fusion values to be in much closer agreement (0–15%) with Kjeldahl values if they added the nitrogen released during a subsequent outgassing period of up to 26 min from an iron bath.

Thermodynamic considerations indicate that conditions for the decomposition of nitrides, even refractory ones such as those of titanium and niobium, can be achieved by vacuum fusion in an iron bath (115,247). From a thorough, excellent study of the problem, Goward (115) concluded that the main deterrent to the attainment of satisfactory vacuum-fusion values from an iron bath is an inability to achieve thermodynamic equilibrium, possibly because of inhomogeneity of composition of the iron–carbon system at the bath–gas interface, where a semisolid network of graphite and carbon-saturated iron has been shown to exist (251). Gerhardt et al. (104) reported recently that they solved the problem by adding a little cerium (see Section IV.F.6) to put the graphite in spheroidal rather than flake form. They also removed the molten sample after each determination with a special furnace and crucible. Fassel et al. (89) obtained satisfactory results using a platinum bath, and their procedure is also described in Section IV.F.6.

Vacuum-fusion values for nitrogen are of interest not only because they constitute additional useful information from a vacuum-fusion analysis, but also because they can be obtained quickly enough to be applied for quality control. Moreover, it is possible that vacuum-fusion values may be useful in determining the form in which the nitrogen is present in steel. Finally, mention should be made of the fact that, for high-silicon (~3%) steels, the vacuum-fusion values have been found to be more accurate than the Kjeldahl values (89,156).

In the case of hydrogen in steel, cooperative programs indicate that the vacuum-fusion technique is not suitable (13,14a). The cause appears to be the combination of the low hydrogen content of steel (normally well under 5 ppm) and a relatively large blank. Fusion at a lower temperature in a tin bath in a quartz container can yield satisfactory results, but the best technique for the removal of hydrogen from steel is hot extraction, heating the sample to a temperature of about 800–1000°C in a vacuum (45,101,141)—a technique described in Section IV.E. On the other hand, it has also been mentioned that a higher carbon content (e.g., >0.5%) can apparently, cause complications, probably by influencing the manner in which the hydrogen exists in the steel, so that some workers (158) have found the temperature of extraction to influence the hydrogen determination on these steels. In such cases, vacuum-fusion results for hydrogen could be of interest.

Because determining the hydrogen by vacuum fusion is usually of no interest in the case of steel, it is possible to have the gases exposed to the copper oxide during the outgassing period. Then $S5$ would be closed and $S2$ opened during the outgassing period, which must be of sufficient duration. The U-trap would be immersed in liquid nitrogen. After the outgassing of the sample, the gases need to be circulated over the copper oxide for only a few minutes. The nitrogen could then be collected as before or pumped out directly if it, too, is of no interest. The CO_2 frozen out would be released for measurement in the same manner as before.

6. Variations in Apparatus and Procedure

One variation consists of the construction of an apparatus with two analytical units (60,124a,270). It is then possible to collect the gases from one sample into one analytical unit while the gases from the previous sample are being analyzed in the other unit. This can be done by connecting the furnace-outgassing diffusion pump (P No. 1 in Fig. 103.6) to two analytical units, as shown in Fig. 103.7. Each analytical unit then contains a diffusion pump such as P No. 2, a copper oxide furnace, McLeod gages, and so on. Both analytical units can be connected to one outgassing diffusion pump (P No. 3) and one mechanical vacuum pump. Substantial savings in time can thus be effected for routine analyses.

Variations are encountered also in the preparation of a specimen for analysis. Some workers (25,165a) prefer an acid or electrolytic etch, followed by washing first with distilled water and then with acetone or some suitable drying solvent, without filing, whereas others avoid

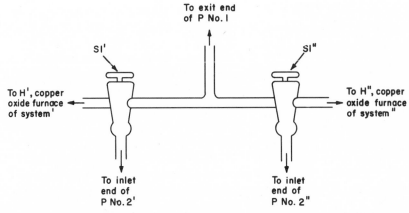

Fig. 103.7. Connections from furnace out-gassing pump to double analytical system.

acid cleaning (112). Kammori et al. (165a) compared various pre-
paratory procedures and concluded that iron and steel specimens are
best prepared by electropolishing in a solution of acetic and perchloric acids. By comparison, they found that filing and abrasion with
silicon carbide paper introduced a few parts per million of oxygen, while
abrasion with emery paper introduced entirely unacceptable amounts of
oxygen.

A mercury lift, essentially a mercury barometer from the bottom end
of which a sample was floated to the top of the mercury column and
into the system, was used earlier for the introduction of a sample, but
for several reasons it is inferior to a vacuum lock of some sort. Although
a crucible plug or cover is frequently used (258), the experience of cooperative groups in the case of a number of metals is that one does
not appear to be necessary. Heating of the crucible can also be accomplished by resistance heating (59,94a,104,180). In addition to platinum,
nickel is often employed for the vacuum-fusion analysis of steel, and
P No. 2 (Fig. 103.6) is often replaced by a Toepler pump (286). A
micro Orsat apparatus has been used for the analysis of the gases (131),
but this technique will probably be found unsuitable for low amounts
of gases. An interesting system designed by Somiya et al. (252) automatically records the pressure indication of an oil (Dow Corning silicone
No. 73 or 74) manometer and Pirani gage and has a computer deliver
the results. It is often useful to follow changes in pressure, either in
the furnace section or in the gas-collecting volume (252), with a suitable

gage such as a Pirani, thermocouple, or Alphatron (National Research Corporation) gage.

An interesting variation for the heating of a sample is levitation melting, a well-known technique which Hickam and Zamaria (140b) applied to vacuum fusion. This technique suspends the specimen in space with electromagnetic forces even after fusion, thus avoiding contact of the specimen with a crucible. Heating is rapid, and a stirring action within the fused mass promotes outgassing. Some special studies are, therefore, possible with a number of metals if a bath or a graphite crucible is not required. Thus surface gases on copper have been studied by Hickam and Zamaria. Also, they report that, because the technique provides rapid quenching of the evolved gases, some hydrocarbons were found to be given off when zirconium was heated to determine the hydrogen content. Since contact with a graphite crucible is so often desired, they have been experimenting with small crucibles that can be levitated along with the sample.

In impulse heating (112a,274a), a large surge of current of about 1000 A is pulsed through the crucible for approximately 12–15 sec to achieve a temperature of $\sim$3200°C very rapidly. Goldbeck (112a) reported good results with this technique for boron, uranium, beryllium, and thorium.

As the result of studies on the determination of nitrogen in steel, Gerhardt et al. (104) concluded that the chief problem in the complete removal of nitrogen from steel during vacuum-fusion analysis was the rapid accumulation of precipitated graphite in flake form (183,251). This precipitated graphite, called "kish," made the bath more viscous, they said, and hindered the release of nitrogen. To prevent this condition, they removed each sample after analysis by means of a spinning crucible and put the precipitated graphite into spheroidal form by adding cerium as a carbon-saturated nickel–cerium alloy along with the sample. The spinning was done around the vertical axis, and the molten metal was caught in a graphite cup surrounding the crucible. The use of a fresh crucible for each determination is in agreement with the writers' experience that the first steel samples analyzed in a graphite crucible usually give higher, more satisfactory nitrogen values. The cerium content of the melt was made about 1% by weight; in the case of titanium-bearing steels it was made somewhat higher. Additional nickel was used with chromium-alloyed steels. The sample weight was less than 1.5 g in each case, and the extraction temperature, achieved by resistance heating, was about 1600°C. They reported complete extraction in only 1.5–3.5 min. Repeated temperature cycling did not bring about the release of any additional nitrogen (103). Satisfactory results were obtained for a series of steels, including 3% silicon steel. Suarez-Acosta (259a) reported that the addition of silver to the bath, in place of cerium, also led to a more complete release of nitrogen from steel.

Fassel et al. (89) found that they could get satisfactory nitrogen values for steels by using a platinum bath. The ratio of platinum to steel was kept at 4:1 or greater, and the sample was introduced with about an equal weight of platinum. An outgassing temperature of 1850°C was usually sufficient, but an increase may be advisable for more highly alloyed steels. Outgassing took place for about 30 min. Others have reported satisfactory results for determining nitrogen in steels by this procedure (14a), and it was studied by a task force of Division I of ASTM Committee E-3. The platinum bath has a low solubility for carbon so that graphite precipitation does not occur and the bath remains fluid. Working with a vacuum-melted steel provided for the steel task force of Division I, ASTM Committee E-3, Carson and Worcester (47b) found that, although the addition of tin before a determination, using a platinum flux technique, was favorable to the determination of oxygen, raising the average from 1.0 to 3.2 ppm, the nitrogen content was lowered by the tin addition from 48 to 39 ppm on the average. Their Pt/sample ratio was 8–10:1. (It might be mentioned that the average for the oxygen content was raised further to 4.3 ppm when an iron bath was employed with a tin addition before each determination, the procedure described for steel.)

Guldner (127,128a) used a xenon flash discharge lamp to decompose thin films in a vacuum and estimated that he got temperatures >5000°C. The gases given off, in the case of a tantalum film, were hydrogen, argon, oxygen, nitrogen, methane, and carbon monoxide, indicating decomposition of the oxide and nitride. When glass was the substrate of the film, the nitrogen values were somewhat low because of an interaction of the nitride with the glass.

More rugged all-metal systems have also been made and reported to be satisfactory (94a,156,180). Greaseless Teflon stopcocks have been found to hold a vacuum (81) satisfactory for vacuum-fusion work. Although greased precision stopcocks continue to be popular, mercury cutoffs are also in wide usage. Proponents of the latter concede that stopcocks appear to be satisfactory, but they believe that the more cumbersome cutoffs add an extra margin of safety by eliminating any possibility of gas emission from the grease (286). Horton and Brady (152a) found indications that the grease makes a small contribution to the blank. Figure 103.8 shows a mercury cutoff in the open position. The

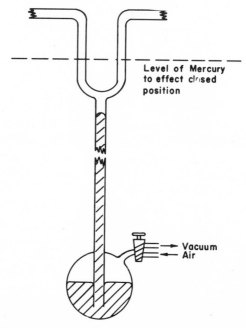

Fig. 103.8. Mercury cutoff valve in open position.

closed position is achieved by raising the mercury to the higher level shown by the dotted line.

Less cumbersome than the preceding is the solenoid-operated mercury cutoff (152a) shown in Fig. 103.9 in the open position. De-energizing the solenoid allows the movable part with the iron slug to drop and effect the closed position. McKinley (201a) used this type of valve to automate a vacuum-fusion apparatus quite effectively—the pushing of a button opening or closing the necessary valves and raising or lowering cold media for traps to bring about a desired condition.

Covington and Bennett (60) got good results with a commercial catalyst, one of the Hopcalite catalysts made by the Mine Safety Appliance Company, to oxidize the carbon monoxide to carbon dioxide and Ascarite to absorb the CO_2. The two preparations are in a measured volume, so that circulation of the gases is unnecessary. Mallett et al. (193), however, report that the Hopcalite–Ascarite mixture also absorbs some of the hydrogen and therefore lessens the accuracy of the oxygen determination.

Magnesium perchlorate (286) and, less often, phosphorus pentoxide

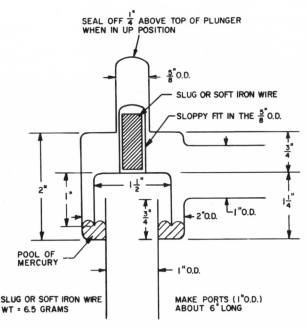

Fig. 103.9. Solenoid-operated mercury cutoff valve in open position. (Courtesy of *Analytical Chemistry*.)

(254) also are sometimes used as absorbents for water vapor in the analysis of the gases. Somiya et al. (252) found the phosphorus pentoxide to be more reliable. The hydrogen can also be removed by diffusion through a silver–palladium thimble (75) (see Section IV.E).

The extracted gases can also be analyzed in a mass spectrometer (6b,198), the analysis being made on all or part (collected in a detachable flask) of the gas. A mass spectrometer does not appear to be justified for routine analyses, but it can be valuable as a research tool, especially when the species of gas present is uncertain. Aspinal (6b) reported a detection limit of 0.1 μg for oxygen and nitrogen, and 0.01 μg for hydrogen, and found a mass spectrometer especially advantageous for materials with low gas contents for specialized purposes. Bagshawe et al. (17b) found a mass spectrometer more sensitive and less involved than the usual gasometric method employing pressure measurements when they worked with low-oxygen (4–50 ppm) steels and nickel-base alloys.

The biggest recent development in the analysis of the gases is the use of the gas chromatographic technique (128a,181a,232). As the sensitivity of this technique improved to a sufficient extent, it was put to use in vacuum-fusion work (92,181,181a,283a) and is reported to provide a rapid and precise analysis. The gas is concentrated in a small volume and then swept by a noble gas through the gas chromatograph. Some workers prefer a hot wire to the thermistor for a detector; either provides a sensitivity comparable to that obtained by the low-pressure gas measurements and fractionation techniques described earlier. More recently, however, a helium ionization detector has been devised which operates on the principle that the introduction of another gas into the helium carrier gas changes the ionization characteristics of the helium. This technique is more sensitive by a factor of ~1000; and, since it is not dependent on thermal conductivity properties, it can provide a good measurement of the hydrogen in the helium carrier, whereas a thermistor or hot-wire detector must choose between good sensitivity for hydrogen (with argon as the carrier) and good sensitivity for nitrogen and carbon monoxide (with helium as the carrier). Lilburne (181a) found the helium ionization detector very satisfactory for vacuum-fusion analyses, particularly for metals with low gas contents. The use of such a detector has been reported for the determination of ~1 ppm of oxygen in samples weighing only 0.2 g (283a). This type of detector, however, involves extra expense and experimental effort.

Kraus (178) reports that he was able to get the partial pressures of the three component gases from the total pressure on the small, closed-end U-tube manometer, the reading on a thermal conductivity gage, and a measurement of the infrared absorption of the gages. The thermal conductivity gage can give the partial pressure of the hydrogen, as explained in Section IV.E on hot extraction, while the infrared absorption measures the partial pressure of the carbon monoxide, the only gas in the system having a dipole moment that is not zero and therefore capable of interacting with the infrared radiation. Holler (144) also used an infrared method to obtain a very fast analysis for control purposes.

Finally, there is a continuing interest in using vacuum-fusion results to determine in what form the oxygen and/or nitrogen is present in the steel. Such knowledge would be of great value in metallurgy, because it could lead to a better understanding and control of the effects of these elements on the properties of the steel. This interest has led to "fractional" vacuum-fusion work, which consists of raising the temperature in the crucible in steps and analyzing the gas evolved at each step. It is then assumed that the more refractory oxides are decomposed and produce CO at the higher temperatures. Although some workers

(94a,271) have great confidence in fractional vacuum fusion, it still appears wiser to view such data as supplementary (187) to information from other sources (e.g., inclusion studies or determination of total oxygen). A knowledge of decomposition temperatures is not sufficient in itself to interpret fractional fusion values, because these values are affected by such factors as fluidity of the bath, gettering, and effect of alloying elements on activity coefficients, and even by whether an analysis is made in a fresh bath (either iron or nickel) or after a number of samples. Nevertheless, additional data and investigations will continue to increase our understanding of fractional vacuum-fusion results, as well as their value. It appears that, even now, division of the evolved gases into "readily evolved" and "less readily evolved" should prove useful.

7. Commercial Equipment

Roboz (226a) gives a detailed review of the commercial equipment available today for vacuum-fusion analyses.

The LECO Model 578-000 (Laboratory Equipment Corporation, St. Joseph, Mich. 49085) employs resistance heating for the fusion and has the dual analytical system designed by Covington and Bennett (60) to speed up analyses. The evolved gases are separated, and their pressures measured with a McLeod gage.

The Metallurgy MS 10 (Associated Electric Industries, Manchester, England; represented in the United States by AEI Corporation, 3 Corporate Park Drive, White Plains, N.Y. 10602) has a mass spectrometer for the analysis of the evolved gases (6b). Heating is done with an induction heater, and the furnace arrangement is similar to the Guldner–Beach type.

The Exhalograph EA 1 (Balzers Aktengesellschaft, Balzers, Liechtenstein; represented in the United States by Balzers High Vacuum Corporation, 1305 East Wakenham, P. O. Box 10816, Santa Ana, Calif. 92702) and newer models EA 2 and EA 3 employ the "spinning crucible" described in Section IV.F.6. The molten contents of the crucible are spun out after each analysis so that, essentially, each specimen is analyzed in a fresh crucible. It will be recalled that this is particularly advantageous for complete extraction of the nitrogen (104,178a). The system is all metal and employs resistance heating. An inductively sensed manometer monitors the gas pressure to indicate completeness of reaction. After admission of a controlled amount of hydrogen, a thermal conductivity measurement is used as a measure of CO plus N_2. The amount of CO is then obtained from the infrared absorption.

The Model VH-8 Evolograph, or Heraeus–Feichtinger vacuum-fusion analyzer (Leybold–Heraeus GMBH & Co., Wilhelm Rohn Str., Hanau, Germany; represented in the United States by Leybold-Heraeus, Inc., Seco Road, Monroeville, Pa. 15146), can be used with either an induction heater or, at the sacrifice of some sensitivity, a resistance heater. It is all metal, save for the furnace section for the induction heater. A five-step temperature programmer promotes reproducibility. Analysis of the gases is carried out with a gas chromatograph employing a thermal conductivity detector. The Model HH 2800 for determining the oxygen and nitrogen contents of metals can heat a specimen to 2800°C and analyzes the gases by measuring infrared absorption and thermal conductivity. The Model VH-9 Evolograph is of a modular design and can be tailored to customer requirements; thus a hot-extraction apparatus for hydrogen determinations can be incorporated.

Hetherington, Palo Alto, Calif., has an apparatus which is designed to deal with low gas contents, such as those found in low-oxygen copper. The evolved gas is measured with a small ion pump and digital integrator. A xenon flash discharge in the sample chamber is designed to first remove the bulk of the surface gas.

8. Some Problems in Vacuum-Fusion Analysis

Before proceeding to the analyses of metals other than low-alloy steels and the modifications that are necessary to apply the vacuum-fusion technique to these metals, some problems in vacuum-fusion analysis will be considered. The emphasis will be placed on the determination of the oxygen content of a metal because, though the values for hydrogen and nitrogen are frequently used, they are usually of secondary interest and can be ascertained reliably by other techniques. This is not to minimize the fact that there is a considerable interest in the determination of nitrogen by vacuum fusion.

It has already been mentioned that the essential task is to remove the oxygen quantitatively out of and away from the molten metal sample and then to measure it with sufficient precision. This is done by reducing the oxides with an excess of carbon to remove the oxygen from the metal as carbon monoxide, which is then transferred to an analytical system for measurement. The analysis and measurement of the collected gases rarely pose serious problems to the analyst. Most of the problems, therefore, center around the reduction of the oxides to form CO and the rapid removal of this gas away from the furnace area. For a large number of metals, there is a considerable area of agreement about the conditions that should be established to determine the oxygen content successfully, but for some metals there is still a need for more work to reach agreement. Unless an isotope dilution technique is applied to specially prepared specimens, the completeness of reaction and gas removal can be checked only indirectly.

The ideal way to check this completeness would be to compare the procedure against a standard of the same composition. However, as explained in Section VIII, "Standards and Comparisons," such standards, which represent the average of the results from a number of laboratories, are rather rare, although their preparation is receiving increasing attention. Synthetically prepared standards (88,152a) can be of great value, even though it can be argued that they are not truly representative of the distribution, form, and behavior of the oxygen in the actual metal, particularly in the case of a metal of very low oxygen content.

Failure to obtain reasonable checks on repetitive analyses of a material may indicate incompleteness of reaction and/or gas removal and calls for a variation.

Finally, interlaboratory agreement is a very good omen, even though it does not guarantee the result. In this connection, the most recent cooperative programs were conducted by Division I for Gases in Metals, ASTM Committee E-3, which had, at one time, about six active task forces and study groups examining techniques and procedures for a number of metals (iron, steel, molybdenum, tungsten, tantalum, beryllium, chromium, copper, and hafnium), including one for nitrogen by extractive methods (8b). Recently Division I was dissolved and its activities were absorbed by Committee E-3.

Normally, reduction of the oxides is brought about by fusing the sample with a "bath" of a suitable metal in the graphite crucible and providing an adequate temperature. The bath should be fluid and have a sufficient solubility for carbon to provide intimate contact of the sample · with carbon. In the case of the iron and steel analyses described, the bath was iron, contributed by the samples themselves after the introduction of some initial conditioning iron, which is advisable to avoid a frequently erratic result with the first sample. At a temperature of analysis of 1650°C, the bath remains fluid for a considerable length of time and the carbon solubility is more than ample. Several reaction mechanisms can be postulated (247), and more than one is probably taking place.

When checking the completeness of reaction by noting when the gas-evolution rate equals the blank rate, it is often found, particularly in the case of steels containing a relatively large concentration of nitrogen, that this equality is not reached in a reasonable length of time. Although all of the oxygen has been reduced, the rate of nitrogen evolution is still high. When the steel is alloyed to an appreciable extent, particularly with a metal such as titanium, it is advisable to check the completeness of reaction by carrying out several determinations at increasingly higher temperatures differing by increments of about 50°C. In increasing the temperature, however, one should bear in mind that an iron bath becomes too viscous rather quickly at elevated temperatures—in less than an hour at 1800°C (200b). This thickening and eventual incapacitation of the bath is caused by the solution and reprecipitation of the graphite, or "kish," in the surface layer of the bath (251). In the case of some metals, this loss of fluidity can lead to incomplete recovery of the oxygen in the metal (200b). Some alloys may be found to yield more oxygen when they are fused in a higher ratio of iron to sample, so that higher dilutions can be tried. This approach is sometimes indicated when a specimen of a known low-oxygen content is analyzed right after the sample in question, and found to give too high a result.

Sometimes a higher-oxygen result is obtained when the bath is diluted with tin (up to 5–10% or more). There has been considerable speculation as to why tin is found to be beneficial, sometimes even essential, in vacuum-fusion analyses (87). It has already been mentioned that tin appears to prevent gettering (26), but in larger amounts it also helps to achieve a greater fluidity of the bath. Also, its volatility may be performing a sweeping-out action to get the gases out of the bath and past surfaces where gettering might occur. Nickel baths have been tried and often found to remain more fluid than iron (21), but are not always as satisfactory (25). In this connection, an iron capsule was found superior to a nickel one for powdered chromium (152a).

The fluidity of an iron bath has also been enhanced by the introduction of a weighed amount of low-oxygen iron along with the sample (200b), a step, however, which can result in a considerable loss of precision if the sample has a low oxygen content. The importance of the fluidity of the bath to the attainment of a complete reaction was demonstrated by the work of McDonald et al. (200b) on titanium, zirconium, and molybdenum. In the case of titanium, they outgassed 10 g of iron before each analysis of an 0.25-g sample; and, after outgassing 1–2 g of tin, they introduced the sample along with 1–3 g of low-oxygen iron. Of course, dilutions of this magnitude severely limited the number of samples that could be run in a crucible. However, insufficient dilution, particularly without the use of tin, can result in the extraction of only a small fraction of the oxygen in the titanium. That Sloman et al. (247) were able to obtain apparently good results at a much lower dilution may have been due to their use of a rather large crucible, because other investigators (59,87,200b) could not duplicate their results. Shitikov and Gederevich (237b) experimented with iron, cobalt, and nickel baths for the determination of oxygen and nitrogen in ferrochrome and reported that they obtained the highest values when they used a nickel bath at 1750°C with tin as an additive.

Though the desirability of fluidity of the bath for complete reaction is generally accepted, Walter (279), using a socalled dry-bath technique, devised a way to determine the oxygen in titanium without a fluid bath. The sample was introduced at a lower temperature with some tin into a crucible containing some small graphite "chips" and outgassed at 1900°C. It was thought at first that the large surface area of the graphite chips made possible the intimate contact of molten metal and carbon, as well as good conditions for escape of the resulting carbon monoxide, which are attributed to a satisfactory fluid bath. Later, however, it was shown that the role of the tin was more vital than that of the graphite chips, which could actually be eliminated (275). A disadvantage of the Walter technique was that, besides requiring an outgassing time of nearly a half hour, the maximum number of samples that could be run in a crucible was only five.

The introduction of platinum as a bath material (119,248,249) constitutes an important event in the analysis of oxygen in metals. Here is a metal which remains very fluid even for extended periods of time at elevated temperatures such as 2000°C and is capable of alloying with a large number of metals. The platinum is then recovered, so that its cost per sample is nominal. Although the solubility of carbon in platinum is relatively low, it is enough to provide the necessary carbon for rapid reduction to take place. The low vapor pressure of the metal is another good feature. Striking improvements from the use of a platinum bath were reported for many metals. In the case of titanium, completeness of reaction was achieved in minutes (60) at a temperature of about 1900°C, and a large number of determinations—Covington and Bennett (60) report 70—were possible in one crucible. Hansen et al. (133) found that they got more reproducible results when the sample was wrapped in platinum foil. This "platinum flux" technique prevented heterogeneity in the bath by ensuring a fresh supply of platinum with each sample. McKinley (201c) made use of an 80% Pt–20% Sn bath for chromium. Others have also made use of a platinum–tin bath for various metals and have reported that, surprisingly, much of the tin was retained in the bath, even after analyses at 1900°C. The eutectic in the Pt–Sn system is at 72% Pt–29% Sn, 1080°C.

Whatever the bath, a temperature of more than 2000°C is rarely used, because it would lead to excessive volatilization of metal and, to a lesser extent, carbon. Larger areas for gettering would result, and in some cases the mouth of the crucible and/or the tube above the crucible could become blocked by heavy deposits. After being released from the bath, the gases must be removed from the furnace section as rapidly as possible to prevent gettering. The use of tin to combat gettering has already been discussed; in the case of refractory metals, notably titanium, it has repeatedly been found to be essential. In the design of the apparatus, it is of paramount importance to make the connection between the furnace section and the outgassing diffusion pump (P No. 1 in Fig. 103.6) as short as possible and large in diameter (at least >1 in.). The faster the speed of the diffusion pump, the better it is.

The blank in an apparatus in good condition comes almost entirely from the furnace and usually consists mostly of carbon monoxide. One of the advantages of a platinum bath is that the blank rate can be kept small over a considerable number of determinations. At least one vendor offers low-oxygen platinum wire and foil for vacuum-fusion work, which should cut down outgassing time and result in a lower blank correction for platinum flux analyses.

In the case of higher-carbon steels, there have been occasional reports of the evolution of small amounts of methane. When this is suspected, it is advisable to analyze the gases on a mass spectrometer or a gas chromatograph. Later, it will be mentioned that in the case of copper the authors found some of the oxygen to be evolved in the form of carbon dioxide. The presence of a little CO_2 was found to occur in other instances also, for example, sometimes in the application of the platinum flux technique with a platinum–tin bath to the analysis of 3% Si steel. Some work with niobium also indicated that significant amounts of CO_2 can be evolved under vacuum-fusion conditions (47b). When the subject was brought up for discussion at the ASTM Symposium on Recent Advances in the Determination of Gases in Metals, June 28, 1966, it was found that a few others had also found some CO_2 in the extracted gases. One instance was the analysis of nickel oxide. It appears advisable, therefore, to have the capability for measuring the CO_2, as well as the CO, evolved.

As might be expected, the analyst must always consider the possible heterogeneity of metal samples. Thus in a steel ingot or casting the specimens near the exterior will probably be found to contain less oxygen than those toward the center, so that the location of the sample will have to be taken into account. Where a cross section is small enough, a pie-shaped wedge may be obtained for a representative sample. Also, in parts of small cross section, metal solidification takes place more rapidly and a smaller number of impurities are involved, so that segregation is less pronounced. When the oxygen level requires a small sample size, segregation may have to be checked carefully by replicate analyses.

9. Specific Metals

The analysis of iron and steel has already received attention. Here it will be mentioned only that in the case of highly alloyed steels, which are usually best analyzed in an iron or iron–tin bath, it is advisable to dilute the bath with iron between determinations. A relatively high sulfur content may require special treatment (124a). Hamner and Fowler (131) found that, when the sulfur was about 0.05%, its effect could be significant, and they devised a way of removing the sulfur in the evolved gases by hydrogenation and freezing. Abe (1), however, reported that he was able to analyze steels with up to 0.7% sulfur, using a commercial Exholograph EA 1. Nickel, apparently, can be treated like iron or low-alloy steel.

Titanium enjoys the distinction of being one of the most intensely investigated metals in vacuum-fusion work. Strong interest in this refractory metal led to the formation of the Task Force for Oxygen for the Metallurgical Advisory Committee on Titanium, a group which did much not only to facilitate the determination of oxygen in titanium but also to promote discussions and exchanges of experiences at a time when a real need for such communication existed. T. D. McKinley, its onetime chairman, has commented that when the task force was formed around 1950 the number of vacuum-fusion people who could be mustered was hardly enough for a good poker game (201c). The unsolved problems at that time were legion, and it was just about then that the number of laboratories undertaking vacuum-fusion work began to rise sharply.

The Walter "dry-bath" technique (279) was the first one generally accepted for oxygen in titanium. Here, a sample of 0.2–0.5 g was dropped with 1–2 g of tin onto some 20-mesh graphite chips into a crucible at ~1200°C; then 4 was heated to 1900°C and outgassed for about 30 min. Later, Venkatesvarlu and Mallett (275) found that the determination could be carried out at 1950°C without the graphite chips, indicating that the tin was more important than a large graphite area.

McDonald et al. (200b) showed that an iron bath, too, could give satisfactory results if sufficient iron ($\sim$10 g) was used to maintain a fluid state and the sample of $\sim$0.25 g was introduced along with 1–3 g of a known low-oxygen iron after the introduction of 1–2 g of tin. Outgassing took place at 1800°C for $\sim$30 min.

Today, titanium is commonly analyzed in a platinum bath (282) at 1850–1950°C, both with and without a prior treatment with tin (though the addition of $\sim$0.3 g of tin between analyses is recommended). Also, there is a general preference for the platinum flux (133) procedure of having the sample wrapped in platinum foil. The ratio of platinum to sample is kept at 7:1. Some analysts like to add a small amount of iron to increase the solubility for carbon, but others do not consider this necessary. With pure titanium or a titanium alloy whose characteristics are known, the outgassing time can be limited to 5 min or less, and up to 70 specimens of $\sim$0.2 g each can be analyzed in a crucible (60). Titanium is no longer considered a "problem" metal, and the task force has been disbanded.

Zirconium, as might be suspected from its position in the periodic table, can be treated like titanium (14a,255a,264a,276). This fact was also indicated by the results of cooperative programs. Temperatures of the platinum bath vary from about 1950° to 2050°C, and the concentration of zirconium is kept below 15%. Apparently hafnium also can be treated like titanium (14a). Work done by the authors showed that the nitrogen values obtained at the higher temperatures with the platinum bath on Zircaloy, a zirconium alloy containing tin and lesser amounts of other metals, are usually in fair agreement with Kjeldahl values. However, Taylor (264a) was unable to get satisfactory results for pure zirconium.

Fassel et al. (88) determined the oxygen in vanadium by vacuum fusion, by inert gas fusion, and by dc carbon-arc, extraction-emission spectrometry and found the results from the three techniques to be in agreement. Furthermore, they were able to recover added amounts of vanadium oxide. The vacuum-fusion results were obtained in a platinum bath with outgassing taking place for 15–30 min at 1900°C. The samples of 0.1–0.4 g were wound with 0.5 g of 12-gage platinum wire, and the vanadium concentration in the bath was kept below 15%. Similar conditions are recommended by Parker (214).

For chromium an iron (214) or iron–tin (152a) bath appears to be most often used, with about 0.5 g of tin added between samples in the iron bath. A temperature of about 1650°C is generally preferred for the extraction, with the chromium concentration kept below 20%. Horton and Brady (152a), conducting analyses on synthetic standards in an iron–tin bath, got the best recoveries at 1777°C, and concluded that even then recovery was not complete but could be determined from a formula that took into account the number of determinations in the crucible. McKinley (201c) obtained consistent results by outgassing at 1650°C with the following bath conditions: Pt/Sn ratio of 4:1; Pt/Fe/Sn in the ratios 4:1:1 and 4:4:1 with additions of 0.5 g of tin between determinations; and iron with 0.5 g of tin added between samples. He kept the chromium concentration below 18%. ASTM Division I had a task force working on this metal (8a,13,14a).

Silicon poses the problem that it attacks the crucible severely; therefore sufficient dilution is necessary. Beach and Guldner (25) found an iron bath, with a ratio of about 15 parts of iron to 1 part of silicon, to give the highest oxygen values consistently. A nickel bath in the same ratio, with outgassing taking place at 1650°C, gave good results at higher oxygen levels, but not at low ones, apparently because the nickel vapor absorbs a small amount of CO. A platinum–iron bath (with 13 g of platinum, 3 g of iron, and 0.5 g of sample) was found to yield only half as much oxygen as the other two baths.

Uranium appears to be analyzed satisfactorily in a platinum bath (214) at 1850–1950°C. A Pt/U ratio of 10:1 seems adequate. The use of tin does not appear to be necessary, but as usual the addition of a little (0.1–0.3 g) tin between samples is considered good practice.

Molybdenum, tungsten, and niobium behave similarly under vacuum-fusion conditions (204) and would be given the same treatment. These metals were studied by an ASTM task force. Beach and Guldner (25) tried various iron, nickel, and platinum baths for oxygen in molybdenum and found that the best recovery and consistency were obtained with an iron–tin bath at 1650°C. The samples were wrapped in 0.3 g of tin sheet, and the molybdenum concentration was kept between 25 and 40% in the bath. Though the iron–tin bath, differing only in minor details from the aforegoing, appears to be the most popular type for these metals (2,190,214), some workers have a preference for a platinum bath containing 10–20% tin at temperatures from 1650°C (190) to as high as 1900°C (13). Harris (134) found that all of the tin was not removed and what was left was present in greater concentration at the bottom after analyses at 1900°C. He also reported that the solubility of molybdenum increases from 10–12% in pure platinum to 35% in the platinum–tin bath. Sample concentration is kept below 30%.

Another method, reportedly not always successful (8a,14a), is the hot-extraction technique of Fagel et al. (82), who reported apparently satisfactory results obtained by heating molybdenum and tungsten samples to 2000°C or higher, without fusion in any bath, and extracting not only all of the oxygen but also a fair amount of the nitrogen by diffusion of the gases out of the solid sample. The oxygen apparently diffused to the surface, where it formed carbon monoxide upon reacting with vaporized carbon. Friedrich and Lassner (98a), determining oxygen in molybdenum, reported that they got best results when they fused in a Ni–20 Fe bath at 1600–1700°C or when they heated without any bath at 2200°C. Attention is again directed to the report that in the case of niobium significant amounts of CO_2 were given off under vacuum-fusion conditions (47b).

Beach and Guldner (25) found an iron–tin bath satisfactory for tantalum, but a platinum bath at 1850–1950°C is considered preferable by others (8a,14a,190,214). Mallett (190) found that the concentration of tantalum in the platinum bath can be as high as 50%; however, some analysts prefer to set a limit of 20%. As in the case of molybdenum and tungsten just discussed, Fagel et al. (82) found that they could apply the hot-extraction technique, without melting, at 2000°C or higher to determine the oxygen (as well as the hydrogen) content of tantalum, and this technique was evaluated by a task force of ASTM Division I. The platinum bath technique described for tantalum is also considered to give satisfactory results for niobium (13,214). Here the extraction is usually done at about 1900°C, and the niobium concentration is kept below 20% in the bath.

Interest in the determination of oxygen in beryllium by vacuum fusion has been rising lately, because some applications call for a very low oxygen content in the metal (13,34). The analysis is particularly troublesome, because the refractory beryllium oxide calls for elevated reaction temperatures (34,273), at which the metal has a high vapor pressure. In addition, beryllium is a good "getter" for the evolved gases. These factors point to the use of a platinum (119) or platinum–tin (214) bath, but the fact that platinum is ten times as dense as beryllium would appear to present difficulties in getting the necessary rapid solution and reaction of the metal. In fact, procedures for the determination of low amounts of oxygen in beryllium are still being studied, although most workers consider a platinum–tin bath, with some platinum foil wrapped around the sample, to be the most satisfactory (14a,119,203,214). Gregory and Mapper (119), who introduced platinum as a bath metal in their study of this analysis, appeared to obtain satisfactory results with a platinum bath at 1900°C with a large dilution factor of about 50 to 1. Parker (214) kept getting erratic results with this procedure, until he used tin as well. In his final procedure for low-oxygen beryllium,

Parker prepared a bath containing 10 g of platinum and 70 mg of tin. He introduced a 50-mg sample and 70 mg of tin when the bath temperature was 600°C, held the temperature at 1400°C for 1 min, and then outgassed for 2 min at 1950°C. Only a small number of samples were analyzed in one crucible before the recovered oxygen began dropping off. Spectroscopically pure tin was used because it gave a more uniform blank, a very important consideration with such a small sample size. The outgassing for 1 min at 1400°C was for the purpose of lowering the tin blank, probably by decomposition of surface oxides.

Taking a cue from an iron flux technique used by Gokcen (111) on powders, Still (258) put the beryllium sample in a steel capsule and carried out the analysis in an iron bath at 1600–1650°C; he got good agreement with the platinum–tin bath technique conducted in Parker's (214) laboratory on one beryllium powder sample. Still found no drop in the oxygen content, even after 22 determinations. The technique also gave consistent results in intralaboratory studies of other samples. However, Still encountered discrepancies with samples that were shown by X-ray analysis to contain crystalline beryllium oxide, and stressed the importance of a complementary X-ray diffraction analysis.

An ASTM task force for the determination of oxygen in beryllium found substantial agreement between different laboratories in the case of a specimen containing 0.5% oxygen (14a). The vacuum-fusion procedure, as well as the inert gas-fusion procedure, used a bath of platinum plus tin with bath/metal ratios varying from 4:1 to 100:1, a small sample size (usually 0.1 g or less), and an outgassing temperature of about 1950°C. The results agreed with those obtained by neutron activation. However, there was considerable disagreement between techniques and laboratories in the case of beryllium containing 500 ppm of oxygen, and the task force made plans to study the problem further in the near future. A special handling technique may have to be derived for the low-oxygen beryllium, which tarnishes in air.

Copper calls for as low a temperature of extraction as appears feasible to avoid heavy deposition of volatilized copper. Temperatures of 1100–1500°C are used, 1200°C being common. Sloman (245) recommends an iron bath at 1550°C when the copper has been deoxidized with phosphorus; however, Horrigan (148) found a platinum–tin bath (80:20) at 1600°C, for this analysis, as well as for that of copper containing nickel or boron, to be much more satisfactory.

Harris and Hickam (135) devised a simple, rapid method for the refinery control of the oxygen in copper, employing fusion at 1150°C in a graphite boat. The furnace is part of the collecting volume, and the pressure, corrected for temperature effects, is a measure of the oxygen content. The gas is found to be almost entirely carbon monoxide. Also, sulfur interference, as a result of the reaction of this element with the oxygen to form sulfur dioxide, is eliminated, because, in the closed system containing the hot graphite, the SO_2 is converted to sulfur and CO.

A real challenge is presented in the determination of the oxygen in OFHC copper, a special, very pure, very low-oxygen copper. In a recent ASTM cooperative program 11 laboratories reported average vacuum-fusion values ranging from 1.2 to 10.1 ppm of oxygen, with an average value of about 3.8 ppm, on a selected piece of OFHC material (14b). Similar values were obtained by inert gas fusion, while 3 laboratories using neutron activation reported <4.7, 3.0, and 15.7 ppm.

A big problem (there is evidence that it may be the main problem) is the preparation of the sample and its subsequent history up to the time of analysis (13,14b). Thus Fleischer (95) found that his results for low-oxygen copper were usually erratic unless he not only stored the samples in a vacuum overnight after filing, but also gave them a preliminary outgassing at 450–500°C for a half hour. He thus cut down the spread on one material from 0.5–2.0 ppm to 0.4–0.8 ppm oxygen. The importance of sample preparation and history is also indicated by calculations of Horrigan and Goward (13), showing that the contribution of one monolayer of oxygen is equivalent to ~1 ppm oxygen.

Recently, one of us (C. C. C.) did some work with C. B. Worcester to compare filing and etching techniques on low-oxygen copper and found the latter to yield similar results with greater convenience and with better precision. In the filing procedure, the samples were filed, heated at 200°C for 1 hr in a vacuum, and then stored in the vacuum system overnight before analysis at ~1350°C. The etching technique, an ASTM procedure (14b) prescribed for comparative tests that have not as yet been reported, called for filing, etching in 1:4 nitric acid for 1 min, rinsing in water, rinsing in methyl alcohol, and drying under an infrared lamp for 15–60 sec. It was found satisfactory to do the drying in air for 60 sec. By either method a precision of ±0.1 ppm on low-oxygen coppers could be achieved if care was taken to obtain a low blank (<20% of the result). An extraction temperature of ~1320°C was found to be quite satisfactory. The results compared well with those of some other laboratories investigating this problem.

The evolution of significant amounts of CO_2 was observed from most copper samples. It appears logical to assume that all of the oxygen in the CO_2 comes from the sample, and the results should be calculated accordingly. However, the work with low-oxygen copper indicated that

the etching procedure described above, followed by extraction at
~1320°C, is less likely to yield measurable amounts of CO_2 than is
the filing–outgassing technique described.

Carson et al. (47) used specially prepared copper melts of very high
purity to evaluate various surface preparations. Analyses were carried
out by vacuum fusion in an apparatus modified to obtain greater sensitivity and precision. The least oxygen pickup (<0.2 ppm) was obtained
with the etching procedure described above, using a 1:1 solution of nitric
acid. Etching yielded much more consistent results than the more cumbersome technique of surface reduction with hydrogen at temperatures
of 200–400°C. However, the most consistent results were obtained when
etching with a 1:1 solution of nitric acid was followed by hydrogen
reduction at 400°C and fusion at 1300°C.

Copper-containing boron has been analyzed for oxygen, using a platinum–tin bath (4:1) at ~1650°C (148).

Savitskii et al. (231) prescribe a nickel bath for the analysis of compact or powdered rhenium, with the temperature raised from 1450 to
2200°C to obtain the nitrogen as well as the oxygen and hydrogen content. The oxygen content undoubtedly could be obtained at a much
lower temperature. They also made use of another technique involving
the insertion of the sample into a degassed graphite capsule which is
then heated to 1900°C.

A platinum bath with a platinum/sample ratio of 2:1 and an extraction temperature of 1850°C or higher was reported as apparently suitable
for the determination of oxygen in yttrium (149). Fassel et al. (87)
concluded from a study that a platinum–tin bath in the ratio of
80 Pt:20 Sn and an extraction temperature of 1700°C gave slightly
higher results, on the average, than did a platinum bath at 1900°C.
For both, the bath/sample ratio was $>10:1$, and the sample was introduced with platinum; a little tin was introduced as well in the case
of the platinum–tin bath. The same workers found these conditions to
be satisfactory for thorium also, although for this metal a platinum
bath with a platinum/sample ratio of $>10:1$ and an extraction temperature of 1950°C were found to be suitable. Here, too, the sample was
introduced with some platinum. For both yttrium and thorium, Fassel
et al. recommend conditioning the bath with a little thorium or yttrium
in preparation for the first determination.

It appears quite safe to say that the platinum–tin bath conditions
of Fassel et al. (87) described above should prove satisfactory for the
metals in the lanthanide and actinide series. It also seems likely that
their platinum bath conditions for thorium should prove satisfactory

for these metals (214), the addition of a little tin between determinations being recommended. For lanthanum, satisfactory results were also reported with a nickel bath at 1900°C (219), while Taylor (264) found the results from an iron bath to agree with those from a platinum bath in the case of plutonium and uranium.

For metals that volatilize rapidly at temperatures well below those required for rapid reaction to take place, the vacuum-fusion technique, if it is to be used, must be modified to remove the volatile metal and to provide suitable conditions for the reduction of the oxide in the residue. Thus Berry et al. (32a) applied the vacuum-fusion technique to a magnesium alloy, Magnox A12, by first removing magnesium by sublimation at 800°C *in vacuo* and then heating the residue in an iron bath at 1700°C. Good accuracy and reproducibility were reported over the range 10–200 ppm of oxygen. Horrigan et al. (149) had a similar solution for determining the oxygen in yttrium fluoride. Here, the sample, wrapped in platinum foil, was heated at a temperature of <650°C to distill off the fluoride; then the temperature was raised to 1850°C for the reduction of the oxides in the residue. The alkali metals, of course, pose a very big problem in this respect. However, Turovtseva et al. (274) did some work with lithium which indicates that the vacuum-fusion technique appears possible for alkali metals. They predistilled the lithium from the graphite crucible at a lower temperature and then carried out a vacuum-fusion analysis of the residue. Anthanis (6a) determined the oxygen and other gases in cesium by refluxing (boiling at 250°C and transferring gases through a tube kept at 450°C) in a vacuum and analyzing the extracted gases with a mass spectrometer. He reported getting the oxygen in the form of CO and H_2O. Gahn and Rosenblum (99a) compared a vacuum-distillation technique with the amalgamation technique and with neutron activation and considered the agreement good.

Dallman and Fassel (65) carried out tests on a number of metals to determine whether suitable nitrogen values could be obtained by vacuum fusion in a platinum bath at 1850–2000°C. The samples were introduced with about 3 times their weight of platinum. Platinum/sample ratios in the bath were the same as those used for helium gas fusion, which are described in some detail in Section VI. A bath of 80 Pt:20 Sn was used for some rare earth metals. Accurate and reliable nitrogen values were usually obtained for chromium, a variety of steels, niobium, vanadium, tantalum, and zirconium. The results for thorium and hafnium tended to be low, while those for titanium and its alloys were invariably low. The results of Dallman and Fassell for the rare earth metals were usually acceptable but not consistently reliable. They concluded that helium fusion, which gave consistently reliable results for all of these metals, was a better technique for determining the nitrogen.

G. ISOTOPIC METHODS

1. Principle

The universally known fact that practically all chemical elements occur as mixtures of isotopes, each with a different mass, has been exploited for the direct determination of oxygen, nitrogen, and hydrogen in organic compounds (122,123). Both the Unterzaucher and the Ter Meulen methods have been adapted (5) to the exchange of ^{18}O-enriched oxygen with alcohols, for example, as a means of determining the oxygen content. For most elements the ratios of the constituent isotopes are constant for the "natural" element when occurring alone or in combination. For example, naturally occurring oxygen contains the nominal masses 16, 17, and 18 in the proportions 0.99757, 0.00039, and 0.00204, respectively (98). Since the building of large-scale isotope-separation plants for atomic energy purposes, it has been possible to obtain compounds in which the isotope ratios are markedly different from the "natural" values. For this reason the method of isotope dilution (97) has become practical and is much used in analyses with radioactive species. It is also possible, however, to employ the method with stable isotopes such as oxygen.

This section is concerned chiefly with isotope dilution as a method for gas analysis, although the last few paragraphs will discuss other aspects of using isotopes for this purpose. In either case it is assumed that mixing two sources of an element which contain different ratios of the isotopes results in the isotopes being distributed at equilibrium in identical proportions in each phase of the final product. Consider source A, containing 1000×10^{16} atoms of natural oxygen. If the ^{17}O is ignored for convenience of exposition, there will be 998×10^{16} atoms of ^{16}O and 2×10^{16} atoms of ^{18}O. Let this material react with source B—154 liter-microns* of oxygen, measured at 25°C and enriched to contain 90 at. % ^{18}O and 10 at. % ^{16}O (again ignoring the ^{17}O). There will be altogether 1098×10^{16} atoms of ^{16}O and 902×10^{16} atoms of ^{18}O. At equilibrium, then, it is expected that every phase of the final reaction mass will contain the oxygen isotopes in the same ratio of

* For the definition of liter-microns see W. S. Horton, "Vacuum Methods," Part I, Vol. 2, p. 1291, of this Treatise.

902 ÷ 1098 = 0.8215, whether the oxygen is finally located in just one phase or is distributed among two or more phases. Thus the uniqueness of this ratio depends not upon whether the reaction went to completion but rather upon whether equilibrium was attained.

Source A, as used in the previous paragraph, corresponds to a 0.1-g unknown metal sample containing 0.266% of oxygen. Source B might be an actual gas. Because the final isotopic ratio is markedly different from both the natural ratio and the enriched ratio, its determination should yield an estimate of the oxygen content of the source, A, or unknown metal sample. For this the data needed are the weight of the sample, S; the quantity of enriched gas used, E; the isotopic ratio of the enriched gas, R_e; and the isotopic ratio of the final gas, R_f. One can show that the percentage of oxygen, P, in the original unknown sample is given by

$$P = \frac{(R_e - R_f)(E/S)}{(R_e + 1)(5796R_f - 12)}$$

if S is in grams and E in liter-microns measured at 25°C. The numerical constants are directly proportional to the absolute temperature of the added enriched gas.

2. Experimental Conditions and Problems

As mentioned above, the use of an isotope dilution method gives the analyst an opportunity to exchange the need for quantitative release of the oxygen, nitrogen, and hydrogen for the need to attain equilibrium. As might be expected, the refractory metals or those with refractory oxides are more likely to be candidates for the isotopic approach. One obvious advantage is that the temperatures required to reach equilibrium within a reasonable time may be rather less than those required to liberate all of the gas for analysis. An obvious procedure, for example, is to prepare an unknown as if for vacuum-fusion analysis and then to add a measured amount of an enriched gas. Now the specimen may be analyzed by conventional vacuum-fusion techniques except for the need to save the product gas for mass spectrometric analysis. With the final isotopic ratio, R_f, available from such analysis, the equation may be used. It is valuable to consider what features of the usual vacuum-fusion analysis are particularly important in this variation. Some of these features are present also in other variations, and hence most problems can be considered in this context.

The blank is no doubt one of the foremost factors which must be under control. Should the blank consist almost solely of hydrogen and nitrogen, however, it would not affect the results for oxygen significantly. In contrast with normal vacuum-fusion methods, the blank is generally kept so low in isotopic dilution as to negate its effect, no matter what its composition may be. Included in the blank, of course, is the vacuum tightness of the system, the degassing of the graphite crucible and the system walls, and the gas content of any fluxing metal added simultaneously with the unknown specimen. The method of computing the influence of a known blank upon the results is obvious. A word of caution is warranted, however, with respect to estimating the magnitude of the blank. With conventional vacuum-fusion methods the recoveries are always preferably near 100%, and as a result, with a little care, blank measurements between individual samples are not always necessary. Since 100% recovery is not required with the isotope dilution method, this is a risky procedure because, following a sample with rather less than quantitative recovery, the blank rate is generally high. For this reason it is advisable to provide a definite degassing period between samples as well as a blank-rate measurement before the next sample is taken. A correction for the blank would require an oxygen and an isotopic analysis of the blank gases in order to be valid and hence is not advisable.

Gettering, a common difficulty with slightly volatile metals, should not be a problem if the desired equilibrium has been attained. As mentioned earlier, the problem of complete gas liberation has been exchanged for the problem of attaining this equilibrium. Thus, even if gettering occurs, the ratio of the isotopes should be the same in the gas phase as in the gettering film. Whether this is true, of course, can be verified only by experiment.

Another problem is that of exchange reactions. These are not problems in vacuum-fusion analysis but are potential sources of difficulty in the isotopic methods. For example, the product gases are generally maintained in the reaction region for a fair length of time in order to ensure statistical distribution of the isotopes. Most often the furnace material is quartz or glass, and the graphite crucible is supported within a quartz closed-end tube. Under the influence of high temperature the oxygen of the gas phase and that of the glass or quartz may exchange, thus diluting the ^{18}O and leading to a high answer for the oxygen content of the specimen. This possibility has been investigated (175) at 800°C with ^{18}O-enriched carbon monoxide. As a result of these experiments it was claimed that at most an error of 0.45% of the oxygen content would occur. Although the fact that the error enters at two points of the calculated answer was neglected, consideration of this fact raises the error to only about 1% of the oxygen answer. It seems unlikely that isotopic exchange is a serious source of difficulty, unless the quartz walls are hotter than 800°C. Oxygen exchange reactions appear to have exceedingly high activation energies, and the reaction rate increases rapidly with rising temperature.

Another problem peculiar to the isotopic method is the determination of isotopic ratio. Since the product gas contains CO, nitrogen, and hydrogen, the mass spectrometric determination is interfered with by the presence of $^{14}N^{14}N$. However, it is standard practice to determine nitrogen from the ^{14}N mass spectrometer peak, and the ^{28}N peak is always in a constant ratio to this. As a result mass peak 28 can be corrected for nitrogen, and the only limitation that then arises is how precisely this correction may be made.

Another feature of the problem is that CO^{2+} contributes to mass peak 14, leading to a need, not for a simple correction factor, but for the solving of simultaneous equations. The amount of CO^{2+} will depend on the accelerating voltage used on the mass spectrometer. Lowering this voltage will reduce the contribution; below the appearance potential of CO^{2+}, the peak should be absent. However, lowering the accelerating voltage also reduces the sensitivity of the instrument to all peaks. Examination of mass peaks 12 and 16 may assist in the differentiation between CO and N_2, but again the sensitivity to these, as coming from CO is low.

An additional source of difficulty may arise if the product gases contain hydrocarbons, particularly methane and ethane. These extraneous gases will also give spuriously increased intensities at the important mass peaks. It appears to be true, however, that in vacuum-fusion procedures the presence of significant quantities of hydrocarbons has not been demonstrated. If one is using the inert gas-fusion method for evolving the product gases, the purity of the inert gas is obviously of great importance. Many of the remarks just made need serious consideration in this case.

One might consider avoiding this problem by oxidizing the carbon monoxide to the dioxide with the usual copper oxide or with iodine pentoxide. In that case mass peaks 44 and 46 would be used to distinguish ^{16}O and ^{18}O. However, before using such a variation a careful investigation of the exchange reaction between the oxidizer and CO would obviously be necessary. The exchange of molecular ^{18}O with copper oxide appears to be significant even at 400°C (4), but no data seem to be available on the oxygen exchange reaction of CO. Another possibility would be to determine the gas density if only CO were involved. This appears to be rather an experimental tour de force, however, for the purpose being served.

For some laboratories the requirement for a mass spectrometer is beyond normal budgetary possibilities. Although mass spectrometric measurements are a commercially available service, the inevitable time delays severely limit the usefulness of this method for routine analysis. It would seem reasonable, therefore, to apply the method under such conditions only as a means of checking the accuracy of other analytical procedures.

The need for isotopic gases is not a drawback. As mentioned earlier, the atomic energy industry has been responsible for the increased availability of these gases. The U.S. Atomic Energy Commission, through the Oak Ridge National Laboratory, publishes a booklet on stable isotopes available there and will provide the names of commercial sources of isotopic oxygen, nitrogen, and hydrogen. If only compounds are available, the gases may be generated as described earlier (151).

The attainment of equilibrium has been a requirement frequently mentioned above. This attainment can be adequately checked only by experimental demonstration that the method is accurate, because the equilibrium must be attained under the actual experimental conditions with attendant problems of "kish," gettering, outgassing, and so on.

3. Variations

In the hypothetical example described earlier the enriched gas was added directly to the unknown specimen. This procedure has been used (136), although not for isotope dilution but rather as a tracer technique for checking recovery from vacuum-fusion analysis. However, the earliest work on the isotope dilution method for oxygen used a "master alloy " (173). In this method the enriched oxygen is added by reaction at elevated temperature to a specimen of metal, usually (but not necessarily) of the type being analyzed. Adequate heating is performed to ensure diffusion throughout the specimen after the gas has completely reacted. The metal–metal oxide combination is then ground to a fine powder for use. To allow for both sources of natural oxygen (i.e., enriched gas and original metal), a sample of the master alloy is reduced with graphite and the $^{18}O/^{16}O$ ratio is determined. Furthermore, it is necessary to know the weight percentage of oxygen in this alloy. If all the enriched gas is absorbed by the metal being used for the purpose and

if the original oxygen content is comparatively low, this figure is cal-
culable if manometric measurements are made during the addition. In
use, a known weight of the unknown is placed with a known weight of
the ^{18}O-labeled metal. The combination is treated as a vacuum-fusion
sample, although lower temperatures may be used. Recirculation of the
evolved gases over the graphite crucible or boat may be desirable to
ensure isotopic equilibrium. The gases are sampled and analyzed isotop-
ically. It is interesting to note that CO_2 as well as CO is mentioned
(173), although even at temperatures as low as 1000°C, the temperature
used, thermodynamic chemical equilibrium with graphite leads to essentially 100% CO (278). The formula given for the percentage of oxygen in the unknown is

$$X\% = \frac{b(m - n)}{a \cdot n} \cdot 100 \tag{3}$$

where b is the weight of the enriched gas used, m and n are the atom
percentages of ^{18}O in excess of normal of the enriched gas and of the
equilibrated gas, respectively, and a is the unknown specimen weight.
It should be recognized that this is a good approximation if m is not
very high. For example, if $m = 15$, there is an error of only about 3%
of the answer. Should m be rather higher, and 90% ^{18}O is available,
the result of equation (3) should be multiplied by the correction factor:

$$F = \frac{1600}{2(m + \sigma) + 1600} \tag{4}$$

where now σ is the normal atom percentage of ^{18}O, and thus $m + \sigma$
is the actual atom percentage in the enriched gas. This variation of
the method has the advantage that enriched gas is added to a metal
only once to provide a sizable quantity of additive. Such an approach
no doubt speeds application of the method.

An obvious variation has been employed to determine the completeness
of vacuum-fusion analysis recovery (136). In this case a measured quantity of molecular oxygen of known isotopic concentration (both ^{17}O and
^{18}O were enriched) was added to niobium metal at between 500 and
600°C. Preheating at 1000°C in vacuum is necessary to remove hydrogen
and other gases; hydrogen seriously interferes with the addition of oxygen to many metals. The enriched isotopes were recovered from the
specimens by the platinum bath variation of the vacuum-fusion technique. Because of complications similar to those discussed above, the
mass peaks used for isotopic analysis of the added gas were due to
singly ionized ^{16}O^{16}O on 32, ^{16}O^{17}O on 33, ^{17}O^{17}O and ^{16}O^{18}O on 34,

$^{17}O^{18}O$ on 35, and $^{18}O^{18}O$ on 36. If it is assumed that two atoms form a diatomic molecule with a probability that is independent of the mass species, six equations are available for four unknowns. For the recovered gas, mass 28 ($^{14}N_2$ and $^{12}C^{16}O$), mass 29 ($^{12}C^{17}O$, $^{13}C^{16}O$, and $^{14}N^{15}N$), and mass 30 ($^{12}C^{18}O$ and $^{13}C^{17}O$) are used.

The point of this procedure, of course, is that recoveries may be checked without concern about the original oxygen content of the material. In the work on niobium described above, recoveries were quite good for both ^{17}O and ^{18}O. Detailed examination of the data, given in Table 103.III, is instructive. These data were interpreted in the paper by quoting individual confidence limits for slope and intercept of a plot of recovery versus added gas. As is well known (194), least-squares calculation of slope and intercept leads to nonindependence of confidence intervals for the two parameters. To make a simultaneous confidence statement for both slope and intercept of the experimental straight line requires a joint confidence region rather than two confidence intervals. The result is that, by using the experimental data, one obtains an ellipse within which one has confidence that the "true" parameters lie.

Figures 103.10 and 103.11 show the ellipses obtained for ^{17}O and ^{18}O, respectively. The curves include the 90% confidence region computed according to Mandel and Linnig. In each case the point (0, 1) for intercept and slope lies within the region, although for ^{18}O it is quite close to the curve. As a result of these two computations it appears reasonable to assume that within experimental error complete recovery has been achieved. One might still object to even the illustrated computa-

TABLE 103.III
Recovery of Oxygen Isotopes from Niobium by Vacuum Fusion[a]

Niobium, g	^{17}O, μg		^{18}O, μg	
	Added	Recovered	Added	Recovered
0.40	0.16	0.18	14.0	13.9
0.27	0.20	0.18	16.9	15.9
0.37	0.20	0.23	17.3	18.5
0.63	0.24	0.24	20.2	20.3
0.41	0.31	0.31	27.0	26.1
0.55	0.38	0.35	33.1	30.0
0.56	0.48	0.47	41.3	40.0

[a] Reproduced from reference 136.

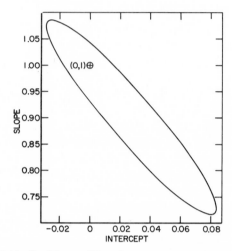

Fig. 103.10. Confidence ellipse for recovery parameters of ^{17}O.

tion by claiming that in the recovered versus added plot the x values (added) are not known without error. The view is taken above that the error in x is rather less than that in y and therefore need not be considered. Should a calculation be desired to avoid this assumption, a more lengthy method (130) is available.

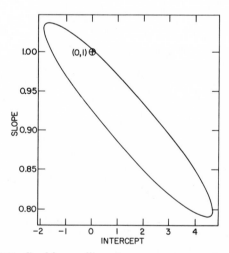

Fig. 103.11. Confidence ellipse for recovery parameters of ^{18}O.

A rather interesting point is raised by comparing the ratios $^{18}O/^{17}O$ for the added and recovered gases. In Table 103.IV the results show that this ratio was more often lower for the recovered gases than for those added. This discrepancy hints that the recovery is not exactly complete although the extent of error must be rather small (say, <2% of the answers). Because "recovery" includes all operations involved, the minor difficulty may lie with the mass spectral determinations. As discussed above, there are problems associated with corrections to the mass peaks for species not containing oxygen. Actually, however, one is impressed not so much with the hint of a small error as with the powerful technique available in isotopic methods to study the procedure.

The use of isotopes is not confined to oxygen, of course, and other impurities have been tested also. For example, hydrogen, as tritium, has been determined in aluminum (107) by hot-extracting the gas and transferring it to an ionization chamber. The output of the chamber is measured with a vibrating-reed electrometer. One notes that as an unknown-determining technique this serves only in the rare cases in which tritium rather than normal hydrogen is somehow caused to be in the specimen. For example, tritium is produced by n, α reaction with ^6Li or by d, 2α reaction with ^9Be (98). The use of the method appears to be of value more when isotope effects are being sought. Then the isotopes can be determined independently. For example, the original reference to the method (266) found differing modes of attack for normal and tritiated liquid water on aluminum powder. The work also reported a comparison between volume measurement and electrometer measurement for which the data, treated similarly to those for Figs. 103.10

TABLE 103.IV
Ratio of ^{18}O to ^{17}O in Added and Recovered Gases

Niobium, g	Added gas	Recovered gas
0.40	87.5	77.2
0.27	84.5	88.3
0.37	86.5	80.4
0.63	84.2	84.6
0.41	87.1	84.2
0.55	87.1	85.7
0.56	86.0	85.1
Average	86.13	83.64

and 103.11, yield intercept and slope of zero and unity within the 90% confidence ellipse. On the other hand, a regular isotope dilution method has been applied to sodium, using tritium (78). Although the technique is experimentally demanding, the concepts are straightforward. The tritium in the gas mixture is determined by Geiger counting. Hydrogen and tritium are stored by absorption in pyrophoric uranium. The recovery data reported look good, although the standard deviation is nearer 0.014 ml of hydrogen at S.T.P. than the 0.006 ml reported. The method was tested with from 0.026 to 0.3 g of sodium containing about 140–1340 ppm of hydrogen.

Another use of tritium, in a semiquantitative way, involves autoradiography (99), wherein the radioactivity of the tritium is allowed to affect an appropriate photographic emulsion. A visual display of the distribution and movement of hydrogen in solid structures can be obtained by this technique. Since it is possible to standardize the emulsion by using homogeneously tritiated specimens, the visual variations may be made quantitative if desired. As a rule, to produce equal optical density, the exposure time for the usual commercial autoradiographic emulsions is inversely proportional to the square root of tritium concentration. Some safety precautions are necessary when using tritium.

Another variation for hydrogen utilizes emission spectroscopy (289) for determining the isotopic dilution of a known deuterium–hydrogen mixture. The spectroscopic principle involved here is that the wavelengths of the hydrogen atom emission lines depend not only on the electron mass but also on the nuclear mass. Whereas, according to simple theory (139), the Balmer line wave numbers are given by

$$\nu = R \left(\frac{1}{4} - \frac{1}{m^2} \right)$$

where m, the principal quantum number, $= 3, 4, 5, \ldots$, and R is the Rydberg constant proportional to the mass of the electron m, consideration of the mutual motion of nucleus and electron about their common center of gravity gives

$$R \propto \mu = \frac{mM}{m + M} \tag{5}$$

where μ is the reduced mass, and M is the mass of nucleus. For hydrogen the effect is greatest since $M \to \infty$, $\mu \to m$. In fact, for hydrogen and deuterium the separations between corresponding lines are between 1 and 2 Å (140). Although, in principle, the same theory applies to oxygen and nitrogen, the amount of isotopic shift is much less. This arises because, as can be easily verified, the shift is essentially proportional to $(M_2 - M_1)/M_1M_2$, where the subscripts refer to the two isotopes. For hydrogen and deuterium this number is 1/2, whereas for ^{16}O and ^{18}O it is 1/144. Therefore, one expects the isotopic shifts for oxygen to be about 1/72 times those of hydrogen. Such small shifts would require a spectrograph of extremely high resolution indeed, in order to be useful.

H. MICROPROBE TECHNIQUES

In some cases there is great interest in determining the local concentration gradients of a gas in a metal, and microprobe techniques for this purpose are improving rapidly. Thus, if a local oxygen level is greater than about 1%, the concentration can be determined with an electron microprobe over very small distances measured in microns. Lower levels of oxygen can be measured, but with poorer resolution, with an ion microprobe. Thus Evans (77a), working with copper, established a working curve over the range 20–2300 ppm oxygen with a resolution of 300–500 μ. The copper surface was bombarded with 14-keV argon ions, Ar^+, in a vacuum, the ejected ions were extracted into a mass spectrometer, and a measurement was made of the $^{16}O^+/^{63}Cu^+$ ratio. The blank was 3–5 ppm, and measurements in the 10–20 ppm (w/w) range were possible.

V. ACTIVATION ANALYSIS

Section IV.G dealt with the use of either stable isotopes or, as in the case of tritium, an added radioactive isotope. During the 1960s there was a steady advance in applying activation analysis to the determination of gases in metals. After the initial paper in 1954 (213a) a steady increase in published work began in 1959 (26a,56b,180a), and by 1970 the total number of papers was about 100 for determining oxygen. Very simply, this technique involves irradiating a specimen with special particles in order to induce nuclear reactions. The idea is to produce a radioactive element not present before, which is then identified and its concentration determined by counting techniques. Under appropriately controlled conditions, the intensity of the radioactive species can be used to calculate the quantity of impurity element being sought. Activation analysis is treated elsewhere (202a), but it is sufficient to real-

ize that the activity (in disintegrations per second) of a product isotope
depends on the number of parent nuclide atoms present in the sample,
the flux of activating particles, the isotopic cross section for the activation reaction, the time of radiation, and the half-life of the nuclide
produced. A reasonable set of guides to activation analysis in general
is provided by Meinke (202a), Lyon (185), and Bowen and Gibbons
(33b). The reference by Pasztor and Wood (215b) provides a good introduction to the special problems associated with applying activation
analysis to the subject of this chapter. A bibliography for the determination of oxygen and nitrogen (as well as other light elements except
hydrogen) has been published (184).

The "special particles" mentioned above may be neutrons, protons,
deuterons, or alpha particles, and these may have more or less thermal
energy. Probably the most commonly used reactions involve irradiation
with neutrons from a nuclear reactor or one of the wide variety of
machines now available. In these a charged particle is accelerated to
an appropriate energy and allowed to strike a target which produces
neutrons. The difference between neutrons from a reactor and those from
an accelerator is that the latter machine permits selection of a fast
neutron of definite energy. High-energy X-rays are also used in "photo-
nuclear" activation analysis. There may be some advantage in using
the $^{16}O(\gamma,n)^{15}O$ reaction because of better sensitivity (124b).

Pasztor and Wood (215b) made a careful and extensive study of activation analysis for determining oxygen in steel. Comparisons were made
with vacuum-fusion analysis and inert gas fusion. Using a Cock-
croft–Walton type of accelerator, they irradiated specimens with about
10^{11} neutrons/sec at an energy of 14 MeV. Samples (which may be
as large as 10 g) were irradiated simultaneously with a Lucite reference
and were then counted to determine the intensity of the ^{16}N gammas
resulting from the n,p reaction with ^{16}O. The ratio of the sample counts
to the reference counts is proportional to the oxygen content of the
sample. Samples with oxygen contents from 0.002 to 0.1% were analyzed
by neutron activation and then cut into several pieces for hot-extraction
analysis. Pasztor and Wood quote reproducibility at the 95% confidence
limit as ±7% for the average of two counts for steels with oxygen
in the 30–300 ppm range. Steels with homogeneous oxygen contents were
said to be significantly better. For high-oxygen-content samples with
low statistical variation the reproducibility was ±2%. Correlation between neutron activation, on the one hand, and either vacuum fusion
or inert gas fusion, on the other, was generally good; correlation with
the latter appeared to be somewhat better, on the whole, than with

vacuum fusion. The difference appears to be related to the obtaining of low values for aluminum-killed steel analyzed by vacuum fusion.

Another isotopic method for oxygen involving nuclear activation has been used for determining oxide film thickness (266). In this scheme the ^{18}O present is activated by protons from a cyclotron through a p,n reaction to produce ^{18}F. The latter is a positron emitter with a half-life of 110 min (266), and its quantity may be estimated by the amount of activity. Although the activity can be computed (or the number of parent nuclide atoms in the sample, given the activity), it turns out to be simpler to use known standards for comparison. This is true in this case because the protons lose energy while traveling through the sample before being absorbed, and the reaction cross section varies with the proton energy. Were it a simple variation the computation might still be useful, but there are resonance peaks and a threshold energy below which no reaction occurs. With anodized specimens of tantalum sheet an essentially linear activity was observed as a function of film thickness between 500 and 2000 Å.

Although Thompson (266) discarded the lowest pair of points (at ~500 Å) when determining the calibration line (apparently by eye), the data do not appear to require this omission. Including all datum points leads to a standard error of fit of 88 Å. As a rough estimate, then, one might say that the method is capable of measuring film thicknesses to within a few hundred angstroms. Factors limiting the sensitivity include the oxygen content of the underlying metal, the available beam current of protons, the beam area, and the irradiation time that one is willing to spend. Actually there are limitations on the amount of time that is useful to spend. Approach to saturation, activation of longer-lived impurities, and the high cost of cyclotron time have been cited as reasons for limiting this time to about one half-life. It is claimed that films between 1 and 250,000 Å may be measured with the bulk oxygen limiting the lower end and the proton range (for 4-MeV protons) the upper end. Actually, the precision data do not warrant expectations much below 200–300 Å at the lower limit. Although one may suspect that the random errors are proportional to the film thickness, the data are insufficient to demonstrate this relationship. The use of 4-MeV protons leads to interferences from the presence of copper, nickel, zinc, and titanium; the last-mentioned interference may be eliminated by using protons below 3.8 MeV.

Another modification (213a) made use of the Harwell reactor to produce ^{18}F in beryllium from the ^{18}O impurity by mixing the specimen with lithium fluoride. Here the bulk impurity was determined by comparison with standards. A similar method (33a) has been claimed to determine as little as 0.01 ppm of oxygen. A similar scheme has been used to analyze gallium arsenide (18). Specimens were wrapped in lithium and placed in a reactor. In both of these cases tritons are produced by the $^6Li(n,T)^4He$ reaction, and these give rise to $^{16}O(T,n)^{18}F$.

Advantages claimed for neutron activation analysis include the possibility of using large samples, the decreased time of analysis, the elimination of some problems encountered in hot-extraction techniques (e.g., incomplete reaction, mechanical ejection or creeping of oxides, gettering), and the nondestructive nature of the analysis. Against these must be balanced mainly the expense of the installation. Although $45,000 has been said (215b) to be the approximate cost of an installation, reference 181c, p. 44, lists costs totaling $57,000. It may be that the claimed savings in operating costs would eventually recover the initial investment.

VI. INERT GAS FUSION AND EXTRACTION

The exclusion of atmospheric gases, the reduction of the partial pressure of the evolved gases, and the removal of the evolved gases can be achieved by sweeping the system with an inert gas as well as by the use of a vacuum. Some small reduction in rates could be expected for any given temperature; however, this has not been found to be a problem and can usually be compensated for by operating at a higher temperature. An increase in the crucible temperature can be tolerated because collisions with the molecules of the sweeping inert gas significantly reduce volatilization and deposition.

Like vacuuum fusion, inert gas fusion takes place in a graphite crucible, where the oxides are reduced to form carbon monoxide, while hydrogen and nitrogen are given off as such. Elimination of the need for a vacuum results in a simpler and more economical apparatus. Also, the operation of the apparatus has been found to be generally more convenient. In situations where a large variety of metals and gas problems are encountered, it may be found advantageous to have both an inert gas-fusion and a vacuum-fusion apparatus.

Not many years ago, the general consensus of opinion was that results from inert gas fusion were comparable to those from vacuum fusion only when the oxygen level was high enough ($\sim$30 ppm or more) (13). Now, however, some workers are reporting for inert gas fusion a precision and sensitivity comparable to that of vacuum fusion (8,63a,146,187,192a). It has also been reported that in some instances the results from inert gas fusion were superior to vacuum-fusion results. One such instance is the determination of relatively large amounts of oxygen (~200 ppm or more) in molten steel "killed" with aluminum (14a).

Though inert gas fusion is normally used for oxygen determinations, satisfactory results have also been reported for nitrogen. Holt and Goodspeed (146) carried out nitrogen determinations effectively for steels, uranium, and zirconium fused in a platinum bath, and Fassel and Dallman (65) and their co-workers at Iowa State University made similar analyses for a number of metals, including chromium, vanadium, zirconium, titanium alloys, rare earths, and thorium, using helium and a platinum–tin bath. Bryan and Bonfiglio (40) determined the nitrogen in a number of steels, using an iron bath at 1900°C with good results.

Holt and Goodspeed also reported good values for hydrogen. In addition, suitable results for hydrogen have been reported from the use of an inert carrier gas in place of a vacuum for the hot-extraction technique (heating the metal to temperatures of 650–1200°C) (218b,234).

Thus a considerable amount of experience with the inert carrier-gas technique has been accumulated with a variety of metals in a relatively short time.

A. HISTORY

Inert gas fusion was first reported in 1940 by Singer (242), who used it to determine the oxygen content of steels. Nitrogen was used as the carrier gas. The evolved CO, after oxidation on hot copper oxide to CO_2, was measured gravimetrically. Smiley (248) significantly improved the method 15 years later to obtain much better sensitivity when he determined oxygen in uranium with a platinum bath, using argon as the carrier gas. Smiley converted the evolved CO to CO_2 with iodine pentoxide (Schütze's reagent), froze out the CO_2 with liquid air, and then measured the CO_2 by reading the pressure when the gas was released into a capillary manometer. This is still a popular way to obtain good sensitivity for the determination of oxygen in a metal by inert gas fusion.

Since Smiley published his paper, many workers have reported results for a variety of metals and of modifications. The modifications are chiefly variations in the method of measurement of the evolved gas or gases. Of particular importance is the introduction of increasingly sensitive gas chromatogpraphic equipment.

B. APPARATUS

The essential simplicity of an inert gas-fusion apparatus is evident from Fig. 103.12. The apparatus is described by Elbling and Goward (76) in connection with the determination of oxygen in zirconium and Zircaloy, but it can also be used for determining oxygen in other metals.

Their furnace section is made of Vycor, and the induction heater is a 2.5-kW, 450-kHz radiofrequency unit, a size commonly used in vacuum fusion. The graphite crucible is insulated with graphite powder, as is usually done, though Smiley (248) supported his crucible on a tungsten stem without graphite insulation and used a 10-kW unit. The connections consist of glass and neoprene tubing. Metal tubing is also used (248). In addition to being passed through sulfuric acid, Ascarite and magnesium perchlorate, the carrier gas, argon, is passed over zirconium chips at 700–800°C for purification.

Sample introduction, which becomes a simple matter, is accomplished with a lock consisting of two hose clamps over a section of neoprene tubing. A small positive pressure of argon keeps out air.

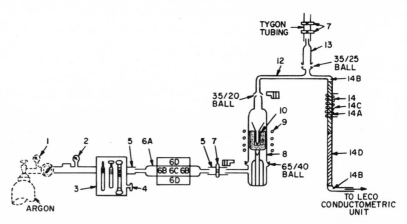

Fig. 103.12. Inert gas-fusion apparatus (76). *1*, two-stage regulator; *2*, third-stage regulator; *3*, purifying train: *A*, sulfuric acid tower; *B*, ascarite and magnesium perchlorate; *C*, flowmeter; *4*, needle valve; *5*, neoprene tubing; *6*, argon purification tube: *A*, nickel tube; *B*, glass wool; *C*, zirconium chips; *D*, split-tube furnace, 700–800°C; *7*, hose clamp; *8*, Vycor furnace tube; *9*, radiofrequency coil; *10*, crucible thimble assembly; *11*, cooling blowers; *12*, sample storage arm; *13*, sample loading port; *14*, oxidizer furnace; *A*, iodine pentoxide; *B*, glass wool; *C*, electrical heating tape; *D*, sodium thiosulfate. (Courtesy of *Analytical Chemistry*.)

The oxidation of CO to CO_2 is carried out here by means of an iodine pentoxide reagent at 130–150°C (248). The reagent does not oxidize hydrogen. In this apparatus, the CO_2 produced is measured conductometrically. The CO_2 is absorbed in an alkaline solution, and the amount absorbed is determined by the change in conductivity of the solution. Calibration of the solution is carried out by analyzing suitable amounts of uranium oxide, silver oxide, zirconium dioxide, niobium pentoxide, or potassium acid phthalate. It is recommended that the calibration be checked at least twice during the day. The conductometric method is in fairly common usage, but its use is diminishing.

Alternatively, the CO is also oxidized over hot copper oxide (148,235,147), as is commonly done in vacuum-fusion analysis. Mention has also been made of Smiley's technique (248) of freezing out the CO_2 in a calibrated capillary manometer and then releasing the gas for measurement after evacuation of the manometer. Holt and Goodspeed (146) report very good sensitivity by this method of measurement for CO_2 as well as for H_2O (produced by oxidation of the evolved hydrogen and measured at 110°C) and nitrogen (absorbed by activated charcoal at −196°C, released by heating the charcoal, and transferred to the manometer with a Toepler pump). A gravimetric method can also be employed (235,147) when the amount of oxygen is sufficiently high by using a tube of Ascarite to absorb the CO_2 issuing from the oxidant. Hydrogen in sufficiently high amounts (280) can also be determined gravimetrically by passing it over hot copper oxide to convert the hydrogen to water and then absorbing the water in a tube of magnesium perchlorate. Moreover, when the hydrogen level is high enough, a titrimetric method can be used by absorbing the resulting water in methanol and titrating with Karl Fisher reagent (234,235).

Satisfactory results for the routine determination of hydrogen in steel have been reported with the use of a katharometer (55a,106,234). This yields a signal, as in the case of a gas chromatograph, when hydrogen is present in the carrier gas.

It is also possible to carry out the gas analysis with a gas chromatograph. One technique (40) consists of absorbing the evolved hydrogen, carbon monoxide, and nitrogen on a molecular sieve at −196°C, then releasing them by heating, and sweeping them through a gas chromatograph after passage through a second molecular sieve at 42°C, where CO and N_2 are separated. When the effort is justified, extraordinary sensitivity can be obtained with a gas chromatograph employing a helium ionization detector, which enabled Hargrove et al. (133a) to work with steel samples weighing only 0.3–10 mg.

C. PROCEDURE

Since the various considerations discussed in Section IV.F on vacuum fusion are applicable also to inert gas fusion, the procedure for the latter will be treated only briefly. Reference is again made to the apparatus of Elbling and Goward (76), shown in Fig. 103.12, and to their procedure for zirconium.

The graphite crucible, usually made a little smaller than for vacuum fusion, is surrounded by lightly packed 200-mesh spectrographic-grade graphite. The lower ball joint is sealed with a silicone high-vacuum grease.

Purified argon is admitted into the furnace assembly and discharged through the sample-loading port at a rate of 0.35 liter/min. The crucible is outgassed at 2500–2600°C for 30 min. With the temperature lowered to 2100°C, the argon is then passed through the iodine pentoxide and into the sodium hydroxide solution. A fresh change of solution is effected, and the change in resistance during a timed period, normally about 10 min, is measured to determine the blank rate. The crucible is subjected to further outgassing if the resistance change exceeds 0.1 Ω/min, which in this case is equivalent to about 2.0 μg of oxygen per minute.

The alkaline solution, which consists of 4 g of sodium hydroxide, 2 g of gelatin, and 3 ml of 2-ethyl–1-hexanol in 18.5 liters of carbon dioxide-free water, is calibrated by analyzing known amounts of a metal oxide. If uranium oxide, U_3O_8, is used, then amounts of 0, 1.00, 2.00, and 3.00 mg wrapped in 0.1–0.15 g of platinum foil are prepared. Suitable amounts of silver oxide, zirconium dioxide, or niobium dioxide can also be used. The platinum foil packages are loaded into the sample-storage arm with proper manipulation of the loading port to prevent air contamination. After the solution in the cell has been changed and the conductance bridge balanced with argon flowing through the cell, a package is dropped into the hot crucible with the aid of a magnetic pusher and swept by argon for 10 min. The change in resistance is noted, and the process is repeated for each of the standards. A plot of the resistance changes after correcting for the blank shows a straight line up to 700 μg of oxygen. Every batch of solution must be calibrated and the curve checked daily with at least two standards.

Zirconium specimens of 0.1–0.3 g are abraded, degreased with a suitable solvent such as trichloroethylene, rinsed with acetone, dried, weighed, and wrapped in a weighed piece of platinum foil. The ratio of platinum to sample is kept at 7:1 or better, separate additions of platinum being made to the bath if necessary. The wrapped samples are loaded into the sample-loading port, and the resistance changes noted for them in the same manner as was done for the standards. The oxygen contents are then determined from the calibration curve.

D. CHARACTERISICS AND VARIATIONS

In general, the problems and variations discussed for vacuum-fusion analyses apply to inert gas-fusion analysis as well.

There is general agreement that inert gas fusion and vacuum fusion give comparable values for larger amounts of oxygen. An increasing number of workers report this to be true for low amounts of oxygen as well when suitable gas chromatographic equipment is used (8). Both techniques have their good points. For inert gas fusion leakage is less of a problem (because the system is under a positive pressure), the apparatus and procedure are simplified, gettering and vapor deposition are diminished, and higher crucible temperatures are possible.

A number of variations were discussed in Section VI.B on apparatus. Therefore, only additional variations will be discussed here.

Although argon is normally used as the inert gas, helium (146,40) is also suitable. Shanahan (234) employed nitrogen to determine the oxygen and hydrogen in steels.

Peterson et al. (221) designed an all-metal system.

Beck and Clark (27) used graphite capsules when volatile samples or products were encountered. Volatile samples can also be dealt with by first evaporating the metal onto a removable cold surface (145).

Holt and Stoessel (147) found that they got more accurate results with macroamounts of oxygen when a perforated lid was kept on the crucible during the fusion process. They also found it advantageous to first outgas the sample at a comparatively low temperature, such as 1700°C, and then at 2000°C, in a platinum bath.

Impulse heating has been reported to be a very satisfactory means of heating the sample (112a,274a). Surges of current in excess of 1000 A are sent through the graphite crucible in 12-sec bursts to reach temperatures greater than 3200°C. The technique is reported to achieve more rapid outgassing at a higher reaction temperature without any insulation around the graphite. Goldbeck et al. (112a) reported good results for oxygen and nitrogen in boron, uranium, beryllium, and thorium.

The relatively high oxygen blanks encountered above 2000°C are caused principally by the reaction of the graphite or carbon black insulation with the quartz thimble to produce carbon monoxide. Dallman and

Fassel (64) found that pyrolytic boron nitride gave an oxygen blank which was low and nearly independent of the temperature. The nitrogen blanks of the two thimbles were comparable.

An interesting variation is the melting of the metals by arcing them in an inert atmosphere and then determining the gas content from the analysis of an aliquot of the gas mixture contained in the closed system (80,283). The technique was developed by Fassel and his co-workers for a variety of metals, using a gas chromatograph for the gas analysis. The sample is placed in a carbon electrode, introduced into a helium chamber along with seven other samples, and then arced for 20 sec with an arcing current of 15 A under a helium pressure of 680 mm Hg. After a gas-mixing period of 60 sec, an aliquot of the gas mixture is taken for analysis. It is estimated that a temperature of about 3000°C is obtained. This temperature was found to be sufficient to release the nitrogen as well as the oxygen from steel (80). By placing the sample on top of a piece of platinum in the electrode, the technique could be extended to refractory metals as well to obtain both the oxygen and the nitrogen contents (283). The same calibration curve for oxygen was used for steel, molybdenum, niobium, terbium, thorium, vanadium, yttrium, chromium, hafnium, tantalum, titanium, and zirconium, while for nitrogen the same curve could be used for all but titanium, which appeared to call for more study.

In a similar variation, Guldner (127) used a xenon flash discharge lamp on a tin metal film over a glass substrate. Temperatures of 5000–8000°C were obtained, and compounds such as tantalum oxide and nitride were found to decompose. The gases were analyzed with a gas chromatograph.

E. SPECIFIC METALS

Again, the criteria and experiences of vacuum-fusion analysis can serve to advantage in inert gas-fusion analysis in considering specific metals, except that higher temperatures can be used. The application of the usual technique of fusion in a graphite crucible with the aid of an induction heater will be discussed.

Steel can be fused in an iron bath at temperatures of 1700–2400°C (40,234,221,14a) or in a platinum bath at temperatures of 1950–2300°C (146,14a). The latter range is preferable if the nitrogen content is desired (147), or possibly if the steel is very highly alloyed. The ratio of platinum to steel in the bath varies from 1:1 to 5:1, and there appears to be a general preference for the flux technique, that is, introducing the sample with some platinum (14a).

Goto et al. (113) fused a manganese alloy steel in a tin bath at 1650°C. Argon was reported by Smith et al. (249a) to be far superior to helium for inert gas-fusion analysis for oxygen in cobalt.

A platinum flux technique, in which the sample is dropped into a platinum bath along with a platinum foil wrapping, is used for refractory metals such as titanium, zirconium, hafnium, and uranium (76,14a). Although the sample is frequently dropped directly into a platinum bath without the accompaniment of some platinum, it is recommended that the sample be introduced into the bath with at least a small piece of platinum foil or wire in intimate contact. The ratio of platinum to sample in the bath should be 7:1 or higher (76) with such metals. Apparently, the presence of tin is not as helpful in inert gas fusion as it is in vacuum fusion; however, its use, either with a sample or between runs or in the bath, is practiced in the majority of laboratories (201) and is recommended for the refractory metals (87). Bath temperatures of 2000–2400°C appear to be satisfactory (76,147,201,14a).

Agreement has not as yet been reached on a method for oxygen in tantalum. A cooperative program conducted by ASTM Division I of Committee E-3 specified the platinum flux technique at 2200° ± 100°C for 5–10 min for the inert gas-fusion method with a minimum Pt/sample ratio of 1:1. The results, which were not considered to be in sufficient agreement, led some operators to believe that the extraction temperature, the extraction time, the Pt/sample ratio, or all three should be increased. A cooperative program was planned which would take these factors into consideration (14a). Kamin et al. (165) reported satisfactory results with a platinum flux technique, using a bath consisting of 60% Pt and 40% Ni with a bath/metal ratio of 6:1 minimum at a temperature of at least 2000°C. They also found the same technique satisfactory for niobium with a minimum bath/metal ratio of 5:1.

ASTM Division I also worked on a method for chromium. The results on a sample containing 50 ppm oxygen were not in sufficient agreement to permit conclusions regarding the methods and conditions used (14a). Among inert gas-fusion operators there appears to be a preference to analyze for oxygen in chromium with a platinum or platinum–tin bath, using Pt/sample ratios ranging from 5:1 to 10:1 and extraction temperatures ranging from 1800° (for a platinum–tin bath) to 2350°C (14a).

In the case of beryllium, there is substantial agreement of Kallman and Collier's inert gas-fusion method (161) with vacuum fusion (using a platinum–tin bath) and with neutron activation for a sample containing 0.5% oxygen (14a). Kallman and Collier wrapped the beryllium sample in about 0.4 g of copper foil and introduced it into the crucible (previously outgassed at 2700°C) along with nickel foil (10 times the weight of beryllium), using an extraction temperature of 2600°C. However, agreement was very poor, both between techniques and between operators using the same technique, when the oxygen content was low (about 100 ppm of oxygen), and this matter was to be studied further by the ASTM Division I on Gases in Metals (14a,87).

Banks et al. (22) determined the oxygen in yttrium, using a platinum flux technique at 2200–2300°C after outgassing the crucible at 2600°C. The ratio of platinum to yttrium in the platinum-wrapped samples and in the bath was about 3:1. The sample weight was selected so as to have about 200 μg of oxygen. The results were found to agree well with those obtained by vacuum fusion and emission spectroscopy.

These conditions, save for a Pt/sample ratio of $>10{:}1$, were found by Fassel et al. (87) to give slightly but unmistakably lower results in their apparatus than did the use of a platinum–tin bath in the ratio of 80 Pt:20 Sn at an extraction temperature of 2300°C. They introduced the sample along with 0.1 g of tin as well as 3 times the sample weight of platinum. They found these conditions satisfactory for thorium also, although for this metal a platinum bath at 2500°C also gave satisfactory results.

For the determination of oxygen in vanadium Fassel et al. (88) used a Pt/V ration of 3:1 with an extraction temperature of 2100°C. They were able to get the same results also with the oxygen content determined by vacuum fusion as well as by dc carbon-arc emission spectrometry. Kamin et al. (165) also obtained satisfactory results by wrapping the sample in platinum foil and using a bath consisting of 60% Pt and 40% Ni with a 1:1 ratio of bath to metal.

To determine the oxygen in zinc, cadmium, and magnesium, Holt and Goodspeed (145) first removed the metal matrix by evaporating it and condensing it on a removable cold surface inside the inert gas-fusion furnace. The oxygen was then extracted from the oxide residue without a flux at 1850°C for zinc, 1900°C for cadmium, and 2100°C for magnesium. The recovery of oxygen was determined to be complete for zinc and nearly complete for cadmium (94 $\pm$ 4%) and magnesium (84 $\pm$ 9%). Accordingly, correction factors were used for these two. A manometric measurement was employed. The sensitivity ranged from 10 ppm for zinc to 50 ppm for magnesium.

In dealing with the determination of oxygen in uranium dicarbide by the inert gas-fusion technique, Smith et al. (250) devised a means of thermally rupturing grains of the sample in the system to get past a highly impervious pyrolytic coating. They then carried out the extraction at 2550°C in a specially designed, current-concentrator furnace.

In a series of successful tests to get nitrogen values which would be comparable with those obtained by chemical methods, Dallman and Fassel (65) reported satisfactory results for a number of metals and steels, using helium fusion in a platinum–tin bath. Their operating conditions are summarized in Table 103.V. The results for unalloyed titanium were about 10% low, but those for a number of titanium alloys were

TABLE 103.V

Extraction Conditions for Simultaneous Determination of Oxygen and Nitrogen in Metals by the Carrier Gas-Fusion Technique (65)

Reaction medium composition,[a,b] wt %	Sample metal[c]	Fusion temperature,[d] °C	Platinum flux ratio	Maximum sample— metal concentration in final fusion melt, wt %
80 Pt:20 Sn	Cu	1725–1775	3:1	10
	Cr	1775–1825		
	Steel	1825–1875		25
	Mo	1850–1950	. . .[e]	20
	Zr		3:1[f]	10
95 Pt:5 Sn	Nb	1950–2050	3:1	15
	V	2050–2150		
	Ta	2150–2250	. . .[e]	20
	Sc, Y, Gd, Tb, Dy, Ho, Er, Lu	2275–2375	3:1[g]	10
	Th	2300–2400	5:1[g]	
	Hf		5:1[f]	
	Ti		8:1[f]	

[a] Not less than 15 g total of the binary bath initially added to the crucible.

[b] Reaction medium preconditioned with 0.2–0.5 g of sample metal or steel.

[c] Samples of 0.05–1.0 g extracted for a total period of 8 min.

[d] All temperature measurements obtained with optical pyrometer (Pyrometer Instrument Co., Bergenfield, N.J., Catalog No. 85) calibrated in accordance with instructions provided by the manufacturer.

[e] Platinum flux not required but normally used to help maintain more dilute sample-metal concentration in the melt.

[f] Samples encapsulated in platinum to prevent direct sample–crucible contact.

[g] Samples encapsulated in platinum to help minimize possible surface contamination by the atmosphere, and to facilitate sample penetration into the melt.

in good agreement, as were their results for all of the other metals. Usually the initial bath weighed 15 g, and 0.2–0.3 g of sample metal was used as a conditioner. The sample was introduced with about 3 times its weight of platinum. Their induction heater operated at 400 kHz, a relatively low frequency which provides a greater penetration of heat into the bath. This is considered advantageous for extracting nitrogen, and it may be an important reason why some workers could not get good nitrogen results for metals other than steels with the LECO Nitrox 6 apparatus, which uses an induction heater of 3–4 mHz (8). Kashima and Yamazaki (166) reported good results for nitrogen in various metals (stainless steel, tungsten, molybdenum, silicon) when they fused the specimens with silicon in an argon atmosphere at <1900°C and measured the nitrogen with a gas chromatograph.

F. COMMERCIAL EQUIPMENT

Of the commercial equipment (226a) available, the LECO Model TC-30 Simultaneous Nitrogen–Oxygen Determinator (Laboratory Equipment Corporation, St. Joseph, Mich., 49085) employs impulse heating (112a), and a separate, small crucible can be used for each determination. The temperature is said to come close to 2500°C. After the gases are passed over hot copper oxide, hydrogen is removed as water by magnesium perchlorate, and a gas chromatographic, thermal conductivity measurement is made of the nitrogen and oxygen (as CO_2), the calculated values being displayed on an electronic digital voltmeter with a sample-weight compensator. The LECO Direct Reading Nitrogen Determinator is trimmed down from the LECO Model TC-30 to obtain only the nitrogen value after the CO_2 is absorbed on Ascarite. Similarly, the LECO Direct Reading Impulse Oxygen Determinator is trimmed down from the TC-30 to yield only the oxygen value after oxidation of the CO to CO_2 with nitrogen as a carrier gas. The LECO Model 734-300 Direct Reading 2½-Minute Oxygen Determinator employs induction heating and helium carrier gas and can reach 2700°C. After oxidation of the gases over hot copper oxide, CO_2 is trapped on a molecular sieve and then released by heating into a gas chromatograph, with a thermal conductivity detector for a measure of the oxygen content.

The Stroehlein Nitrogen–Oxygen Analyzer (Dinometer) (Ströhlein Fabrik Chemischer Apparate & Co., Liechtenstein; represented in the United States by Brinkmann Instruments, Inc., Westbury, N.Y., 11590) employs a graphite resistance heater. The oxygen content is determined with an elegant infrared gas analyzer calibrated against pure CO. Then, after oxidation over Schütze's reagent, the CO is removed as CO_2 on a molecular sieve, and the nitrogen in helium carrier gas is measured by thermal conductivity. Values are read from digital displays. A new crucible, moved into place by an electrically operated lift, is used for each determination.

A Japanese instrument, "Coulomatic O" (274d), employs argon, Schütze's reagent, and a coulometric titration of the evolved CO_2 with barium chlorate solution at a controlled potential.

De La Rue Frigistor, Ltd., Langley, Bucks., England, sells a hot-extraction apparatus designed for hydrogen determinations (218b). Argon is used as the carrier gas, and the cumulative signal from the measurement of the thermal conductivity is related to the hydrogen content. The equipment, called the Hytest, has no glass parts and is described as rugged. The company also sells sealed steel capsules containing known amounts of hydrogen for calibration of the instrument. Such capsules are also available from the British Welding Institute (56a) and from the U.S. National Bureau of Standards.

VII. SPECTROGRAPHIC METHODS

A. PRINCIPLE

The long-established and successful use of spectrographic methods for trace analysis, together with the usual speed of these methods, has spurred attempts to apply them to gases in metals. Although the practice of emission spectroscopy is treated elsewhere in this Treatise (233) and in the general literature (52), the basic principles warrant repetition, particularly as they bear on the subject of this chapter. Practical details of general emission spectroscopy will not be treated; only techniques peculiar to determining oxygen, hydrogen, or nitrogen in metals will be discussed. Infrared absorption has been used to measure oxygen in silicon (213b) and in germanium (205a). Because this technique is not yet widely used and is still under investigation for both metals (268b), it will only be mentioned here.

If two electrodes (two rods electrically connected to a source of power) have a strong current passed across the gap between them, the emitted light may be directed into a spectrograph, wherein a grating or a prism, perhaps in optical arrangement with mirrors, disperses the light and separates the component wavelengths. This latter light may now be recorded photographically or electronically in such a manner that both the component wavelengths and the corresponding intensities can be measured. The resulting spectra are of at least three kinds: line, band, and continuous. The first is generally the result of transitions among electronic energy levels of atoms or monatomic ions. Band spectra are generally due to electronic transitions among energy levels of diatomic or polyatomic ions or molecules, together with vibrational and rotational energy changes occurring simultaneously. The continuum results from ionizaton and the differing kinetic energies of the liberated electrons.

This refers to the ultraviolet and visible region of the spectrum, for in the infrared, far and near, pure rotation, vibration, and rotation–vibration spectra may be obtained without electronic transitions. The continuous spectra appear as a general darkening on photographic plates and are due to the kinetic energy, generally of atoms from dissociated molecules. Spectrometers with low dispersion will sometimes show the lines of band spectra so close together that they may be misidentified as continuous. The spectra from the electrodes described above will depend on the material of the electrodes, the material on and between the electrodes, and the kind and amount of power applied. For example, graphite is a common electrode material, and usually air is not kept away from the electrode region. As a result it is exceedingly common to have the cyanogen bands appear in the spectra as a result of reaction between the carbon and nitrogen.

The usual spectrographic procedure is to prepare a mixture of the analytical specimen and graphite. This is placed in the cuplike end of a small vertical graphite rod which acts as one electrode. Another graphite rod, generally sharpened to a point like that of a pencil, serves as the second electrode, mounted vertically with the point aimed downward toward the sample electrode. (It should be mentioned that many analyses are performed by sparking directly between the upper electrode and a metal—or at least conducting—specimen held in an electrode holder below.) An arc or spark is struck between the electrodes, and the spectrum recorded as usual. Now the record will be characteristic not only of the electrodes and the atmosphere but also of the specimen in the cup electrode. The intense heat generated vaporizes the specimen and excites the atoms, ions, and molecules to higher than normal energy levels. On returning to the ground state the exciting energy is re-emitted, giving rise to the spectra. Since the energy levels—electronic, rotational, and vibrational—are quantized, these returning transitions appear as discrete lines. Kinetic energy is tantamount to being unquantized, and therefore the corresponding transitions appear as continuous in wavelength.

Now, since the wavelength of a spectral line depends on the difference in energy levels:

$$\lambda = \frac{hc}{E_2 - E_1} \tag{6}$$

(h = Planck's constant, c = velocity of light, E_2 and E_1 = energy of the two levels), and these are specific to the atom, ion, or molecule involved, identification is performed by determination of wavelength. Furthermore, the intensity of the line, as exhibited by darkening of the photographic plate, is correlated in a positive sense with the number of atoms emitting, and these are, in a carefully developed analytical method, correlated in a positive sense also with the number of these atoms in the specimen. Thus one needs merely to choose an appropriate line (wavelength corresponding to a species of interest), measure the intensity of the line as produced under controlled conditons, and, comparing this with standard results obtained as nearly as possible under identical conditions, interpolate the inferred unknown concentration. Sometimes this is simple; sometimes it is difficult indeed.

To apply this general method to gases in metals, then, one would expect to look for the spectra of either oxygen-containing molecules or ions, say, or of oxygen itself—and similarly for nitrogen and hydrogen. Immediately it is apparent that simple procedures will not work because of the presence of the elemental gases, moisture, and carbon dioxide in the air's atmosphere. A controlled atmosphere chamber will be necessary around the electrodes and the specimen. Furthermore, the electrodes, if made of graphite, will be suppliers of these sought elements also unless they are "degassed" in some way. Once these problems are satisfactorily solved, the question arises of which spectral line is to be used for the determination. Will it be representative of a monatomic or of other species? Although the most extensive and successful work in this method has been performed with atomic oxygen lines, some success has been claimed with band heads of TiO_2 (288). With hydrogen, too, an atomic line appears to be more useful. However, nitrogen has been determined spectrographically using the CN band system at 3883 Å, although an atomic line has since appeared to be more useful.

The experimental variables in emission spectroscopy are under only moderate control. Conditions in the arc between the electrodes are generally a bit erratic, and a common way of counteracting some of this undesirable variability utilizes an "internal standard." Generally, an elemental spectral line is chosen from among those present in the matrix (the substance is added, if necessary) for which the excitation energy and the wavelength are about the same as for the line of the element being determined. Keeping the amount of this auxiliary element constant at a level to give a nicely measurable line generally provides a measure of the exposure conditions. Thus, whereas in ordinary spectrographic measurements the analyst plots $\log I$ versus $\log m$ (or $\log c$), where I is the line intensity and m the mass of the unknown element (or c its concentration), in this case he plots $\log (I_x/I_{is})$ versus $\log m$. Here the subscripts x and is refer to unknown and internal standard, respectively. In either case, when conditions are reasonably ideal, a straight line results from known x's, which may be used as a calibration.

B. HISTORY

Russian workers (195) appear to have been the first to attempt spectrographic determination of gases in metals, starting with oxygen and nitrogen in steel (83). Oxygen, nitrogen, and hydrogen analyses have been tried by these means in titanium (102,262). However, by 1956, V. A. Fassel and his co-workers at the Iowa State College had successfully developed a procedure for the determination of oxygen in plain carbon steel (93). Later, the procedure was applied to titanium and its alloys (90), to oxygen in zirconium, niobium, and yttrium, and to hydrogen in titanium. These latter determinations are described in an excellent article (91) reviewing the techniques that this group has developed. It is probable that these latter techniques are the most reliable of the spectroscopic methods at this time, and some general details in regard to them will be presented here as exemplifying the solution of the problems that are involved.

C. PROBLEMS

In order to avoid the interference of atmospheric gases, an airtight arc chamber which can hold a number of electrodes within it is provided. These may be rotated into position under the second electrode as needed. The chamber is flushed with purified argon. All electrodes are pre-arced for thermal degassing. The chamber is outgassed by a conventional high-vacuum system before admission of the gas. The order of these procedures is as follows. Empty electrodes are placed in the holder within the chamber which is then closed and evacuated. Pure argon is admitted to about $\frac{1}{5}$ atm, and the electrodes are arced at 35 A for 5 sec. The chamber is re-evacuated, the arc stopped, and the next electrode moved into position. Upon sufficient evacuation, fresh pure argon is led in and the process

repeated until all electrodes have been degassed. The chamber, as shown
in Fig. 103.13, is water cooled and this cooling is utilized during this outgassing. When all electrodes are degassed, the chamber is opened and the
previously prepared specimens are placed in the electrodes; then the
chamber is closed, quickly evacuated, and refilled with pure argon to
about $\frac{5}{6}$ atm. The amount of argon in the chamber is carefully controlled
by water cooling for temperature regulation and by measuring the pressure carefully—the volume, of course, is fixed. The sample electrodes
are of the undercut type shown in Fig. 103.14 in order to reduce thermal
conduction from the specimen pedestal. As a result, it is estimated that
temperatures as high as 3000°K have been attained. The metal melts
and generally dissolves the graphite wall, thus acquiring carbon for the
reduction. There appears to be sufficient surface tension to retain the
molten globule on the tiny electrode platform that remains.

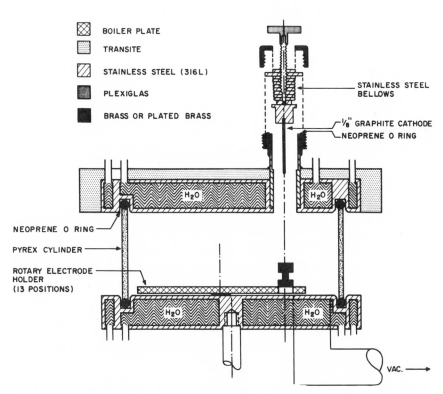

Fig. 103.13. Arc chamber for spectrographic method. (Courtesy of *Analytical
Chemistry.*)

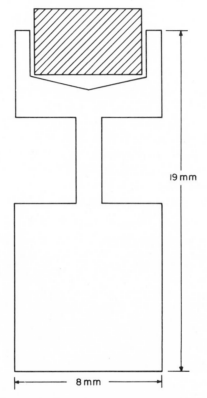

Fig. 103.14. Generally used electrode for spectrographic method. (Courtesy of *Spectrochimica Acta.*)

In the procedure just described it is noted that efforts have been made to reduce the blank; carbon has been provided, as well as a high temperature. These latter conditions, as in vacuum-fusion or carrier-gas techniques,* will generally lead to the evolution of CO, N_2, and H_2. In this case, however, these will now be in an argon atmosphere subjected to a high-intensity electric arc and will emit spectra. The argon will also emit a spectrum, thus providing an internal standard. The only general problems remaining involve the choice of appropriate spectral lines and standardization. Table 103.VI shows the line choices made by the Iowa State chemistry group.

These techniques appear to handle the problems likely to occur with

* Cf. Sections IV.F and V of this chapter.

TABLE 103.VI
Spectral Lines Used for Gas Analysis in Metals (91,163)

Element sought: line	Internal standard: line
Oxygen: O–7772 Å	Argon: A–7891 Å
Nitrogen: N–8216 Å	Argon: A–8053 Å
Hydrogen: H–6563 Å	Argon: A–6753 Å

metals in general. Specific problems with certain metals will be treated in the appropriate sections.

D. A VARIATION

There is a variant of the spectrographic technique which is, in fact, not spectrographic at all. However, its inclusion at this point is justified because of the use of the dc arc normally employed in the spectrographic technique described above. The method utilizes a chamber and arc source similar to that for the spectrographic method, with similar degassing operations, but uses helium as the inert gas and passes the resulting gas mixture through a gas chromatograph (80).

By this scheme oxygen and nitrogen have been determined in steels. By using a commercially available gas chromatograph, with a molecular sieve column to separate carbon monoxide and nitrogen in the stream of helium, 0.003–0.1 wt % oxygen and nitrogen are said to be measured with a relative standard deviation from 2.5 to 10%, depending on the concentration. Actually, the precision data presented in the original paper are insufficient to support the argument of a variable *relative* precision. Rather, the data suggest an unchanging relative standard deviation of about $3\frac{1}{2}$% in the range 0.008–0.04%. The method appears quite applicable, however, from 0.004 to 0.1%.

E. SPECIFIC METALS

1. Iron and Steel

The general procedure described earlier is essentially that used for iron and steels. The electrode and the specimen are as shown in Fig. 103.14. The cylindrical specimen weighs about 1 g. The spectral lines for oxygen, nitrogen, and argon are as shown in Table 103.VI. The nitrogen–argon spectral line combination shown there is a change (163) from the earlier pair (N 8216 Å–Ar 8179 Å), since it was found that the Ar 8179 Å line was a Rowland ghost of Ar 8115 Å. Although with the spectrograph grating of the Ames Laboratory the original lines served as a good pair for the purpose, this would not be universal and the alternative choice has been made to avoid using a ghost line.

Svet and Kozlova (262b) have used a "steelometer," a Russian version of the steeloscope but with photomultiplier detection, for determining oxygen in carbon steel between 0.005 and 0.1%. The 15-mm-diameter specimen is sharpened to a cone at one end and sparked against a pointed tungsten counterelectrode in a flowing argon atmosphere. The claimed reproducibility error is 10% relative.

2. Titanium

As discussed earlier in this chapter, some refractory metals have proved to be difficult for the vacuum-fusion analyst because of the high melting points and the refractory nature of the oxides and carbides. Metals like titanium tend, for example, to form solid masses in the vacuum-fusion crucibles when the carbide products reach even minor concentrations. The attendant difficulties have been solved in this case by the use of platinum as flux, as bath, or as both. In the spectrographic method the same problems are solved in essentially the same way. In Fig. 103.15, the electrode assembly used (90) for titanium and its alloys

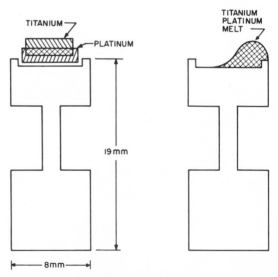

Fig. 103.15. Spectrographic electrode for refractory metal samples. (Courtesy of *Analytical Chemistry*.)

is shown. The electrode is seen to be modified to include a small platinum cup which holds the sample. When arced, the platinum and metal form a molten metal solution which remains on the electrode and absorbs enough carbon to permit the CO-producing reaction. Fassel and Gordon showed that platinum weighing about 5 times the titanium specimen is needed in order to obtain the maximum quantity of oxygen. These authors investigated iron, tin, iron plus tin, tin saturated with carbon, and nickel as bath materials also, finding, as with vacuum-fusion, that none of these was successful. The oxygen, nitrogen, hydrogen, and argon lines used are shown in Table 103.VI. Titanium and its alloys appear to be the only metals to date which have been analyzed for hydrogen by the spectrographic technique.

3. Zirconium

Although extensive details are not available beyond the general directions already discussed for refractory metals (91), some data have been reported on a comparison of the method with vacuum-fusion results (84). These are presented in Table 103.VII. The data suggest two conclusions: (1) the random errors of at least one method increase with increasing oxygen content, and (2) the algebraic differences do not show significant correlation with the level of oxygen content. Each of these conclusions is reached by a rank correlation method (72). The first conclusion, already well accepted in spectrographic analysis, emphasizes that one should consider a constant percentage error rather than a constant error when discussing the results. The second conclusion, although

TABLE 103.VII
Oxygen in Zirconium

Sample	Spectrographic	Vacuum fusion
Q138	0.12	0.13
WCE–12	0.096	0.097
1	0.17	0.19
2	0.15	0.14
3	0.13	0.13
4	0.08	0.077
5	0.36	0.34
6	0.42	0.36
7	0.18	0.21

not sufficient, is a necessary condition if the two methods are to exhibit agreement with one another. One would like to test the results by the method of Mandel and Linnig (194), but the proportionate random error rules out this possibility. However, using the sign test, (71) one finds as many positive as negative deviations, suggesting the identity of mean values for the samples by the two methods. This, in conjunction with conclusion (2) given above, is strong evidence that the methods agree.

4. Vanadium

This metal has been the subject of not only individual work (86) but also comparative study among laboratories and methods (88). One of the most important conclusions of the individual study on the spectrographic method is that the rate and degree of evolution of the oxygen content are sensitively dependent on the environmental conditions of the graphite electrodes. Most important appear to be sample size, electrode geometry, and arcing current. The usual oxygen and argon spectrographic lines are used.

Examination of the data from both studies indicates that errors are proportional to level; the spectrographic results appear to exhibit a coefficient of variation of about 6%. Between 0.03 and 0.20% oxygen content, the data exhibit quite good agreement. At the extremes beyond this range, however, there are suggestions of disparity in the results.

5. Niobium

For this metal the graphite electrode is used, either alone or with platinum, depending on the presence or absence of alloying constituents (e.g., $\sim 1.0\%$ Ti) (79). Although precision data are reported, accuracy data are merely mentioned as being concordant with the preparation of standards and vacuum-fusion results. In the range 0.04–0.4% oxygen the coefficient of variation is about 4%. Although the Ar 7891 Å line is universally used, either the O 7772 Å or the O 7775 Å line can serve for oxygen, with the former more sensitive to low oxygen concentrations than the latter.

6. Lanthanum

Lanthanum, which has yielded to a nickel bath approach for oxygen determination by the vacuum-fusion technique, can apparently be treated spectrographically by an adaptation similar to that made for the platinum bath techniques (92). A nickel cup fits inside the graphite electrode cup, and the lanthanum metal sample fits into the nickel. Standards are made by reacting specimens with oxygen gas at 520°C and annealing for several hours to ensure homogeneity by diffusion. Corrections to the unknowns and standards must be made, of course, for the oxygen content of the nickel, and to the standards for their prior oxygen content. The spectral line used is the O 7772 Å, as given in Table 103.VI, but Ar 7372 Å is reportedly employed for the internal standard. Determinations appear to be feasible in the 0.08–0.50 wt % range of oxygen.

7. Yttrium

This element does not seem to have been discussed individually in the literature. From its listing in tables of representative data, however, it apparently has been analyzed for oxygen by the platinum bath technique, using procedures similar to those for titanium, zirconium, and niobium (91,85).

VIII. STANDARDS AND COMPARISONS

A. "COOPERATIVE" MATERIALS

It will have been noticed that the various methods discussed in this chapter include both comparative and absolute types. The first of these, of course, require standards for calibration before an unknown can be evaluated. The latter, at least in principle, do not require such standardization; calibration is performed with the analytical apparatus by some direct physical measurements that may be used to compute the gas concentration from first principles. As an example, consider a vacuum-fusion apparatus with which the gas content of samples is determined volumetrically. That is, let the amount of determined constituent be computed by the gas laws from a measurement of the pressure in a calibrated volume. In such a case the McLeod gage, calibrated (150) absolutely, and the volume, also calibrated absolutely, provide a direct measuring system. If 100% "recovery" is presumed, there is no essential need for standards in order to obtain the desired answer for an unknown determination.* On the other hand, consider the spectrographic method, whereby not only must intensity of spectral lines be calibrated but also

---

* The question of recovery has, of course, been treated earlier in the chapter. There is no intention to beg this question here; the aim is merely to illustrate the principles under discussion.

the relation between intensity and quantity of oxygen (or other constituent) requires determination. Having once demonstrated complete recovery and given an adequate physical calibration, the absolute or direct methods have no need for standards, whereas the relative methods cannot proceed without them.

It should be recognized that the difference between an absolute and a comparative method often lies only in the manner with which the analyst proceeds. An analyst using a vacuum-fusion apparatus, for example, may dispense with physical calibration of the volume and of the McLeod gage if adequate standards are available. Presume, for instance, that a known specimen is analyzed and the McLeod gage measurement corresponding to the oxygen content is given by l_s mm. This will be the depression of the mercury in the gas-trapping capillary of the gage when it is used as a quadratic scale (150). Since the specimen weight, m, is known as well as the oxygen content, p, in parts per million, $pm \propto l^2$ as long as the same volume is always used. The apparatus constant may then be computed as

$$k = \frac{p_s m_s}{l_s{}^2} \qquad (7)$$

When an unknown sample is analyzed, therefore, and l_x is the corresponding mercury depression in the gage, the unknown oxygen content becomes

$$p_x = \frac{k l_x{}^2}{m_x} \qquad (8)$$

where k is the constant previously determined, and m_x is the weight of the unknown specimen in the same units that were used for the calibration.

Now, even though some methods may not require standards in principle, primary and even secondary standards are of great value. An absolute calibration of the type described above measures everything but the "recovery" of the method. Research on the analysis of a metal not previously tried is severely hampered without a primary standard. By "primary standard" in this context is meant material with "known" gas content; that is, known with certainty! Although, in a sense, "knowing with certainty" is contrary to scientific objectivity, some means are possible for producing material known with "reasonable" certainty. But care must be exercised. A primary standard should have the constituent present in the same manner that it is to be present in unknowns. It is not sufficient, for example, to use oxide powder added to metal.

Use of such standards led to wrong conclusions with a vacuum-fusion analysis of zirconium; it gave complete but misleading recovery (36). Analysis of more realistic standards made by known addition of oxygen gas to individual metal specimens showed the other procedures to be unsatisfactory. Use of a platinum bath technique then was successful (36). Specimens prepared in this manner are nearly ideal, although slow to produce. Metallographic comparison of cross sections of such specimens with those from commercial material should exhibit strong resemblance if such primary standards are to serve. It must be recognized, of course, that some materials like zirconium will soak up oxygen and become homogeneous, whereas others will only form an outer oxide or oxide-rich layer in a reasonable length of time.

Primary standards may be made in such fashion for some materials and are invaluable for research. Standards may be available for similar purposes if a variety of methods differing in principle can be used for analysis of homogeneous representative material. Such material is the only reasonable primary standard which can be distributed to other laboratories, or within a given single laboratory, for checking methods. Distribution of material analyzed by one method only would be presumptuous and subjective. On the other hand, were the aim solely to discover whether two laboratories obtained the same answer with the same representative material, such a procedure would be justified. In fact, cases may arise in which a consumer laboratory knows from experience that material analyses made by a specified laboratory in a specified way will serve to differentiate acceptable from unacceptable material. This may be true even if the recovery of the analytical method is unknown. During an era of rapid advancement in the use of heretofore unavailable metals this is often the only practical way to proceed. The analysts and, in particular, those using the analytical results, however, must forever keep in mind this distinction between an "acceptability comparison" and a supposedly absolute analysis. Failure to do so will lead to confusion when alternative methods of analysis are introduced.

An example of primary standards for distribution is given by the "cooperative steels" prepared by the U.S. National Bureau of Standards, distributed to cooperating laboratories, and analyzed by a variety of methods (268). These steels and one iron, eight in all, range from about 0.002 to 0.11% oxygen, 0.004 to 0.014% nitrogen, and 0.00022 to 0.00033% hydrogen. There was some reluctance among analysts to use the nitrogen values, and the materials are no longer available. However, for a long time these were the only reference materials available. The study which formed the original source of these materials was quite extensive and involved 35 cooperating laboratories. For the oxygen results by vacuum-fusion analysis 15 laboratories participated, and "best values" and "acceptable ranges" were chosen after a technical consideration of agreement and disagreements.

Currently (May 1971) many ferrous materials are available through the NBS Office of Standard Reference Materials (42b): 17 chip form steels, 1 granular form steel, and 5 chip form irons. These are certified for nitrogen content. Four steels and 1 iron in solid form are certified for oxygen content, and some of these are described in an NBS *Miscellaneous Publication* (202b). Nitrogen values are available as supplementary information. Also available from NBS are standards of unalloyed titanium: numbers 352, 3, and 4, containing hydrogen at three levels. These have been described by Sterling et al. (255b). Standard reference materials 355 and 6 are titanium-base samples containing oxygen. Steel samples with known amounts of oxygen can be obtained from the Bundesanstalt for Material Prufung (BAM), Unter den Eichen 87, 1 Berlin 45, Germany. Standards for the determination of hydrogen in steel have been developed by the British Welding Research Association and are available from Fosico Teknik, Ltd., Long Acres, Nechells, Birmingham, England.

So far these are the only known materials available or potentially available for wide distribution. Other reference materials are usually available only in small quantity and are limited to the cooperating laboratories. One such set consisted of titanium specimens which were analyzed by a variety of methods and laboratories. Much of this kind of activity was handled by Committee E-3, Division I, for Gases in Metals, of the American Society for Testing Materials. The division sets up a task force for studying a particular aspect of gases in metals when sufficient laboratory interest has been indicated. Participation in such a task force is very beneficial to the group as a whole and to the individuals. This has proved to be an effective way to catalyze communication between active workers in the field.

B. STATISTICAL ASPECTS

Although various chapters of this Treatise (230,287) cover some general statistical concepts, the vital part that interlaboratory comparisons and comparisons with standards play in the gases-in-metals field warrants some pointed remarks here.

Probably the technique most often used in this field and others of analytical chemistry is to check results obtained with so-called known standards. Such standards are hard to come by, and "known values" are even more difficult to obtain. For some metals, such as zirconium and titanium, direct addition will at least give individual pieces with known amounts. However, let us assume that such a set is available; probably the most common method of assessing the results is to compare the "found" with the "added" (or "known"). At times a table is presented with the comment that agreement is quite good. This subjective statement is sometimes justified when the results obviously indicate only very small differences, such as 1 or 2 units in the last of two significant figures for magnitudes greater than, say, 30. On other occasions the device resorted to involves taking algebraic differences between the corresponding members of each pair and showing that the average value of the algebraic difference is zero within experimental error, commonly using Student's t-test (70). The latter approach is not, in general, a wise one, however, for reasons understandable by reference to the following diagrams.

In Fig. 103.16 a common situation of lack of agreement is presented.

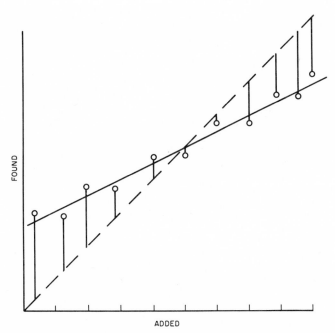

Fig. 103.16. Data comparison for verification of a method. A possible type of deviation. See also Fig. 103.17.

The solid line represents the true relation between the added and the found constituent. The small circles represent the experimental results obtained from an attempt to establish the true relation. The dashed line represents the ideal analytical situation, where "found" = "added" throughout the range of experimentation. The deviations from the solid line represent random errors.

Note that the vertical lines connecting the experimental points and the dashed line represent the differences which are used for calculation in the t-test described. It is quite obvious how the test can fail to expose an actually unsatisfactory method. Two factors enter into this case, both of which tend to weaken the test. First, approximately half of the vertical lines go upward. Algebraically this means that the calculated average difference tends toward zero. Second, the lengths of the vertical lines vary from short, near the intersection of the two found–added relations, to long at the extremes. This means that the calculated estimate of standard deviation will be larger than the appropriately calculated one. Now, because the calculated $t = \sqrt{n}(\bar{d}/s)$ (where n is the number of experimental points, $\bar{d}$ is the average algebraic difference, and s is the estimated standard deviation), we see that this test applied in these circumstances is very likely to give an unjustifiably low value of t. In other words, this method of testing the data is insensitive to this particular kind of systematic error.

Of course, Fig. 103.16 represents only one of many possible deviations from true recovery. Figures 103.17a, b, and c show others, where for clarity illustrative experimental points have not been included. Straight line relations are pictured, although there is no guarantee that these are always appropriate. The cases illustrated in Fig. 103.17 are more likely to be detected than the one in Fig. 103.16, all other things being equal. Rather than dwell on the various possibilities, it suffices to say that the deficiency of the t-test for this purpose lies in its inability to detect certain important alternative hypotheses. It is powerful only against one alternative hypothesis, that is, that $\bar{d} \neq 0$. What is desired is a test that is powerful against a combined double hypothesis; that $A \neq 0$ and $B \neq 1$ in the equation

$$f = A + Ba \qquad (9)$$

where f represents the found amount, and a the added (or known) amount. We wish to test the combined hypothesis $A = 0$ and $B = 1$ against the alternatives just mentioned. As indicated earlier, this is handled by the method of Mandel and Linnig (194), wherein an elliptical area is calculated in the coordinate system for estimates of A and B.

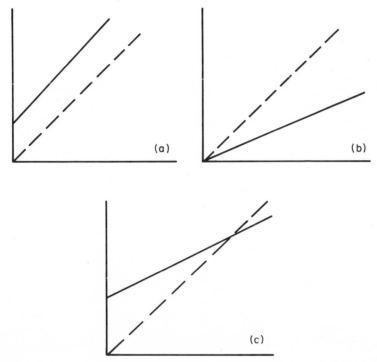

Fig. 103.17. Data comparison for verification of a method. Other possible types of deviation. See also Fig. 103.16.

If the point (0, 1) lies within the ellipse, then one states, with the appropriate confidence level for which the ellipse was computed, that the method is valid. For other applications of this confidence ellipse the reader is referred to the original paper. It is worth mentioning, however, that even this approach is limited by the need for the data to satisfy four requirements:

1. Random errors may exist only in the "found" answers.
2. These errors must be normally distributed.
3. The error from each f value must not be correlated with that from any other.
4. The true standard deviation for f's at one value of a must be the same size as that for the f's at any other value of a.

These demands are not really too stringent. The last three are probably applicable quite often if the range of values is not too large [although it was found in one study (50) of chemical analytical data that only 10–15% of some 250 distributions were normal]. Obviously some random error exists in the "added" numbers used, but it is likely that if this corresponds to less than one-fourth the error in the "found," no serious effect will arise.

The subject of interlaboratory comparisons should be included in this section because a reasonably valid program and interpretation of the data are often lacking. Since this is a subject which may be enlarged ad infinitum, no attempt will be made to cover it extensively. Reference may be made to a paper by Pierson and Fay (222), wherein many salient points are made about statistical and general matters and which contains a number of references helpful in regard to statistical matters pertaining to this particular kind of experimentation. Touched only lightly in this reference, however, is the vital importance of homogeneous samples for use in the comparisons when the determination of gases in metals is involved.

The first standard reference materials (268) were very carefully prepared, examined, and selected with regard to homogeneity. For each composition a portion of a single large ingot was selected so as to obtain material free from vertical segregation. Horizontal segregation was avoided by hot-rolling into a rod. The idea was to aim for a situation in which a complete cross section of the rod represented a complete cross section of the ingot. Adjacent samples were also expected to be identical. Samples were taken at intervals along the rod which upon analysis indicated reasonable uniformity. Metallographic specimens showed inner-to-outer differences for a cross section but no difference along a rod. Vacuum-fusion analyses representing (1) complete cross sections and (2) core areas verified the inner-to-outer variations. Complete cross sections were recommended as the appropriate sample, but pie-shaped sections have worked satisfactorily in most cases. This attention to homogeneity of specimens is of prime importance.

When apparent homogeneity is not attained or is suspect, the statistical design of the experiment may take this factor into account. For example, one round robin was conducted with a bar of zirconium which was first analyzed along its length by the supplier and found to have an oxygen content varying linearly from end to end. When this material was passed out to co-operating laboratories, the positions of the assigned specimens were noted and the comparison was evaluated by use of the positional variation originally found and that deduced from the comparison results. As long as the positional variation is not too pronounced, this procedure appears to be satisfactory. It would be appropriate to use the analysis of covariance technique to remove this effect from the statistical treatment of the data.

REFERENCES

1. Abe, Y., "Determination of the Gas Content in Metals," *Proc. Balzers 3rd Conf., 1967*, pp. 61–62, February 1969.
1a. Adcock, F., *J. Iron Steel Inst. (London)*, 115, 369 (1927).
2. Advisory Group for Aeronautical Research and Development, AGARD, NATO, *Working Paper* M33, 1962.
3. Alleman, G., and C. J. Darlington, *J. Franklin Inst.*, 185, 161 (1918).
4. Allen, J. A., and I. Lauder, *Nature*, 164, 142 (1949).
5. Anbar, M., K. Dostrovsky, F. Klein, and D. Samuel, *J. Chem. Soc. (London)*, 1955, 155.
6. Anderson, H. J., and J. C. Langbord, *Anal. Chem.*, 35, 1093 (1963).
6a. Anthanis, T., Cesium Analysis Study, *Tech. Doc. Rept.* ML-TDR-64-200, Air Force Materials Laboratory, Wright–Patterson Air Force Base, Ohio, August 1964.
6b. Aspinal, M. L., *Analyst*, 91, 33 (1966).
7. ASTM, *Methods for Chemical Analysis of Metals*, Philadelphia, Pa., 1960, p. 503.
8. ASTM, Committee E-3, Meeting of Division I, Gases in Metals, St. Joseph, Mich., Apr. 4–5, 1966.
8b. ASTM, Committee E-3, Meeting of Division I, Gases in Metals, Washington, D.C., Apr. 2–4, 1968.
9. ASTM, *1965 Book of ASTM Standards*, Part 32, Philadelphia, Pa., 1965, pp. 66–68.
10. *Ibid.*, pp. 615–617.
11. *Ibid.*, pp. 674–677.
12. *Ibid.*, pp. 739–741.
13. ASTM, Committee E-3, Meeting of Division I, Gases in Metals, St. Joseph, Mich., Nov. 21, 1963.
14a. ASTM, Committee E-3, Meeting of Division I, Gases in Metals, Watertown, N.Y., Nov. 23–24, 1964.
14b. ASTM, Committee E-3, Meeting of Division I, Gases in Metals, San Diego, Calif., Apr. 4–6, 1967.
14c. Aston, R. J., and C. Ross, "Determining the Gas Content in Metals," *Proc. 3rd Balzars Conf., 1967*, pp. 58–60, February 1969.
15. Austin, G. W., *J. Iron Steel Inst. (London)*, 86, 236 (1912).
16. Babko, A. K., A. I. Volkova, and O. F. Drako, *Tr. Komis Ps Analit. Khim., Akad. Nauk SSSR*, 12, 53 (1960); *Zh. Khim.*, 1960, 22.
17. Bach, B. B., J. V. Dawson, and L. W. L. Smith, *J. Iron Steel Inst. (London)*, 176, 257 (1954).
17a. Bach, B. B., R. R. Willis, and R. B. Reid, *Metallurgia*, 74, 191 (1966).
17b. Bagshaw, G., H. B. Headridge, R. Pemberton, E. D. Rawsthorne, and D. F. Wilson, *Metallurgia*, 82, 237 (1970).
18. Bailey, R. F., and D. A. Ross, *Anal. Chem.*, 35, 791 (1963).
19. Baker, W. A., *J. Inst. Metals*, 65, 345 (1939).
20. Baker, W. A., *Metallurgia*, 40, 188 (1949).

21. Banerjee, N. G., *NML Tech. J. (Jamshedpur, India)*, **2**, 10 (1960).
22. Banks, C. F., J. W. O'Laughlin, and G. J. Kamin, *Anal. Chem.*, **32**, 1613 (1960).
23. Barraclough, K. C., *Murex Rev.*, **1**, 305 (1954).
24. Bauer, G. A., *Anal. Chem.*, **37**, 300 (1965).
25. Beach, A. L., and W. A. Guldner, Symposium on Determination of Gases in Metals, *ASTM Spec. Tech. Publ.* **222**, pp. 15–24, 1957.
26. Beach, A. L., and W. G. Guldner, *Anal. Chem.*, **31**, 1722 (1959).
26a. Beard, D. B., R. G. Johnson, and W. G. Bradshaw, *Nucleonics*, **17**, 90–94, 96 (1959).
27. Beck, E. J., and F. E. Clark, *Anal. Chem.*, **33**, 1767 (1961).
28. Beeghly, H. F., *Ind. Eng. Chem., Anal. Ed.*, **14**, 137 (1942).
29. Beeghly, H. F., *Anal. Chem.*, **21**, 1513 (1949).
30. Beiter, E. W., F. P. Byrne, W. F. Harris, M. I. Mistrik, and R. H. Wynne, Paper presented at Pittsburgh Conference on Analytical Chemistry and Applied Spectroscopy, 1957.
30a. Berger, H., *Neutron Radiography*, American Elsevier, New York, 1965.
31. Bergstrasser, K. S., and G. R. Waterbury, *U.S. At. Energy Comm. Rept.* **LAMS-1698**, 1954.
31a. Berstrasser, K. S., G. R. Waterbury, and C. F. Metz, "Determination of Trace Amounts of Oxygen Added to Metallic Sodium," *U.S. At. Energy Comm. Rept.* **LA-3343**, 1965.
32a. Berry, R., J. A. J. Walker, and R. E. Johnson, *U.K. At. Energy Authority Rept.* **DEG-111 (C)**, 1960.
32b. Blake, P. D., and W. I. Pumphrey, *Brit. Weld. J.*, **9**, 594 (1962).
33a. Born, H. J., and N. Riehl, *Angew. Chem.*, **72**, 559 (1960).
33b. Bowen, H. J. M., and D. Gibbons, *Radioactivation Analysis*, Oxford University Press, London, 1963.
33c. Bradley, H., and R. B. Hand, "The Determination of Oxygen in Lithium by Vacuum Distillation," *Rept.* **GESP-277, R69-NSP-11**, Nuclear Systems Programs, Missile and Space Division, General Electric Co., Cincinnati, Ohio, July 1, 1969.
34. Bradshaw, W. G., *Rept.* **LMSD-2312**, Lockheed Aircraft Corp., March 1958.
35. Bradstreet, R. B., *Kjeldahl Method for Organic Nitrogen*, Academic, New York, 1965.
36. Brady, J., and W. S. Horton, unpublished work, 1954.
37. Bramley, G. E. A., and T. Raine, *J. Iron Steel Inst. (London) Spec. Rept.* **25**, p. 87, 1939.
38. Brandt, J. L., and C. N. Cochran, *J. Metals*, **8**, 1672 (1956).
39. Brown, G. P., A. Dinardo, G. K. Cheng, and T. K. Sherwood, *J. Appl. Phys.*, **17**, 802 (1946).
40. Bryan, F. R., and S. Bonfiglio, *J. Gas Chromotog.*, **2**, 97 (1964).
41. Cailletet, M. L., *Compt. Rend.*, **61**, 850 (1865).
42a. Caldwell, D. O., *Rev. Sci, Instr.*, **23**, 501 (1952).
42b. Cali, J. P., "Catalog of Standard Reference Materials," *Natl. Bur. Std. (U.S.) Spec. Publ.* **260**, July 1970 ed., Government Printing Office, Washington, D.C., 20402 (75 cents).
43a. Carney, D. J., *Gases in Metals*, American Society for Metals, Cleveland, Ohio, 1953, pp. 100–114.

43b. Carney, D. J., J. Chipman, and N. J. Grant, *J. Metals,* **188,** 397 (1950).
44. Carson, C. C., *Gen. Elec. Rept.* **DF49SL116,** Schenectady, N.Y., Nov. 2, 1949.
45. Carson, C. C., *Anal. Chem.,* **32,** 936 (1960).
46. Carson, C. C., and B. J. Alperin, *Trans. Am. Foundrymen's Soc.,* **70** (1959).
47. Carson, C. C., S. Foldes, and C. B. Worcester, "The Determination of Oxygen in Low-Oxygen Copper," paper presented at Meeting of Metallurgical Societies, Cleveland, Ohio, October 1970.
47a. Carson, C. C., and D. D. Rossettie, unpublished work.
47b. Carson, C. C., and C. B. Worcester, to be published.
48. Cassler, C. D., and G. R. Fitterer, *J. Metals,* **17,** 655 (1965).
49. Champeix, L., R. Darras, and J. Duflo, *J. Nucl. Mater.,* **1,** 113 (1959).
50. Chancey, V. J., *Nature,* **159,** 339 (1947).
51. Chubb, W., G. T. Meuhlenkamp, F. R. Shober, and D. A. Schwope, "Mechanical Properties of Zirconium and Its Alloys," in B. Lustman and F. Kerze, Jr., eds., *The Metallurgy of Zirconium,* McGraw–Hill, New York, pp. 516, 518.
52. Churchill, J. R., *Ind. Eng. Chem., Anal. Ed.,* **16,** 653 (1940).
53. Codell, M., and G. Norwitz, WAL Report No. 401/229 (Frankford Arsenal Report No. **S-4128**), Aug. 20, 1954; *Anal. Chem.,* **27,** 1083 (1955).
54. Codell, M., and G. Norwitz, *Anal. Chem.,* **28,** 106 (1956).
55. Coe, F. R., "Hydrogen in Steel," *J. Iron Steel Inst. (London) Spec. Rept.* **73,** p. 111, 1961.
55a. Coe, F. R., *Steel Times,* **196,** 456 (July 1968).
56. Coe, F. R., and N. Jenkins, *J. Iron Steel Inst. (London) Spec. Rept.* **68,** p. 229, 1960.
56a. Coe, F. R., N. Jenkins, and D. H. Parker, *Anal. Chem.,* **39,** 982 (1967).
57. Colin, R. H., and G. H. Snelling, *Elec. Furnace Steel Proc.,* **15,** 37 (1957).
58. Consolazio, G. A., and W. J. McMahon, *O.T.S. Rept.* **P.B. 131121,** Oct. 28, 1955.
59. Cook, R. M., and G. E. Speight, *J. Iron Steel Inst. (London),* **176,** 252 (1954).
60. Covington, L. C., and S. J. Bennett, *Anal. Chem.,* **32,** 1334 (1960).
61. Crane, L. T., *J. Chem. Phys.,* **36,** 10 (1962).
62. Crawley, R. H. A., *Anal. Chim. Acta,* **7,** 63 (1952).
63. Cunningham, T. R., and R. J. Price, *Ind. Eng. Chem., Anal. Ed.,* **5,** 27 (1933).
63a. Dallman, W. E., F. D. Snell, and L. S. Ettre, "Carrier Gas and Vacuum Fusion Methods," in *Encyclopedia of Industrial Chemical Analysis,* Vol. 8, Wiley–Interscience, New York, 1970, pp. 613–693.
64. Dallman, W. E., and V. A. Fassel, *Anal. Chem.,* **38,** 662, (1966).
65. *Ibid.,* **39,** 133R (1967).
66. Dardel, Y., *Metals Tech.,* **15,** Tech. Publ. 2484 (1948).
67. Darling, A. S., *Platinum Metals Rev.,* **2,** 16 (1958).
68. Dawson, J. V., and L. W. L. Smith, *J. Iron Steel Inst. (London) Spec. Rept.* **68,** p. 219, 1960 (plus 2 plates).
69. Derge, G., W. Peifer, and B. Alexander, *Trans. AIMME,* **162,** 361 (1945).
70. Dixon, W. J., and F. J. Massey, Jr., *Introduction to Statistical Analysis,* 2nd ed., McGraw–Hill, New York, 1957, pp. 124–127.
71. *Ibid.,* p. 280.
72. *Ibid.,* p. 294.
73. Dorner, S., and K. Kummerer, *Atomkernenergie,* **9,** 167 (1964).
73a. Dulski, T. R., and R. M. Raybeck, *Anal. Chem.,* **41,** 1025 (1969).

74a. Dushman, S., *Scientific Foundations of Vacuum Technique*, 2nd ed., J. M. Lafferty, ed., Wiley, New York, 1962, p. 491.

74b. *Ibid.*, p. 544.

75. Eborall, R., "The Determination of Gases in Metals," *J. Iron Steel Inst. (London) Spec. Rept.* **68**, p. 192, 1960.

76. Elbling, P., and G. W. Goward, *Anal. Chem.*, **32**, 1610 (1960).

77. Elwell, W. T., and D. M. Peake, *Analyst*, **82**, 734 (1957).

77a. Evans, C. A., Jr., *Anal. Chem.*, **42**, 1130 (1970).

78. Evans, C., and J. Herrington, *Anal. Chem.*, **35**, 1907 (1963).

79. Evens, F. M., and V. A. Fassel, *Anal. Chem.*, **33**, 1056 (1961).

80. *Ibid.*, **35**, 1444 (1963).

81. Fagel, J. E., Jr., General Electric Co., Cleveland, Ohio, personal communication.

82. Fagel, J. E., Jr., R. F. Witbeck, and H. A. Smith, *Anal. Chem.*, **31**, 1115 (1959).

83. Falkova, O. B., *Izvest. Akad. Nauk SSSR, Ser. Fiz.*, **19**, 149 (1955).

84. Fassel, V. A., private communication.

85. Fassel, V. A., *J. Iron Steel Inst. (London) Spec. Rept.* **68**, p. 103, 1960.

86. Fassel, V. A., and L. L. Altpeter, *Spectrochim. Acta*, **16**, 443 (1960).

87. Fassel, V. A., W. E. Dallman, and C. C. Hill, *Anal. Chem.*, **38**, 421 (1966).

88. Fassel, V. A., W. E. Dallman, R. Skogerboe, and V. M. Horrigan, *Anal. Chem.*, **34**, 1364 (1962).

89. Fassel, V. A., F. M. Evens, and C. C. Hill, *Anal. Chem.*, **36**, 2115 (1964).

90. Fassel, V. A., and W. A. Gordon, *Anal. Chem.*, **30**, 179 (1958).

91. Fassel, V. A., W. A. Gordon, and R. J. Jasinski, "Spectrographic Determination of Oxygen, Nitrogen, and Hydrogen in Metals," in *Progress in Nuclear Energy. Ser. IX: Analytical Chemistry*, Pergamon, New York, 1959, p. 230.

92. Fassel, V. A., W. A. Gordon, and R. W. Tabelling, "Emission Spectrographic Determination of Oxygen in Metals," in Symposium on Spectrochemical Analysis for Trace Elements, *ASTM Spec. Tech. Publ.* **221**, 1958, pp. 3–22.

93. Fassel, V. A., and R. W. Tabelling, *Spectrochim. Acta*, **8**, 201 (1956).

93a. Fedorov, A. A., and A. M. Krichevskaya, *Zavodsk. Lab.*, **34**, 1425 (1968).

94a. Feichtinger, H., H. Bachtold, and W. Schuhknecht, *Schweiz. Arch. Angew. Wiss. Tech.*, **25**, 426 (1959).

94b. Fischer, W. A., and W. Ackerman, *Arch. Eisenhüttenw*, **36**, 643 (1965).

94c. *Ibid.*, **36**, 695 (1965).

94d. Fischer, W. A., and W. A. Ackerman, *Arch. Eisenhüttenw.*, **37**, 43 (1966).

94e. *Ibid.*, **37**, 692 (1966).

94f. Fitterer, G. R., C. D. Cassley, and V. Vierbicky, *J. Metals*, **20**, 74 (1968).

95. Fleischer, J. F., personal communication.

95a. Franklin, N. F., and D. Potts, *Plating*, **54**, 717 (1967).

96. Frazer, J. W., and C. W. Schoenfelder, *U.S. At. Energy Comm. Rept.* **UCRL 4944**, 1957.

97. Freidlander, G., and J. W. Kennedy, *Introduction to Radiochemistry*, Wiley, New York, 1949, p. 291.

98. *Ibid.*, p. 301.

98a. Friedrich, K., and E. Lassner, *J. Less Common Metals*, **13**, 171, (1967).

99. Foster, L. M., A. S. Gillespie, Jr., T. H. Jack, and W. W. Hill, *Nucleonics*, **21**, 53 (1964).

99a. Gahn, R. F., and L. Rosenblum, *Anal. Chem.*, **38**, 1016 (1966).

100. Gardner, R. D., W. H. Ashley, and P. Bergstresser, *Los Alamos Sci. Lab. Rept.* **LA-3225,** 1965.
101. Geller, W., and H. T. Sun, *Arch. Eisenhüttenw.,* **21,** 423 (1950).
101a. Gentry, C., and R. Schott, *Anal. Chem.,* **42,** 7 (1970).
102. Gerasimova, N. G., T. F. Ivanova, N. S. Sventitskii, G. P. Startsev, K. I. Tagarrov, and M. E. Trantovius, *Izvest. Akad. Nauk SSSR, Ser. Fiz.,* **19,** 147 (1955).
103. Gerhardt, A., personal communication.
104. Gerhardt, A., T. Kraus, and M. G. Frohberg, Pittsburgh Conference on Analytical Chemistry and Applied Spectroscopy, February 1966.
104a. Gertsman, S. L., D. K. Farschow, and J. C. Pope, *Foundry,* **98,** 78 (1970).
105. Gilbert, T. W., Jr., A. S. Meyer, Jr., and J. C. White, *Anal. Chem.,* **29,** 1627 (1957).
106. Gill, W. E., *J. Iron Steel Inst. (London),* **201,** Part II, 960 (1963).
107. Gillespie, A. S., *Anal. Chem.,* **32,** 1624 (1960).
108. Glass, J. W., C. M. Larsen, and J. M. Scarborough, *Anal. Chem.,* **38,** 942 (1966).
108a. Glushko, Y. V., and B. A. Shmalev, *Zavodsk. Lab.,* **32,** 14 (1966).
109. Goerens, P., *Metallurgia,* **7,** 384 (1910).
110. Goerens, P., and J. Paquet, *Ferrum,* **12,** 57 (1915).
111. Gokcen, N. A., *Trans. AIME,* **212,** 93 (1958).
112. Gokcen, N. A., and E. S. Tankins, *J. Metals,* **14,** 585 (1962).
112a. Goldbeck, G. C., paper presented at Annual Meeting of Division I, ASTM, Committee E-3, Washington, D.C., Apr. 2–4, 1968; Goldbeck, C. G., S. P. Turel, and C. J. Rodden, *Anal. Chem.,* **40,** 1393 (1968).
113. Goto, H., S. Ikeda, A. Anuma, and T. Shimanuki, *Japan Inst. Metals J.,* **27,** 558 (1963); *Rev. Met. Lit.,* **21,** Abstr. 02890S (1964).
114. Goutall, E., *Metallurgia,* **7,** 340 (1910).
115. Goward, G. W., *Anal. Chem.,* **37,** 117R (1965).
116. Graham, T., *Proc. Roy. Soc. (London),* **15,** 502 (1866–67).
117. Gray, N., and M. C. Sanders, *J. Iron Steel Inst. (London),* **137,** 348P (1938).
118. *Ibid.,* **143,** 321P (1941).
119. Gregory, J. N., and D. Mapper, *Analyst,* **80,** 225 (1955).
120. *Ibid.,* **80,** 230 (1955).
121. Griffith, C. B., and M. W. Mallett, *Anal. Chem.,* **25,** 1085 (1953).
122. Grosse, A. V., S. G. Hardin, and A. D. Kirshenbaum, *Anal. Chem.,* **21,** 386 (1949).
123. Grosse, A. V., and A. D. Kirshenbaum, *Anal. Chem.,* **24,** 584 (1952).
124a. Guernsey, D. C., and R. H. Franklin, Symposium on Determination of Gases in Metals, *ASTM Spec. Publ.* **222,** pp. 3–14, 1957.
124b. Guinn, V. P., personal communication.
124c. Gulbransen, E. A., "Gas–Metal Reactions of Zirconium," in *The Metallurgy of Zirconium,* B. Lustman and F. Kerze, Jr., eds., McGraw-Hill, New York, 1955, pp. 593, 597–599.
125. Gulbransen, E. A., and K. F. Andrew, *J. Electrochem. Soc.,* **101,** 348 (1954).
125a. Gulbransen, E. A., and K. F. Andrew, *Electrochem. Tech.,* **5,** 471 (1967).
126. Guldner, W. G., *Talanta,* **8,** 191 (1961).
127. Guldner, W. G., *Anal. Chem.,* **35,** 1744 (1963).
128. Guldner, W. G., and A. L. Beach, *Anal. Chem.,* **22,** 366 (1950).

128a. Guldner, W. G., and R. Brown, "The Application of a Xenon Discharge Lamp to the Analysis of Thin Metal Films," in *Measurement Techniques for Thin Films,* B. and N. Schwartz, eds., Electrochemical Society, 1967, pp. 82–101.

128b. Hagemaier, D., H. Halchak, and G. Basl, *Materials Evaluation,* **27,** 193 (1969).

129. Hague, J. L., R. A. Paulson, and H. A. Bright, *J. Res. Natl. Bur. Std.,* **43,** 201 (1949).

130. Halperin, M., *J. Am. Stat. Assoc.,* **56,** 667 (1961).

131. Hamner, H. L., and R. M. Fowler, *J. Metals,* **4,** 1313 (1952).

132. Hampel, C. A., ed., *Rare Metals Handbook,* Reinhold, New York, 1954.

133. Hansen, W. R., M. W. Mallett, and M. J. Trzeciak, *Anal. Chem.,* **31,** 1237 (1959).

133a. Hargrove, G. L., R. C. Shepard, and H. Farrar, IV, *Anal. Chem.,* **43,** 439 (1971).

134. Harris, W. F., "The Determination of Nitrogen and Oxygen in Niobium," in *Technology of Columbium (Niobium),* B. W. Gonser and E. M. Sherwood, eds., Wiley, New York, 1958, pp. 57–59.

135. Harris, W. F., and W. M. Hickam, *Anal. Chem.,* **31,** 281 (1959).

136. Harris, W. F., M. W. Hickam, M. H. Loeffler, and D. H. Shaffer, Trans. *AIME,* **218,** 625 (1960).

137. Hartcorn, L. A., and R. E. Westerman, *U.S. At. Energy Comm. Rept.* **HW-74949,** November 1963.

137a. Heistand, R. N., *Anal. Chem.,* **42,** 903 (1970).

138. Herwig, D. W., *Stahl Eisen,* **33,** 1721 (1913).

139. Herzberg, G., *Atomic Spectra and Atomic Structure,* Prentice–Hall, Englewood Cliffs, N.J., 1937, Chapter I.

140. *Ibid.,* p. 183.

140a. Hetherington, J. S., U.S. Patent 3,498,105 (Mar. 3, 1970) (Cl. 73–19; Gol$_a$).

140b. Hickam, W. M., and J. F. Zamaria, *Westinghouse Engr.,* **122,** July 1968.

140c. Hill, J. H., C. J. Morris, and J. W. Frazer, *U.S. At. Energy Comm. Rept.* **UCRL-14959,** 1966.

141. Hill, M. L., and E. W. Johnson, *Trans, AIME,* **221,** 622 (1961).

142. Hobart, E. W., Jr., and S. Kallman, ASTM, Committee E-3, Division I, paper presented at Symposium at Atlantic City, N.J., June 28, 1966.

143. Hoekstra, H. R., and J. J. Katz, *Anal. Chem.,* **25,** 1608 (1953).

144. Holler, P., *Arch. Eisenhüttenw.,* **34,** 425 (1963).

145. Holt, B. D., and H. T. Goodspeed, *Anal. Chem.,* **34,** 374 (1962).

146. *Ibid.,* **35,** 1510 (1963).

147. Holt, B. D., and J. E. Stoessel, *Anal. Chem.,* **36,** 1320 (1964).

148. Horrigan, V. M., personal communication.

149. Horrigan, V. M., V. A. Fassel, and J. W. Goetzinger, *Anal. Chem.,* **32,** 787 (1960).

150. Horton, W. S., "Vacuum Methods," in *Treatise on Analytical Chemistry,* I. M. Kolthoff, P. J. Elving, and E. B. Sandell, eds., Interscience, New York, 1961, Part I, Vol. 2, p. 1288.

151. *Ibid.,* p. 1296.

152a. Horton, W. S., and J. Brady, *Anal. Chem.,* **25,** 1891 (1953).

152b. Horton, W. S., and J. Brady, unpublished work, 1954.

153. Howe, H. M., *The Metallurgy of Steel,* 2nd ed., Scientific Publishing Co., New York, 1891, Vol. I, pp. 108, 114.

153a. Hoyt, J. L., C. R. Porter, and C. D. Wilkinson, *Trans. Am. Nucl. Soc.,* **10,** 2 (1967).

154. Hunter, J. P., U.S. Pat. 2,663,561 (Dec. 11, 1956).

155. Hurst, J. E., and R. V. Riley, *Proc. Inst. Brit. Foundrymen,* **42,** A185-188; Discussion, 189 (1949) (Paper 941).

156. *Iron Steel Inst. (London) Spec. Rept.* **62,** "The Determination of Nitrogen in Steel," 1958.

157. Iwaji, I., K. Takashi, and M. Yoshida, *Bull. Chem. Soc. Japan,* **33,** 1340 (1960).

158. Izmanova, T. A., and Yu. A. Klyachko, *Khim. Nauk. i Prom.,* **2,** 528 (1957).

159. Johnson, W. H., *J. Iron Steel Inst. (London),* **23,** 168 (1874).

160. Jordan, L., and J. R. Eckman, *Ind. Eng. Chem.,* **18,** 279 (1926).

161. Kallman, S., and F. Collier, *Anal. Chem.,* **32,** 1616 (1960).

161b. Kallman, S., E. W. Hobart, H. K. Oberthin, and W. C. Brienza, Jr., *Anal. Chem.,* **40,** 332 (1968).

162. Kallman, S. R. Liu, and H. Oberthin, Air Force Materials Laboratory, *Tech. Rept.* **AFML-TR-65-194,** Wright-Patterson Air Force Base August 1965.

163. Kamada, H., and V. A. Fassel, *Spectrochim. Acta,* **17,** 121 (1961).

164. Kamada, J., H. Yamamoto, and Y. Iida, *Japan Analyst,* **2,** 36 (1953).

165. Kamin, G. J., J. W. O'Laughlin, and C. V. Banks, *Anal. Chem.,* **35,** 1053 (1963).

165a. Kammori, O., N. Yamaguchi, and R. Suzuki, *Nippon Kinzoku Gakkaishi,* **32,** 1190 (1968); *Chem. Abstr.,* **70,** 522, Abstr. 83998a (1969).

165b. Karp, H. S., L. L. Lewis, and L. M. Melnick, *J. Iron Steel Inst. (London),* **200,** 1032 (1962).

166. Kashima, J., and T. Yamazaki, *Bull. Chem. Soc. Japan,* **42,** 542 (1969).

167. Kass, S., and D. B. Scott, *J. Electrochem. Soc.,* **109,** 92 (1962).

167a. Kawamura, K., S. Watanabe, T. Otsubo, and S. Goto, *Fuji Seitetsu Giho,* **17,** 1968; *Chem. Abstr.,* **70,** 522, Abstr. 83995x (1969).

168. Kempf, H., and K. Abresch, *Arch. Eisenhüttenw.,* **14,** 255 (1940).

169. *Ibid.,* **17,** 119 (1943).

170. *Ibid.,* **17,** 261 (1944).

170a. Keough, R. F., "Modifications to the Determination of Hydrogen in Sodium by Amalgam Refluxing," paper presented at the 13th Conference on Analytical Chemistry in Nuclear Science and Technology, Gatlinburg, Tenn., October 1969.

171. Kern, W., and G. Brauer, *Talanta,* **11,** 1177 (1964).

172. Kindt, B. H., G. B. Rickman, and A. M. Palmer, *Gen. Elec. Rept.* **DF55SL115,** Schenectady, N.Y., June 22, 1955.

173. Kirshenbaum, A. D., and A. V. Grosse, *Trans. Am. Soc. Metals,* **45,** 758 (1953).

175. Kirshenbaum, A. D., R. A. Mossman, and A. V. Grosse, *Trans. Am. Soc. Metals,* **46,** 525 (1954).

175a. Kirtchik, H., *Anal. Chem.,* **37,** 1287 (1965).

175b. Klopp, W. D., C. T. Sims, and I. R. Jaffee, "Vacuum Reactions of Niobium During Sintering," in *Technology of Columbium (Niobium),* B. W. Gonser and E. M. Sherwood, eds., Wiley, New York, 1958, pp. 111, 116.

176. Klyachko, Yu. A., and T. A. Izmanova, *Zavodsk. Lab.,* **25,** 396 (1959); *Khim, Nauka i Promy.,* **2,** 528 (1957).

176b. Klyachko, Yu. A., L. I. Levi, and O. M. Borisova, *Zavodsk. Lab.,* **32,** 273 (1966).

177. Klyachko, Yu. A., and M. M. Shapiro, *Zavodsk. Lab.*, **23**, 140 (1957).
178. Kraus, T., *Schweiz. Arch. Angew. Wiss. Tech.*, **28**, 506 (1962).
178a. Kraus, T., *Rept.* **6**, Balzers High Vacuum Corp., Santa Ana, Calif., 1966.
179. Kroll, W. J., W. F. Hergert, and L. A. Yerkes, *J. Electrochem. Soc.*, **97**, 258 (1950).
179a. Kulkarni, A. D., R. D. Johnson, and G. W. Perbix, *J. Inst. Metals*, **99**, 15 (1971).
180. Laboratory Equipment Corp., St. Joseph, Mich.
180a. Lapteva, A. S., E. A. Bondarevskaya, and E. N. Merkulova, *Zavodsk. Lab.*, **34**, 1310 (1968).
180b. Larina, O. D., L. A. Domogatskikh, and S. A. Sdvizhkova, *Sb. Tr. Tsentr. Nauchn.-Issled. Inst. Chernoi. Met.*, No. 66, 62 (1969).
180b-1. Lbov, A. A., and I. I. Naumova, *At. Energy (USSR)*, **6**, 330 (1959).
180c. Levi, L. I., and Ya. A. Kitaev, *Zavodsk. Lab.*, **35**, 1438 (1969).
181. Lewis, L. L., and L. M. Melnick, *Anal. Chem.*, **34**, 868 (1962).
181a. Lilburne, M. T., *Analyst*, **91**, 571 (1966).
182. Littlewood, R., *Can. Met. Quart.*, **5**, 1 (1966).
183. Lumley, E. J., *Anal. Chem.*, **34**, 657 (1962).
184. Lutz, G. J., "Determination of the Light Elements in Metals: A Bibiliography of Activation Analysis Papers," *Natl. Bur. Std. (U.S.) Tech. Note* **524**, May 1970.
185. Lyon, W. S., Jr., *Guide to Activation Analysis*, Van Nostrand, Princeton, N. J., 1964.
186. Ma, T. S., and G. Zuazaga, *Ind. Eng. Chem., Anal. Ed.*, **14**, 280 (1942).
187. Maekawa, S., Y. Nakagawa, and M. Shudo, *Tetsu To Hagane*, **44**, 1382 (1958).
188. Malikova, E. D., and Z. M. Turovtseva, *Tr. Komiss, Anal. Khim., Akad. Nauk SSSR*, **10**, 91 (1960), through *Anal. Abstr.*, **8**, No. 3587 (1961).
189. *Ibid.*, **10**, 103, (1960), through *Anal. Abstr.*, **8**, No. 3600 (1961).
190. Mallett, M. W., *Talanta*, **9**, 133 (1962).
191. Mallett, M. W., A. F. Gerds, and C. B. Griffith, *Anal. Chem.*, **25**, 116 (1953).
192. Mallett, M. W., and C. B. Griffith, *Trans. Am. Soc. Metals*, **46**, 375 (1954).
192a. Mallett, M. W., and S. Kallman, "Vacuum Fusion, Vacuum Extraction, and Inert Gas Fusion Techniques," in R. F. Bunshah, ed., *Modern Analytical Techniques for Metals and Alloys*, Wiley–Interscience, New York, 1970, Part I, Vol. III, Chapter 3.
193. Mallett, M. W., D. F. Kohler, R. B. Iden, and B. G. Koehl, *Tech. Rept.* **WAL TR 823/5**, Battelle Memorial Institute, 1962.
193a. Malyshev, V. I., Z. M. Turovtseva, and N. F. Litvinova, *Zh. Analit. Khim.*, **20**, 1214 (1965).
194. Mandel, J., and F. J. Linnig, *Anal. Chem.*, **29**, 743 (1957).
195. Mandel'shtam, S. L., and O. B. Falkova, *Zavodsk. Lab.*, **16**, 430 (1950).
196. Marot, J., *Rev. Met.*, **52**, 943 (1955).
197. Martin, J. F., personal communication.
198. Martin, J. F., J. E. Friedline, L. M. Melnick, and G. E. Pellissier, *Trans. AIME*, **212**, 514 (1958).
199. Martin, J. F., R. C. Takacs, R. Rapp, and L. M. Melnick, *Trans. AIME*, **230**, 107 (1964).
200. Masson, R., and M. S. Pearce, *Trans. AIME*, **224**, 1134 (1962).
200a. Mayer, V., "Determining the Gas Content in Metals," *Proc. 3rd Balzers Conf., 1967*, pp. 56–57, February 1969.

200b. McDonald, R. S., J. E. Fagel, and E. W. Balis, *Anal. Chem.,* **27,** 1632 (1955).
201. McDonnell, H. L., R. J. Prosman, and J. P. Williams, *Anal. Chem.,* **35,** 579 (1963).
201a. McKinley, T. D., personal communication.
201b. McKinley, T. D., *Trans. AIME,* **212,** 563 (1958).
201c. McKinley, T. D., Pittsburgh Conference on Analytical Chemistry and Applied Spectroscopy, 1959.
202. Meacham, S. A., and E. F. Hill, *U.S. At. Energy Comm. Rept.* **ADPA-183.**
202a. Meinke, W. W., "Activation Analysis," this Treatise, Part I, Vol. 7, Chapter 79, 1963.
202b. Menis, O., and J. T. Sterling, "Standard Reference Materials: Determination of Oxygen in Ferrous Materials—SRM 1090, 1091, and 1092," *Natl. Bur. Std. (U.S.) Misc. Pub.* **260-41,** Washington, D.C., 1966.
202c. *Metallurgia,* p. 79 (February 1970).
203. Meyer, R. A., S. B. Austerman, and D. G. Swarthout, *Anal. Chem.,* **35,** 2144 (1963).
204. Mikhailov, G. V., Z. M. Turovtseva, T. S. Khaltov, and V. I. Vannadskii, *Zh. Anal. Khim.,* **12,** 338 (1957).
205a. Millett, E. J., L. S. Wood, and G. Bew, *Brit. J. Appl. Phys.,* **16,** 1593 (1965).
205b. Milner, O. I., and R. J. Zahner, *Anal. Chem.,* **32,** 294 (1960).
206. Müller, F. C. G., *Stahl Eisen,* **4,** 69 (1884).
207. Murray, W. M., and S. E. Q. Ashley, *Ind. Eng. Chem., Anal. Ed.,* **16,** 248 (1944).
207a. Nakagawa, Y., and Y. Shiga, *Tetsu To Hagane,* **53,** 549 (1967); Abstract 77901, *J. Iron Steel Inst. (London),* **206,** 1283 (1968).
208. Namiki, M., Y. Kakita, and H. Goto, *Talanta,* **11,** 813 (1964).
209. Naughton, J. J., and H. H. Uhlig, *Ind. Eng. Chem., Anal. Ed.,* **15,** 750 (1943).
210. Newell, W. C., *J. Iron Steel Inst. (London),* **141,** 243P (1940).
211. *Ibid.,* **152,** 333P (1945).
212. Orsag, J., *Rev. Met.,* **52,** 237 (1955).
213a. Osmond, R. G., and A. A. Smales, *Anal. Chem. Acta,* **10,** 117 (1954).
213b. Pajot, B., *Solid State Electronics,* **12,** 923 (1969).
213c. Pargeter, J. K., *J. Metals,* **20,** 27 (1968).
214. Parker, A., "The Determination of Gases in Metals," *J. Iron Steel Inst. (London) Spec. Rept.* **68,** p. 64, 1960.
215a. Parry, J., *J. Iron Steel Inst. (London),* **1881,** 183.
215b. Pasztor, L. C., and D. E. Wood, *Talanta,* **13,** 389 (1966).
216. Pearce, M. L., *AIME Trans.,* **227,** 1393 (1963).
217. Pepkowitz, L. P., and W. C. Judd, *Anal. Chem.,* **22,** 1283 (1950).
218. Pepkowitz, L. P., and E. R. Proud, *Ind. Eng. Chem., Anal. Ed.,* **21,** 1000 (1949).
218b. Perriton, R. C., and F. R. Coe, *Metallurgia,* **78,** 43 (1968).
219. Peterson, D. T., and D. J. Beernstein, *Anal. Chem.,* **29,** 254 (1957).
220. Peterson, D. T., and V. G. Fattore, *Anal. Chem.,* **34,** 579 (1962).
221. Peterson, J. I., F. A. Melnick, and J. E. Steers, Jr., *Anal. Chem.,* **30,** 1086 (1958).
222. Pierson, H. R., and E. A. Fay, *Anal. Chem.,* **31,** 25A (1959).
223. Ponomarev, A. I., and A. A Astanina, *Zh. Anal. Khim.,* **14,** 234 (1959).
224. Ransley, C. E., *G.E.C. J.,* **11,** 135 (1940).
225. Ransley, C. E., and D. E. J. Talbot, *J. Inst. Metals,* **84,** 445 (1956).

226. Read, E. B., and L. P. Zopatti, *Repts.* **MIT-1038** and **AECD-2798**, Feb. 14, 1950.

226a. Roboz, J., *J. Chem. Educ.*, **48**, A9 (1971).

226b. Rommers, P. J., and J. Visser, *Analyst*, **94**, 653 (1969).

227. Rooney, T. E., and A. G. Stapleton, *J. Iron Steel Inst. (London)*, **131**, 249 (1935).

228. Sachs, K., "The Determination of Gases in Metals," *J. Iron Steel Inst. (London) Spec. Rept.* **68**, p. 243, 1960.

229. Sachs, K., and M. Odgers, *Nature*, **197**, 373 (1963).

230. Sandell, E. B., "Errors in Chemical Analysis," this Treatise, Part I, Vol. 1, Chapter 2.

230b. Sannier, J., and J. Leroy (Commissariat a l'Energie Atomique), *Rapp.* **CEA R-2957** (French), 1966.

231. Savitskii, E. M., H. E. Chuprikov, and H. H. Glavin, *Zavodsk. Lab.*, No. 8, 957 (1962).

232. Schmidt-Collerus, J. J., and A. J. Frank, *WADD Tech. Rept.* **60-482,** March 1961.

233. Scribner, B. F., and M. Margoshes "Emission Spectroscopy" This Treatise, Part I, Vol. 6, Chapter 64.

234. Shanahan, C. E. A., *J. Iron Steel Inst. (London) Spec. Rept.* **68,** pp. 75–92, 1960.

235. Shanahan, C. E. A., *Rev. Met.*, **58**, 55 (1961).

235a. Shemelev, B. A., and T. V. Ryutina, *Zavodsk. Lab.*, **35**, 422 (1969).

236. Shields, B. M., J. Chipman, and N. J. Grant, *J. Metals*, **5**, 180 (1953).

237. Shirley, E. L., unpublished work.

237b. Shitikov, V. S., and N. A. Gederevich, *Zavodsk. Lab.*, **33**, 1713 (1967).

238. Short, H. G., *Analyst*, **75**, 335 (1950).

239. Shvaiger, M. I., *Zavodsk. Lab.*, **26**, 1223 (1960).

240. Silverman, L., and W. Bradshaw, *Rept.* **NAA-SR-2633,** North American Aviation, Inc.

241. Silverman, L., *Anal. Chem. Acta*, **8**, 436 (1953).

242. Singer, L., *Ind. Chem., Anal. Ed.*, **12**, 127 (1940).

243. Sloman, H. A., *J. Iron Steel Inst. (London) Spec. Rept.* **25,** p. 43, 1939.

244. Sloman, H. A., *J. Inst. Metals*, **71**, 39 (1945).

245. *Ibid.*, **71**, 391 (1945).

246. Sloman, H. A., *J. Iron Steel Inst. (London)*, **182**, 307 (1956).

247. Sloman, H. A., C. A. Harvey, and O. Kubaschewski, *J. Inst. Metals*, **80,** 391 (1952).

248. Smiley, W. G., *Anal. Chem.*, **27**, 1098 (1955).

249. Smiley, W. G., Symposium on Determination of Gases in Metals, *ASTM Spec. Tech. Publ.* **222**, p. 25, 1957.

249a. Smith, G. A., W. C. Lenahan, and D. S. Macleod, *Metallurgia*, **79**, 121 (1969).

250. Smith, M. E., J. M. Hansel, and G. R. Waterbury, *Anal. Chem.*, **37**, 782 (1965).

251. Smith, W. H., *Anal. Chem.*, **27**, 1636 (1955).

252. Somiya, T., S. Hirano, H. Kamada, and I. Ogahara, *Talanta*, **11**, 581 (1964).

253. Speight, G. E., and R. M. Cook, *J. Iron Steel Inst. (London)*, **160**, 397 (1948).

254. Speight, G. E., and E. W. Gill, *Metallurgia*, **55**, 155 (1957).

255a. Stanley, J. K., J. von Hoene, and G. Wiener, *Anal. Chem.*, **23**, 377 (1951).

255b. Sterling, J. T., F. J. Palumbo, and L. L. Wyman, *J. Res. Natl. Bur. Std.*, **66A**, 483 (1962).

256. Stevenson, W. W., and G. E. Speight, "7th Report on the Heterogeneity of Steel Ingots," *J. Iron Steel Inst. (London) Spec. Rept.* **16**, p. 65, 1937.

257. Stevenson, W. W., and G. E. Speight, *J. Iron Steel Inst. (London)*, **143**, 326P (1941).

258. Still, J. E., "The Determination of Gases in Metals," *J. Iron Steel Inst. (London) Spec. Rept.* **68**, p. 43, 1960.

259. Ströhl, G., *Z. Anal. Chem.*, **179**, 259 (1961).

259a. Suarez-Acosta, R., "Determining the Gas Content in Metals," *Proc. 3rd Balzers Conf., 1967*, pp. 40–41, February 1969.

260. Sucha, M., *Hutnicke Listy*, **15**, 729 (1960).

261. Sullivan, T. A., *U.S. Bur. Mines Rept. Invest.* **5834**, 1961.

262. Sventitskii, N. S., K. A. Sukenkho, P. P. Galenov, O. B. Fal'kova, M. C. Alpatov, and K. I. Taganov, *Zavodsk. Lab.*, **22**, 668 (1956).

262b. Svet, V. I., and I. K. Kozlova, *Zavodsk. Lab.*, **33**, 715 (1967).

262c. Swinburne, D. G., and P. J. Melia, *Metallurgia*, **81**, 37 (1970).

263. Task Force on Oxygen-Panel of Methods of Analysis of the Metallurgical Advisory Committee on Titanium, "Recommended Inert Gas Fusion Method (Platinum Flux Technique) for the Determination of Oxygen in Titanium Metal and Alloys," June 8, 1962.

264a. Taylor, R. E., *Anal. Chim. Acta*, **21**, 549 (1959).

264b. Tedmon, C. S., Jr., personal communication.

265. Thomas, D. E., in *The Metallurgy of Zirconium*, B. Lustman and F. Kerze, Jr., eds., McGraw–Hill, New York, 1955, Chapter 11, Part II, pp. 29, 617–618.

266. Thompson, B. A., *Anal. Chem.*, **33**, 583 (1961).

267. Thompson, J. F., and G. R. Morrison, *Anal. Chem.*, **23**, 1153 (1951).

268a. Thompson, J. G., H. C. Vacher, and H. A. Bright, *J. Res. Natl. Bur. Std.*, **18**, 259 (1937).

268b. Thurber, W. R., "Determination of Oxygen Concentration in Silicon and Germanium by Infrared Absorption." *Natl. Bur. Std. (U.S.) Tech. Note* **529**, May 1970, Government Printing Office, Washington, D.C., 20402 (30 cents).

269. Tighe, J. J., A. F. Gerds, E. J. Center, and M. W. Mallett, *Battelle Memorial Inst. Rept.* **EMI-799**, Dec. 29, 1952.

270. Torrisi, A. F., and J. L. Kernahan, *Anal. Chem.*, **23**, 928 (1951).

271. Torterat, P., L. Backer, and E. Herzog, *Rev. Met. Mem. Sci.*, **57**, 493 (1960).

271a. Turkdogan, E. T., and H. E. Fruehan, paper presented at 76th General Meeting of AISI, New York, May 23, 1968.

272. Turkdogan, E. T., and S. Ignatowics, *J. Iron Steel Inst. (London)*, **185**, 200 (1957).

273. Turovtseva, Z. M., and L. L. Kunin, *Analysis of Gases in Metals*, Academy of Sciences, USSR, Moscow (Trans. Consultants' Bureau, New York), 1959.

274. Turovtseva, Z. M., and N. F. Litvinova, *Proc. 2nd Intern. Conf. Peaceful Uses At. Energy*, **28**, 593 (1958).

274a. Turovtseva, Z. M., and A. M. Vasserman, *Zh. Analit. Khim.*, **20**, 1359 (1965).

274b. Ul'yanov, A. I., *Zavodsk. Lab.*, **34**, 1442 (1968).

274c. Vasil'ev, D. M., A. I. Mel'ker, and V. V. Trofimov, *Zavodsk. Lab.*, **33**, 1758 (1967).

274d. Vasserman, A. M., Yu. N. Surovoi, L. L. Kunin, and V. A. Mansurov, *Zavodsk. Lab.,* **34,** 919 (1968).

275. Venkatesvarlu, C., and M. W. Mallett, *Anal. Chem.,* **32,** 1888 (1960).

276. Venkatesvarlu, C., and M. W. Mallett, *Talanta,* **5,** 283 (1960).

277. Von Oberhoffer, P., and H. Schenk, *Stahl Eisen,* **47,** 1526 (1927).

277a. Walker, J. A. J., E. D. France, and W. T. Edwards, *Analyst,* **90,** 727 (1965).

278. Walker, P. L., Jr., F. Rusinko, and L. G. Austin, "Gas Reactions of Carbon," in *Advances in Catalysis,* Academic, New York, 1959, Vol. XI, p. 135.

279. Walter, D. I., *Anal. Chem.,* **22,** 297 (1950).

280. Walter, R. J., and H. G. Offner, *Anal. Chem.,* **36,** 1779 (1964).

280a. Weatherburn, M. W., *Anal. Chem.,* **39,** 971 (1967).

281. Wells, J. E., and K. C. Barraclough, *J. Iron Steel Inst. (London),* **155,** 27 (1947).

282. Wilkins, D. H., and J. F. Fleischer, *Anal. Chim. Acta,* **15,** 334 (1956).

282a. Wilkinson, C. D., and G. F. Hoffman, *Precision Metal,* **27,** No. 2, 39 (1969).

282b. Willis, R. R., *Metallurgia,* **79,** 129 (1969).

283. Winge, R. K., and V. A. Fassel, *Anal. Chem.,* **37,** 67 (1965).

283a. Wood, D. F., and G. Wolfenden, *Anal. Chim. Acta,* **38,** 385 (1967).

284. Wood, G. B., *J. Electrochem. Soc.,* **110,** 867 (1963).

285. Wrzesinska, E., *Prace Inst. Hutniczych.,* **10,** 180 (1958).

285a. Yamaguchi, N., T. Sudo, and O. Kammori, *Bunseki Kagaku,* **18,** 13 (1969); *Chem. Abstr.,* **70,** 522, Abstr. 83996y (1969).

286. Yeaton, R. A., *Vacuum,* **2,** 115 (1952).

287. Youden, W. J., "Accuracy and Precision: Evaluation and Interpretation of Analytical Data," this Treatise, Part I, Vol. I, Chapter 3.

288. Yudowitch, K. L., *Armour Res. Foundation Rept.* **NP-4486,** July 25, 1952.

289. Zaidel, A. N., *Zh. Tech. Fiz.,* **25,** 2571 (1955); U.S. Atomic Energy Comm. Rept. **TR-3162.**

SECTION F: Principles of Instrumentation

Part I
Section F

Chapter 104

GENERAL CONCEPTS OF INSTRUMENTATION FOR CHEMICAL ANALYSIS

By Gordon D. Patterson, Jr., E. I. du Pont de Nemours & Company, Wilmington, Delaware

"The most powerful instrument in the cause of science is the human mind."

Contents

I. CONCEPTS OF MODERN ANALYTICAL INSTRUMENTATION

While justifiably resisting the inclination to regard instruments as ends in themselves, the competent analytical chemist recognizes the advantages to be gained from adequate familiarity with all tools useful in his work. Electrical instruments especially have assumed such a strong role in analytical chemistry that separate consideration* is necessary in any comprehensive treatment of analysis. Their complexity immediately raises the question, What is adequate familiarity when the analyst plugs in his instruments and begins to turn the dials?

Instrumentation, like analytical chemistry itself, is a discipline which spans the boundaries separating the fields of science. Wildhack (195), under the title, "Instrumentation in Perspective," very early joined others in viewing instrumentation as a field unto itself and spoke, even in 1950, of the "science of instrumentation," agreeing with Condon's (38) affirmative answer to "Is There A Science Of Instrumentation?" Condon emphasized the common elements in the problems, devices, and methods of measurement. Wildhack claimed that the progress of science and technology "can be greatly accelerated by focussing specifically on the wider exploitation of existing techniques of measurement and on the improvement and development of new instruments and methods for measurement and control."

Ten years later Klopsteg (99) went even further, saying that "instruments are unifying elements which help self-centered disciplines shed their isolationism." His further expositions in his presidential address to the American Association for the Advancement of Science revealed a philosophy worth rereading. Strengths which come from innovation in joining together old things in new ways, Klopsteg said, are especially present when the scientist's interest in an instrument leads him to extend his curiosity beyond his immediate field. A scientist's use of instruments may be akin to a lawyer's use of the law. Experimental science operates through the use of instruments, because they are essential tools for measuring the physical properties on which every chemical analysis depends.

Müller, in his Beckman award address (135), traced the course of

---

* This chapter emphasizes concepts only and refers the reader to appropriate literature sources for technical details such as graphical and diagrammatic descriptions.

instrumentation in analytical chemistry, starting with the early efforts at gadgeteering. He advocated a unified and general approach to the instrumentation of any phenomena and claimed that instrumentation is largely responsible for recent technological growth.

An appreciation for the historical development of electrical and optical components and also for their basic theory will assist the user of an instrument to make it perform as he wishes. Blind acceptance of the meter or recorder reading or the computer typeout without knowledge of the factors affecting such readouts can lead to serious errors and misinterpretations. Furthermore, instruments often have built-in flexibility which enables them to perform tasks not apparent to the uninformed user. It is the purpose of these chapters to consider mainly the nonchemical, or "hidden," factors in instrumentation. Also included are brief descriptions of laboratory instruments and their applications that illustrate key points and information about their use to help analysts perform more effectively (see also 125,129,149).

An adequate understanding of the fundamentals of analysis includes emphasis on possibilities and limitations. These in turn rest heavily on instrumentation theory, as well as on chemical theory and practice. Trends toward specialization in narrowly defined techniques tend to inhibit this thorough understanding, resulting in analytical chemists not properly prepared to meet a "real" problem head-on. Thus the viewpoint in these chapters will aim to be selective rather than specialized and complete.

The reader will find that, mainly since 1930, the analytical chemist has adopted the tools of physics for *manipulating energy*, especially radiant and electrical energy, into, through, or out of his sample. In some respects, the early growth of this trend was spotty and illogical. Furthermore, analytical electrical instrumentation is only part of a much larger total technology. The instrument industry consists of more than 3000 firms in the United States alone, producing annually more than $10 billion worth of hardware. Its growth rate has averaged 5 times that of industry as a whole. A comprehensive view of the whole field can result in a recognition of merits (and demerits), and consequent pertinent choices, in meeting specific chemical problems.

A. PHYSICS

The tools of physics are useful simply because it is the magnitudes of physical properties and their effects on energy that the analyst uses in arriving at his quantitative answer. Like the physicist, the modern analyst finds electronic components playing an important part in the "proper manipulation" of the various forms of energy producing his answers, both quantitative and qualitative. However, this section will attempt to place both electronic and radiant-energy devices in their proper perspective, along with other physical parameters to be encountered. A survey of components behind the little "black box," intended to remove the mystery surrounding such terms as "diodes," "triodes," "transformers," "transducers," "signal/noise," "response times," "amplifiers," "scalers," triggers," "feedback," "bridges," and "oscillographs," will be followed by illustrations of how these concepts are put together for useful chemical applications.

B. WHY USE INSTRUMENTS?

Instruments are used for exactly the same reasons that other tools are used. They either can do a job which is otherwise impossible or can do it better (easier, faster, cheaper, more accurately) than is otherwise possible. The modern scientist uses any apparatus to achieve observations beyond the range of his unaided senses and to create powers of manipulation greater than those in his own bare hands. Of course, the question of economics enters into both the original decision to resort to an electrical instrument and the choice among several available. The "cheapest" apparatus available for a given job may not always be the one whose initial purchase price is lowest.

Lewin (104), in his discourse on "Proper Utilization of Analytical Instrumentation," said that the "bedrock upon which all of our chemical knowledge and capability rests is the reliability of the data supplied from analytical instrumentation." It behooves chemists to know the characteristics, potentialities, limitations, and idiosyncrasies of all their tools, especially the "black-box" instruments. He also said correctly that colleges and universities often offer courses grossly misnamed "Instrumental Analysis" just because they use black boxes. All chemical analysis is instrumental in nature, since it is a physical property whose magnitude is finally measured.

Actually, one of the most perplexing problems facing analytical chemists is how to know in advance and without prior experience what a new device can (or should) do. Not only the newer techniques but also many older ones rest upon shaky, or at least incompletely defined, theoretical grounds, so that selecting the proper set of characteristics necessary in an instrument becomes too often a matter of shrewd guessing.

One temptation is to buy or build maximum versatility. Unfortunately, this quality is usually accompanied by complicated operation, increased maintenance, and possible compromise of sensitivity. Instrument manufacturers generally attempt both to describe the fundamental characteristics and capabilities of various models and to anticipate, before design, the major analytical variables of interest.

Proper provision for accessory modifications will always be a factor in making the "right choice." In research, economic justification for a specific approach may best be answered in terms of long-range considerations because of the very unpredictability of analytical problems of the future. The purchaser will always be torn in at least two directions (versatility versus simplicity), and the compromise must be based on an adequate "feel" for the basic aspects of the instrument as well as the chemistry of the analytical problems anticipated.

II. HISTORICAL REVIEW

Although it is neither possible nor practical to identify the first instrument in history, it is useful to recollect the stages which formed the foundations for present-day instrument technology. Historically, simple instruments and tools like the lever and the pulley, which multiplied man's muscles, were followed by mechanical power, which multiplied his energy. The third stage may be considered that of *measuring* instruments and automatic controls which broadly multiply man's capacity by taking over routine and repetitive tasks of observation.

An interesting survey of the historical aspects of the development of measuring instruments is contained in Klopsteg's AAAS presidential address (99), in which his picture of instruments as the indispensable tool of science was supported not only by statistics on the number of Nobel Laureates (112 out of the 138 in physics and chemistry from 1901 to 1960) whose works were particularly dominated by instruments, but also by an eloquent narration of pre-Nobelian science landmarks. These include the efforts of such notables as Democritus, Aristotle, Galileo, Torricelli, Hooke, Thompson, Rutherford, Hales, Volta, Coulomb, Cavendish, and van Leeuwenhoek, whose leadership in using *experimental* evidence to support theory added importantly to the development of the instruments and the *experimental* techniques eventually to be identifiable as a separate science.

Other narratives are given in, for example, Michel's *Scientific Instruments in Art and History* (126); this account includes antique instruments, illustrated in vivid color to demonstrate that they were more than functional. Composed of costly and beautiful materials, richly embellished, they were works of art and of masterful craftsmanship. This work records Pascal's first adding machine (1642), Giovanni Polim's first automatic calculating machine (1709), the first machine for frictionally generating electricity (1663), and other landmark instruments. Wheatland's *The Apparatus of Science at Harvard, 1765–1800* (192), describes a remarkable collection of historical instruments of Harvard University. Photographs of a wide range of instruments significant to astronomy, physics, and chemistry in the laboratory are given.

Going back to the ancients, we find that Alexandrian astronomers were responsible for the first major efforts in precision instrument-making, which peaked with the work of Ptolemy and then faded with the rest of classical science. Again, when Greek science passed to the Arabic-speaking peoples, it was astronomy which provided the impetus for developing better instruments. These bursts of interest and those which followed in European countries up through the fourteenth century never succeeded in establishing any sustained tradition in instrument-making. Although the beginnings of a renaissance in this craft occurred in Germany in the late fifteenth century, it remained rather limited in its applications to astronomy (and later surveying).

The scientific revolution of the seventeenth century had as one of its main features the development of new tools (instruments) for the scientist, who thereby at last became able to check both old and new theories and to open rich worlds to his experience. There developed a transition away from strict custom-inventing instrument construction, tailored to the specific demands of those able to pay. This transition was accelerated by a sudden new popularity of science through publications in books and journals that began to have an audience even among amateurs. Demands for mass production followed promptly.

A. ANALYTICAL BALANCES

In man's historical advances in techniques of measurement, progress in measures of weight followed developments in measures of length. The earliest instrument of interest to analysts would be the balance. The earliest commercial use of weighing probably occurred about 2500 B.C. in the pre-Aryan Indus civilization of northern India and perhaps to a limited extent in Mesopotamia. In Egypt the use of the balance in ordinary trade apparently did not begin until 1350 B.C., although pictures of balances in use by Egyptian goldsmiths date back to about 2500

B.C. Weighing does not appear to have extensively preceded the use of gold, although the oldest known weights are limestone specimens found in prehistoric graves at Naqada in Egypt, and all weights were of polished stone until about 1450 B.C.

Before the introduction of the metric system in Europe in the nineteenth century, various nations differed not only from each other, but also among their own principal cities, and different standards applied for various kinds of merchandise (one for bulky materials, one troy weight for gold, another for silver, etc.). The preservation of specimen weights and measures as material reference standards, inscribed with the king's name and deposited in the principal temples, was initiated at least as early as 3000 B.C.

Records show that by A.D. 780 coiners in a Muslim mint could weigh to within a third of a milligram. "To reach such accuracy it was needful to use the finest chemical balance, with closed case, double weigh the glass weights against each other, and read a long series of swings of the balance" (147).

There were only a few exceptions to the early alchemist's ignorance of precise balances. One consisted of pictures in Thomas Norton's *Ordinall of Alchimy*, dating from 1477, which shows the earliest known representation of a balance in a glazed case.

By the sixteenth century the typical assayer's equipment included three separate balances of different capacities and sensitivities as good as 0.1 mg. By that time knife-edges were protected by means of beam-lifting devices. Significant modifications in design were introduced during 1770–1775 in balances built by John Harrison (English clock-maker) for Henry Cavendish. He added such features as triangular structure of the beam (for rigidity), movable nuts to allow control of the center of gravity, triangular knife-edges, and the beam-lift at-rest arrangement. Lavoisier benefited from similar advances introduced in somewhat smoother fashion by Megnie. The late eighteenth and early nineteenth centuries saw significant growth in the balance-maker's trade, and the successful experimental work of many of the chemists of this period was due in large degree to the precision balances then becoming available.

An interesting account of the development of the chemical balance insofar as British science is concerned was written by Stock (172). Part of a series of essays, this monograph starts with the work of Black and Lavoisier and the emergence of the concept of quantitative studies needing scientific balances. Stock details the usage of balances from the sixteenth to the nineteenth centuries, including the swan-neck beam balance of Ercker (1574), the application of agate planes as bearings in balances for Henry Cavendish, and the work of Thomas Robinson, first British manufacturer of precision balances.

Many centuries were necessary for the acquisition of the necessary constructional skills to fabricate parts for measuring instruments requiring precise positioning, later to be utilized by analytical chemists. Among these parts were screws. Short screws, both coarse and fine, of metal or wood were commonly used in scientific instruments after 1650, but long lead screws, less frequently employed, were expensive and often inaccurate. This and other mechanical features so necessary in measuring instruments had to await the perfection and wide availability of precision lathes and gear-cutting machines, which accompanied the development of machine tools in the nineteenth century.

B. TEMPERATURE-MEASURING DEVICES

Crude forms of temperature measurement can be traced back to very early times. The expansion of matter by heat was described by Philo of Byzantium (second century B.C.) and Heron of Alexandria (first and second centuries B.C.), although measurement of the amount of heat does not seem to have been their objective. G. della Porta (Naples, 1589) devised a simple thermometer involving the expansion of air when heated by fire, the air displacing water from a retort. Galileo, however, is commonly credited with the first real thermometer (about 1592), utilizing wine in a stem also containing air which expanded when heated. He used the term "degrees" and described the bulb as having a capacity of "three or four drinking glasses."* Sealed thermometers independent of atmospheric pressure and having fixed points capable of accurate comparison were not developed for at least another century.

The word "thermometer" itself is first found in a description by a Jesuit priest, Father Leurechon (1624). It was after the experiments of Torricelli in 1643 that the barometer and the thermometer came into popular use. The early history of both of these instruments parallels the rise of glass-blowing craftsmanship. The rest of the seventeenth century saw increasing recognition of the thermometer's potentialities, particularly in the medical profession, although it was not until 1714 that G. D. Fahrenheit of Danzig devised the scale used today. Fahrenheit, who had begun to make instruments in Amsterdam in 1706, was among

---

* A gift of one of these to an English doctor evoked the reply, "Your gracious gift of rare Italian wine has been gratefully received. When shall I expect the receipt of your new instrument?"

the first to fill the need for standard instruments for measuring temperature. Professor A. Celsius (Uppsala, 1742) and his colleagues introduced the Celsius (or centigrade) scale used in scientific work in most of the world. In 1948 the General Conference on Weights and Measures recommended adoption of the name "Celsius" to designate this scale in preference to the other names. Further interesting historical discussion on the development of the thermometer is contained in Middleton's book (127).

C. POTENTIOMETERS AND OTHER ELECTROANALYTICAL COMPONENTS

The potentiometer, which dates from Poggendorff's first description in 1841, is the oldest precision current- and voltage-measuring device. Rapid development and refinement followed in Germany by such men as Brooks, Wenner, Diesselhorst, Hausrath, Feussner, Lindeck, and Rothe. The Feussner–Brooks high-internal-resistance design, dating from 1890, offers measuring accuracy of ±0.005% in the range above 100 mV. The Diesselhorst–Hausrath instrument, first designed in 1908, features low internal resistance for accuracies of ±0.003% when differences of a few microvolts are to be measured. The cascade, or Kelvin, potentiometer subsequently developed is intermediate between the two, being suited for measurements in the millivolt or tenths-of-millivolt range with accuracies of ±0.01% or better. These instruments greatly affected the growth of electrochemistry and associated electrical methods of analysis.

Nevertheless the slow pace of the development of work on the dissociation of electrolytes in water, ranging from the efforts of Arrhenius at the end of the nineteenth century to G. N. Lewis' research on ionic activity, predicted the historical difficulties in establishing acceptable pH scales on an absolute basis. Wallis' account (187) mentions the pressures during the 1950s to study many electrode systems, emphasizing continuous pH-recording systems for plant control. Along with this came increased concern in regard to the pollution of rivers.

Efforts for a simple concept of pH in all environments shifted to the production of reliable instruments for a variety of conceivable purposes. Needs for measuring pH, at extremes of temperature and pressure in particular, also developed with demands to check on corrosion by high-temperature water in the atomic energy industry, to monitor sulfur dioxide solutions for the paper industry at temperatures up to 130°C, to check sterilization of pharmaceutical culture media at 125°C, to monitor underground water pH and other aspects of petrochemical processing, and so forth. Much of the successful development of the glass electrode, now able to operate in conditions of up to 150°C and 1000 kg/cm² pressure, stems from the work of Eisenman et al. (48a), who considered that the fundamental factor controlling the selectivity of glasses to ionic solutions is the free energy of association of the exchanging ions to the oxygen ions in the glass lattice.

Oddly enough, one of the earlier forerunners of *modern* electroanalytical instrumental analyses was connected, not with conventional photometric and titrimetric techniques, but with Heyrovsky's polarography (1922), which was introduced to the United States in 1932, subsequently was reviewed with enthusiasm by Kolthoff in 1939, and eventually earned Heyrovsky the Nobel prize in 1959. He disclosed the quantitative and qualitative behavior of a dissolved electrolyzable sample, as the applied voltage varies to produce a sigmoidal plot of current passing between the reference electrode and a polarizable electrode formed (usually) by a series of mercury drops. This led to a wide extension of an already significant electrochemical science of both inorganic and organic materials.

According to Bottom (20), Julius E. Lilienfeld patented the invention of the solid-state amplifier more than two decades before it was rediscovered and made into a practical device. The first of three relevant patents was filed on October 8, 1926, under the title "Method and Apparatus for Controlling Electric Currents" and was granted as U.S. Patent 1,745,175 on Jan. 28, 1930. From the description and the circuitry Bottom concludes that the device was what we now call an NPN transistor. Two factors may be responsible for the delay of nearly a quarter-century in practical follow-up: the nonavailability of good semiconducting material, and inadequate theories to explain the operation of the devices and to guide further development and application.

D. SPECTROMETRY

Starting with Wollaston's observation of dark solar lines in 1802, the nineteenth century saw significant development of the understanding of spectra. McGucken (121) has emphasized the gain in *understanding* more than the progress in apparatus, but the former aspect of the history of spectroscopy is, in many ways, as relevant as the development of its instrumentation.

Learner's paper (102) describes the improvements in spectrograph design during the 50-year period from 1918 to 1968. Before 1918, most of the familiar instruments had already been developed, namely, those of Littrow in 1862, Rowland in 1883, Abney in 1886, Ebert in 1889, Wadsworth in 1896, Paschen in 1902, and Eagle in 1910, all well described in the standard texts.

The period since 1918 has seen mainly improvements in the performance and understanding of spectrographs developed *before* that time (except for the grazing incidence and X-ray spectrographs), including vacuum techniques, photodetectors, data-processing devices, and diffraction gratings. Among other things Learner regrets the occasional rediscovery of the properties of a plane grating in convergent light (e.g., Monk in 1928, Gillieson in 1949, Richards et al. in 1957).

The field of X-ray spectrometry has benefited from the early start given it by the valuable (and not accidental, as sometimes supposed) discovery of X-rays by Röntgen, who received the first Nobel prize in physics for this achievement. Müller (134) and others have pointed out that Röntgen knew exactly what he was doing. He compares Röntgen's genius, skill, and originality in this work to the similar curiosity and thoroughness of Henri Becquerel, whose fogged photographic plates were observed by an *un*fogged mind in his work with uranium compounds.

E. RECORDERS

Recorders, now taken so much for granted, have been briefly reviewed by Keinath (93). R. von Voss (1908) built the first rectilinear current recorder with linear scale and ink recording. A mirror-type light-beam oscillograph having a natural frequency of 6000 Hz was in production in Germany in 1910. Five years later, a high-speed moving-coil electrocardiograph was available. High-sensitivity potentiometer recorders were an American development during the 1920s. The Japanese developed the first 12,000-Hz galvanometer oscillographic recorder in 1934.

The perfection and widespread use of *X-Y* multichannel recorders has come since 1942. Along with it, there has developed a proliferation of edge-wise or profile instruments, round instruments with 90° or 30° deflection, small ones and big ones, disk charts, strip charts, all of different appearance and maintenance problems. Modifications such as automatic standardizing and scale expansion have also become available.

F. HISTORICAL TRENDS

A perusal of the mid-twentieth-century history of instrumentation reveals numerous examples of mechanizations of manual operations, sometimes very crude. For example, an early semiautomatic combustion device for elementary analysis was described by Verdery (186). It kept a microburner moving back and forth along the combustion tube by mounting it on a car of a toy electric train, which was kept moving on its track by means of a modified worm-gear connected to the motor of an old record player. There are descriptions in the literature also of fraction collectors ranging from a simple merry-go-round to very complex precision-timed positioning devices.

Another trend readily discerned is the movement of laboratory techniques into factories. Early, rather naive attempts were often unsuccessful because of lack of recognition both of limitations of existing components and of special needs imposed by continuous (or intermittent) measurements. In time, a whole new technology evolved to speed up this transition from laboratory to plant so that lead time can now be strikingly low when analysts and engineers work together in meeting all requirements.

A case in point may be cited in the development of process refractometers. The utility of the refractive index as a physical property for analysis of two-component systems was recognized for many decades. Not until 1950, however, did several process firms build their own continuous refractometers, and 5 more years passed before significant commercial devices were offered for sale. Contrast this with the existence in 1957 of five instrument-makers offering process gas chromatographs, only 14 years after the first U.S. commercial laboratory instrument was developed, following closely after Martin and Synge's first suggestion of the gas–liquid variant of chromatography (119).

It is clear that applicability to process control will continue to accelerate the perfection of analytical instrumentation because of economic factors. Conversely, it is perhaps ironic to note the delayed progress of laboratory automation, for example, the long years of waiting for versatile automatic titrators and recording balances of adequate precision and sensitivity, in spite of the excellent groundwork reported in the technical literature by academic researchers.

III. GENERAL DEFINITIONS AND CLASSIFICATIONS

A. INSTRUMENTATION

As for defining the word "instrumentation," there is Wildhack's (195) position that the "science of instrumentation may be recognized without being very exactly defined, having very fuzzy edges like most other branches of science." Klopsteg said (99) that generally "an instrument is an amplifier for sensory perception and a means for measuring whatever attributes of a physical entity are susceptible of quantitative treatment." Wildhack characterized the instrumentation field first by its relation to the other sciences, second by its content, and third by its goals. He viewed instrumentation as a horizontal field, encompassing segments of practically all other sciences. Just as the whole is often greater than the apparent sum of its parts, so is the science of instrumentation greater than the sum of the individual phases inherent in its fields, for instance, of electricity, mechanics, electronics, and optics.

Further useful discussions regarding the content of instrumentation science are well presented in such papers as McNish's "Fundamentals of Measurement" (122), Huntoon's "Concept of a National Measurement System" (81), and Howlett's "International Basis for Uniform Measurement" (79). Howlett contends that "without continual improvement in the precision and accuracy of measurement, progress in science and industry would be at first critically handicapped and, in due course, stopped."

The concept of an *analytical instrument as any tool designed to perform laboratory measuring operations better than manual skill alone* is a broad one that may be interpreted to include everything from spatulas to nuclear magnetic resonance spectrometers. Conversely, the assumption that an analytical instrument necessarily contains electrical circuitry excludes many useful tools that deserve consideration. The view taken here is that an *instrument for chemical analysis comprises any assembly of components intended to perform operations contributing to the qualitative and quantitative analysis of a sample.*

Acquiring the *"component* approach" early will help in understanding the functioning of the multiplicity of instruments which the modern chemist will encounter. Likewise the essential idea of measurement *operations* may seem obvious to some but is useful to emphasize, because it permits arranging the broad subject into a logical classification incorporating understandable categories. Only then can the laws and generalizations required for proper utilization be formulated.

Analytical instruments, then, are mechanical, optical, or electrical devices used in performing certain operations in the identification and measurement of the material under observation. *Instrumentation* comprises the science and technology of making and using the tools known as instruments.

These definitions, of course, tie instruments to the operation of *measurement,* which, itself, is not simply defined. Dingle (45) has suggested that it is "any precisely specified operation that yields a number; that is, measurement is related to an operation performed." The theme of a symposium on basic problems published in the book *Measurement: Definitions and Theories* (34) was that measurement has forced its way into almost every field of study for the reason that precise description usually requires it. This is most true in the physical sciences, and analytical chemistry accomplishes its contribution entirely through the use of instruments in their broadest sense.

Furthermore, in general, we may say that measurement per se requires the disturbance, by some form of energy, of the system being measured. It is the ensuing result of this disturbance of the energy that gives us the quantitative (and often qualitative) output information desired. Usually the effect of this disturbance on the sample is insignificant, as in many radiant energy methods. In other cases, particularly those involving electrical currents or heat passing through the system or those requiring chemical or physical changes, the sample producing the information may not be precisely the same as that originally taken. Recognition of this possibility will enhance both the validity of the choices made and the conclusions drawn. Hence an understanding of the chemical aspects of an analytical problem, combined with adequate knowledge of the functioning of the tools to be used, constitutes the chief contribution an analyst can make in the planning phases of solving many perplexing problems in modern technology.

As for defining units, the reader is referred to the *ASTM Metric Practice Guide* (2), McNish's paper (122), and U.S. National Bureau of Standards publications (182).

B. OPERATIONS IN INSTRUMENTAL ANALYSIS

The steps in an instrumental analysis follow the same logical sequence of operational *systems* used by an analyst conducting a conventional "wet" analysis: sampling system, selectivity system, and reference system. In addition the instrumental analysis may involve further operational systems for electrical signals evolving from the measurement: primary and secondary transducer systems, operational amplification and modification circuits, and data and program control systems. All are cited in examples described in other chapters of this Treatise.

The necessity for a sampling system can hardly be overemphasized, and the requirements for an adequate one, in terms of representing the intended population, must always be kept in mind. Instruments functioning wholly or partly as sampling devices can fall prey to the same pitfalls of invalid assumptions and conclusions that manual sampling techniques can.

In fact Kaiser (91) says that the automation of sampling and sample preconditioning is more complicated than the automation of analytical measurements, even though necessary. Kaiser's theoretical considerations and experimental results indicate, for example, that a discontinuous system works better than a continuous, automatic sample-preconditioner. The moving of premounted samples into place, as has been done by Bakker et al. (11) and by Johnson et al. (89) for infrared spectrometers, can be simple and reliable. Some aspects of sample treatment will be considered further in Section VIII.

The selectivity system refers to means for isolating the physical properties most directly related to the sought-for constituents. For example, instruments may carry out actual physical separations incorporating such techniques as phase changes, chemical reactions, and/or alteration of the physical environment of the specimen. Physical separations may often be successfully obviated by converting the constituent in question into a form wherein some useful (i.e., measurable) physical property results. A chemical reaction resulting in color generation is an example. Furthermore, the choice of transducer and its parameters will often provide adequate or enhanced selectivity.

The reference system, of course, is *always* present by implication, if not actually in existence at the time of routine measurement. Many physical properties are by their nature not completely predictable measures of the constituent of interest, and hence standards are used in preparing calibration curves. Working standards are necessarily used when comparison, standard addition, or null-balancing techniques are part of the procedure. The peculiarities and special problems encountered in referencing or verifying (calibrating) instruments, especially the finding of ways to make them do their own referencing, are worth extensive consideration.

Final outputs must, in general, be of a form visible to the human eye, preferably unaided by auxiliary or optical means. These outputs nearly always incorporate mechanical movements of one sort or another; either the indicator needle or the galvanometer moves, or the inked recording pen moves up and down scale, or the typewriter types out. (An exception might appear to be the cathode-ray oscilloscope, but even here a permanent record is produced by operation of a photographic camera.) Such systems, usually calling for rotational or linear motions, encounter inertial effects, as well as frictional wear, which influence response speeds, sensitivities, and accuracies.

Lewin (103) proposed a useful, further operational classification of the principles of instrument design, based on the number of ways in which the basic elements (or components) of an instrument can be assembled to follow various designs having certain inherent advantages and disadvantages:

1. Single channel, direct deflection. An amplified signal (only one) drives a meter directly.

2. Single channel, bridge balance. A single amplified signal is balanced in a bridge or potentiometer circuit whose slide-wire position is read.

3. Double channel, direct ratio. The use of a double-beam (or other pair of signals) ratio technique eliminates drift and long-term instability of circuitry.

4. Double channel, null balance. The servomechanism principle cuts the dependence on linearity and long-term constancy of detector responses.

Instruments then are hybrids derived by using such diverse fields as electronics, mechanics, and optics. Even statistics and psychology cannot be ignored in the proper design, choice, and utilization of the tools we call instruments.

C. COMPONENTS

Another useful classification of the components which comprise useful electrical instruments consists of a separation into "active" and "passive" elements. Active elements may be considered to consist of those components which actually produce electrical energy. Among these would be included batteries, power supplies, and certain transducers. Among passive elements might be included those components which merely control the flow of energy derived, but which may be viewed as if they were the source of it (e.g., tubes, transistors, certain transducers, resistors, capacitors, inductors, switches, and relays of various types).

1. Transducers

A transducer may be thought of as a converter of some physical phenomenon into a useful output. In the context of analytical chemistry the "physical phenomenon" of interest is one related to the quantitative chemistry of a sample, and the "useful output" is one which gives valuable information, in a form very often electrical, for the identification and quantitative measurement of the desired constituent(s). Wildhack

(195) defines a transducer as "a device that responds to one physical entity, or change therein, by producing a change of some condition, factor or entity."

Because *electrical* transducers offer speed, stability, versatility, sensitivity, and selectivity greatly in excess of the capabilities of most mechanical transducers, they are the ones most frequently encountered. In the most literal sense, of course, an electrical signal per se is of no consequence to the observer; it must itself be transduced into a mechanical or otherwise visible form before it acquires utility to the observer.

Ample means exist to indicate electrical signals of wide ranges and characteristics. The electrical transduction occurs at the point where a physical characteristic of the desired constituent is converted (transduced) into an electrical signal. Once in this form, a wide choice exists for the instrument designer (or user) to add selectivity, amplification, correction, and/or stabilization, with high reliability and speed, before further conversion into a visible form. It goes almost without saying that, if the visible form is required only for further use in controlling a chemical process, it is logical to use the electrical signal to adjust mechanical devices to control independent process variables. The detailed aspects of "closing the loop" for automatic process control are outside our scope, although appropriate examples will be mentioned to illustrate situations which add to the impetus for translating analytical information into electrical terms.

Transducers in general lack chemical specificity, although there are notable exceptions (e.g., certain electrodes). Although the necessary specificity is gained in traditional wet analysis by chemical means (reactions causing phase changes, color changes, etc., and leading to separations or other alterations permitting measurement), it is advantageous to circumvent them when circumstances permit. Selection of the strategic physical property to be transduced, coupled with the appropriate transducer, *often* enables this to be done. Here, again, it is essential both to understand the physical–chemical behavior of the system to be studied and to acquire adequate knowledge of the transducers available and of the complete significance of their electrical signals.

The primary transducer is one which obtains the electrical signal containing the desired qualitative and quantitative information. It almost never operates alone, that is, there must be associated components which both control its environment and generate the signals to be modified by the physical properties of the desired constituent. Occasionally the transducer acts as its own signal generator (e.g., the thermocouple), but the measurement of its potential depends on a reference junction as well as a precision potentiometer or null-balancing recorder. In another general case, the signal to be modified by the desired constituent may be external to the transducer, as in the usual measurement of radiant-energy absorption.

2. Modification and Transmission Circuitry

The modification circuitry changes the signal of the primary transducer in some way to enhance its usefulness and transmission to some other location. Biologists and physicists have very actively adopted standard modular components of flexible, stable, versatile design. These are characterized by high gain (10^3–10^6 or more) and are often used with external feedback loops. Users of instruments want them to produce numerical data in terms and units directly applicable without further calculation. Hence computational steps, often utilizing "operational" amplifiers (in terms of *mathematical* operations), when appropriate, may be incorporated in the instrument system as finally assembled.

Interchangeable general-purpose amplifiers and associated circuitry not only ease maintenance and trouble-shooting problems but also permit adaptation of a given setup to a variety of measurement situations with simple alterations. Hence relatively straightforward circuits exist for performing such operations as inverting, adding, integrating, or differentiating, and for functioning as a single-polarity response (switch), peak follower, primary standard voltage source (stable for various loadings), or signal generator of various wave forms.

If the information gained in the transducer is needed at a remote location, it must be transmitted over some line of communication. A simple example is the mercury-in-glass thermometer, where a change in volume is the response of the mercury to the change in temperature. If the thermometer has an extremely long stem, this information may be transmitted the required (short) distance. Instead, the mercury's expansion may change its amount in a stem surrounded by a coil, so that the impedance of an electrical circuit is changed and the information thereby indicated at a longer distance away.

IV. CONSTRUCTION ASPECTS OF INSTRUMENTS

In building or using an instrument, it is essential to consider carefully the aspects (both mechanical and electrical) of precision, accuracy, and sensitivity. Assuming an adequate utilization of fundamental statistics, the analyst who knows the quantitative capabilities of the components comprising his instrument will be in a position to judge the end result of its operation intelligently, in terms of precision, accuracy, and sensitivity. Of course, interactions may well complicate such conclusions, and compromises are often required between what is ideal "on paper" and what is attainable in practice, particularly when cost factors are concerned.

A. MACHINING, MATERIALS, DRAWINGS

Users of instruments should be at least generally aware of a number of construction and design aspects (50), which will be mentioned briefly in Sections IV.A–IV.H. For example, the accuracy attainable in machining operations enables one profitably, without excessive cost, to take full advantage of modern machine-tool developments. Terms which are in common use in engineering practice, such as "nominal size," "tolerance limits," and "allowances," should be understood by those concerned with imperfections of workmanship. Such machining operations as lathe turning, plane milling, surface grinding, center grinding, lapping, spinning, and location of holes are important in the construction of a properly built instrument.

Another consideration involves the properties of materials used for general construction of instruments. These would include varieties of wood and of metals, including iron, mild steel, carbon steel, alloy steels, cast iron, and stainless steels. Alloys of copper and aluminum are also widely used. A variety of plastic materials, including such polymers as poly(phenol-formaldehyde), poly(methyl methacrylate), polystyrene, polyethylene, nylon, polyacetals, and polytetrafluoroethylene, have properties which should be understood well enough to permit their proper use in the construction of instruments. Knowledge in regard to materials for springs, including the elastic properties of steel, phosphor bronze, beryllium copper, and fused silica, is often useful. Developments in new materials, especially composites, constitute one of the most rapidly moving areas of technology. For example, a special glass has radioactivity levels less than 6% of those of comparable ordinary glasses; therefore, it is better suited to the manufacture of photomultiplier tubes. Advantages to the instrument-user of awareness of such developments are obvious.

In order to minimize the confusion of complex circuitry, it is common to use a block diagram which is a graphical representation of the functions (blocks) and of the flow of information (arrows). There are no general rules which apply, but usual practice is to have (1) lines with arrows representing the flow of information and (2) boxes placed in appropriate locations of the system labeled according to function.

In the preparation of drawings, a proper understanding of assembly and detail drawings, methods of projection, types of lines, scales, sections, screw threads, dimensions, and tolerances, as well as requirements for proper reproduction of drawings, are subjects for study. Situations involving constrained motion and the constraints to be used will often be encountered.

Complex electrical circuitry requires standard codes and symbols for use on drawings, in order to speed maintenance and cut down-time. Uniformity of expression always helps.

Wiring diagrams should be simple, while still including all essential details. Ideally they should distinguish between the electrical functioning of the devices and the physical layout of the electrical system. Thus there will be two diagrams: (1) the elementary or schematic diagram and (2) the interconnection or machine diagram.

B. MEASUREMENT OF SMALL DISPLACEMENTS

Physical measuring instruments often call for the magnification of small displacements. Sometimes the displacements in themselves are the object of interest, or they may be merely an intermediate step, as in movements of a galvanometer coil. It is useful to distinguish between two limiting cases: first, where the force producing the displacement is so great, compared with that required to keep a light-weight mechanism in contact with the moving surface, that the displacement is not affected by such contact; and second, where the lightest conceivable contact will cause a change in the displacement. Examples of these extremes might be cited as the thermal expansion of a solid, and the movement of a balance beam, respectively. Failure in many electrical instruments having meter movements can occur because of damage in pivot and jewel bearings and delicate control springs. Newer instruments eliminate these by using a taut ribbon suspension.

Among various means of obtaining magnification of small displacements, the most common ones are mechanical, optical, and electrical. Mechanical means require contact with the displaced body, and examples include lever magnification, rolling cylinder and related methods, parallel deformable strips, ratchet and toothed wheels, and dial gages. Magnification by optical methods, varying from optical levers to simple- and compound-lens microscopes and telescopes, is conveniently used to render small movements more easily visible and thus facilitate their measurement. When the optical system is used to project an image on an opaque scale, the accuracy depends to a great extent on suitable fiduciary marks.

The most common optical device employed for magnification is the *optical lever*, as employed in galvanometers. The principle is that light impinging on a mirror, when the mirror is rotated through angle θ, is reflected through an angle of 2θ.

Three techniques may be used to observe the angular movement of light beams: (*1*) the lamp and scale, in which the image of the lamp is thrown, by means of a lens, onto a mirror attached to a rotating part, permitting position readings to as low as 0.1 mm (assuming good-quality mirrors); (*2*) a telescope and scale, whereby an image of the scale is formed in the focal plane of a telescope provided with a cross wire, and the position of the cross wire on the scale image is then read by means of an eyepiece; and (*3*) the autocollimator, in which light from a small bulb is reflected past a cross wire down the tube of the telescope to the objective. Since the cross wire is at the principal focus of the objective, the light rays from each point in the cross wire leave as parallel beams. These beams then fall on a plane mirror whose rotation is to be measured, whence they are reflected back to form an image in the plane of the cross wire.

Interference methods may be used to measure small displacements accurately; these are as a rule quite simple and do not require very expensive apparatus. Furthermore, they are much more accurate than many methods in more general use, except for extremely expensive ones. There are several arrangements in which small movements of a glass plate produce large movements of a system of interference bands.

The *thermal relay* (131) was an early version of the photoelectric relay. Both of these have their sensitive elements (e.g., two thermocouple junctions), placed with such geometry that displacement of a light beam affects their electrical output. The thermal relay is used to amplify moderate galvanometer deflections but suffers appreciable response delays. The photocell has a very fast response and has come into wide use for both small and large amplifications of light-beam displacements.

The choice between a photovoltaic and a photoemissive cell (or photocell) depends partly on whether output to a pen recorder is desired. In spite of the lower output of the photoemissive cells, their higher impedance makes their signals much easier to amplify. One of the common objections to photovoltaic cells has been that of fatigue, but this can be considerably reduced by the insertion of an infrared absorbing filter in the light beam. Photoresistive cells are also coming into use because of their ruggedness and small size, coupled with considerable sensitivity.

A final category concerns electrical methods for measuring small displacements which may have some advantages in certain situations, for example, where the apparatus must be quite compact. This includes the capacitance or inductance change of an oscillating circuit produced by the mechanical displacement to be measured. The object undergoing displacement may form one plate of a capacitor that is in an oscillator circuit whose frequency then is measured as it forms a beat frequency against another oscillator. Similarly, for inductive sensing of displacement, the moving body is the armature and the current change resulting from its effect on inductance is measured directly with a microammeter. Remote reading is possible, and, where desirable, the results can be recorded continuously on either a pen or a photographic recorder.

C. SENSITIVITY

Much stress has been laid on the need for sensitivity in instruments. In the case of direct-reading instruments this may be expressed most simply in terms of scale divisions per unit change of the quantity being measured. However, a useful criterion of performance of an instrument is better obtained by considering the least change in the quantity measured which can be detected by the instrument at the desired accuracy and signal/noise ratio. Nothing is gained by making the scale divisions much smaller than would correspond to this amount. Furthermore, scale divisions are occasionally too small for easy reading by the eye, necessitating some magnifying device.

A useful expression of sensitivity, suggested by Young (197) and used by Kieselbach (97) in evaluating detectors for gas chromatography, merits wider application. It utilizes the unit pQ_0, the negative logarithm of the concentration in moles per liter corresponding to twice noise level, as the minimum detectable signal. Parenthetically, noise is one of the basic difficulties encountered with all electronic devices: even in the absence of an input signal, an output occurs which is called "noise." It occurs because all charged particles, including electrons, induce electrical signals in output circuits arising from random cathode emission and from random variations in their velocity.

The really pertinent consideration regarding noise is the signal/noise ratio (discussed in Section V.D). Noise, whether it arises from thermal resistance (Johnson noise), from tube plate or grid current fluctuations, or from the shot effect, can be minimized by low temperatures, by high-quality constant-performance components (including tight, vibration-free connections and well-soldered joints), and by narrow-bandwidth, sharply tuned amplifiers. Relaxation of the time-constant or response-time requirement, to a degree consistent with negligible distortion of the desired signal, is also helpful.

D. STABILITY

The question of stability attains paramount importance if the indications of the instrument are to be magnified, and it is worthwhile to consider how good this stability really must be. In the case of a coil and mirror in a galvanometer, the restoring couple per unit twist of the coil is not affected by the stretching force applied in holding taut the spring. Hence the periodic time of torsional oscillations also will not be affected. If, however, the arrangement is made to vibrate like a stretched spring, the time of vibration is decreased by increasing the stretching force. The net result is that a stretched suspension is much less sensitive to disturbances producing transverse vibrations than is an unstretched system. If this system is perfectly balanced, then movement of the points to which the suspension wires are fastened at top and bottom cannot cause rotation of the coil.

The cathode-ray tube has certain obvious advantages as a detector and recording device. However, it is well to realize the kinds of stability problems which can develop. The fact that a potential of several hundred volts is required to deflect the spot to full scale in the ordinary CRO (cathode-ray oscilloscope) immediately imposes a need for amplification when the signal is only a few millivolts. Conventional amplifiers can be used, although they should be made symmetrical, so that variations in velocity of the electron beam (and, therefore, sensitivity) will not accompany changes in the potential applied to the deflection plates.

Magnetic-deflection CROs do not have this limitation, but the amplifier output must produce an adequate current (rather than hundreds of volts). There then may also be certain frequency restrictions due to effects on the time constants of the deflection coils.

E. ERRORS

The subject of instrumental errors has been considered in detail by many workers, and their publications should be consulted for detailed treatments. It must be recognized first that an instrument cannot possibly be constructed so that all of its parts have exactly the desired shapes and dimensions, even when the best methods of construction are utilized. It is important, therefore, to consider the extent to which the allowable errors of construction may be minimized by suitable design and use.

Many measuring instruments use some kind of pointer moving over a scale, and the quantity observed is supposed to be related in a known manner to the movement of the pointer. Errors due to the instrument, as distinct from those due to the user, arise on account of inaccuracies in the supposed relation (quantity measured versus movement of the pointer). Although this relation often cannot readily be pre-estimated, it may be determined by calibration. In this case, apart from the observer's errors made in the process of calibration and subsequently in using the instrument, the only errors will be those due to erratic behavior, which show up if the very same instrument reading is not always obtained for a given value of the quantity being measured. Causes of such erratic behavior may include external disturbing influences, unduly large frictional forces where moving parts are involved, backlash, and indefiniteness of location of the parts. These causes are generally due to faulty design rather than inferior workmanship. For example, when sliding of surfaces is present in an instrument, erratic errors may be too large to accept. In such a case, rolling motion or motion depending on elastic deformation should be employed in the design.

Errors in an instrument, other than those of calibration or those resulting from external causes, may often be investigated by determining the shape of the hysteresis loop. To do this, all external conditions which would affect readings must first be kept constant. The input quantity which the instrument is to measure is increased stepwise, with readings taken at each step. Then this quantity is reduced so that a complete cycle has been traversed.

A plot of instrument readings against the measured quantity will usually show a closed loop, which should be reproducible when the instrument is taken through several identical cycles. If the hysteresis loop is not repeatable, the instrument is not in proper condition and should be examined for correction, looking for such suspicious things as grit, damage to working surfaces, or defective springs. Such a hysteresis loop will reveal the presence of backlash by showing straight portions in the loop parallel to one of the axes. In addition, however, if the ascending and descending parts of this graph become straight only after appreciable increments of one of the measured quantities, even after backlash has been overcome, it may be that elastic deformation of the driving parts of the mechanism has taken place. This effect should perhaps be expected, but always should be small.

F. SYSTEMATIC ERRORS

In instruments in which the relationship between measured quantity and scale reading is calculated by a relation involving a knowledge of the dimensions of the instruments and not entirely by calibration, systematic errors may enter when the shapes and dimensions of certain parts of the apparatus are not as assumed. These errors will generally arise from imperfections of workmanship, and their effect on the measurement depends to a considerable extent on the design. Their probable existence should be kept in mind when planning arrangements of parts. Clever design can minimize or perhaps completely eliminate such effects.

An interesting example of faulty design, causing variable systematic error, was described by Westneat et al. (193), who discovered the cause of excessive nonlinearity in the low-light-level response of a widely used photoelectric colorimeter. Heat from the pilot light, used to illuminate the balancing galvanometer and the slide-wire scale, caused thermal potentials in the slide-wire, thereby converting it to a thermocouple with the hot junction at the midposition and the cold junctions at the ends. Thermal emf's of up to 0.12 mV, opposing the normal potential developed by the photocells, were generated.

Whenever rectilinear motion of a sliding part relative to the frame of an instrument is involved (e.g., the movement of a micrometer screw or scale in which the motion is coupled to some other part of the instrument), errors will enter because of the assumption that the ways which guide the motion are aligned to be perfectly true. When using a traveling microscope to measure a distance between two points, the assumption mathematically depending on parallel motion may be valid if the microscope moves parallel to itself as it travels back and forth across the scale. However, if two positions of the microscope for two different displacements are *not* exactly parallel, there is a small angle between the directions of the microscope axis at the two positions, and the difference of the scale reading will depend on the sine or cosine of this angle.

Measuring *screws* enter into the construction of many kinds of apparatus used in scientific work; hence it is important to realize that a screw in its mounting may be responsible for larger errors than would at first be expected. For the instrument designer the most important errors involving screws are those of pitch.

Parallax error is an observational error made when reading the position of a pointer with reference to a scale when the pointer does not lie in the plane of the scale. It is avoided when the observer's eye is lined up with a line passing through the pointer perpendicular to the surface of the scale. Often a mirror is mounted with its reflecting surface in the plane of the scale, so that the pointer and its image may be properly aligned.

G. ISOLATION OF APPARATUS FROM DISTURBING INFLUENCES

Instruments located on tables or other supports are subject to vibrations, of local or more distant origins, whose magnitudes will vary greatly, according to conditions such as the ground on which the building stands and the type of building construction. Vibrations of local origin can be very severe because of such factors as heavy highway traffic, moving machinery in the building, and flexible flooring. White (194) complains that manufacturers too seldom advise the consumer regarding the ideal or minimum-acceptable environment.

Although few generalizations can be made to suit particular situations, it can be said that *natural* periods of vibration of walls and floors are usually a small fraction of a second (caused perhaps by footsteps of persons walking nearby). These are often superimposed by lower frequencies caused by slow-running machinery. Such local disturbances can be minimized by utilizing piers, supported directly on the ground, and having no direct contact with the building.

The early work of du Bois and Rubens (48), in connection with the moving-magnet galvanometer, applies to mechanical disturbances on a freely hanging mass suspended by a flexible fiber, where rotation around the vertical axis is of concern. Briefly, these and other analyses indicate that *symmetrical* construction of moving parts about their axis of rotation will prove useful in apparatus having suspended or pivoted rotating parts. It should be remembered also that supporting frames are best built with high rigidity. It is for this reason that cast iron is preferred over mild steel or aluminum, especially when the frequencies to be encountered are high. Adequate damping of moving parts can likewise reduce sensitivity to disturbances.

When a sensitive instrument cannot be built sufficiently symmetrical to achieve independence, the instrument may be mounted to prevent vibrations from having a harmful effect by:

1. Reducing the coupling between the instrument and its surroundings.

2. Damping out the reduced disturbances in the mounting itself.

3. Supporting the instrument by means of materials which do not fully transmit mechanical waves.

Techniques include vanes or dashpots immersed in oil baths, air-filled cushions, coupled oscillators, and spring mountings. Novel simplicities, like mounting on tennis balls or alternate layers of felt and stones, have advantages in offering partial freedom from both vertical and horizontal displacements.

Extensive advice on the isolation from vibration of knife-edge balances has been offered by Macinante and Waldersee (113), who reviewed the available literature on balance mountings and classified them into four groups: independent pillar, floor, bench, and wall mountings. They conclude with the plea that early consideration should be given to vibration-isolation requirements when planning a new laboratory. They also point out the absence of reliable data on the permissible levels of vibration for balances, making this largely a matter of guesswork and trial and error.

Disturbances from thermal effects are often overlooked in situations in which these effects are potentially serious. Use of materials having low coefficients of thermal expansion will minimize fluctuations resulting from thermal changes in dimension. If measurement of length, including distances between lines on a photographic plate, is the "heart" of the measurement, then use of a mild steel for the screw or scale, followed by an appropriate calculated correction for the temperature effect, can be simpler than other alternatives. Also used is an arrangement for "self-compensation" utilizing differential expansion. Unfortunately these may set up strains giving rise to greater distortion than would occur if strictly homogeneous material were used.

Macurdy (114) has considered the large instabilities caused by small thermal gradients on precise balances and has reported on the effects of various types. Poulis and Thomas (150) explained the spurious mass changes, observed when microbalances are operated at low pressures, as really resulting from pressure differences arising from temperature inhomogeneities. They discussed ways to reduce such disturbances.

In radiant-energy-measuring detectors such as the bolometer, two identical fine strips of platinum may form two arms of a Wheatstone bridge. If one strip is exposed to the incident radiant energy and the other shaded, both will react equally to undesired changes such as those produced by bridge current fluctuation and ambient air-temperature changes, thus leaving the bridge balance unaffected, insofar as the strips are truly identical.

In other cases of thermal distortions, paired compensating thermopiles are useful. They must be identical and connected in opposition in a common enclosure. Still another alternative is to use a *calculated* correction when the thermal behavior of all the components involved is known or when feedback servo arrangements may be incorporated to obviate separate computation. Tuning of alternating-current amplifiers to filter out unwanted responses has proved especially successful in chopped-beam infrared spectrophotometers. Nearly complete freedom from zero drift, caused by changing room temperature, can be obtained.

Temperature control of the room and/or apparatus housing is an alternative, if perhaps expensive, choice. Thorough air circulation is essential if thermal gradients are to be minimized. Insulation (including *no* windows) to lag the cyclic fluctuations inherent in any on–off temperature control system is also desirable.

H. TESTS FOR STRAIGHTNESS, FLATNESS, AND SQUARENESS

Engineers and the gage and tool industry have devised many useful methods for testing flat surfaces and determining the relationships between one surface and another. Users of instruments might well be aware of some of these techniques, especially when optical measurements, as in spectrophotometry, are to be made, because they affect the specifications and design characteristics of components.

Two instruments are in common use for measuring the inclination of a surface—the autocollimating telescope and the spirit level. The best-known method for giving quantitative results involves the observation of interference fringes. An optically polished glass reference surface is laid on the surface to be tested (both being as clean and dust-free as possible). The contour fringes which then can be made to appear as a result of having a thin air film between two semireflecting surfaces demonstrate very effectively the shape of the unknown surface, that is, straight parallel fringes indicate flat surfaces. Deviation from straightness of one fringe width indicates a departure from flatness of one-half the wavelength of the incident light, conventionally the Hg green line of 5460 nm.

Standard reference plates and straightedges with various degrees of accuracy are available. Generally the standard is assumed perfect in any comparison to be made, but there are good ways of calibrating standards by internal measurements.

When one is concerned with the squareness of components of an instrument, the problem is highly dependent on the conditions. Squareness tests for two flat surfaces usually call for comparison with a try-square. When this is done carefully, a gap of 2.5 μ (micrometers) is observable between blade and surface. Use of a micrometer-dial gage, or an autocollimator mounted on a carriage which can be traversed, permits greater sensitivity and quantitative observation. By comparing traversed measurements of the unknown part with those of a substituted square, considerable accuracy can be achieved.

The squareness checking of two rectilinear motions, such as found in measurements of photographic plates to give Cartesian coordinates or of the stage micrometers used on many microscopes, can be approached in several ways. One of the best is to move the carriage and to base the measurements on the use of a square and a dial gage or equivalent indicator. Errors in the square itself can be canceled out by running it both ways, that is, reversing its position and averaging the two sets of readings thus obtained.

V. THE ROLE OF ELECTRONICS IN ANALYTICAL CHEMICAL INSTRUMENTATION

It is largely as a result of rapid advances in *electronics* that the practicality of making analytical measurements automatically by instrumental means has increased so greatly over the past several decades. The introduction of new types of circuit components, as well as improvements in existing types, has led to new flexibilities in circuit design. Of special interest are a variety of elements with useful nonlinear characteristics. These developments not only have led to welcome improvements in the accuracy and reliability attainable in instrumental analysis, but also have extended the range over which many variables can be measured accurately, and have made feasible previously impossible measurements.

This discussion of electronics will not attempt to substitute for a detailed study of the subject; rather it will serve as an introduction to some of the concepts and terminology which the conscientious analytical chemist will want to find in appropriate reference sources reported and commented upon here and elsewhere (see General References on Instrumentation and Electronics at the end of this chapter, as well as the Selected Patent References). Some of the problems that electronics specialists have had to solve in meeting the needs of chemical instrumentation will also be related.

In addition to the general literature and patent references given at the end of this chapter, good sources for recommendations on aspects of electronics worth studying are found in such papers on the *teaching* of chemical instrumentation as those by Tabbutt (175), Ewing (53), and Eisner (49). Lewin's reports (105) on amplifiers and oscillators as electronic tools for the analytical chemist are valuable also. The scientist wanting to study in this field is also fortunate in having a variety of sources of training kits for student use from many laboratory supply houses and other companies directly in the electronics or instrumentation business.

A. TRANSDUCERS

We introduced the concept of transducers in Section III.C.1. To do its job, a transducer must *reproducibly* convert an input function, whose magnitude is desired, into a second variable, such as displacement, voltage, or force. The nature of the mathematical relationship between the two variables (input and output) is predetermined by the operating principle and functioning of the particular type of transducer. The output may be used in any of several ways. It may be employed directly as a means of observing the variable, or it may be used to provide a signal to additional elements which amplify, convert, or otherwise modify the response.

In modern instrumentation, transducers that provide an electrical signal, such as voltage, current, or frequency, are generally most flexible in their applicability because of the ease with which such signals can be monitored, amplified, and transmitted. For this reason, transducers whose output is fundamentally mechanical are frequently coupled with mechanical-to-electrical transducers. Nevertheless, purely mechanical devices, such as pneumatic sensors and amplifiers, are sometimes better suited for particular applications.

When an instrument is designed or selected for a special application, it is important that critical consideration be given to the functioning of the primary transducer. Some questions that should be posed are the following:

1. How is transducer output related to the variable whose value is required?

2. Is the transducer unaffected by extraneous environmental conditions throughout the operating range of interest?

3. Is the transducer action sufficiently free from drift so that the desired accuracy can be obtained?

4. Does the action of the transducer adversely affect the variable being measured?

5. What is the useful operating life under the service conditions anticipated?

6. What services (power, coolant, etc.) must be provided?

The suitability of a given transducer for a particular application will generally be established by the answers to questions such as these. In some cases, environmental control, such as thermostatting, may be required to obtain a satisfactory performance level.

The following are examples of commonly used transducers:

1. Thermometer. Converts molecular kinetic energy first to a volume expansion and then to an observable linear displacement.

2. Phototube or photomultiplier. Converts light quanta into electrical current.

3. Differential transformer. Converts the position of its core relative to its windings into an alternating voltage, thus providing a measure of displacement.

4. pH electrode. Converts hydrogen-ion activity to an electrical potential.

5. Voltmeter. Transforms electrical potential to position of a pointer. The common voltmeter can be looked upon as a *series* of transducers also: the voltage is transformed first into a current which produces a magnetic field; this is transformed into a mechanical force which finally positions the pointer.

B. BASIC ELECTRICAL QUANTITIES

These basic quantities—voltage, current, frequency, charge, resistance, capacitance, inductance, and so on—are defined in any standard physics textbook. A number of simple relations also exist between various of these quantities, and one quantity can be electrically converted to another with more or less complex equipment. There are limitations to the accuracy with which electrical signals can be transduced. Some of these are inherent in the physical principles involved; other practical limitations result from the use of components that are less than perfect.

Where *low* signal levels are measured, "noise" must frequently be contended with (this is discussed in Section V.D). The commonly used measure of the seriousness of noise is the signal/noise ratio (S/N). Noise may be introduced by the system being measured or by the transducer system. It may, at least in part, be an inherent characteristic which cannot be eliminated. Under these circumstances, special circuitry may be recruited to "enhance" the S/N, untangling a small signal from a large noise background. Other descriptive terms, more or less "basic," will be introduced as they occur in the following discussions.

C. BASIC ELECTRICAL COMPONENTS AND BUILDING BLOCKS

Components available to the circuit designer range from the traditional standard cell, resistor, capacitor, inductor, photocell, and vacuum tube (long available and well described in physics textbooks, as well as in many of the references listed at the end of this chapter), to the more specialized, modern components such as semiconductor devices (diodes, transistors), power supplies, amplifiers, oscillators, triggers, scalers, meters, potentiometers, bridges, oscilloscopes, and servomechanisms. The following sections will assume the reader's knowledge of the most fundamental concepts (cited as "traditional" above) and will instead expand on selected items among these, as indications of the level of understanding likely to be helpful to the chemist using his "black box."

1. Transistors

Transistors and some types of diodes often depend on the physical and electrical characteristics of semiconductor materials. One definition of a semiconductor is "an electronic conductor with resistivity in the range between metals and insulators, in which the electrical charge-carrier concentration increases with increasing temperature, at least over a certain range. Certain semiconductors possess at least two types of carriers: negative electrons and positive holes."

Typical materials used in transistors include indium, germanium, and antimony; other materials such as gallium, silicon, and arsenic are also employed. When in the highly pure state very little conduction can occur, and the resistivity would be too high in a perfect single crystal, as there are not enough impurities available to act as charge carriers. On the other hand, semiconductor properties can be obtained, in the case of germanium, by so-called doping with traces of antimony to replace germanium atoms in the crystal lattice, leaving one unsatisfied bond (because antimony has a valence of 5 compared to 4 for germanium). The resulting material is then called "negative" or "N type." Conversely, in the case of indium, having a valence of 3, doping it into a germanium crystal lattice leaves fewer electrons available and a "hole" is said to exist. This then is "positive" or "P type" germanium. "Holes" can be considered as "receptacles" into which electrons can drop and temporarily remain in their course through the material.

If we combine P and N materials, the result is known as a "PN junction," which gives diode action, meaning it will conduct in one direction and not in the reverse. By varying the amount of doping, the characteristics of the junction and the efficiency with which it will conduct or cut off electric current can be readily controlled. Transistors have a number of advantages over vacuum tubes:

1. Low power consumption. A direct result of transistors' "cold" emission of electrons, this endows transistors with clear superiority over electronic vacuum tubes in weight, size, power supply, and heating problems.

2. Low anode voltage. Transistors require only 10–15 V, compared to the 100–300 V required on tubes, reducing power supplies to a fraction of the weight and size required by tubes.

3. Miniature size. The transistor itself is miniature, less than one-tenth the size of an equivalent vacuum tube.

4. Long life. The transistor's longer life permits reliability to be built into electronic instruments, paving the way for applications inconceivable for the vacuum tube.

5. Low cost. Basically the transistor is a simple device, easy to construct. The requirements of extreme purity and delicate assembly were only temporary stumbling blocks in achieving economic production to make transistors competitive with vacuum tubes. The former have essentially supplanted vacuum tubes for a wide range of applications, except for high-power, high-temperature, or high-frequency situations.

In regard to the applications of transistors, these devices can, for example, replace other systems as stepping switches, relays, or commutators, performing a function traditionally done mechanically by rotary-contact or mercury-jet switches. A transistor commutator has a number of advantages over mechanical types:

1. It is inertialess and can operate at frequencies above 10 kHz, where the phase shift is small.

2. It is immune to shock and vibration, and is not microphonic.

3. It can easily match low-impedance sources such as thermocouples.

4. It has a long life.

5. It has a dynamic range of 100 decibels (dB).

A major disadvantage has been the careful selection required for adequate matching and freedom from noise.

When transistors are used as amplifiers, their ultimate limits of operation are governed by the effect of parameter variations, such as changes in the current appearing at the collector junction when zero current flows in the emitter, and by noise. The effect of parameter variations and nonlinearity is most pronounced at low temperatures and can be reduced (similarly to the situation for vacuum-tube amplifiers) by operating the transistor in its linear region and by using negative feedback. *Feedback* then becomes *essential* to the operation of low-level, instrumentation-quality transistorized signal amplifiers.

Transistors have had the reputation of being noisier than vacuum tubes. Noise, in terms of the noise figure (*NF*), is the ratio of the noise power out of an amplifier to the noise power out of an *ideal* amplifier with only thermal noise in the source resistance. The frequency region most favorable for operating a transistor (*NF* at a minimum) occurs from about 1 to 100 kHz. In this region, the *NF* is determined primarily by thermal and shot noise. The *NF* rises above 100 kHz because of decreasing transistor gain with increasing frequency. It also increases with *decreasing* frequency—a characteristic of transistors commonly called "1/f noise"—which is due to a surface phenomenon. Careful control of humidity and sources of contamination during manufacture can result in low noise figures.

Transistors are finding use as analog-to-digital converters, so commonly required in data logging. The requirements of such applications for commutators having long life, high reliability, speed, and compact size have led to extensive use of transistors.

2. Zener Diodes

Although the original incentives for solid-state devices in electronics were related to reduced maintenance costs, the zener diode power supply competes sharply in terms not only of cost, but also of performance. This applies not only to the supplies that deliver operating power, but to reference voltage supplies influencing the accuracy of many electrical measuring instruments, as well.

The zener diode is a variant of the silicon-junction diode, consisting of a semiconductor formed of two types of silicon having different electrical properties and forming a PN junction; it has a very high ratio of forward-to-reverse resistance and, therefore, is usually used as a rectifier to block the current flow in one direction. However, if a voltage applied to the diode in the reverse direction is gradually increased, the current remains extremely small until the breakdown voltage (zener breakdown) is reached. At this voltage and under controlled conditions a nondestructive breakdown of the high reverse resistance will occur, and current increases rapidly. In the region of breakdown, the voltage drop across the diode will be very nearly independent of the current, depending only on the very small reverse resistance of the diode. The usual zener diode is fabricated so that this breakdown will occur in a particular voltage region (e.g., 3–200 V, more commonly 5–12 V).

The zener breakdown phenomenon is a field emission effect which takes place when there is a high electric field in the depletion layer. Based on a tunneling action, it has a negative temperature coefficient. This breakdown is similar to the leakage in insulators and occurs at lower voltages as the temperature is raised. The other factor responsible is the avalanche breakdown, occurring when the field is increased enough that the initial carriers gain sufficient energy to create new electron-hole pairs. Avalanche breakdown, characterized by a positive temperature coefficient, predominates in diodes above 40 V.

Silicon junctions which are more heavily doped exhibit breakdown at lower voltages; for example, at 5 V the zener breakdown predominates, resulting in a negative temperature coefficient. At around 5.5 V this coefficient passes through zero, providing one method for the selection of stable diodes.

Control of the zener diode temperature to within narrow limits is desirable for best performance. In the case of compensated assemblies, temperature control to $\pm 0.2°C$ may require that the quality of the heat sink joining the zener junction to the compensating junction maintain the difference to less than $0.002°C$. Along with these considerations is the important factor that zener diodes have a thermoelectric power (in the case of silicon, about 400 $\mu V/°C$). It becomes evident that the thermal resistances surrounding the diode case and its leads should be *fixed* relative to a heat sink. Silicone grease has been used to exclude air between the case and the heat sink, and the leads may be soldered to feed-through terminals mounted in the heat sink.

To keep fluctuations in temperature of the diodes and the heat sink to a minimum, a proportional controlled oven may be considered necessary, along with temperature indication to show when the warmup interval is over. Such needs can be met by using a photoconductive pickup on a bimetal thermometer to actuate a silicon transistor serving as a heating element. In this case an illuminated photoconductive cell may provide base current for the transistor, which, in turn, raises the temperature of the aluminum chamber containing the reference diode. When the operating temperature is reached, the photocell is partially shaded by the thermometer pointer and the base current is reduced. In spite of the inherent slow response of bimetallic systems, good stability can be achieved, and control to $\pm 0.1°C$ is practical for a 10% change in ambient temperature.

a. Temperature Measurement with Zener Diodes

For the single-PN-junction device, an impressive method can be used to determine very small changes in temperature. A constant-current source supplies the diode in the forward bias direction, and the resulting voltage drop is measured with a differential voltmeter or potentiometer. Since the change in voltage ranges to about 2000 $\mu V/°C$, a resolution of less than $0.01°C$ can be easily obtained. It is particularly important that this measurement indicates the change at the junction exactly and accounts for all the thermal paths and thermal capacities that otherwise tend to average out some of the external fluctuations in ambient temperature. Thus the junction itself acts like a thermocouple having a very high sensitivity and low resistance without causing any added heat loss. Such a technique should be particularly useful in observing the effects of oven cycling in the case of on–off controllers and also in checking line-voltage influence on proportional controllers.

b. Circuit Requirements

Resistors must have good stability; this requirement is particularly important in regard to the resistor which determines the current flow into the zener. Voltage dividers are required across the output to drop the level to some integer value, such as 5 V, or to a level near 1.018 for direct comparison to a standard cell. If the zener voltage is only slightly higher than the desired output, the quality of the resistor is not quite so critical.

A constant-current source is required to develop a stable voltage output from a zener reference diode. The simplest way to achieve this is to use a voltage source which is large compared to the zener voltage, and a series resistor with a value large in comparison to the incremental resistance of the zener diode.

To achieve even closer regulation, a cascaded arrangement can be used wherein a first zener diode, having a higher voltage rating than the second, acts as a partially regulated voltage source. Further cascading with still another zener becomes less effective, however, consuming much greater energy. One circuit to handle this problem is called an impedance cancellation bridge; in this, a portion of the input is sampled and subtracted from the zener output. With such an arrangement, regulation factors of 1000 to 1 can be obtained over a limited range of input voltage for which the curvature of the zener characteristic is negligible.

Another circuit, said to be just as effective as the impedance cancellation bridge, involves a transistor current source and has the advantages of a common ground and a low output impedance.

c. Load and Overload Currents

Load currents can be taken from zener power supplies in almost any amounts without damage. The output may even be shorted, and then recover completely when the short is removed and the diode regains its thermal equilibrium. Reverse currents may be more of a problem, since they tend to flow directly into the zener. Moderate amounts can be tolerated provided the junction temperature ratings (generally 150–200°C) are not exceeded. If a very large reverse current flows, even for a fraction of a second, the junction can be destroyed, developing a short circuit.

d. Zener Merits

The advantages of zener diodes as voltage references include the following:

1. Large load currents are delivered without damage.
2. There is immunity to shock, vibration, and position.
3. The temperature coefficient is adjustable to zero.
4. The higher output reduces thermal emf effects.
5. Output impedance is lower.
6. The range of operating temperatures is wide.
7. There is a long shelf life.

Possible disadvantages of zener diodes would include the following:

1. Power is required, even at "no load."
2. Regulation against power-line changes requires elaborate circuits.
3. Diode selection is tedious.
4. Voltage levels may differ too widely from unit to unit.

In general, zener diodes are considered rugged, shippable, and easily incorporated directly into a measuring circuit so designed that a particular voltage will be derived to a specified load over a wide range of input voltage and temperature. Hence, in most recording and controlling equipment of modern vintage, zener diodes replace standard cells as reference voltages. They also serve as standards in applications requiring accuracies of about 0.01%, and possibilities exist for accuracies of 0.001% or even better. The former hand-carrying of saturated standard cells is becoming unnecessary with the impressively higher accuracies achievable with zener diodes.

There are two major differences between zener diodes and standard or other chemical cells serving as sources of reference emf:

1. Zener diodes require a source of operating current, whereas traditional standard cells do not.

2. The operating voltage of zener diodes is 5–12 times that of the standard cell.

3. Operational Amplifiers

As implied in the name, "op amps" perform operational transfer functions, having been designed originally for analog computers. The early models were chopper-stabilized vacuum-tube circuits, but by the mid-1960s improved solid-state components made possible compact transistorized amplifier modules costing less than $25.

Because of their tremendous versatility in so many applications of value to chemical instrumentation, operational amplifiers warrant consideration here. An electronic amplifier in general is a device which makes a small signal larger, performing a function essentially of multiplying the input signal by a constant. The operational amplifier, in conjunction with additional external circuitry, is unusually flexible in being able to perform a variety of mathematical operations, such as summation, subtraction, and integration and differentiation with respect to time.

Functionally speaking, an operational amplifier is a device which, by means of negative feedback, is capable of processing a signal with a high degree of accuracy, limited primarily by the tolerances in the values of the external passive elements used in input and feedback network.

Electronically, an operational amplifier is simply a high-gain amplifier designed to remain stable with large amounts of negative feedback from output to input. In more precise terms, an operational amplifier consists of a wide-band amplifier (dc up through several thousand hertz) of very high gain (10^4 or higher) which is especially designed to be stable and highly free from drift. Also, it is inverting, that is, has a negative gain to provide for stabilization when feedback is employed.

a. USE OF FEEDBACK

There are several well-known techniques for selecting some percentage of an output signal and feeding it back to the input. Voltage feedback, the most common one, derives the feedback signal by tapping off in *parallel* with the load. As its name implies, this feedback signal is proportional to output *voltage*. It is also possible to derive feedback by tapping off in *series* with the load; in this case the feedback voltage is proportional to the output *current* and is thus commonly called current feedback.

Selection of the type of feedback really is an impedance problem. A study of the situation reveals that voltage feedback lowers the output impedance, whereas current feedback raises it. Thus the designer must determine whether one type of feedback tends to stabilize the output impedance more than the other.

Since the primary functions of the operational amplifier are achieved by means of negative feedback from the output to the input, this requires that the output be inverted (180° out of phase) with respect to the input. The conventional symbol on wiring diagrams for the operational amplifier is the triangle. The output is the apex of the triangle, while the input is the opposite side. The negative feedback occurs through a resistor, capacitor, inductor, or network of nonlinear impedance, and is applied from the output to the input. When the input and output impedances are both pure resistances, the system will behave as an inverting amplifier of gain equal to the ratio R_2/R_1, R_2 being the resistance in parallel with the amplifier and R_1 being that in series with it. The accuracy of multiplication is determined by how well the ratio R_2/R_1 is known and by the gain of the amplifier relative to this value.

By suitable choice of a constant value of R_1 and the application of a known fixed voltage, resistances of unknown value can be substituted for R_2 (in parallel with the amplifier), and the instrument becomes an ohmmeter which reads out the value of the resistance in a linear manner. On the other hand, if a conductometric cell serves as the input impedance, R_1, the output voltage thereby measured will be linearly proportional to the conductance of the solution in the cell. This system then becomes a useful direct-reading conductivity meter for automatic recording of conductometric titrations and so forth. In another arrangement, a dc voltage of about 1 V can be applied and the input impedance, R_1, may be a photoconductive cell (e.g., cadmium selenide), permitting the output voltage to be used in photometric measurements such as colorimetry.

Obviously thermistors and other resistance-type transducers sensitive to temperature change can be substituted for R_1 or for R_2 for appropriate thermal measurements. Alternatively, by placing an electrolysis cell in place of R_2 and by using a fixed resistance as R_1, a constant current of magnitude equal to the value E_1/R_1, the input voltage divided by the resistance, will be applied to the electrolysis cell. This circuit, in other words, acts as a constant-current supply whose magnitude is readily set at any desired level simply by altering the value of E_1 or R_1. It is important that the magnitude of current flowing through the electrolysis cell be essentially independent of the value of the potential drop across the cell. Hence this becomes a very useful technique in carrying out coulometric titrations and in chronopotentiometry.

b. CHARACTERISTICS

Four characteristics of operational amplifiers should be known before attempting to utilize these devices in practice. First, a dc offset exists in all amplifiers, having a value which drifts with time. This offset can be considered as equivalent to the insertion of a small potential source between the summing point and the input to the operational amplifier. Therefore, the operational amplifier circuitry must contain an adjustable bias source to effectively cancel this dc offset potential. Generally, this cancellation is accomplished by connecting the input of the circuit to ground and adjusting the bias until the voltage output is zero. The amount of drift depends on the particular operational amplifier and the stability and regulation of the power supply; the use of chopper-stabilized amplifiers greatly minimizes the magnitude of such drift.

Second, the amplifier input current should be as small as possible. With suitable operational amplifier circuitry, this value can be made as low as 10^{-12} A.

A third factor is the noise generated by the operational amplifier; for many commercial units this value is well below 1 mV and is generally greater than for the chopper-stabilized types. Since a large portion of the noise measured in the output voltage may arise from the external circuitry, suitable attention should be devoted to appropriate shielding, and to the proper design of these operational amplifiers, especially those portions of external circuits which are connected to the input point.

The fourth characteristic consists of the phase shift and frequency response of the operational amplifier. To provide stability in the enclosed feedback loop, the frequency response of commercial operational amplifiers is purposely designed so that the gain of the amplifier decreases beyond a fixed cutoff frequency. Although this characteristic seldom creates a problem in chemical instrumentation, its presence should be kept in mind if very fast response (less than 1 msec) is desired.

c. Limitations

The accuracy of all operations is limited by the open-loop gain of the amplifier, which determines how closely the device is capable of holding the minus input null. An amplifier with infinite gain would provide a null of exactly 0 volt, and the impedance at the minus input (using feedback) would be exactly 0 ohm. With *finite* gain the minus input does not quite null and does not appear as 0 ohm, and there will be a definite output signal error in addition to the error introduced by the tolerances of input and output impedances.

The amplifier gain varies with frequency, and it is important to know what the effective value may be for the frequencies or signal-frequency components being used. The error introduced by the gain becomes greater with frequency, and for accurate measurements the allowable ratio of voltage output to voltage input must be reduced as higher-frequency information is processed.

During integration with a tube operational amplifier, any grid current flowing in the minus input will be integrated along with the current through the voltage input, except when this current is bucked out through a dc pass from output to input. It is not usually practicable to try to adjust a wide-band operational amplifier for "zero" input current, since this condition is not as stable as is a fixed value of grid current appropriate to the input tube type and amplifier design. In low-frequency amplifiers using electrometer tubes as input elements, extremely low values of input grid current can be obtained with good stability. In wide-band units, higher values must be tolerated. Once the grid current has been set to a known value, its effect on a given integrating operation can be calculated; and as long as its value is very small compared to the average value of the input current coming in, the effect of grid current can be largely ignored.

Any operational amplifier is limited in the amount of current and voltage it can deliver to its feedback network and to any external load with good linearity. If these limits are exceeded during any part of an operation, the accuracy of at least that part of the operation will be impaired.

Highly linear operations using precision input and feedback components will be accurate only if the source impedance of the signal is quite small compared to the impedance of the input component, or if the value of the input or feedback component is trimmed to allow for the impedance of the signal source. When trimming of components is not practical or the signal source impedance is not linearly resistive, it is usual to process the signal first through a gain-of-one high-input-impedance, low-output-impedance amplifier, so as to allow a low-impedance signal source for the desired operation.

d. Transistors and Vacuum Tubes in Amplifiers

Tubes and transistors are often used together to achieve a particular result, and it is useful to recognize that the two are in many ways complementary. Vacuum tubes and transistors should be considered as separate entities; each is peculiar to its own mode of operation. Even though the two are different in concept, however, there are analogies between them. Transistor and vacuum-tube data, such as curves depicting the behavior of these devices under very closely defined conditions, demonstrate such analogies but are more useful to the designer than to the user. Hence we will not attempt to duplicate here what is readily available in tube manuals.

It can be shown that the gain of a linear transistor amplifier is set by external conditions, and similar reasoning can be applied also to vacuum tubes. The current that is produced in the plate circuit of a tube by a voltage signal acting on the grid is best studied by postulating that the plate current can be replaced by a generator having an internal resistance. Also, a vacuum-tube amplifier can be viewed in terms of the constant-current form by replacing the voltage generator in the constant-voltage form with a current generator shunting the internal resistance. Such approaches appear to be valid, but not very useful. Good electronic textbooks classify vacuum-tube amplifiers into several types according to the following (105):

1. Electron transport mechanism.
2. End use.
3. Grid bias.
4. Circuit configuration.
5. Signal frequency.
6. Interstage coupling.

Circuits in which tubes or transistors are used exhibit a gain that depends on the ratio of impedances. Of course, the gain equation can be derived for both devices in terms of mutual conductance; also, the transfer curves of both devices have striking similarities (i.e., the plate current versus grid voltage in the case of tubes, and the collector current versus base-to-emitter voltage in the case of PNP transistors). It has been said that the cathode-follower (grounded plate) is identical in operation with the common-collector configuration, the two differing only in concept. A similar argument has been expressed for the common-base amplifier and the grounded-grid amplifier and, likewise, for the common-emitter amplifier and the grounded-cathode amplifier, allowing for different input impedances for the two devices.

For further commentaries on the chemical uses of operational amplifiers, such as summation, integration, or differentiation, or on their application as standard voltage sources or differential amplifiers, see Reilley (153). He also describes a few multiamplifier (or consecutive) circuits in which, for example, one integrator followed by another yields a response that is the *product* of the transfer functions of each circuit. Another use is the analog computation of a variety of differential and algebraic equations, such as linear simultaneous equations.

4. Selecting an Optimum Power Supply

The versatility of modern power supplies permits dc outputs to be maintained precisely at constant levels, or to be adjusted quickly and accurately to satisfy the power requirements for a wide range of applications. In other words, since power supplies in modern instruments are no longer limited to the old rule of "battery substitutes" and electronic B+ sources, the technology and terminology associated with these versatile components have become more complex. The following guidelines, while not exhaustive, are typical of the kind of considerations a chemist might develop, for an awareness in building circuits.

In most analytical applications the voltage and current requirements, as well as the output mode (i.e., constant voltage or constant current), are known or can be readily determined.

a. Voltage Output

A power supply with a voltage rating about 10% higher than that required is recommended. This provides the necessary flexibility for testing and adjusting the analytical system.

b. Current Output

For optimum performance the current rating of the selected power supply should equal or exceed the maximum instantaneous (not just the average) output demanded by the load. Protection circuits on the better supplies limit the output current to a preselected value, thus protecting both the power supply and the load from damage.

c. Output Mode

Most power supplies are designed to provide either a constant-voltage or a constant-current output. Some, however, are designed to operate in *either* mode and to switch automatically from one to the other at a preselected point in the output range.

d. Secondary Factors

Other guidelines include such power supply characteristics as load regulation, ripple, line regulation, and drift. These terms, expressed in millivolts, milliamperes, or percentage of the power supply output, indicate the source and the magnitude of small, undesirable changes in the current or voltage output of the power supply. In general, specifications should require these characteristics to be at the minimum levels consistent with other characteristics, for optimum operating performance.

e. Miscellaneous

If the distance between the power supply and the load is 2 meters or more, remote sensing through separate leads may be recommended to maintain a constant-voltage output at the load. Furthermore, overvoltage prevention may be desirable as a useful device for protecting delicate loads from inadvertent overadjustment of front panel controls, accidentally opened sensing or programming leads, or failure of components within the supply.

5. Differential Transformers

A differential transformer is one of the most useful devices for converting a mechanical displacement into an electrical signal and therefore is the basis for a large variety of transducers which measure not only displacement but also pressure, force, temperature, and other properties. Unless displacement is the primary variable to be measured, the differential transformer must be connected to some auxiliary mechanical system which converts the quantity being measured into a displacement proportional to that quantity. For example, in a pressure transducer, a simple diaphragm has its center attached to the movable element in the differential transformer. This diaphragm thus comprises the auxiliary mechanical system which converts the measured quantity (pressure) into a corresponding displacement.

A differential transformer is usually composed of a primary winding, a pair of secondary windings, and a movable core. The core motion changes the mutual inductance between the primary and each secondary winding, with one secondary becoming more tightly coupled to the primary, while the other secondary becomes more loosely coupled. If the two secondary windings are connected in series opposition, then the open-circuit output voltage will be the difference between the voltages developed in each secondary winding. In a symmetrically constructed transformer it should then be possible to find a core position near the center of the available core travel where the secondary voltages are equal and the output is thereby zero.

As the core is displaced from its null position in one direction, some net output voltage will appear. Displacement in the opposite direction will also produce a net output, but of the opposite phase. If the direction of the displacement is needed, the transformer output must feed into a phase-sensitive indicating or recording system.

The *ideal* differential transformer (1) has purely inductive coupling between primary and secondaries, (2) is perfectly symmetrical so that the neutral core position results in a zero net output, and (3) has coupling which is exactly proportional to core displacement. In practice, such perfections are not achieved, and a sensitive indicator connected to the transformer actually displays an output that goes through a minimum as the core is moved through its range, but not through zero (usually because of harmonics and stray capacitance coupling).

Capacitance coupling between primary and secondaries will not produce opposing voltages across the two windings; therefore, even when the core position provides equal and opposite inductive-coupled voltages in the two windings, the capacitively coupled voltage remains. Stray capacitance coupling is also a function of frequency, and any harmonic in the primary excitation voltage is exaggerated in the capacitively coupled output voltage.

Usually harmonic difficulties can be kept negligible by appropriate choice of excitation level and the use of a good excitation voltage source. Actually, capacitance coupling can often be ignored, especially if the sensitivity of the indicating system is not set as too extreme. If necessary, there are other ways to balance out this stray capacitance coupling, such as adding a small adjustable capacitor in the circuit.

Ordinarily a transformer output voltage is not exactly in phase with the primary excitation voltage. However, if it is desired to deliver an output voltage to the indicating or recording system which *is* in phase with the excitation voltage, then an appropriate load can be connected across the output terminals of the transformer which will produce the desired phase shift. Quite elaborate networks can be interposed between the transformer output and the indicating system input circuits, but the reasoning involves extensive consideration of the properties of the indicating system and the presence or absence of suppression controls.

6. Electrometers

The following discussion on electrometers is included because many of their aspects bear on electronic circuitry in general and because they have assumed a dual role of unusual importance as a result of (1) the need to handle smaller and smaller signals in the newer analytical techniques, and (2) increasing concern for lower concentrations.

An *ideal* voltmeter would have infinitely high input impedance, so that in making the voltage measurements it would draw *no* current. The electrostatic voltmeter nearly has this desired "infinite-impedance" characteristic for dc measurements, because it uses the force generated between well-insulated charged plates to indicate the voltage. Unfortunately so much charge is required to repel the two plates that the electrostatic voltmeter has low sensitivity for dc measurements, and, although its input resistance is high, it takes considerable transient current from a voltage source to charge its plates.

Ordinary dc voltmeters *do* draw significant currents from the sources they are measuring. A voltmeter with a $10\text{-}\mu\text{A}$ meter-movement has internal resistance of *ca.* $100{,}000 \ \Omega/\text{V}$, while common dc vacuum-tube voltmeters have a $10\text{-}M\Omega$ input resistance. Current drawn by a voltmeter can produce a significant voltage drop in the internal resistance of the source being measured and a resulting error. Compensation for voltmeter current often can be made; but when the potential of a 1000-pF capacitor is being measured, no realistic reading can be obtained even with a vacuum-tube voltmeter (VTVM).

The term "electrometer" usually means a "high-input-impedance dc voltmeter." When such a voltmeter is used to measure the voltage drop across a high resistor, the combination becomes a very sensitive pico-ammeter, of which there are several types.

a. THE ELECTROMETER VACUUM-TUBE VOLTMETER (VTVM)

This is a highly refined dc VTVM. The signal to be measured is applied directly on the grid of an input vacuum tube (electrometer tube), and measurement is made of the change in current through the tube, such currents being amplified before being measured.

The conventional VTVM has a resistance voltage divider of about 10^9 Ω at its input for selecting the various ranges. When a 1-V signal is applied to the input, 10^{-9} A flows from the source to ground through the 10^9-Ω divider. In many applications this is excessive loading. In addition, the input tube grid acts as a source of current of about 10^{-9} A additionally flowing into the input divider and into the source being measured.

The electrometer VTVM is designed with special insulation around its input conductors, and the electrometer tube is built to have greater than 10^{14}-Ω resistance from the grid to the other electrodes. Thus the current flowing from the tube grid is less than about 10^{-14} A, and the input impedance is greater than 10^{14} Ω, which is 10^5 times higher than the value for a conventional VTVM.

The high insulation resistances are achieved through the use of poly-tetrafluoroethylene and silicone-varnished glass. The low grid current is obtained by having a very high vacuum in the glass envelope, low electrode potentials to minimize gas ionization, low cathode emission current, and the proper grid potential to assure that all the emission current goes to the screen and plate, not to the control grid. Furthermore, the tube is operated in darkness to minimize photoemission from the grid, and the envelope exterior is varnished to minimize leakage currents.

Vacuum-tube voltmeters and picoammeters display some *zero drift,* incompletely understood but resulting principally from changes within the input tube (thus the drift cannot be reduced by negative feedback in the circuit). For example, 1 mV/A is the normal zero drift of a well-designed VTVM; picoammeters can achieve zero drifts below 5×10^{-15} A/hr.

Since the zero drift is due to the direct-coupled amplifier in a VTVM, several types of *modulators* have been developed to overcome it. A modulator generates an ac signal which is proportional to the signal applied to its input. The ac signal can be amplified by capacitor-coupled amplifiers with less than the 1-mV/hr drift of the vacuum-tube electrometer.

b. Vibrating-Reed Modulators
(Vibrating-Capacitor Modulators)

These devices have an extremely high input impedance, and the drift is only 10 μV/hr. Typical vibrating-reed voltmeters have sensitivities from 1 mV to 10 V full scale. Because of the low zero drift, high input impedance, and freedom from grid current, full-scale ranges as low as 10^{-15} A are practical. This superior performance is expensive, however, both in price and in equipment complexity, compared with a vacuum-tube electrometer.

c. Mechanical Choppers

These devices, using metal-to-metal contact, are especially suitable for voltage measurements below 10 mV. Full-scale sensitivities of 1 μV or less can be obtained with instruments in which the chopper is followed with a transformer to increase the signal voltage before it is mixed with the tube noise of the first amplifier stage. The limit of usable sensitivity is set principally by the Johnson (thermal) noise generated in the resistance of the signal source.

Electrometer-tube voltmeters are preferred to chopper voltmeters for general-purpose use in ranges above about 2 V, because of their higher input impedance and frequency response, and essentially negligible zero drift in this range. In ranges between 1 V and 10 mV each type has its special advantages. Below 10 mV, however, the chopper voltmeter is superior because it has less drift and tube noise.

d. Effect of Johnson Noise on Electrometer Performance

The full-scale sensitivity of a voltmeter-shunt picoammeter (a voltmeter measuring the voltage drop across a shunt resistor) is E/R, where E is the full-scale voltmeter sensitivity and R is the shunt resistance in ohms. A given current can be measured by using a wide selection of shunt resistors within the ability of the voltmeter to read the voltage.

Johnson noise in current measurement is the noise voltage generated in the shunt resistor; it increases as the square root of the resistance increases. In selecting a voltage/resistance ratio to measure a given current, the higher the resistance value, the greater the voltage and the less will be the interference from Johnson noise; 10^{-11} A measured at 10 μV and 1 MΩ has 10^3 times as much Johnson noise in its measurement as 10^{-11} A measured at 10 V and 10^{12} Ω. This is the reason why it is desirable to measure with as large an input impedance as possible.

Because of input-impedance limitations, chopper picoammeters must measure small currents at low potentials, and the Johnson noise is a serious problem. In a 1-MΩ resistor, for example, peak-to-peak Johnson noise (about 4 times the rms value) is about 0.7 μV, which is appreciable on the 10-μV scales used in measuring 10^{-11} A.

In addition to the Johnson noise limitations, chopper picoammeters are affected by thermal dc emf's (inadvertent thermocouples) in the input circuitry, and by the random current generated by the charge transferred when the chopper contacts break. They are further limited in frequency response because of the carrier frequency generated by the chopper, as well as by limited chopper service life.

The basic point of superiority of the chopper over the vacuum-tube electrometer picoammeter is less zero drift. Because of the thermal noise and charge-transfer current, however, choppers are less stable than vacuum-tube picoammeters for currents smaller than 10^{-8} A; for higher currents (up to milliamperes) choppers are just as stable as vacuum-tube circuits.

e. Photoconductive Modulators

Photoconductive modulators are similar to chopper modulators but do not suffer from charge-transfer currents or limited contact life. However, they have more inherent noise than mechanical choppers and greater dc thermal potentials. They provide a more voltage-sensitive voltmeter than the electrometer-tube type, but offer no particular advantage as picoammeters in measuring currents of less than 10^{-12} A.

f. Insulation Leakage and Resistance

In measuring voltages accurately (including measuring a current by means of voltage drop across a resistance), the insulation leakage resistance of the sample holders, test leads, and measuring voltmeters must be several orders of magnitude higher than the resistance of the source of the voltage. Although this is true for all impedance levels, it is especially important in electrometer work, since many common "insulators" have a resistance comparable with that of the high impedance source. For this reason it pays to use high-resistivity materials like polytetrafluoroethylene, polystyrene, polyethylene, sapphire, or quartz. Many less inexpensive insulating materials [e.g., phenolics, poly-(vinyl chloride), nylon, or the acrylics] have too low a volume resistivity or have surface characteristics inappropriate for electrometer work.

In input connections, the best of insulation can be spoiled by being covered by dust, dirt, powdered graphite, solder, flux, or films of oil or water vapor. Constant attention should be paid to keeping insulation clean and dry. The conductor-to-conductor resistances of two rubber-insulated, flexible test leads lying on a wooden bench top can be as low as 10^9 Ω, much too low for electrometer high-impedance work.

g. Shielding

Electronic voltmeters and picoammeters are so sensitive that they are affected by stray electrostatic fields (voltages as high as 30,000

V can be generated simply by rubbing a polystyrene comb against a woolen suit). Therefore shielded inputs are recommended for most electrometer measurements. Shields can be built to enclose (*1*) just the circuit being measured, having shielded leads running to the electrometer, (*2*) the circuit being measured, plus the electrometer itself, or (*3*) the circuit being measured, the electrometer, and the person making the measurement—in other words, a screened booth.

h. CIRCUIT RESPONSE TIME

Electrometer measuring setups may have response times of many seconds. Although a typical input capacitance may be only 10–30 pF (and this is negligibly small at conventional impedance levels), when 30 pF shunt 10^{12} Ω (as is common in electrometer applications), the time constant is 30 sec and accurate readings cannot be made until 4–5 time constants have elapsed. Thus even extremely small capacitances can make a high-resistance circuit very sluggish.

Negative feedback is used to shorten the time of response of instruments measuring high-impedance circuits. For example, a feedback resistor is connected so that the capacitance of the input circuitry does not shunt the input, and the time constant of the circuit is only about 0.001 as long as when the 1-V voltmeter measures across a resistor with capacitance shunting. This reduction comes from the reduction of the input impedance (internal) of the picoammeter by negative feedback or from the fact that only 1 mV is required at the input for full-scale output instead of the conventional 1 V.

i. UNUSUAL SOURCES OF NOISE

Generation of unwanted charges by friction of moving components in the input circuit of an electrometer can have an appreciable effect because of the characteristically low capacitance and good voltage sensitivity of the instrument. Therefore, care must be taken to provide mechanical rigidity to prevent conductors and insulators from sliding over each other and to keep stray surface capacitances from changing.

Cable noise can also cause trouble in current measurements. When a cable is moved, a shield braid slides over the polyethylene or polytetrafluoroethylene surrounding the central conductor, generating charges which introduce spurious potentials in the essential conductor. So-called low-noise cable can reduce this difficulty, being made with the insulation under the shield braid, coated with graphite powder to provide a conducting flexible equipotential cylinder around the insulator and a conducting path for any charges generated.

D. NOISE AND SIGNAL CONDITIONING

1. Noise

Noise generally may be broken down into two types: broad-band "white" noise (so called because of the analogy to white light) and low-frequency noise (referred to by some as "pink" noise), in which the power varies inversely with the frequency. All resistors, tubes, and semiconductors exhibit both types of noise to some extent.

At a given temperature any resistance component of an impedance generates random wide-band noise because of thermal agitation of electrons. The *noise power* generated is proportional to resistance, temperature, and bandwidth of the circuit. Current flowing in a resistor (especially a carbon resistor) produces a low-frequency noise called "excess noise" or "current noise." Its quantity varies widely with the type of construction of the resistor. This noise power occurring in the low-frequency region varies inversely with frequency.

a. Tube Noise

Tubes have "shot noise," so called because the effect of pouring electrons from cathode to plate is considered analogous to the noise of pouring buckshot into a barrel. It is a combination of the effects of cathode temperature and resistance and the fact that current flow, consisting of discrete charges, is subject to random fluctuation, a matter of statistical probability. The flow of grid current likewise is subject to random fluctuation. The resulting output noise and frequency distortion depend on the impedance into which the grid "looks."

Tubes also have a low-frequency noise, commonly called "flicker noise" which is not nearly so predictable as the wide-band shot noise. This kind of noise is incompletely understood and is characterized by unusual spectral characteristics, but for many electronic devices it dominates thermal or shot noise only for frequencies below about 100 Hz. It is generally found where electron conduction occurs in granular semiconductor material, or where cathode emission is governed by the fusion of clusters of variant atoms on the cathode surface. Almost all electronic elements exhibit "flicker" noise to some degree. Even galvanometers exhibit flicker-noise-like effects. The power distribution of flicker noise varies inversely with frequency; it is predominant (in tubes) over broad-band noise below about 1–10 kHz. It really becomes serious in vacuum-tube amplifier performance below 50 Hz. Therefore, dc measurements are to be avoided if S/N is a problem.

b. Transistor Noise

Electrons crossing any semiconductor "barrier" generate wide-band noise which is proportional to the current and to circuit bandwidths. Low-frequency flicker noise is generated in transistors as well as in vacuum tubes, but is more readily controlled.

The wide-band shot noise has a minimum for any semiconductor carrying a given value of current indicated by the equation:

$$I_{rms} = (3.2 \times 10^{-19}If)^{1/2}$$

where I is the dc collector current, and f is the bandwidth. The constant is twice the charge of an electron (1.6×10^{-19} coulomb). In a transistor operated at 1-mA collector current, the minimum wide-band noise over a 1-MHz bandwidth would be about 18 nA rms at the collector. In the case of low-frequency noise for transistors, the same problem applies as with vacuum tubes in that flicker noise is not mathematically predictable. So-called turnover points, as low as 100 Hz, can be obtained below which flicker noise increases at 6 dB/octave. The turnover point is one below which low-frequency noise exceeds the broad-band value.

c. Resistance Noise

Thermal or "Johnson" noise *power* is proportional to temperature, resistance, and bandwidth. The rms noise *voltage* is proportional to the square root of the absolute temperature, T, multiplied by the resistance, R, and the effective bandwidth, f, of the system, in the equation:

$$E_{rms} = 2(1.38 \times 10^{-23}TRf)^{1/2}$$

The constant shown is Boltzmann's constant, expressed in meter-kilogram-second units (joules/degree K). A simpler formula for use at room temperature is:

$$E_{rms} = (1.65 \times 10^{-20}Rf)^{1/2}$$

The values obtained will be essentially the same from $-15°C$ to $+65°C$. In usual practice it is convenient to use bandwidths of -3 dB in such calculations, on the assumption that the total attenuated noise outside the -3 dB points is about equal to the "rolled-off" noise inside these points.

A 1-MΩ resistor at room temperature generates about 130 μV of rms noise over a 1-MHz bandwidth. However, when such a resistance is shunted by a capacitance of about 10 pF, the noise output above 16 kHz will be rolled off at 6 dB/octave. The approximate noise level at a bandwidth of 16 kHz would then be 16 μV rms for the noise voltage, or about 50 μV peak to peak over a short span of time.

Because resistor noise is typical of all broad-band noise, white noise levels are often specified as "equivalent noise resistance," which allows specification of noise in terms applicable to any bandwidth. The assumed temperature is 25°C, and for any particular bandwidth the rms voltage can be calculated.

d. CURRENT NOISE

Composition and deposited-carbon resistors, to a greater extent than wire-wound or high-grade metal-film resistors, generate low-frequency noise proportional to the applied voltage. This noise stems primarily from the current density. In most resistors this current noise, "excess noise," or "1/f noise," as it is variously called, is about equal to thermal noise at 100 kHz and is negligible abouve 1 MHz. The noise power of current varies inversely with frequency.

A commonly used specification is "noise figure" (*NF*), defined as the ratio of the signal/noise ratio at the collector to the "available" signal/noise ratio at the base; this is normally expressed in decibels and does not include flicker noise. The "available" signal-to-noise ratio involves the noise of the optimum driving resistance; that is, if a transistor exhibits lowest noise when driven by a very high impedance, its noise figure may be very good but its actual noise contribution quite high.

Noise specifications at best are only a general guide, and in-circuit evaluation is the only way, for transistors as for tubes, to evaluate limits of achievable performance, especially for low-frequency noise. Only to the extent that one can *predict* the nature of the signal he wishes to measure and also delineate the characteristic of rejectable nonsignificant signals can substantial improvements in sensitivity and bandwidth be made.

2. Signal-to-Noise Enhancement

Chemistry, as well as the physical sciences in general, in striving for greater accuracy and the detection of finer small-effect phenomena, approaches ever more closely to the fundamental limitations imposed by nature. These limitations arise for the following reasons:

1. All matter not at absolute zero has thermal fluctuations, and such noise has properties derivable from thermodynamic and statistical–mechanical principles.

2. Charge, light, and energy are quantized. Therefore current in transistors, vacuum tubes, and photocells flows by virtue of the uncorrelated transfer of single electrons and hence is statistical.

Although quite analogous to Brownian motion, thermal noise in resistors was not recognized until comparatively late. This "Johnson noise," named after its discoverer in 1928, is a "fundamental" noise in all physical measurements and is present in addition to other noise from more common sources, such as 60-Hz line pickup, building vibrations, and variations in ambient temperature, considered elsewhere in this chapter. The environmental disturbances can, theoretically at least, be reduced to an arbitrarily small value.

In spite of the chemist's concern with noise as a limitation of detectability in his measurement work, he may be unaware of relatively simple techniques which would allow improvement in the results. Coor (39) has made a valuable contribution to this aspect of instrumentation problems in a very concise discussion. Much of the literature on S/N ratios requires abstract mathematics, yet most of the results are simple to understand and should be readily applied to many experimental situations. In his paper Coor reviews the properties of the various sources of noise confronting a chemist making sensitive physical measurements. He then considers various ways of avoiding unnecessary noise in achieving an electrical signal most representative of the quantity being determined. Examples of typical areas in which a chemist may be faced with tiny signals immersed in noise include (*1*) electron spin resonance, (*2*) nuclear magnetic resonance, (*3*) nuclear quadrupole resonance, (*4*) microcalorimetry, (*5*) differential thermal analysis, (*6*) all spectroscopy, (*7*) photochemistry, (*8*) chromatography, and (*9*) electrochemistry.

One must always contend with the possibilities that the noise contaminating a desired signal may be *inherent* in the physical process, arise from the probing action itself, or be picked up from the environment. Consider the familiar Nernst equation applied to the measurement of pH by a glass microelectrode:

$$E_{\text{H}} = -2.303 \frac{RT}{nF} (\text{pH}) + K$$

The equation can represent only the *average* potential at the electrode. In a microelectrode, the concentrations of ions in the critical area fluctuate because they are present in only finite quantities. This microscopic fluctuating concentration shows up as a corresponding fluctuation in the electrode potential; the smaller the electrode size and the lower the concentration, the greater will be the influence of this statistical fluctuation.

A more precise measure of the "true" pH for a glass microelectrode might be obtained by taking an average of many measurements of the fluctuating cell potential over a period of time. Although in theory any arbitrary degree of precision might be achieved in this way, there must also be considered the noise properties of the probe and the electronic system used in making the measurement. The effects of glass-membrane resistance, flicker noise in the glass electrodes and (in amplifiers) extraneous signal pickup from the environment, temperature-fluctuation noise in constant-temperature baths, and more—all must be considered in predicting the effectiveness of increasing the averaging time as a means of gaining greater precision in simple dc measurements.

Good resistors carrying current can be said to have little noise in excess of the Johnson noise, in spite of the facts that electrical conduction occurs by means of electrons and that charge is quantized. However, if a diode (or photocell, transistor, triode, etc.) is introduced into the circuit, the passage of current becomes *governed* by a random, single-event, electron emission process, whereas *metallic* conduction is a large-scale correlated drift phenomenon, with many discrete changes taking place all at once. Therefore current through the single-event devices is subject to large statistical fluctuations, whereas metallic conduction is not. The "shot noise" due to random emission of electrons from cathodes or semiconductor junctions was predicted by Schottky in 1918. All active amplifying devices used in electronics—the transistor, photocell, vacuum tube—exhibit "shot noise," an important factor in determining the overall performance of an amplifying/measuring system. However, in vacuum tubes, so-called space charge effects introduce a smoothing action on this random conduction process and thereby reduce the apparent "shot noise" otherwise expected.

Miscellaneous sources of noise, coming mainly from the environment, include 60-Hz and higher harmonics from power lines, radio stations, motor and switch sparks, corona, and so forth. Fortunately, except for power line frequencies, which are very sharp lines at 60, 120, 180, and 240 Hz, the region between 10 and 10^6 Hz is comparatively quiet of environmental noise and can be considered a good one in which to have information appear.

a. NATURE OF THE DESIRED ELECTRICAL SIGNAL

One of the first considerations in system specification for improving S/N is the nature of the information to be handled. The simplest source is a dc instrument in which the quantity of interest is steady or varies very slowly with time, and the information carrying the quantitative magnitude may be handled as a dc signal. Other systems may carry this information as the amplitude of sinusoidal or square-wave signals. These ac signals may result from modulation of the phenomena being measured or from the nature of the transducer readings. In the latter case a steady or dc input to the transducer is converted by some chopping or modulating action into an ac signal, as is done in a vibrating electron-capacitor electrometer.

Sometimes in designing a system there is a choice as to where the band of information will lie. In atomic absorption spectroscopy the changes in transmission of resonance radiation through the flame can be determined by a dc measurement, or the resonance radiation can be chopped or alternated between the flame path and a reference path at a suitable frequency (e.g., 400 Hz). By chopping, the band of information is removed from very low frequencies (dc) to near the chopping frequency. This permits the use of ac amplifiers, avoiding the inherent drift and flicker noise in dc equipment. Furthermore, it enables the use of other techniques for improving S/N.

After the frequency at which the information is to appear has been chosen, the question arises of system bandwidth. If the quantity being chopped or modulated has a steady or dc value, the bandwidth is very narrow. However, if the quantity is changing with time, the bandwidth required to carry the information will be greater. This is necessary for faithful reproduction of the original time-dependent quantity, since every element component of the system through which the information passes must have a bandwidth large enough to pass all components of the original signal. This should not be overdone, however, since most noise sources increase with wider bandwidths.

b. DYNAMIC RANGE

Dynamic range is a concept often overlooked in system design. Any measuring system will have its own irreducible noise level and drift, imposing lower limits on the signals that can be handled. At the other end, the system will always overload, becoming nonlinear at sufficiently large inputs. Coor (39) defined dynamic range in two ways: (1) the ratio of the signal producing full-scale output to the signal corresponding to the inherent system noise and drift, and (2) the ratio of the noise level producing overload to the signal that corresponds to full-scale output. The first definition is more useful in quiet systems, whereas the second serves better where the input S/N is less than unity.

c. Noise Figure

Although any physical system containing some quantity to be measured has a certain irreducible noise associated with it, to transduce, amplify, and measure the desired signal with the least possible increase of the noise associated with such processing is desirable. Doing this requires that the transducer, amplifier, and other equipment which might be used degrade the initial S/N as little as possible. Therefore it is useful to have a quantitative measure of the S/N degradation which occurs in passing information through some system. One of these is the so-called noise figure, defined as:

$$NF = 10 \log \frac{S_i/N_i}{S_o/N_o}$$

where S_i/N_i is the ratio of the available *input* signal and noise powers, and S_o/N_o is the ratio of the available *output* powers, expressed in decibels. A perfect signal-processing system has an NF value of 0 decibels, while one that degrades the S/N ratio of the signal source by a factor of 2 has a 3-db NF.

There are a number of other ways to measure and designate noise. These can be confusing if not well specified whenever quoted.

Marzetta, of the U.S. National Bureau of Standards, advocated the concept of "equivalent noise temperature" in his comparisons (120) of thermal noise performance for a variety of low-frequency signal detectors. He pointed out that the terms "noise voltage" or "current," "noise power," "noise factor," "noise figure," and "noise temperature" quite commonly express the performance of signal detectors. However, in attempting to compare the noise limits of various types of detectors, the use of noise voltage or current is misleading, since two devices with identical noise figures could have quite different noise voltage or currents if their source resistances differed greatly. The noise factor or its logarithmic equivalent, the noise figure, could be used, but such dimensionless terms make it difficult to visualize the relative advantage gained by noise reductions, as the theoretical limit is approached in several modern devices.

Marzetta defines the equivalent noise temperature in the equation

$$T_n = T_{290}(F - 1)$$

where F is the noise *factor* equaling the ratio of the total noise power of a device (including signal source) to the noise power in the signal-source resistance. Then an amplifier, having a noise figure of 3 dB

($F = 2$), has an equivalent noise temperature of 290°K (which means the amplifier generates as much noise power as its associated signal-source resistance generates at 290°K). Marzetta's log-log plot of the thermal noise performance of low-frequency signal detectors, showing equivalent noise temperatures over 7 decades versus signal-source resistance in ohms over 17 decades, includes data for galvanometers, electrometers, chopper-transformer voltmeters, and operational amplifiers of four types.

d. System Grounding

In designing a system for minimum noise, system grounding can be one of the most frustrating problems encountered by a chemist. In most laboratories several electrical grounds are available: the water or gas pipes, the third (neutral) conductor in the 60-Hz power distribution system, or, less frequently, a building ground-bus. For small signals *none* of these grounds suffices, since the usual ground system contains hundreds of millivolts between two different points. The chassis of electronic equipment should be grounded to only *one* point of a distribution system, and *all* shields should be referred to this point. Often more extensive precautions will have to be taken because of such factors as ac magnetic fields from transformers, which might induce potential differences, even if the chassis has zero resistance. Avoiding these ground-loop voltages is often difficult.

E. DECISIONS INVOLVED IN ELECTRONICS DESIGN

1. Direct- Versus Alternating-Current Systems

From the standpoint of signal/noise ratio, it is important to decide whether the electrical signals to be measured and handled will be ac or dc. In general, the *original* signal from the system under observation will be not ac, but will have some steady or slowly varying value. It is necessary to provide in the apparatus for some means of "zeroing the system" in such a way as to leave as much of it as unchanged as possible. The ideal way would be to reduce only the phenomenon of interest to zero, leaving the transducer or probe and all following equipment unchanged. To avoid errors from noise and drifting, the zeroing should be done frequently.

If the quantity of interest and the output of the transducer are dc, the simplest way to make the measurement is to amplify and record the signals in dc form. If the signals are small, the system is likely to drift and to be subject to all the other undesirable characteristics of 1/f flicker and environmental noise.

If the system could be zeroed very frequently, say 100 times per second, several advantages would follow: (1) the rapidly corrected zeros would easily keep up with system drift, (2) ac amplifiers and ac processing techniques could be used, avoiding 1/f noise, (3) it might be possible to have *only* the quantity of interest chopped, interfering elements being unaffected. This would allow the chopped signal to be separated out by some signal-processing arrangement after amplification.

2. Choice of Transducers

It is natural to choose the transducer which gives the largest signal with the smallest amount of interference and without sacrificing linearity, accuracy, and stability. If the signal is to be chopped or modulated, or comes through because it is inherently non-dc, then the frequency response of the transducer should also be considered. As far as the signal/noise ratio is concerned, it is important that the transducer has a good NF and a high-signal-level output.

The photomultiplier, often used in photochemistry and spectroscopy, is efficient in converting weak light into electrical signals, has very high frequency response, and has high inherent gain with large dynamic range. It has a low NF, being very near 0 decibel; and when pushed for measuring exceedingly weak light signals, the NF can be further improved by cooling the photomultiplier to low temperatures, even though its gain is strongly temperature and voltage dependent. Occasionally hysteresis effects may be troublesome, since exposure to strong light can cause a shift in operating characteristics from which it may take hours to recover.

Another transducer, commonly used to detect radiation in spectrographic equipment, is the bolometer. This usually consists of a small, blackened semiconductor element used as a resistance thermometer. Some of the difficulties in using bolometers include the following: (1) they have an appreciable Johnson noise since their resistance is usually high; (2) since they are semiconductors and are used with a standing current through them, the "flicker" noise problem occurs to a rather large degree; and (3) they are fairly slow in response and are not suitable where chopping may be used at a high frequency for the purpose of avoiding 1/f noise. Hence chopping at some low frequency, usually in the neighborhood of 10 Hz, is generally used with bolometers. Semiconducting diodes, in the range of wavelengths where they exhibit good sensitivity, give much better NF's than bolometers.

3. Amplifiers

Many of the characteristics useful in selecting amplifiers have already been discussed. Since amplifiers are usually employed to process the lowest signal levels, it is especially important to consider carefully their effect on the S/N. They should have sufficient dynamic range and bandwidth, but no more than necessary to carry the required information, since the *narrower* the bandwidth, the lower will be the noise at the output.

Because the noise occurring in tubes and transistors is dependent on the source impedance to which they are connected, the NF of an amplifier becomes a complicated function of source impedance, frequency, input-device characteristics, and other factors. For any given amplifier, there is an optimum source impedance for obtaining the best NF. Transformers can be used to match source impedances to the inputs of given amplifiers to attain better NF, but there is a risk of magnetic pickup and some loss of NF because of Johnson noise in the winding resistance itself.

Small-signal amplifiers sometimes have trouble with power-line-frequency pickup, a difficulty mentioned in Section V.D.2.d on grounding. One useful technique for avoiding pickup problems of this sort is to use differential amplifiers in the lowest-level stage of the instrument. A good ac amplifier has a common-mode rejection greater than 10^6 for frequencies below 500 Hz. Therefore the proper use of differential amplifiers can greatly reduce spurious environmental pickup of all kinds.

4. Signal Processing

What is generally understood as signal processing amounts to a cleanup job on the information after it has been amplified to a level where there is little further danger that noise will degrade it; that is, enhance the signal and reject the noise. Also, signals which are ac may have to be converted into dc to drive a recorder.

With dc measurements, often a simple low-pass filter is used to smooth the amplified signal. Averaging over longer periods of time can *sometimes* improve the S/N, although there is danger that dc drifts, "flicker noise," and similar effects will cancel the theoretical improvement.

If the signal of interest is at a single frequency, *tuned* amplifiers which respond only to a small band of frequencies can be used, thereby excluding broad-band and discrete noise occurring elsewhere than at the signal frequencies. The narrow band of frequencies which passes through such tuned amplifiers can then be detected (i.e., changed from ac to dc) and averaged in a low-pass filter system. The minimum narrowness of the bandwidth is practically limited by frequency-drift problems. *Any* drift of the signal frequency, compared to the response frequency of the narrow filters, would make it difficult if not impossible to use them.

The way to get around this problem is to have a clean reference signal at the same frequency as the signal of interest, derived in some way from the modulation or chopping process, such that this "reference" is phase-locked to the small signal to be measured. This technique allows one to use very narrow bandwidths for random noise rejection without frequency-drift problems. This depends on the fact that the "cross correlation" of random noise with a periodic signal tends to zero with increasing averaging time, whereas the "cross correlation" of the signal with a replica of itself yields a constant independent of integrating time.

So-called lock-in amplifiers, or synchronous detectors, are instruments utilizing the techniques of cross correlation. For best S/N, a detector should have as narrow a bandwidth as possible, within limits set by the rapidity with which the amplitude of the initial unknown signal can be allowed to change and also by the observation time required to make the measurement. In going to this very narrow bandwith, provision must be made to prevent the center frequency of the detector from drifting off the carrier frequency, and this is what the lock-in or synchronous detector does.

This kind of detector can also be thought of as a switch, driven by the reference signal so that it is connected to one phase of the detected signal for 180° of the signal and then to the other phase for the remainder of the cycle. The result effectively is full-wave rectification of the *signal*, hence becoming a dc component, whereas *noise* will not impose any net direct current on the output. If sufficient filtering is assumed, the output will be a dc voltage proportional to the original ac signal, with noise averaging to zero.

Signal conditioning of other types would include circuitry for range changing, system calibration, output standardization, and excitation sources (for passive transducers).

VI. INSTRUMENT DESIGN CHOICES

After the obvious primary considerations of specificity, accuracy, precision, and initial cost are studied, a potential procurer of an instrument should look at predicted maintenance costs. If the instrument has been on the market long enough, suppliers often offer a preventive maintenance contract which may reveal the likely percentage of down-time and interrupted operations to be expected. The profitable and competitive nature of the instrument business has meant that new developments reach the market very rapidly, and the rush to the market may result in equipment reaching the user before its "bugs" have been eliminated. In making choices, both user and designer should benefit from the following discussion.

A. SIMPLIFICATION

Two bits of advice offered by Lewin (103) bear repeating:

1. Always choose the *simplest* mechanics, optics, and electronics permitted by the resolution and sensitivity required.

2. Obtain data at the slowest rate allowed by the job requirements and the stability of the sample. Fast scans are invariably accompanied by noise and/or distortion.

In connection with the first point it is unfortunate that the need for devices to perform complicated chores often leads to excessively complex instruments. However, a significant example *of the reverse* is the electrolytic analyzer for determining water, invented by Keidel (92). It measures the electrical Faradaic current required to electrolyze water in a gaseous sample flowing through the analyzer. The simplification embodied in this instrument is striking when compared with alternative approaches based on infrared, Karl Fischer reagent, dew point, and so on.

Baker, in discussing the meaning of measurement (10), has commented further on this aspect of simplicity. He complains of "enlarging volumes of data generated by proliferated instrumentation, in which the machinery is becoming so elaborate that it is impractical in reporting results to describe exactly what the experimental conditions of measurement were."

Simplification of a different sort can be accomplished by reducing the versatility built into a device. When the intended application can be so specifically defined as to justify such limited utility, it is undoubtedly desirable. On the other hand, "corner cutting" to reduce *initial* costs can prevent usage in solving unpredicted *future* problems, especially when the instrument is to be applied in research.

B. THE HUMAN FACTOR

Design is easiest but least successful when it lacks anticipation of all factors conceivably to be encountered in the lifetime of an instrument.

Beals (15) emphasized selected details of the human factor in instrument-panel design, dealing with placement, grouping, size, resistance to motion, direction of motion, lettering, pointers, and selection of scale divisions. His advice, intended for aircraft instruments, really applies to instrument panels in general. Maddock (116) discussed the several conditions leading to the best design of scales for indicating instruments. These included particularly the numbering systems, the size and spacing of graduation marks, and the relationship between reading distance and scale length.

The following tongue-in-cheek collection of "Finagle's Universal Laws" (5) offers food for thought for users, designers, manufacturers, and purchasers (like most exaggerations, it contains some truth in every point):

1. In any calculation, any error which can creep in will do so.

2. Any error will be in the direction of most harm.

3. In any formula, constants (especially from engineering handbooks) are to be treated as variables.

4. The best approximation of service conditions in the laboratory will not begin to meet the conditions encountered in actual service.

5. The most vital dimension on any plan or drawing stands the greatest chance of being omitted.

6. If only one bid can be secured on any project, the price will be unreasonable.

7. If a test installation functions perfectly, all subsequent production units will malfunction.

8. All delivery promises must be multiplied by a factor of 5.0.

9. Major changes in construction will always be requested after fabrication is nearly completed.

10. Parts that positively cannot be assembled in improper order will be.

11. Interchangeable parts won't.

12. Manufacturer's specifications of performance should be multiplied by a factor of 0.5.

13. Salesmen's claims for performance should be multiplied by a factor of 0.25.

14. Installation and operating instructions shipped with any device will be promptly discarded by the Receiving Department.

15. Any device requiring service or adjustment will be least accessible.

16. Service conditions as given on specifications will be exceeded.

17. If more than one person is responsible for a miscalculation, no one will be at fault.

18. Identical units which test in an identical fashion will not behave in an identical fashion in the field.

19. If, in engineering practice, a safety factor is set through service experience at an ultimate value, an ingenious idiot will promptly calculate a method to exceed said safety factor.

20. Warranty and guarantee clauses are voided by payment of the invoice.

Webb (189) advises that successful maintenance depends on good instrument purchase specifications, careful equipment selection, sound installation standards, personal supervision of installation, information feedback to designers, and close user–maker cooperation.

A very concise pamphlet (183), issued by the U.S. Navy Electronics Laboratory, which should be required reading for all designers of electronic equipment, pleads for increased consideration of the relationship between man and machines. Its clarification of functions in which man and machine, respectively, excel follows.

Men excel in their ability to:

1. Detect minimum amounts of visual or acoustic energy.

2. Perceive patterns of light or sound.

3. Improvise and use flexible procedures.

4. Store large amounts of information over long periods subject to relevant recall at appropriate times.

5. Reason inductively.

6. Exercise judgment.

7. Use imagination and insight.

Machines excel in their ability to:

1. Respond rapidly to control signals.

2. Apply great force smoothly and precisely.

3. Perform repetitive tasks reliably.

4. Store information briefly and *erase* completely.

5. Reason deductively and compute.

6. Handle highly complex operations involving many tasks at once.

Man's superior adaptability to changing demands will remain a fundamental reason for including human elements in the operation and maintenance of mechanical and electronic systems (27).

C. CRITERIA IN SELECTING AN INSTRUMENT

In deciding what kind of measurement technique to depend on for solving a problem, the chemist may do well to construct a chart similar to the ones often presented by review authors in various journals. One of these, by Melnechuk (125), tabulates the names of the methods on one axis and the various attributes across the columns. The attributes include such items as sample limits, time requirements, specificity, quantitative and qualitative characteristics, sensitivity, and precision. In this way a bird's-eye view can be gained of all the possibilities and their merits pin-pointed.

Once a chemist has selected an instrumental approach to solving a particular problem, he must make further decisions regarding which of many instruments will most nearly suit his needs. There really is little difference, in the way of going about this, from the basic approach best used in making other chemical decisions. However, a reminder of this fact, in the form of a conscious stepwise consideration of all factors, can be very helpful. The analyst should have in mind answers to such questions as the following:

1. What kind of sample is involved (physical state, temperature, corrosion and safety aspects), and how much sample will there be?

2. What is the approximate composition, including matrix and minor constituent(s) of both primary and secondary interest?

3. Are the complete concentration ranges of all components known, along with their chemical and physical properties?

4. What kind of information is desired? (This would include precision and speed, as well as the form in which the data should be reported.)

5. Must the specimen be preserved, or can the technique involve its destruction?

6. What apparatus is already available, or what would it cost (in money and effort) to obtain it? (Similar questions regarding reference or calibration *standards* are also pertinent.)

7. What personal background and experience is involved, including advice or consulting availability?

Intelligent compromise is often the secret in arriving at the final choice. An alternative second choice will profitably be kept in mind. Finally, personal preferences and experience should weigh heavily, especially if major technical skills are necessary.

Smith (161), in a useful commentary entitled "Specify What You Need," suggests that the prospective purchaser ask himself such questions as:

1. What do you need?

2. What will be adequate?

3. What will be *more* than adequate?

4. What will be needed to completely describe what is required—words, graphic media, or both?

He cites several examples in which incomplete specifications having sizable loopholes led to troublesome operation, because of wrong choices made by the vendor or his suppliers. An error of omission by the specification writer can lead to a serious production delay and expensive reprocurement in order to solve a problem after the fact.

Further attention is called to the pitfalls of extraneous tolerances. In this connection Smith's example is the use of title block tolerances on engineering drawings where standard tolerances for fractional, decimal, and angular dimensions are permanently printed on the title block. This may lead to inappropriate, too-tight restrictions on purchased components for which relaxed requirements should be permissible. The extra expense and delayed delivery that may result can be avoided by careful checking of the reliability of blanket tolerances when a design calls for purchased items like meters, relays, capacitors, connectors, shafts, or gears. Another expense can result from reluctance to ease up on tolerances originally established for prototype models; here conservative choices are originally justified but may contribute to unnecessary expense in later production models. It has been said that, as the tolerance on a dimension is decreased below ± 0.015 in., the cost on the part rises exponentially.

At each step, of course, consulting advice, either in the person of an expert or through the literature, should be sought.

D. STANDARDIZATION OF INSTRUMENTS

Since chemists are interested in the chemical meanings of information issuing from instruments, the standardizing problems are basically not different from those encountered in other areas of analytical chemistry. Techniques, however, will differ greatly and often involve more complex relationships. The ultimate test of an instrument's performance will always be: Does it give a precisely correct indication when used to measure the composition of a known standard sample? Furthermore, if adjustments force a quantitatively correct output for one standard, will these same adjustments hold true for similar standards having different compositions?

Analytical chemists in certain fields are fortunate in having standard samples available for this kind of standardization, and there the procedure to follow is clear cut and is well described in other portions of this Treatise. When appropriate standard samples are unavailable or when physical property magnitudes themselves are part of the objective, means to standardize instruments and their components in terms of basic physical (electrical, dimensional, photometric, radiation, mechanical) parameters are necessary.

Standardization benefits particularly when the terms used in various languages are clearly translatable. The utility of multilingual dictionaries, such as the one for molecular spectroscopy developed by the International Union of Pure and Applied Chemistry's Commission on Molecular Structure and Spectroscopy (84), is self-evident.

The existence of the U.S. National Bureau of Standards Electronic Calibration Center is of great value to analytical chemists and other scientists. Housed in a wing of the Radio Standards Laboratory at Boulder, Colorado, it provides government, industry and the military services with access to an increasing range of primary electronic standards. The mechanism involves calibration of interlaboratory standards for such quantities as voltage, power, and impedance. These interlaboratory standards are available for assuring the accuracy of reference and working standards used in individual laboratories.

The Office of Basic Instrumentation was established at the National Bureau of Standards in 1950 with the responsibility of coordinating and administering a program of research and development. It was charged with a policy of cutting across traditional division lines. In spite of its small size it served as a research, reference, and consultation center for government agencies and their contractors. Drawing upon the competence of existing groups, it aided in originating projects to solve recognized problems and to probe particular fields not under adequate study elsewhere. Standardization is invariably an objective of all of these activities.

The complexity which has accompanied the rapid growth of electrical instruments has become so great that branching chains of measurement are necessary to extend the primary material standards to the shops and laboratories where each precision step is vital. The large number of links in such a chain of events involves the highest practical level of accuracy at each step to achieve the quality desired in the ultimate product, be it chemical or instrumental. The whole system is based on the fundamental units of length, mass, and time—the meter, kilogram, and second—as well as on the basic electrical units—the absolute ohm and the absolute ampere. The former (ohm) is derived from the absolute henry (based on an inductor of accurately known dimensions). The latter (ampere) is established in terms of the magnetic force on an accurately dimensioned, current-carrying coil, measured with a current balance.

To the extent that present electrical standards depend at least partly on mechanical standards, their precision will be seriously limited in comparison to the standards of length and frequency. The practice of depending on particular material examples of a standard cell, a resistor, or a capacitor inconveniently maintained at central standardizing laboratories is yielding to the development of better electrical standards, which should eventually have a quantum basis.

At this point it is well to be reminded again that precision measurements of all types depend heavily on proper environment. This would include availability of a well-lighted area free of mechanical vibration and shielded, when necessary, from radio and other electrical interference. It should include also regulated, well-filtered, well-grounded power lines. Calibration checking of standards usually requires dust-free air of controlled temperature and humidity.

Periodic conferences on standards and electronic measurements both reassure allied technologies of continuing high-level concern and permit innovating modifications to keep abreast of needs. The chemist who thinks in terms of chemical *quantities* can ill afford to ignore the standardizing steps which ensure the validity of the equipment he uses to measure his chemical quantities. The objective of official groups responsible for policies on standardization in the United States (27) is to provide the procedures, materials, and techniques to encourage the trend away from old prototype standards and toward the basing of all measurement standards on natural physical constants, as commonly available as possible.

McNish (123) has emphasized that the standards which physically embody three of the fundamental units (length, time, and temperature) are subject to change, in line with continual efforts to make all measurement units more effective tools. For example, the man-made meter bar was replaced as a length standard by the wavelength of an energy transition in krypton-86, and McNish predicts that this in turn may be replaced by a length standard based on the laser. Such a system avoids rigid centralized control.

The U.S. National Bureau of Standards thus functions as a scientific rather than a regulatory body. It purposely lacks authority to endorse or approve other standards laboratories. Rather, it strives to promote the competence and effectiveness of such laboratories by providing calibration and certifying standards and information intended to be useful in designing and planning new laboratories and in developing procedures and equipment.

McPherson (124) has described the requirements for local standardization laboratories. They include those concerned with temperature and humidity control, as well as freedom from vibration and dust. Economy of space, provision of adequate personnel and equipment, and coordination with NBS services are other problems discussed by McPherson. Many commercial standards laboratories of high competence have been organized with careful evaluation of performance through round-robin interlaboratory checks. Their selection is best carried out in the same manner as that used for any firm rendering professional services.

Standards of several types exist. These include the National Prototype Standards, on which all measurements of our present system are based (involving arbitrarily agreed definitions). Reference standards are derived from national ones to establish working standards which are intended for everyday use.

Chemists have ample access to reference standards for voltage (6) and resistance (8). For use in laboratories the Weston cadmium standard cell has long been the basic working reference standard for voltage. More recently, secondary-standard voltage references with 0.01% accuracy and greater flexibility have become available. Some of these permit significant current flows (e.g., 5 mA) without sacrifice of the specified 0.01% accuracy. Resistance standards include such materials as manganin, constantan, gold–chromium, and nichrome. Bridge techniques are most commonly used for measurement comparisons of resistance materials. Standards-quality resistors of an accuracy of 0.01% or better are commercially available in values from 10^{-5} to 10^{5} Ω and higher.

VII. THE PRODUCTION AND HANDLING OF DATA

It is well to keep in mind that the main product of the proper utilization of instruments is a set of *data* from which some individual(s) may make decisions or take actions of a wide variety. The ways in which data can be "handled" concern both the "producer" of such data and the "consumer." Superficially, data handling may be thought of as involving the replacement of the conventional array of recorders with computers and digital outputs. But this description is inadequate since data handling encompasses functions and operations beyond the capabilities of conventional systems and the limitations of human handling.

The primary function of a data system is to scan a large number of information points and to present the derived information in an organized and generally reduced form through some kind of readout device.

The term "conversion to reduced form" refers to a number of mathematical operations which may be performed on data before their presentation to the operator. These may range from general calculations of almost any simple process-control function, all the way to extensive statistical evaluations and trend analyses. These systems make possible rapid studies of processing parameters and economic analyses of what is occurring during a production or semiworks operation. The presentation of essentially complete information in a predigested form can be considerably more meaningful for studies involving analyses or process control. Computers may enter directly during the data collection or may later take the data and perform additional mathematical operations, or both.

A. JUSTIFICATION OF DATA SYSTEMS

Analysts concerned with routine data reduction or with the use of mechanical techniques to assist in their own statistical studies should undertake to keep informed about trends in this field. The merits of digital versus analog systems and the pros and cons of versatile and specialized computing devices are among the aspects to be considered. Details such as the means of storing data (e.g., punched paper or magnetic tape) and the form of data output (typewriter, recorder trace, X-Y plotting, etc.) will bewilder the analyst who is uninformed about the choices to be made.

Automatic data handling seems best adapted to large installations where considerable cost is involved in the basic equipment, and it is in such cases that economic justification is easy. Justification of data reduction equipment is heavily affected by the high purchase cost per measurement point and the expense of stocking spare parts. Costs may be allotted to four major areas: design, purchase cost, operating cost, and maintenance.

In view of the large investments involved in most data-handling systems, they seem most applicable to new plants or to old processes undergoing extensive modernization. However, an understanding of the possibilities should contribute to more reliable studies, both of analytical research and of processes which are unusually dependent on analytical measurements.

Chemical processes which require the handling of large numbers of data in chart form or in manually read logs of data points would experience immediate advantages from introducing at least partially computerized adaptation of data-handling systems. The real justification is cost reduction, through substantially improved operating efficiency and better quality control. In general such an achievement requires a very thorough coverage of process information, and this in turn entails both more and better information-collecting devices, namely, analytical instruments.

It follows that, although data handling may be predicted to have a key part in the development of modern processes, it can never be any better than the sensing elements and process analyzers which are available. The making of intelligent choices among these, together with the development of improved devices and the adaptation of totally new techniques, must not only continue but also accelerate in coming years if satisfactory progress is to be achieved.

B. RELIABILITY IN DATA SYSTEMS

The foremost problem in obtaining satisfactory data-handling installations is certainly that of reliability. While the consumer considers the responsibility for reliability to lie primarily with the manufacturer, maintenance is ordinarily the problem of the user. Manufacturers have responded to these needs by converting heavily to solid-state devices, high-quality components, and minimum numbers of vacuum tubes and moving contacts. Their circuits have been designed to be operative at conservative ratings, and designs are established with plug-in units, self-standardizing circuits, built-in automatic checking, and trouble-shooting features.

Systematic maintenance, generally involving preventive maintenance programs, seems to be essential for consistent reliability in the operation of equipment of this sort. The maintenance skills required constitute no small problem and usually involve retraining already-skilled men. Furthermore, data presentation must be both simple and readily understandable by operating personnel, without sacrificing the total information content which supervisors, analysts, and process engineers require in order to make "back-deductions" whenever after-the-fact studies become necessary.

C. INDICATING AND RECORDING DEVICES

We have emphasized that the experimental end result, as far as the analytical chemist is concerned, is the readout or indication of the values he is trying to measure. The pressure gage, electrometer, or length of mercury in the thermometer stem are examples, but increasingly the representation is digital or continuous recording, giving a permanent record of the measurement as a function of time or of some controlled parameter. The moving pen on circular or strip charts is perhaps the most common, and multirange and multichannel recorders are commonplace when numerous conditions and/or sampling locations are involved.

The modern high-speed strip-chart recorder dates commercially from the late 1930s, when the crystal pen-motor drive first was incorporated into analog recorders. About this same time the first of the multichannel recorders also appeared. During the 1950s, position feedback was introduced, considerably reducing inaccuracies in recorder tracing. Subsequently solid-state devices, including integrated circuits, coupled with a much more sophisticated recorder technology, proved feasible. Units capable of operating at very high frequencies, with rapid response and on a multiplicity of channels, came later.

Some circuits, whose rapid response permits following *high-frequency* phenomena, thereby require oscillographic or high-speed chart recorders which have a higher-frequency capability. Many such units use mirror galvanometers for recording. Being low-mass devices with very high sensitivity, mirror galvanometers achieve a minimum response time. They can record high-frequency and square-wave data with much more fidelity than most pen-writing recorders, although the high-speed feedback galvanometer encroaches on the lower ranges of the oscillographic recorder.

When the need is for recorded digitized data, either punched-paper tape (for slow speed) or magnetic tape (for higher frequencies) will serve the purpose. However, French (59), among others, has described a device even for digitizing oscilloscope recordings onto punched tapes.

Sometimes one recorder is not enough, and authors have often claimed originality in adding an auxiliary or "slave" recorder to some standard instrument, such as an infrared spectrometer (185). Lloyd (109) summarized ways to obtain "erasable trace" records on chart recorders for continuous display of data not permanently needed. Techniques for doing this are highly developed for cathode-ray oscilloscopes but have been more limited for mechanical recorders.

Recorders can be broadly classified into three basic types: (*1*) galvanometers (direct recording, light-beam recording, chopper-bar recording, and light-beam servo-operated recording), (*2*) null-balance potentiometer recording, and (*3*) cathode-ray tube plus camera recording.

1. Basic Construction of Recorders

The fundamental components of any recorder may be regarded as consisting of an amplifier, a deflection system, a writing mechanism, and the chart record. The amplifier may not only raise the amplitude or power of the input signal, but also convert the signal into a more useful form, such as "chopping" direct current into alternating current or rectifying or demodulating an ac signal. The power of the amplifier is then used to move a writing mechanism, which may consist in the movement of a pen holder, the deflection of a mirror from which a light beam is reflected, or a motor which drives the writing instrument. The writing mechanism produces the record on a chart, which is usually driven at a constant rate so that the data are spread out on a linear time axis.

Recorders for laboratories are available in a wide variety, covering various capabilities requiring a large or extremely small (down to picowatts) power from the input signal. Rates of chart travel range from inches per day to hundreds of feet per minute, and every kind of chart ruling, from linear to hyperbolic functions, can also be obtained from commercial sources.

2. Writing Mechanisms

Several kinds of writing mechanisms are found in common use. By far the most prevalent is the pen-and-ink device utilizing either a low-viscosity ink with capillary-action feed or a high-viscosity ink in a ball-point pen. Unfortunately, the latter requires a relatively high pressure contact of the pen against the paper. In some cases with the capillary-feed pen mechanism, the ink cartridge is a hermetically sealed unit to minimize evaporation and spillage. Disposable cartridges are often used and provide the versatility of various colored inks.

Another technique utilizes electrochemical reactions. So-called inkless writing systems may use zinc-coated paper and a metal stylus. Current flowing from a power supply through the metal contact where it touches the paper oxidizes the metallic zinc coating at that point. Thus a transparent trace is produced on an opaque background, and the thickness of the trace can be controlled through the magnitude of the current.

In another method a fine stainless steel wire stylus touches the electro-sensitive paper, causing an electric current to pass and generate ferric ions. These ions then react with a dye at the surface of the paper, producing a sepia-colored trace. Electric writing styli are lighter than the usual pen-and-ink mechanisms and hence can respond more rapidly to fast-changing signals. The electric techniques avoid the irritations of plugging of the pen capillary and evaporation and splashing of ink, common to pen-and-ink mechanisms. On the other hand, the electric techniques require special papers that are up to ten times more expensive than ordinary chart papers.

Photographic papers (also expensive) are used with galvanometer recorders, where a light beam is the deflecting agent. They have obvious advantages of high speed and freedom from inertial and frictional interaction with the deflection mechanism. Some types require loading in a dark room and developing and fixing by standard photographic techniques, although other papers which are insensitive to visible light, but turn blue on exposure to ultraviolet light, can eliminate these trips to the dark room.

3. Direct-Writing Recorders

The simplest and cheapest recorders are those in which a writing mechanism is directly attached to a standard D'Arsonval meter movement, so that the needle deflection creates a written record. These recorders are relatively inexpensive and can record data in any application in which a milliammeter or voltmeter is involved. Since the meter coil moves the pen, the minimum current that can be reliably recorded is limited by the inertia and friction due to its mass. Therefore these recorders are generally not used for sensitivities better than 1 mA full-scale deflection, although specifications sometimes list sensitivities as great as 50 μA full scale. The earlier models of these recorders involved a pen traveling the arc of a circle on the chart paper, yielding curvilinear graphs. Because of the strong desire for rectilinear readout, considerable effort was expended in devising a system to produce a rectilinear tracing, while still using the intrinsic curvilinear deflection mechanism.

4. Galvanometer Recorders

The heart of all galvanometer-type recorders is obviously the galvanometer itself. Galvanometers provide a flat frequency response up to more than 60% of their natural frequency, but sensitivity decreases as frequency increases. Through the addition of calibrated amplifiers, amplifier gain can be used to compensate for lost galvanometer sensitivity, and the recording frequency can still be extended.

In direct-recording galvanometer recorders, the stylus or pen is attached to a coil in the field of a permanent magnet. They may use either (1) an ink pen, (2) a stylus tracing on carbon-coated film, (3) a heated stylus and heat-sensitive paper, (4) a metal stylus and electro-sensitive paper, or (5) a chopper bar that presses the stylus against the paper, using, for example, a typewriter ribbon to make the impression.

Galvanometer light-beam recorders involve a galvanometer which moves only the tiny mass of a mirror, making possible an increase of the frequency-response ceiling from 1000 Hz for the direct-recording types up to approximately 5000 Hz, which is about the practical limit of light-beam recorders. The beam makes its impression on photosensitive film or paper. Both rapid wet- and rapid dry-process techniques without chemicals nowadays give almost immediate readout.

5. Potentiometer (Servomechanism) Recorders

One of the more significant factors in accelerating the growth of chemical instrumentation has been the application of the servomechanism principle to the design of recorders, thereby reducing the restrictions, so prevalent in direct-writing recorders, involving force developed on a meter coil whose magnitude must be large enough to produce the deflection. Of course, light-beam galvanometers can handle extremely small signals, but they are expensive, delicate, and unsuited for wide, routine use.

Servo-operated null-balance recorders have basic advantages of high sensitivity and independence of lead length. The sensitivity results from the inherent amplification in the servosystem; independence of lead length is realized by canceling the input signal so that no signal flows at balance. These two advantages are gained at the expense of response time, and potentiometers generally cannot operate at speeds faster than about 0.25-sec full-scale pen travel, limiting the response to signals of less than 1 Hz.

The principle is no different from the servomechanism principle previously described for amplifiers. If a potentiometer is unbalanced, an unbalance signal current flows in a direction dependent on the direction of the potentiometer unbalance—that is, if the slide-wire contactor is above the balance point, the current goes in one direction; if below the balance point, the current flows in the opposite direction. This unbalance current is converted to alternating current by an electromechanical chopper and fed into an amplifier. The amplifier output is an ac voltage which is applied to one of the sets of coils of a phase-sensitive reversible motor. The other set of coils (the rotor) is connected to the line ac and always has a fluctuating magnetic field associated with it.

Hence the rotor turns only if both stator and rotor are energized by alternating signals that are out of phase with each other. The direction in which the motor turns depends on the phase of the ac output of the amplifier. The shaft of the rotor is linked mechanically to the slide-wire contactor so that shaft rotation drives this connector along the slide wire. A mechanical linkage is arranged so that the direction of drive of the sliding contactor is always that which will reduce the amplifier output signal.

In this manner deviation of the potentiometer from balance produces a signal that initiates several events leading to motion in the slide-wire contactor in a direction to reduce the signal. The contactor, therefore, continually sets the balance point, since that is the only position in which the motor is not energized.

This servomechanism principle is applied in recorders in several forms. In potentiometric recorders the deflection system is the servomotor, and a pen is driven across a chart of paper, that is, a strip chart, by the same drive shaft that causes the potentiometer slide-wire contactor to seek the balance point continuously. Since the servomotor can deliver large torques, the pen-writing mechanism can be quite large and heavy.

In bridge recorders the error signal is the unbalance voltage of a Wheatstone bridge. The same kind of servosystem balances the bridge and drives a pen.

Through the use of negative-feedback-stabilized amplifiers and high-speed servomechanisms, recorders are available which respond to input signals with a time delay of less than 0.1 sec, giving stable, low-noise linear deflections for inputs down to microvolts.

Direct-writing D'Arsonval recorders are suitable for currents in the milliampere range, for potentials of the order of volts, and for dc to low-frequency ac signal variations. Servomechanism recorders are preferable for much smaller signals. Light-beam galvanometer (oscillographic) recorders are appropriate for low-level, high-frequency applications. The following sections discuss the appropriate recorders to be used with various kinds of signals of three different types.

a. Large Signal, Low Source Impedance

A low source impedance means that appreciable currents (milliamperes) can flow from the source without significantly affecting the source voltage. It may be desirable to monitor, for example, the voltage applied to a lamp in a photometer, a heater in a bath, electrodes in an electro-deposition cell, or the coil of an electromagnet. Here the current likely to be drawn by any recorder is generally negligible compared to the current already flowing. Therefore a direct-writing recorder can be used; and since these devices are essentially moving-coil meters with a writing attachment, they would be applied with all the factors in mind that govern the use of ammeters or voltmeters. That is, an *ammeter* is connected in *series* with the load, and its effective resistance (parallel resistance of coil and shunt) must be small compared with that of the load. A *voltmeter* must be connected in *parallel* with the load, and its effective resistance (coil plus multiplier) must be large relative to the load.

In some cases small signals from sources of either high or low impedance are fed into amplifiers which are designed to give an output suitable for amplification to direct-writing D'Arsonval recorders.

b. SMALL SIGNAL, LOW SOURCE IMPEDANCE

A thermocouple is an example of this type of source. For example, a copper–constantan couple generates only 41.5×10^{-6} V/C difference in temperature between the hot and the cold junctions, and the total resistance of the couple is of the order of ohms (depending on wire diameter). If the thermocouple is in good thermal contact with its surroundings and the surroundings have a large heat capacity, the current drawn from the thermocouple will have only a small effect on its voltage (otherwise appreciable errors in temperature measurement can occur when significant currents are drawn from thermocouples). This kind of signal produces only small torques in the usual electromagnetic meter movements, requiring a very sensitive movement if a direct deflection technique is to be used.

Here, then, an oscillograph (galvanometer recorder) would be appropriate. Preferably, the signal may be balanced against a standard voltage in a potentiometer circuit, so that no current at all would be drawn from the signal source at balance. A servopotentiometer recorder, with a high-gain feedback-stabilized amplifier for balance detection, would be used in this case.

Other examples of small signals from low impedance sources include outputs from strain gages, resistance thermometers, thermal conductivity detectors, and barrier layer photocells.

c. LARGE OR SMALL SIGNAL, HIGH SOURCE IMPEDANCE

There are many experimental situations in which the internal impedance of the source is so large that the recorder must make its measurement without drawing more than extremely minute currents. High-impedance signal sources include the outputs of pH-sensing electrodes, Geiger-Müller counters, ionization chambers, phototubes, photomultipliers, and high-resistance conductivity cells.

In such cases it may be necessary to employ an electron amplifier, or equivalent circuitry, between the transducer and the input to the recorder. Often this preamplifier will be constructed as a part of the recorder.

6. Papers and Charts for Recording Potentiometers

Strip charts may be driven either by a drive sprocket in the instrument and holes in the chart paper, or by a friction drive. For applications in which time accuracy is required, and for long-duration runs, a sprocket drive is preferred because even a slight variation in paper thickness may allow some slip to affect the accuracy of paper movement in a friction-drive assembly. Also, most papers are dimensionally sensitive to changes in humidity. A chart may shrink or expand from its original printed dimensions because the relative humidity decreases or increases, or because paper tension in the recording instrument varies from that in the press on which the chart was printed.

Generally, chart papers are printed at 50% relative humidity and can contract up to about 0.5% from their printed dimensions when humidity drops from 50 to 20%; expansions of 0.5% occur for humidity increases of 50 to about 85%. Charts are often furnished with an end-of-roll warning, which either can be a warning stripe running a short distance along the edge of the inner end of the chart roll, or can be wording printed on the chart near the inner end of the roll.

The core on which the chart is wound can be notched at one end or both ends, for use in braking the supply roll or driving the takeup roll. Although charts can be wound on almost any size of core desired, an excessively small core can be a disadvantage because the difference in diameter between an almost-full roll and an almost-empty roll can be so great that clutch problems develop between the supply roll and the takeup roll.

Charts may be wound with the printed side in or out and may have the round hole on either the right or the left side, depending on where the greater accuracy is required.

7. Recorder and Indicator Performance, Errors, and Interference

In addition to the magnitude of the signal and the internal impedance associated with it, the rate of variation of the signal with time (frequency) exerts a determining influence on the type of recorder that can be effectively used, if rapid variations are to be faithfully recorded. Hence an oscilloscope would be considered in cases where signals are encountered, for example, in high-speed scanning spectrophotometers, kinetics measurements of fast chemical reactions, shock-tube experiments, some mass spectrometers, and electrode mechanism studies. With slowly varying signals such as the outputs of colorimeters, polarographs, furnaces, titrimeters, and gas chromatographs of some types, direct-writing or servomechanism recorders are adequate.

It is worth remembering that there is an inherent time constant characteristic of any writing mechanism, and the input signal is damped or distorted by any recorder to an extent determined by the ratio of the frequency of the signal variation to the time constant of the recorder. Hence, in recording spectrophotometers, the absorption bands of sharp peaks tend to distort much more than broad peaks, because the effective frequency of the input signal for a sharp peak is large, whereas that of a broad peak is small.

For testing pen recorders, a device by Thompson (180) which is primarily a chromatographic peak simulator has been found useful.

Aaker (1) has discussed how potentiometric recorder performance affects chemical analysis. In his paper he examines four principal elements in potentiometer recorder performance: dynamic accuracy, deadband, linearity, and interference rejection.

Dynamic accuracy determines the capability of an instrument for tracing peak height without overshoot and for following small rapid peaks. Recorder dynamic performance is critical in regions of maximum trace curvature, where acceleration demands are greatest. Since maximum accleration occurs at the top of the peak, good dynamic characteristics are required to trace the peak height accurately without overshoot. Very demanding accelerations occur also at the initiation and the termination of a trace.

Deadband is the range through which a measured quantity can vary without initiating response (American Standards Association C39.4). It is easy to measure and provides a good test of the components of the recorder. The six main factors contributing to deadband are (*1*) the servomotor torque, that is, the driving force behind the pen; (*2*) the system inertia and friction; (*3*) the feedback slide-wire, whose resolution sets the lower limit to the deadband; (*4*) the amplifier gain, which, if too small, may force the amount of error signal required to move the pen to be too large; (*5*) interference rejection (excessive interference can create deadband); and (*6*) source resistance (any external circuit resistance effectively attenuates the error voltage, thereby increasing deadband). Deadband inhibits the ability of the recorder to monitor faithfully the signal from the detector. It may be observed on the recorder trace as a stepping movement at the trailing edge of the trace or as a rounded or flat peak. On a slow-moving peak, the recorder pen will fall short of the actual peak by an amount equal to one-half of the deadband. Similarly, the pen may fail to return to the actual base line by an amount equal to one-half the deadband.

Insofar as *linearity* is concerned, it can be no better than that of the slide-wire. Linearity is limited by the resolution of the slide-wire, its mechanical construction, and its resistance-loading effects. The usual slide-wire is wire-wound and will have a resolution dictated by the number of convolutions. A 1500-turn slide-wire has 0.06% resolution; a linearity of 0.1% is commonly available.

Interference, defined by the American Standards Association (C39.4) as "any spurious voltage or current appearing in the circuits of the instrument," will occur either as transverse interference, "a form which appears as a voltage between measuring circuits," or as longitudinal interference, "a form which appears between the measuring circuits and ground."

Transverse interference may be caused by inductive coupling, much as in a transformer, where the primary would be the interference source and the secondary would be the input leads. Longitudinal interference is introduced primarily by capacitive and resistive coupling in the circuits of the analytical instrument. It is harmful when it is converted to transverse interference by returning to ground through the measuring circuit of the amplifier. One way in which this may occur is through leakage capacitance from the recorder circuits to ground. Another route to ground could be the leakage capacitance to the input transformer secondary. The resulting current produces an undesirable signal in the amplifier.

The most obvious means to minimize the effect of transverse and longitudinal interference is the inclusion of a filter in the input circuit of the recorder. Such filters are usually RLC or RC type, tuned to 60 Hz, so as to attentuate the 60-Hz signal introduced at the recorder input. This transverse-interference rejection depends on the attentuation achieved by the filter. Guard shields may also enclose portions of the circuit to serve as electrostatic barriers.

Interference appearing at the amplifier input, if of the same phase as the signal, will be amplified and behave as if it were a legitimate signal. On the recorder trace it may appear as a zero shift. Even if the interference appearing at the amplifier input is out of phase with the signal and, therefore, does not cause pen movement, it is still looked upon as input by the amplifier itself. This may force the amplifier to operate at an inefficient point on its characteristic curve with reduced amplifier gain. Deadband is increased, and operation becomes more sluggish.

Precautions worth taking during operation to minimize interference include the use of well-shielded cables to extend the guard which may already be in the recorder; physical removal of any obvious sources of interferences from the recorder; and elimination of ground loops by tying the input circuit to ground at one point only. Grounds are rarely at precisely identical voltages, and any voltage differentials inadvertently connected through multiple grounding will pass through the recorder quite readily.

In discussing accuracies for dc indicating instruments, Stolar (173) categorizes ten kinds of error which can be caused by faults in the mechanical construction of the instrument and which can plague recording indicators as well: conditions such as temperature, stray magnetic fields, and magnetic mounting panels; mishandling due to mechanical shock and electrical overload; and exposure to extreme environmental service conditions, such as vibration, temperature, shock, and humidity. Various comments on these ten types of error follow.

a. FULL-SCALE ERROR

This exists if the actual full-scale current or voltage differs from the rated full-scale values for the instrument. Such variations typically are caused by necessary manufacturing tolerances for spring torque, flux density, and turns on the moving coil of an indicator.

b. BALANCE ERROR

This exists if the pointer moves when the movement is rotated. Moving elements of pivoted instruments, if supported on jeweled bearings providing a low-friction support, have their static balance necessarily accomplished by the use of adjustable balance weights along two mutual perpendicular axes. Since these cannot be mechanically perfect, some error will exist.

c. FRICTION ERROR AND ROLL

Turning actions of pivots in jewel bearings cause some friction which will appear as a readout error in an instrument having a relatively low torque/weight ratio. Friction acts as a retarding torque and is usually eliminated by tapping the instrument lightly. In instruments used in a vertical position there may be a tendency for the pivot to roll up the side of the jewel, yielding an additional error known as "roll." Here again "roll" is a function of torque/weight ratio and can be minimized by lightly tapping.

d. Scale Error

Data for the design of scales are obtained by applying fixed values of current or voltage to a group of instruments and measuring the resulting angular deflection, which may vary as much as 2° from instrument to instrument. The printed scales are designed by computing the average angular deflection for each cardinal point in a group of instruments tested. Differences which occur in the markings of the printed scale are known as scale errors.

e. Spring-Set Error

When springs are deflected for extended time intervals, they may *not* return immediately to the original zero position when current is removed; this is sometimes referred to as spring-set error, due to mechanical hysteresis. Such a pointer may require several hours to return all the way to zero, indicating that it suffers both temporary and (semi) permanent spring-set error. The temporary spring-set error is usually about four times as large as the permanent spring-set error, depending on the type of spring material and the angular deflection.

f. Magnetic Error

This error is caused by impurities located within the moving element, which may be magnetized by the instrument magnet, and acts like hysteresis, in that readings taken by deflecting the pointer up-scale will differ from those taken by deflecting the pointer down-scale. It is most apparent in instruments of low torque/weight ratio and relatively high air-gap–flux density.

g. Temperature Error

Both the torque spring and the magnet are affected by temperature. The torque on the spring decreases with an increase in temperature, whereas the magnetic flux declines with increasing temperature. Hence the two effects are partially compensating insofar as their additive effect on the current required to produce a given deflection is concerned.

h. Fall-over Error

Since a certain amount of end play must exist between the moving element and the pits of the jewels, the moving element can tilt slightly in the jewel; this shows up as a mechanical error designated as fall-over. With proper jewel settings, this error can be kept to a minimum.

i. External Field Error

Magnetic fields coming from any force outside the instrument cause a so-called external field error. Instruments using core mechanisms are nearly immune to external field effects because of inherent self-shielding. This error is usually temporary in that removal of the magnetic field eliminates the cause. Nevertheless, large external magnetic fields can produce permanent instrument errors of this type.

j. Magnetic Panel Error

Unshielded instruments can be affected by the proximity of a magnetic panel because of the change produced in the stray-field pattern of the instrument magnet. This type of error can be compensated for by the simple expedient of calibrating the instruments when mounted in the panel specified.

8. Oscilloscopes

The oscilloscope is a logical extension of the strip-chart recorder, if the X-Y recorder is considered as a natural transition. One can imagine an oscilloscope screen replacing the graph paper, and the electron beam substituting for the pen, the big difference being that the graph drawn by the oscilloscope beam disappears almost immediately in conventional types. Of course, storage-tube oscilloscopes have the ability to retain the written image for indefinite lengths of time, and even most nonstorage types of oscilloscopes can present a fixed visual display *if* the signal is quantitatively repetitive.

For the chemist, the essential working components of an oscilloscope are (power supplies have been discussed elsewhere) (*1*) the cathode-ray tube (CRT), (*2*) the amplifier, (*3*) the time base, and (*4*) the trigger circuit.

a. The Cathode-Ray Tube

This can be considered the end result in a series of transductions. For example, in an experiment in which a sudden change in temperature occurs, a thermistor can be used to convert this to a change in voltage transmitted to an oscilloscope. Within the oscilloscope, the voltage change is amplified and applied to the cathode-ray tube, where it is converted to a visual display on the screen of the tube.

This action of the CRT starts with the *electron gun*, where a stream of electrons is produced. Such a gun contains an indirectly heated cathode which produces a cloud of electrons; these are then attracted down the neck of the conical tube to the screen by a potential of several thousand volts between the screen and the electron gun. The measurement and the control operations consist of elements within the gun focusing the beam to a small dot where it strikes the screen.

The screen consists of a thin layer of fluorescing material, usually cadmium or zinc sulfide deposited on the inner glass surface. A steady stream of electrons striking the screen causes momentarily a dot of fluorescence. If the construction of the tube is symmetrical, this dot occurs directly in the center of the tube. Hence the dot on the screen is similar to a resting pen on a recorder, awaiting signals.

In the oscilloscope such signals are applied by two pairs of *deflection plates*, contained within the tube neck and located midway between the screen and the electron gun. Sufficiently high voltages applied to a pair of deflection plates generate an electrostatic field between them; and if the applied voltages between the two pairs are unequal, the stream of electrons deflects toward the more positive plates. If a 40-V potential difference between the plates moves the beam 2 cm, the "deflection factor" of these plates is 20 V/cm. It is not uncommon for the deflection factor of the vertical plates in the same tube to be considerably different from that of the horizontal plates.

Many CRT's have deflection-plate connections directly accessible on the neck of the oscillograph tube, and one simply removes the internal connections and carefully tapes them to clips on the new signal connections. In the usual applications only the vertical plates are connected to external signals, with the internal horizontal time-base amplifier used for horizontal deflection.

Electrical signals on their way to the CRT screen may suffer slight distortions and phase shifting, especially when passed through attenuators and amplifiers. These effects are usually negligible, however, especially when high-frequency oscilloscopes are used. On the other hand, in a low-frequency instrument it is likely that the observed signal will be a poor replica of the original. In this case one might benefit by connecting directly to the oscilloscope plates; or if some higher-frequency amplifier is available, this could be substituted to precede the connections to the plates.

Alternatively, one can intensify the trace at precise points by using a pulse of the correct polarity and interval. This can be especially useful to co-ordinate some exterior event with the waveform to be shown on the oscilloscope.

Different CRT screens have different *persistencies*. Various compounds or phosphors continue to fluoresce or persist for different lengths of time after being struck by electrons, extending from several milliseconds to as long as 30 sec. The choice of persistency depends on the type of experiments planned, with slow-scan work using a long-persistence CRT phosphor.

A long-persistence CRT used in repetitive scanning tests can give the impression of a stationary image; an example is a cathode-ray polarographic analyzer in which the scan is to repeat every 7 sec, corresponding perhaps to drop times. If a screen with 10 sec of persistence is used, the resulting polarogram appears stationary.

b. ELECTRONICS IN OSCILLOSCOPES

The *vertical amplifier*, often called the *Y*-axis amplifier, receives the signals to be measured, converting them to a form that can be displayed by vertical deflection on the electron beam of the CRT. Most simply, it consists of a fixed-gain amplifier with a calibrated variable attenuator at its input and the vertical plate of the CRT at its output. The signal to be displayed is applied to the vertical amplifier, through appropriate probes or cables, in the form of a voltage or current having been transduced from temperature, pressure, vibration, light intensity, or other source.

The sensitivity control of the oscilloscope vertical amplifier is usually calibrated in volts per division of deflection. For example a 10-V attenuator setting requires that a 10-V signal be connected to the input to produce a 1-division vertical deflection on the CRT graticule. Customarily, the calibrated attenuator is labeled the "volts-per-division" switch and contains switch positions, extending from the most sensitive to the least sensitive, in a repeated sequence, that is, 10, 5, 2, 1 "V/div." In many amplifiers the most sensitive position of the volts-per-division switch also indicates the sensitivity of the fixed-gain amplifier when the actual attenuation is unity.

Chemists using oscilloscopes generally need to have the waveform displayed on the CRT an exact representation of the signal applied. However, unless the user is familiar with the bandwidth of his instrument, he may be seeing a poor reproduction. Bandwidth, in an oscilloscope vertical amplifier, means the sine-wave frequency at which the displayed signal is 70% (−3 dB) of the applied signal. A good rule of thumb is to use an amplifier that has a bandwidth at least 10 times the frequency of the signal of interest.

In some situations a chemist may need to be concerned with the response of the oscilloscope amplifier to a step function or square wave. Despite the square-sided appearance of a step function, a finite time is still required to change from one voltage level to another. This *rise time* is the time interval between the 10% and 90% amplitude points on the rise of the step function.

Of course wide bandwidths in oscilloscopes are not always desirable. High-frequency noise often obscures the signals of interest, and the penalty of unduly wide bandwidth becomes apparent. An example would be the amplifier used in a dropping mercury electrode in polarography. To amplify the small currents requires high sensitivity, yet the column of mercury acts as an antenna, and considerable noise pickup will be evident with wide-band oscilloscope amplifiers. Hence it is necessary to reduce the bandwidth to a point where small polarographic currents actually can be observed. For this reason some amplifiers have a bandwidth control permitting adjustment over a considerable range. Also, when using operational amplifiers, bandwidth can be reduced by a small capacitor placed across the feedback network.

The time base (usually designed as the X-axis) establishes the time value of each horizontal division on the oscilloscope screen. Internally, the voltage change produced by the time base is a linear ramp that rises to some value and then rapidly falls to zero. Because of the appearance of this wave it is called a sawtooth. It is usually developed in a sawtooth sweep generator by integrating a constant current to produce a linear rise. Since the rate of rise of the integrator is related to both current and capacitance, it can be changed to different values by altering the value of either the resistors or the capacitors involved. Calibrated time-base integrators have been designed to extend from 10 sec per division to 1 nsec per division. Such integrator output is applied to a horizontal amplifier, whence it goes to the horizontal deflection plates.

c. Triggering

When a repetitive signal such as the sine wave is applied to an oscilloscope and displayed on the screen, it appears as a stationary waveform, assuming proper phosphor persistence and repetitive rate. This ability to superimpose repetitive signals *exactly* is an important characteristic of oscilloscopes. What makes it possible is that each scan of the time base can be initiated by a trigger impulse derived from the measurement signal.

Actually, a triggered oscilloscope is one in which each horizontal sweep is initiated by a signal other than the time base. Usually the trigger signal is received from the vertical amplifier, although it can come from an external source. Obviously, in the case of a single sweep, the trigger signal can be applied by actuating a front-panel switch. Chemists probably use single events more than repetitive signals requiring triggering;

for example, for the capture of a fleeting transient in a flash-photolysis experiment, the sudden application of a high-energy flash produces chemical changes that can be followed electrochemically on the oscilloscope by using the pulse of light to trigger the sweep. The events that follow immediately in the electrolysis cell can be recorded from the screen by camera or recorded intact on a storage CRT.

In chronopotentiometry, a constant current is suddenly applied to the electrolysis cell, and the resultant potential is monitored. The change in potential upon application of the constant current can be used to trigger the time base. Since a trigger signal can be applied from an external source, it is possible also to couple electrical signals from mechanical devices. For example, in a dropping mercury electrode assembly, one may want to trigger an oscilloscope whenever a drop is dislodged. In some cases a mechanical striker "knocks off" the drop and simultaneously closes a microswitch to provide the necessary pulse triggering the oscilloscope circuit.

Time-delay circuits are also available to offset the triggered time base to suit the application.

d. PLUG-IN AMPLIFIERS

Many modern oscilloscopes accept so-called plug-in units for the X and Y axes, and many vertical amplifier and time-base combinations are possible. As one example, the dual-trace amplifier can accept two signals which are then displayed simultaneously on the cathode-ray tube. The time relationships between the events in the two traces are immediately apparent and may indicate cause-and-effect relationships. The dual traces are usually accomplished internally by switching each channel alternately (e.g., at 100 kHz) into the output amplifier. This is less expensive than the dual-beam arrangement in which two separate cathode-ray tubes are contained in one glass envelope, although the results are similar.

Other special plug-in units involve differential amplifiers, having the ability to amplify and display the difference between two signals. This technique is sometimes useful in canceling or considerably reducing background or residual currents, including noise and line-frequency hum. This ability to cancel smilar signals is called common-mode rejection. A number of special plug-in operational amplifiers, sometimes containing programmable amplifier units, are available.

Special oscilloscopes include the storage oscilloscope, digital readout, and sampling and signal-averaging oscilloscopes. Each of these has special capabilities not possessed by conventional instruments, although their description is beyond the scope of this discussion (140).

e. Appraising Oscilloscopes

A few remarks concerning the appraisal of oscilloscope specifications and performance will be helpful.

(1) Display Geometry

A horizontal trace should coincide with the horizontal scale markings on the graticule. Misalignment generally indicates a need to reorient the CRT (or scale), but can be caused by inadequate shielding or the presence of a strong magnetic field. Likewise, the CRT deflection perpendicularity should be good. That is, the horizontal plates should deflect the beam in a direction truly perpendicular to that of the vertical-deflection plates: deviation should be less than 1° near the center of the screen; typically, a 1° error is a displacement from perpendicular of 1 mm in 5.7 cm.

(2) Trace Bowing

Beam deflection may deviate from a straight line when a trace appears near the outer limits of the useful screen area. A CRT may be tested by using sets of horizontal or vertical lines or by manually positioning the beam rapidly back and forth horizontally and vertically to all four sides of the useful area. Bowing tolerance will depend on the CRT type and its operating voltages. A high-quality instrument will not deviate from parallel lines by more than 1 mm on the edges, or by more than 0.5 mm at the top and bottom. To minimize bowing, some systems have adjustable voltages on special electrodes in the CRT.

(3) Amplititude Linearity

The amplititude linearity of a dc-coupled amplifier can be checked by observing any change of amplitude of a small signal while positioning the display through the useful scan area. For ac-coupled deflection plates, compression is checked by changing the input signal in measured increments, and observing whether the displayed signal amplitude changes exactly correspondingly. Similar comments apply to horizontal linearity.

(4) Range of Sweep Speeds

Whatever the total range of sweep speeds may be, continuous coverage usually is provided. Most of it occurs in discrete calibrated steps similar to those on a voltmeter. A variable sweep–time/division control may also be provided to offer continuous coverage between steps; this may or may not be time-calibrated. Such a variable control is convenient for spreading or compressing a display, or for letting some portion of the waveform cover a particular number of divisions. If calibrated, the variable control allows fractional time measurements also to be made without using subdivisions on the scale.

(5) Time-Base and Sweep Accuracy

This is usually specified in terms of the permissible full-scale timing error for any calibrated sweep. For example, an accuracy of 3% means that the actual full-scale period of any sweep should not be more than 3% greater or less than indicated. Magnified sweeps may well have poorer accuracy ratings than unmagnified sweeps since the magnification is generally achieved by reducing amplifier feedback.

This sweep linearity problem is an important factor often overlooked. Equal changes in voltage on the horizontal plates *should* produce equal changes in spot position. Several forms of nonlinearity may exist in a scope, so there has been no simple way to specify linearities in terms which are significant for all cases. Manufacturers recommend that experience or careful side-by-side comparison is most productive, keeping in mind that the fastest sweeps are usually the most nonlinear, with the most prevalent problem being slowness at the beginning and at the end of the sweep.

(6) Sweep Magnification and Delay

Sometimes it may be desirable to display parts of wave forms occurring considerably later than suitable sweep-triggering signals. Such waveforms can always be displayed on sweeps which last long enough; but if the duration is short compared to that of the full sweep, an accurate examination is difficult. Thus portions of sweeps may be magnified by increasing the gain of the horizontal amplifier and positioning the display so that the desired portion is on screen. Another way is to generate suitably delayed sweep-triggering signals so that fast sweeps may be triggered just before the instant when the signal to be examined occurs.

(7) Vertical Signal Delay

Whenever a sweep is to be triggered by the display signal and the entire leading edge is to appear, the arrival of the signal at the vertical deflection plates must be delayed by a delay cable or delay line somewhere in the signal path. The amount of delay time required slightly exceeds the time that it takes the trigger circuit to start the sweep and the sweep to become linear, or the time that it takes for the CRT beam to be turned on, in case that is longer. The need for signal delay is greatest in the faster scopes, where the time required for sweeps to start and the beam to be turned on occupies a significant portion of the fastest sweep.

(8) Transient Response

This reflects the fidelity with which a deflection system displays a fast-rising step signal. The most common distortions affecting transient response are overshoot, ringing, and reflection from impedance discontinuities in the vertical signal-delay line. These distortions make "clean" step signals seem to have spikes, squiggles, or bumps when they actually do not. Transient fidelity is essential but is seldom specified except in terms of permissible overshoot.

Transient response does not remain good indefinitely without some adjustment, and vacuum tubes are the most frequent culprit for degradation of transient response. Termed "cathode interface," this defect shows up an excessive overshoot after several hundred hours of usage. "Interface" may be due to a loss of vacuum-tube gain, for all but the high-frequency signal components. Transient response, of course, is related to rise time.

(9) Rise Time and High-Frequency Response

An important qualification to be considered in specifying a scope is adequate rise time or high-frequency sine-wave response. Rise time is more important for faster scopes, and passband (or bandwidth) is the more frequently considered specification for slower scopes. The two are mathematically related when fast step signals produce little or no overshoot or ringing. The product of rise time and frequency response should be a factor lying between 0.33 and 0.35 when transient response is optimum. Factors larger than 0.35 indicate overshoot exceeding 2%, while those above 0.4 probably indicate more than 5% overshoot.

Ideally scopes should have a vertical system capable of rising in about 0.2 the time that the fastest step signal applied rises. In such a case, the rise time of the signal (as indicated on the scope) will be in error by only 2%, assuming that sweep timing and linearity are perfect. Yet vertical deflection systems with a rise time no better than equal to the fastest-rising applied signal are often considered adequate—a conclusion which may or may not hold true, depending on the accuracy needed.

If one wishes to check a rise-time specification, a square-wave step signal is applied to occupy almost the full vertical scale; the rising portion of the signal then should be displayed at nearly a 45° slope. This can be done only if the fastest sweep is able to move the beam horizontally, nearly equal to the full vertical scale in a time interval equivalent to the use time of the vertical deflection system.

In situations requiring accurate rise-time measurements involving the fastest sweep, a useful figure of merit for the adequacy of that sweep is

$$M = \frac{T_r}{T_d}$$

where M = figure of merit, T_r = vertical system (amplifier) rise time, and T_d = time per division of the fastest (horizontal) sweep. Figures of merit greater than 1 seldom are found in oscilloscopes having rise times of less than approximately 30 nsec. Figures of merit greater than about 6 offer no particular advantage. Usually most rise-time measurements are made, not to determine *actual* rise time, but to determine whether certain limits are met or exceeded.

(10) Deflection Factor (Sensitivity)

Sensitivity, like frequency response and rise time, is one of the prime factors determining the suitability of a scope. Achieving the utmost in sensitivity requires some sacrifice in bandwidth because of the greater background noise associated with wide-band, high-gain amplifiers. With high amplification, background noise may be evident in a display. When it is, the amount of noise should be specified so that a signal/noise ratio can be predicted. Oscilloscope deflection factors should be stated as peak-to-peak voltage rather than rms voltage, since only sine-wave amplifiers can read directly from a scope calibrated in rms volts. When noise figures are stated in terms of rms voltage, they should be multiplied by 5 or 6 to determine the approximate amount of peak-to-peak deflection that the noise might produce on the screen.

(11) Position Drift

High-gain dc-coupled amplifiers may drift appreciably. After warmup, the maximum amount of drift is often specified in terms of millivolts (or microvolts) per hour. The amount of position change represented by such a drift obviously depends on the deflection factor selected. How long the balance and position controls, which interact somewhat, continue to operate properly without requiring internal adjustments may be important, although nearly impossible to predict accurately. As in many another of these considerations of performance specifications, experience will constitute the best basis for sound judgment.

(12) Input Impedance

The input impedance of most oscilloscopes is specified by input resistance and input capacitance, with input resistance generally falling between 0.1 and 10 MΩ (except in very wide-band instruments, where the input resistance matches the impedance of coaxial cables). Input capacitance is usually between 20 and 50 pF. To minimize the effects of circuit-loading imposed when connecting a circuit or device to the input, the input impedance should be as high as possible. For high-frequency or fast-rising signals, the input capacitance imposes more loading than the input resistance does. It is often recommended that a coaxial cable be used between a scope input and the test point, but the added capacitance of the cable further increases loading. Passive attenuator probes are used commonly, instead of plain coaxial cables, to reduce such loading effects. These probes may attenuate loading by a factor as great as the signal attenuation that they impose (up to 100 times).

(13) Differential, Balanced, or Push–Pull Inputs

Push–pull signals may be introduced into the vertical deflection system of an oscilloscope if the input circuits are suitably designed. Amplifiers of this type are commonly called differential or balance amplifiers. Providing a feature beyond mere accommodation of push–pull signals, they include an ability to cancel or reject, very effectively, signal components equal in amplitude and phase that appear at both inputs.

This feature amounts to a difference amplifier, since essentially only the difference between two signals is amplified. The degree to which unwanted, common-mode signal components can be rejected is termed the "common-rejection ratio." A rejection ratio of 100 indicates that the amplitude of the displayed common-mode signal components is only 1% of what might be displayed without the differential input.

Significant rejection of common-mode signals much larger in amplitude than the signals to be displayed requires difference inputs with *very* high rejection ratios. Most difference amplifiers do not maintain their maximum rejection ratio for weeks at a time, although adjustments usually allow re-establishment of maximum rejection when needed. Differential amplifiers usually reject low-frequency signal components better than high-frequency components, and hence the rejection ratio should be specified for a given frequency.

Other considerations outside the scope of this discussion are the following:

1. "Summing" inputs and switched inputs.

2. Sweep-triggering synchronization details.

3. Triggering level and slope selection (this system meets the need occasionally to *prevent* sweeps from being triggered by low-amplitude signals, such as noise).

4. Dc- and ac-coupled triggering.

5. Automatic triggering and triggering sensitivity.

6. Curve-tracing (*X–Y*) scopes.

7. Writing rate (because short-duration, nonrecurrent signals must be photographed, it may be important to know how fast a CRT spot can move and still be photographed).

D. APPLICATIONS OF DATA-REDUCTION RECORDING

Recording systems that go far beyond mere transcription into analog or digital terms border on being computers and are not considered "instruments" in the context of this chapter. Likewise, on-line systems, often with computer control of one or more instruments, will not be discussed at length, although references to selected key applications incorporating complete computer operation of analytical experiments will be cited.

Much of the activity leading to automatic data reduction occurred as the result of two trends. First, the success achieved by instrumental analysis has bred its own problems, in that instruments can generate more data in shorter times than users can assimilate; and, second, the prices of computer and data-reduction hardware, particularly since the advent of integrated circuits, have dropped to the point of reasonable consideration for use with analytical instruments.

The teaming of chemists using instruments with computer programmers and operators has resulted. The data can follow a variety of routes; for example, they can be sent off-line to a central processing unit via punched paper, magnetic tape, or punched cards. Alternatively, a small, usually inexpensive computer can be dedicated for use solely with a single analytical instrument. A third option involves a number of instruments, either all of the same type or of different types, being multiplexed into a larger dedicated computer, or into a large computer not dedicated but available through time-sharing. All of these alternatives have certain requirements in common, involving getting the analog data into digital form for readout or subsequent computation.

Several authors have discussed in quite general ways digital techniques for data processing in analytical chemistry. Savitsky (156) reviewed the computer handling of spectra, particularly for infrared spectrometers, and described digital recorders tailored to the spectroscopic problem, pointing to the necessary link between the computer and the instrument. The emphasis was in two directions: (1) direct computational processing of the spectroscopic data to improve signal/noise ratios, extraction of band parameters, automated quantitative analysis, and a study of band shapes; and (2) logical processing of spectra to extend the utility of correlation procedures for spectral retrieval in qualitative identification.

Childs, Hallman, et al. (31) reviewed aspects of the use of digital computers in the particular areas of statistical treatment, X-ray analysis, spectroscopy, mass spectrometry, gas chromatography, and electroanalytical chemisty.

1. Chromatography

Walsh et al. (188), in work reported in a paper entitled "Evaluation of Chromatography Column Parameters by Digital Computers," departed from earlier efforts, using digital devices mainly to facilitate the handling of large numbers of data, such as those from several chromatographs, on a time-sharing basis. In demonstrating the power of a computer to provide qualitative information, they claimed that on-line evaluation of column performance by the chromatographer greatly improved the ease and effectiveness of utilization and selection of various types of columns by following an on-line computation of the Height Equivalent to a Theoretical Plate for each component of a mixture. They were able to evaluate columns in regard to their efficiencies toward component types of various polarities. Other parameters were also found useful in practical ways, enabling the evaluation of a gas chromatographic column *in situ*.

Hancock and Lichtenstein (71) discussed practical aspects of on-line data processing, focusing on reducing and interpreting sampled data from a chromatograph. They discussed the utilization of data filtering, peak detection, and peak analysis for the improved resolution of components in poorly resolved peak combinations, finding the Gaussian method of complex-peak analysis superior to older methods.

Davis and Kipiniak (43) described a computer-controlled system for serving 15–32 chromatographs built around a digital computer, keeping peripheral and interface hardware as simple as possible. Their computer system included programs for the following functions:

1. Sample switching.
2. Temperature control and programming.
3. Calibration.

4. Maintenance check.

5. Signal integration.

6. Overlap peak resolution.

7. Relative thermal response tables.

8. Concentration calculations.

9. Display.

They concluded that the use of a stored-program digital computer as an integral part of a chromatograph system, taking over all its control, computing, and display functions, was a sound design principle, both technically and economically.

2. Infrared Spectroscopy

Many papers have described efforts to collect infrared spectral data digitally. Savitsky (156) studied the digitizing and digital handling of infrared data, reminding the reader that the original Coblentz spectra were all digitized, being taken point by point and laboriously plotted. In his studies, Savitsky devoted considerable effort to data smoothing to minimize the effects of noise in a spectrum. His use of least-squares procedures led to smoothing the data and finding derivatives, simplifying the mathematical approaches necessary to achieve high resolution in qualitative evaluations of infrared spectra of considerable complexity.

Neal et al. (139) described an analog computer which will perform the following functions: plot spectra of any required size, convert percentage transmittance signals from the spectrometer to absorbance units, and integrate absorption bands.

Stine et al. (170) reported on various applications of digital techniques to infrared spectroscopy, pointing out that the ability to store and process the enormous volumes of spectral data now produced by spectrometers has opened up vast new possibilities for automatic manipulation of infrared data. Their discussion of digital techniques for corrections or modifications of raw spectral data is particularly useful: (*1*) 100% and zero-line corrections, (*2*) wave number corrections, (*3*) ordinate (per cent *T*) corrections, (*4*) slit-function corrections, (*5*) tracking error corrections, and (*6*) signal-to-noise improvements.

Much of what Stine et al. said applies to spectral data in general, not just infrared. For example, before obtaining a sample scan, data on the 100% (background) and zero lines may be stored in digital form and later used to correct the sample spectrum for aberrations which do not constitute true spectral information. They pointed out that this can be especially valuable when backgrounds are severely distorted by beam clipping, by polarizing accessories, or by other factors.

They categorized 11 computer manipulations of digitized spectral data: (1) per cent T to absorbance conversion, (2) differentiation or integration, (3) complex multiple-band resolution, (4) quantitative analysis, (5) automatic searching, (6) interpretive routines, (7) differential analysis, (8) synthesis of hypothetical spectra, (9) physical studies, (10) replotting, and (11) peak tabulation.

3. Emission Spectroscopy

An early electrical spectrographic calculator involving an electrical circuit based on analog computer principles was described by Epstein (51). This represented an early attempt to replace the mechanical or graphic aids previously used by spectroscopists to reduce the labor required for calculating results. He gave the complete circuit and parts list for the construction of the computer.

Weber (190) described a similar application. However, he used a digital computer and divided his problems into two types: (1) regression analysis, correlation analysis, and improved precision of a spectral line; and (2) calculations of the concentration of the given element from densities of the spectral lines measured on a photographic plate.

Aslund and Cronhjort (9) described how a digital computer was used in combination with an emission spectrometer to determine the chemical compositions of steels. The mathematical model, which was derived for the relations between composition and the intensities of the spectral lines, considered both overlapping and matrix effects. Glover et al. (61) described the automatic transcription and transmission of spectrographic results, which represented an extension of the degree of automation in the routine analysis of iron and steel samples by spectrography.

The development of small, special-purpose analog computers, to perform the necessary scaling, linearization, and interelement corrections automatically, made it possible to print out accurate element concentrations without an intermediate conversion step. As confidence grew in the more widely used spectrographic analyses for quality control, the need for further automation became apparent, particularly in the dissemination of results. An adequate system identifies each set of results by appropriate number, date, and time, and transmits these results automatically to the various points at which they are required. In addition, the results can be displayed in the manner best suited to the place involved and in the printout format required.

4. Absorption Spectroscopy

Digital recording of spectra on magnetic tape in computer format was described and illustrated by Brackett (23). The density of points required and the resulting volume of data were considered in terms of needs for computation and for spectral representation. Kendall (95) studied "peak smearing" in spectrometers by analog simulation. The effects of finite slit widths and other factors influencing resolving power can be described mathematically in terms of convolution transforms. Kendall developed a device which can be used to compute and predict the effects of various spectrometer design parameters on observed spectra.

Malmstadt and Crouch (117) developed systems for the automatic direct readout of reaction-rate data after determining that the *rate* of change of one physical quantity with respect to another is indispensable in many technical investigations. Their system replaces tedious point-by-point or recorded plotting of the value of one quantity versus another.

The direct readout of rate information, although having obvious advantages, has been difficult because of unwanted noise fluctuations accompanying many typical experimental rate signals. The Malmstadt–Crouch paper described three systems for rate measurements, based on comparison-measurement principles, with components readily assembled from inexpensive commercial modules providing high precision and accuracy: (1) rate measurements by a manual comparison technique, (2) automatic rate measurements by a servo-comparison technique, and (3) automatic rate measurements by an all-electronic comparison technique.

The experimental demonstration of the direct-readout rate meter was carried out on the specific enzymatic determination of glucose, based on the use of the enzyme glucose oxidase to catalyze the oxidation of glucose to gluconic acid plus hydrogen peroxide. A subsequent reaction, wherein the hydrogen peroxide oxidizes iodide to triiodide, can be followed by various techniques—in this case, spectrophotometrically by following the absorbance of triiodide at 360 nm.

Hannon et al. (72) developed hardware to implement the computer control of a single-beam monochromator. The unique feature here was the circuitry used to position and step the wavelength settings to yield very precise wavelength positioning in an open-loop configuration, with a minimum of computer usage.

A spectrophotometric system for data acquisition over a wide range was described by Grant (65). The operations of the computer included regulation of wavelength setting, determination of the system gain, analog-to-digital conversion of the output signal, and positioning of the sample and detector. It allowed for studies of electroreflectance, fluorescence photoconductivity, and ordinary reflection and transmission measurement.

5. X-ray Data

Where image clarity was needed in rapid X-ray diffraction studies, Kennedy (96) used electron intensification of the image of the diffraction pattern to expedite the orientation of crystal specimens by inspection of the transient diffraction patterns. Cole (37) demonstrated computer control of X-ray equipment as an example of laboratory automation. He pointed out that, particularly for some kinds of X-ray diffraction and fluorescence, instruments should be operated under varying degrees of computer control, not merely to accumulate and store data, but also to obtain the necessary information on the sample more quickly. The need for the experimenter to interact with the computer can become as important as the instrument–computer interaction. This requirement constitutes the largest difference between implementing laboratory automation and process-control automatic instrumentation.

Hamilton (70) talks about "the revolution in crystallography," reviewing how automation and computers have made X-ray structure determination a routine laboratory procedure. He concludes that for any molecule that exists a complete geometrical structure determination can be carried out in a short time. Even the determination of a protein structure forms only part of a doctoral thesis, since the advent of more efficient diffractometers and the further development of automatic methods for computer solution of structures.

6. Titrimetry and Electroanalytical Techniques

A semiautomatic subtractor of titration curves with a versatile curve-plotting device was described by Stevens and Duggan (169). It employed a slide-wire for converting distances into voltages for routine subtraction to yield a difference titration curve. The same apparatus could be used also for curve subtraction in spectroscopy. Another apparatus described by Zenchelsky and Segatto (199) aided in the selection of the end point in a thermometric titration by obtaining the first and second derivative curves. They used mechanical amplification and included a filter to increase the apparent signal noise ratio.

Brown et al. (26) described instrumentation for digital data acquisition in voltammetric techniques, including dc and ac polarography. The instrument they preferred incorporated a combination of known electrochemical control and signal-conditioning components, connected to a commercial digital data acquisition system. Its purpose was the digital readout of current and potential signals on punched cards for voltammetric techniques involving relatively long experiment times. This subject will be discussed further in Section VIII.C.

7. Miscellaneous

Johnson et al. (88), in a discussion of the logical and timing requirements and the control circuitry for computer-assisted spectroscopy, used a magnetic resonance spectrometer to exemplify the components necessary in any computer–instrument interface, demonstrating adequately flexible and open-ended control programs. Soulen (164) described the calculation of thermogravimetric data by electronic digital computer.

Hagan and de Laeter (68) noted the use of a mass spectrometer data-processing system in which digitized isotopic information was fed on-line to a time-shared, general-purpose computer. They outlined a program applicable to many isotopic problems requiring accurate peak selection.

Spragg (166) coupled a photoelectric scanner of an analytical ultra-centrifuge to a high-speed digital computer. Use was made of the theory of radar-pulse reception to show how the uncertainty in the measured pulse amplitudes could be improved, utilizing two-way information exchanges between the ultracentrifuge and the processor.

Müller and Lonadier (137) developed an early multipoint plotter, using a multipoint printout recorder and a sliding translatory potentiometer to devise a convenient data-plotting apparatus. Through appropriate circuit modification, other functions, such as squaring, square rooting, taking logs, or plotting hyperbolas, could be performed.

Hill (75) described the development, at the U.S. National Bureau of Standards, of a number of modular transistorized digital circuits that have been used to automate many data-recording and preliminary-processing tasks encountered in the scientific operations of the NBS. These versatile building blocks can be connected systematically to form digital circuits which accept raw data from experimental equipment and transpose them into forms suitable for input to high-speed electronic computers. In these systems the output can be (*1*) fed directly to a computer, (*2*) recorded on a medium (paper tape, magnetic tape, etc.) suitable for computer input at a later time, or (*3*) used to drive display equipment that keeps the experimenter informed about the status of his experiment.

Savitsky and Golay (157), in their discussion of "Smoothing and

Differentiation of Data by Simplified Least Square Procedures," reported that the operations of differentiation and filtering are especially important, both as ends in themselves and as preludes to further treatment of the data. They considered numerical counterparts of such analog devices as RC filters. They concluded that the method of least squares can be effectively used, without excessive complexity of a computational nature, to achieve considerable improvement in the information obtained. In their study they utilized convolution of the data points with properly chosen sets of integers.

E. COMPUTERS IN THE LABORATORY

Perone (144), during a symposium on computer automation of analytical gas chromatography, gave a very concise but complete introduction to digital techniques and the fundamentals of on-line computers in chemical laboratories. The paper included general considerations of what a digital computer is, its component parts and their functions, as well as the principles of programming, considerations of computer size, and the pros and cons of off-line and on-line computers.

In discussing the problems of computer automation and the initial steps of justification and recognition of problems requiring a computer, Perone recognized that the basic comparison to be made is the one between a dedicated small-computer system and time-shared access to a larger computer system. As for operational problems, one of the more serious is often overlooked in "going on-line" with a multiple-data-input processing system, namely, that the packaged system must necessarily include restrictions on the type of input data which it can successfully handle, thereby imposing a certain amount of *conformity* on all input stations and their mode of operation. "Odd-ball" or "bootleg" experiments are thereby discouraged.

The earliest applications of computers to chemical instruments involved analog devices. In the late 1940s petroleum companies utilized such devices for analysis of infrared spectra of multicomponent hydrocarbon mixtures requiring simultaneous equations with many variables. It was very soon recognized that digital computers offer advantages for most instrumental applications. These include high accuracy, large memories, flexibility, and well-developed software.

Methods of digitally collecting instrumental data off-line have the following advantages:

1. They can be used with nearly all instruments having conventional readouts without extensive modification of the instruments.

2. The data can be processed on the computer during off hours or, if necessary, immediately.

3. The data can be transported manually to a computer, reducing the investment involved in a signal line going from the instrument to the computer.

4. The software and interfacing problems are small, since long-term usage already has solved many of them.

In many instrumental applications, however, the following disadvantages of the off-line method are important:

1. Too much time may be required to hand-digitize the data.

2. Human error is possible in transposing the information from instrument to computer.

3. Results are relatively slow.

4. No possibility exists for later computer control of the instrument.

Smythe (163) discussed the development and influence of autoanalyzers and data processing in analytical chemistry and gave a good review of the early history. He predicted increasing usage of automated chemical analysis methods for the more numerous routine determinations and commented on trends and predictions among the following categories: classical wet methods, spectrophotometry, spectrofluorimetry, emission spectrometry (including flame spectrometry and atomic absorption), X-ray fluorescence spectrometry, mass spectrometry, nuclear magnetic resonance spectrometry, electroanalysis, chromatography (gas, paper, and ion exchange), and radiochemical analysis.

Smythe predicted that automation in the average industrial laboratory will continue to be confined to the more numerous routine determinations, that is, the types of which more than about 20 are done per day. He said that it is too much to expect a chemist to understand *all* aspects of the instrument–computer complex, and anticipated a maximum return in situations where a major instrument or group of instruments could be used to full capacity by teams consisting of chemists, instrument or electronic engineers, and mathematicians. In view of the high investment cost of such instrument–computer complexes, Smythe cited the obvious conclusion that such laboratories would be used continuously, instead of just 8 hours per day.

Spinrad, in his discourse (165) asserting that on-line computers provide new freedom in the design and conduct of experiments, analyzed quite generally the steps involved in going from a classical experiment to a fully automated one, in broad context not limited to chemical analytical techniques. He pointed out that the first step in automating a laboratory procedure occurs when the experimenter decides to generate his data in machine-readable form. This involves incorporating analog-to-digital converters in the output apparatus, with results appearing directly on punched tape or cards or on magnetic tape. He may or may not observe the progress of the experiment, inserting his judgment during its course to change conditions or alter such factors as the rate of data collection.

Spinrad's discussion of a partially automated experiment described the experimenter preparing a series of "macrocommands" that designate one run on the apparatus. Along with the appropriate auxiliary program, these commands enter into a computer, where they are elaborated into "microcommands," detailing the individual commands required to drive the experimental apparatus. As the run proceeds, the outputs are generated on another paper tape, which subsequently is fed into a computer along with the data analysis program, with printed or plotted results presented to the experimenter at the end of the computer's run. The experimenter then makes certain conclusions and prepares a new program of macrocommands to execute a follow-up experiment.

Clearly, *if* the scientist is able to state quantitatively the mental processes employed in going from the results to input commands for the next experiment, these steps likewise can be computer programmed. *If* this is done, the data analysis program and the input preparation program can be *joined* to form a "decision-making" program. In short, then, a master program would accept data from run i, process them, apply the programmed decision criteria, and then output the instructions for run $i + 1$. Spinrad then proceeded to consider the concept involved in "going on-line" and described the implications in terms of the additional apparatus needed to achieve an automated laboratory.

Of course, the artificial separation of hardware and software in planning computer applications obscures the important aspect of systems philosophy and procedures; it is a fact that systems design requires simultaneous development of both hardware and software apparatus and procedures, with strong interplay between the two. Costs can be considerably influenced by choices as to whether functions will be performed through software or hardware.

Another aspect of cost has developed through recognition of the differences between laboratory instrumentation and "process control." First, it is evident that laboratory automation is less expensive than process automation, because there is much less emphasis on reliability and the ensuing need for component redundancy. Second, laboratory automation demands much greater flexibility; an industrial process is a reasonably stable enterprise in which changes of any significance occur on time scales of months or years.

Anderson (3) defined fast analyzers as analytical instruments which provide output data in a form and at a rate suitable for direct input into a computer. He mentioned particularly the workloads of hospital clinical laboratories, increasing at an estimated 15% per year and making almost mandatory new means for mechanizing analyses, such as the autoanalyzer developed by Skeggs. This device incorporates a continuously flowing, air-segmented liquid stream carrying a sample past a series of stations where reagents are added or where processes of heating, dialysis, incubation, filtration, or absorbance measurement are performed. Such systems can *enormously* increase the number of analyses which can be done in a short time and on a single sample.

Similar devices, which introduce discrete samples into individual glass or plastic reaction vessels, move these vessels past stations where additions, reactions, and measurements can occur at rapid rates. Such machines can create enormous record-keeping workloads, with estimates that personnel in clinical laboratories might devote from 20 to 70% of their time to manual data taking and recording. A difficulty in adapting computers to assist in this kind of problem occurs, however, because of (*1*) the rates at which data are generated, (*2*) the rates at which an analyst can evaluate data, and (*3*) the rates at which computers can reduce data to final form. These rates are not all well matched.

Anderson distinguished between automation and mechanization. He required that automation involve some method of monitoring a reaction through its entire course to ascertain such things as proper preparation and condition of reagents, correctly controlled temperatures, proper compensation of temperatures, and proper timing of absorbance or other transducer measurements.

Anderson stated that the objective of interfacing a general analytical system with a computer in such a manner that the latter controls the system completely has not been fully reached. He pointed out that the basic problem with conventional analyzers, which are slow (about 20–120 sec between data points), has been that data are stored on strip charts, paper or magnetic tape, or punched cards, and that complex interfacing is usually required. Inherent in such slow systems is the chemical or electronic origin, inhibiting the efficient use of the computer, which ought to have its data fed in directly at a *high* rate of speed, with all signals suitably identified. *If* data signals for 15–90 analyses can be produced in a small fraction of a second, then electronic or other drifts will be minimized within that set.

The exciting prospect of repeating all readings on each sample many times, averaging them, and doing the necessary statistical analyses and simultaneously deciding how many readings should be obtained for the proper statistics looms when this concept of complete automation is considered. Usual analytical systems involve serial analyses of samples. Alternatively, analyses could be done in parallel, with all reactions, additions, and other steps occurring for all samples at the same time.

Of course, the interval required for radiant energy absorption measurements must be very small compared to the reaction time. These computer-oriented requirements dictate the directions in which work on fast analyzers should proceed.

VIII. SELECTED APPLICATIONS OF INSTRUMENTATION IN LABORATORY CHEMICAL ANALYSIS

Instruments for laboratory use function in one or both of two ways. One is the direct transduction of chemical quantities into numerical or analog form, and the other is the mechanization of laboratory operations, either for continuous measurement or for discrete analyses. The analytical chemist, whether specializing in a single technical area or broadly concerned with analyses in general, has come to depend, to a greater degree than most scientists, on a wide variety of these instruments and machines. This section reviews several examples embodying the techniques and principles involved. None of these attempts to be complete in any particular subject area (see other chapters in this Treatise for detailed accounts) but emphasizes the instrumentation viewpoint.

Instrumentation know-how and know-why includes all the potentialities and characteristics not only of the components that may be used, but also of the systems formed when they are assembled and put to work together. As instruments find increasingly widespread and indispensable application in routine laboratories, there has been a strong trend to make them look externally simple. This often-deceiving surface simplicity requires compensating internal complexity. Certain powerful, new instrumental functions are found again and again in analytical apparatus. Three of particular importance are high-impedance amplification, inverse feedback stabilization, and servoelectromechanical recording. These auxiliary aids, plus dozens of important new, improved transducers, have become sufficiently established to make possible certain broad generalizations.

Measuring instruments (contrasted with other "instruments," better considered *machines,* which perform solely mechanical operations) may be regarded as consisting of three block components: (*1*) the transducer (or detector), which receives a signal as input and converts ("transduces") it over to some other form, nearly always electrical; (*2*) the operational circuitry to modify the transducer signal into some other form suitable for readout, for example, conversion from direct to alternating current by chopping, transformation of voltage into current or vice versa, amplification of power or voltage, simple computation, such as differentiating, referencing, storing, accumulating, and discriminating (as in pulse-height analysis), and inversion (back to direct current for conventional output indication); and (*3*) the readout device, which, through an indicator or recorder, makes a sensible record for the operator to observe visually.

Not only does each block component have its own more or less well-characterized performance characteristics, but also interactions may occur which affect performance in ways that are often not simply predictable. The same complications present in the new interdisciplines of systems engineering are often encountered when using the "black box." Especially when several "black boxes" are put together in a "brown box," the limitations imposed are easily forgotten as the switch is thrown and the knobs are turned.

While the design of complex instruments requires the detailed application of the mathematics of circuit theory, proper use of instrumental tools can result from a qualitative knowledge of their characteristics. Indeed this is necessary for decisions made daily in regard to whether to accept or reject an answer, or whether to use this or that instrument, or in which mode to use a particular device when a choice exists.

Inseparable from the considerations vital in the selection of an instrument are those dealing with the proper utilization of the equipment already chosen. The following points are pertinent.

1. Calibration. Standards should duplicate as closely as possible the samples to be measured. Hence the interpolation (known or unknowingly assumed) between sample and reference may be correct.

2. Sensitivity. A good rule of thumb is to operate so that some noise or "pen jitter" comes through. This is useful evidence of readouts which are undistorted by friction or inertia. Likewise, the time constant (if available) should be set to be the smallest value which does not produce excessive noise.

3. Scanning speed. The slowest readout gives the most reliable data (in cases where the events being recorded either are constant or are occurring quite slowly). In fact, nonscanning measurements are to be preferred whenever the situation permits.

4. Checking procedures. Simple, *in situ* standard signal input techniques are a desirable part of any well-designed measuring tool, and instruments are no exception. An instrument freshly turned on must have an adequate warmup, followed by complete input circuit checking for assurance of equipment sensitivity, stability, and response time.

The laboratory chemist finds, in addition to the two main categories of measurement instruments and automatic machines, a further useful breakdown of the former into continuous and intermittent types. Continuous-measuring instruments are finding wide utility in laboratory work, but their greatest value still lies in process applications (both process development and production monitoring).

A. SAMPLE PREPARATION

Of course, even batch analytical instruments may be speed-limited by the bottleneck of sample preparation, and the method of choice may well be influenced largely by the need for fast sample handling. Thus, for metallurgical analysis, a solution-flame emission procedure may be as satisfactory as a pin- or disk-spark emission technique. However, the latter is sometimes chosen in spite of its higher investment cost because the time required for sample preparation is so much shorter. Likewise, the use of dry powdered-cement samples for quality control is made possible by using X-ray spectrometry rather than other techniques which require dissolving the samples. The time saved can permit product shipment to the user several hours earlier than would otherwise be possible.

When separations are necessary and the older, traditional techniques such as precipitation, electrodeposition, or volatilization would be called for, the use of ion-exchange columns and highly selective extraction procedures is helpful in modern instruments. Higher selectivities are also available in masking and/or chelating agents and indicators, making automation and the use of measuring instruments feasible in ways which once were considered impractical.

Automatic sample treatment may be very important to the achievement of really meaningful results from the measuring instrument, although this topic cannot be discussed in detail here. This importance should be kept in mind as we now review just two examples—gravimetry (balances) and titrimetry—of instrument applications to laboratory chemical analysis, chosen from among the numerous possible categories.

B. GRAVIMETRY (BALANCES)

No property of matter is more central to quantitative science than is the concept of mass. None has been both so routine and so nonroutine; hence the dichotomy of automatic instrumentation—the early high-priority efforts toward precision and reliability, contrasted with the needs, developing after 1945, for speed.

Excellent reviews of the principles inherent in the construction of both conventional analytical balances and electrically operated automatic and recording balances were published by Lewin in 1959 (106) and by Hirsch in 1967–1968 (76). Gordon and Campbell's review (64) of automatic and recording balances is a good source of similar information, including a description of the automatic recording balances commercially available as of 1960.

In 1965, Thomas and Williams (179) reviewed the various types of thermodynamic, kinetic, and structural information used in establishing specifications governing the operation of vacuum microbalances.

1. Early Automatic Recording Balances

Although automatic recording balances were known as early as 1910, and recording thermobalances by 1935, it was not until the early 1950s that chemists and instrument manufacturers began to exploit fully the possibilities inherent in the automation of weighing procedures. Before 1950, the tedious point-by-point recording of sample weight as a function of its experimental environment was common practice. A few chemists did construct automatic recording balances for their own use to record weight changes due to specific phenomena under direct investigation (e.g., animal metabolism, the drying of enamels, the dehydration and decomposition of mineralogical samples, the determination of particle size by sedimentation or of the specific surface of powders by the BET gas absorption method, etc.).

The commercial availability of the Chevenard thermobalance in France in 1950, followed by Duval's work on determining the thermal stability of analytical precipitates, turned the serious attention of the chemist and the instrument manufacturer to general exploitation of the automatically recorded weight-change technique. As a result there are now many commercially available recording balances which find use in various fields of physicochemical study.

In general, recording balances have the same sensitivity and range as their manual counterparts. Their operation involves both null-point and deflection techniques, and with one exception—the differential thermobalance—they are single-channel instruments, that is, weight is rendered directly as a function of time, temperature, humidity, or other variables.

2. Beam-Deflection Principles

In the case of beam-deflection-type balances, a suitable transducer—electrical, optical, mechanical, or a combination of these—is utilized to determine the degree of deflection and to activate a recording device calibrated directly in units of weight. Balances utilizing the null-point technique depend on the transducer to detect deviation of the balance beam from its null position and to activate a restoring force to return the beam to that position; the magnitude of this restoring force is recorded directly, or through an appropriate transducer, in units of weight.

Use of a conventional beam is not really necessary; in many cases, the beam is replaced by a cantilevered support, torsion wire, helical spring, or strain gage. In all cases, deflection or deformation is related to weight and recorded as such. Use of a beam is also avoided in one balance (35) which makes use of the principle of electromagnetic levitation, utilizing an electromagnet to suspend an oil-damped sample pan in mid-air and relating the current required to maintain the suspended material in the null position to the weight of the material. The null detector is a shutter intercepting the light beam incident upon two photocells.

Many techniques have been utilized for detecting beam deflection and converting it into recorded weight. Kuhlmann's first automatic recording balance (101) utilized a mirror mounted on the center of the beam of a conventional analytical balance, the mirror reflecting light from an appropriate source onto photographic paper mounted on a rotating drum. The deflection of the light trace on the developed paper was proportional to the change in weight of the material being examined.

Rabatin and Gale (152) devised an automatic recording sedimentation balance utilizing a sensitive spring to weigh the particles settled on the pan. As the weight on the pan changed, a shutter intercepted the light reaching a photocell.

Small range is an inherent limitation of deflection-type recording balances, for the range of the balance is limited by the weight required for complete beam deflection. Many techniques have been employed to circumvent this limitation. For example, the effective range of such a balance may be expanded by devices which arrest the beam at the point of maximum deflection and add a restoring weight to the balancing pan by various electromechanical means (16,101).

Alternatively, the hydrostatic principle may be utilized. A rod, suspended from the nonsample end of the balance beam, partially immerses in a nonvolatile liquid; the deflection of the beam becomes a function of the weight of the sample, and the linear range of the deflection is a function of the diameter of the rod and the density of the liquid (44,181).

Linear-variable differential transformers have been extensively employed as transducers for deflection-type balances; the transformer armature is suspended from one end of the beam, and the transformer output is then proportional to beam deflection (29). This technique has been used to convert ordinary beam-type analytical balances to the recording mode (146) and, in conjunction with the hydrostatic principle, to make them equally effective in both the microgram and the gram range (29).

Linear-variable transformers may also be used to indicate the deflection of a cantilevered beam, of a torsion-type beam (19,145), and of a spring-deflection balance (78). A commercial version of the last type records automatically changes in the weight of a sample as a function of temperature in controlled atmospheres or in vacuum. The temperature may be programmed or kept isothermal between ambient and 1000°C, and the range of the balance is determined by the characteristic of the spring. Changes in sample weight from 0 to 200 mg can be measured with 1% accuracy, or better, with sample weights up to 10 g.

Strain gages have also been used to indicate beam "deflection" (14); in this case the deflection is very minute. One end of the beam is attached to the floor of the balance by the strain-sensitive wire gage so that, as the sample undergoes a change in weight, the resistance of the gage also changes. The gage is part of a conventional bridge whose output is used to feed a self-balancing recording potentiometer.

Bowman and Macurdy (21,115) developed a special photoelectric device to automatically follow beam position in a high-precision balance, indicating it to 10^{-4} in. It was developed primarily as a research tool for gaining better understanding of fundamental balance errors. It included a fixed-filament lamp whose output was first reflected from a rotating mirror; then a simple arrangement allowed the angle of tilt to be set at any value up to 1 minute of arc. The mirror was rotated by a 3600-rpm motor, so that the reflected light beam oscillated at 60 Hz with a peak-to-peak amplitude adjustable from 0 to 2 minutes of arc. The light beam was reflected back to the photocell, producing an image of the lamp element on the plane of the photocell slit. The filament image oscillated at 60 Hz through a distance of about 0.19 in. When the slit was located in the center of this vibration, it was illuminated 120 times per second. However, if the slit was displaced from the center of vibration, a remedial feature came into play. The electrical output of the photocell contained a 60-Hz component whose phase and magnitude depended on the direction and amount of displacement. This 60-Hz component was filtered, amplified, and applied to the control winding of a two-phase motor driving the lead screw in the proper direction to re-enter the slit.

Thus the photocell followed the swinging light beam to within 4×10^{-5} degree in the angular position of the balance arm. This photoelectric reader allowed the U.S. National Bureau of Standards to accelerate research on precision balance operations by 5–10 times, shortening the interval during which major errors due to small changes in temperature and humidity might occur.

3. Null Balances

Many techniques have been employed in the construction of null-type recording balances to supply the force necessary to restore the beam to its equilibrium position. The initial approaches simulated manual operation of the balance and involved the incremental, automatic addition of weights to the balance pan (141,174) or the use of a cam and spring arrangement to reposition the beam. In the latter case, the rotational position of the cam was recorded directly in units of weight. In most cases, electrical contacts were used to indicate the position of the beam with respect to the null point.

The deficiencies inherent in this technique, due to the incremental addition of weights, are obvious. They were overcome to a large extent by the use of the chainomatic principle (52,82,136), with the range and sensitivity of the balance being varied by suitable choice of chain length and link weight. Both photometric (82,110,136) and capacitance (90) methods have been employed to indicate beam deflection, a portion of the detector signal being used to activate servomotors positioning the chain.

Magnetic methods were also widely used to restore the beam to its equilibrium position. In 1895, Angstrom (4) described a null-type balance with a magnet suspended from one end of the beam into a solenoid coil. The coil current was manually adjusted to return the beam to its equilibrium position. This current was then measured with a mirror galvanometer and related to weight. This technique has been widely automated (e.g., for moisture sorption measurements), and photoelectric, capacitance, inductance, strain gage, and other transducers have been used as the null detectors.

The D'Arsonval galvanometer principle has been applied by Cahn in a commercially available null-type electronic microbalance (28). In this instrument the galvanometer indicator functions as the beam, and the torque produced by the unknown weight is offset by the magnetic torque of an electric coil. Since the beam is always returned to its null position, the current is proportional to the torque and the setting of the balancing potentiometer is proportional to the sample weight. A phototube–shutter–light-source system serves as a null detector and operates through a servosystem to restore the beam to its equilibrium position.

The torsion wire has been used in place of a beam in several instances. Twisting of the wire is detected by the motion of the pan system, and a restoring force is applied either by actuation of a magnetic coil counterbalance or by mechanical twisting of one end of the wire.

4. Recording Thermobalances

Longchambon (111) utilized both the beam mirror and a galvanometer mirror to produce a simultaneous, photographically recorded trace of sample-weight change and furnace temperature in an apparatus wherein the sample was suspended from the beam into a furnace. Brefort (24) devised a photographically recording thermobalance in which a shutter attached to the balance beam partially obscured the light source from a photocell; the photocell current, which varied with the degree of beam deflection, was recorded by means of a mirror galvanometer reflecting a beam of light onto photographic paper. In several commercially available balances the photocell is replaced by more modern phototubes, whose current output or potential can be directly recorded by faster and more convenient nonphotographic techniques.

A Mettler recording deflection balance utilizes a conventional single-pan projection balance; light is reflected from the beam-mounted mirror to a pair of matched photocells mounted on a servomotor-driven carriage.

Deflection of the beam causes unequal illumination of the photocells, which actuates the servo-driven carriage in the direction required to achieve equal illumination. Since the photocell carriage is mechanically coupled to a strip-chart recording system, the record is a direct indication of beam deflection. Six current-carrying recording styli are used, operating at different full-scale sensitivities. In this manner, a 200-mm chart can have an effective range of 1200 mm and corresponding weight ranges of about 0–1200 mg or 0–1200 g.

The Chevenard thermobalance (30), consisting of a wire-suspended balance beam, records changes in weight photographically by the deflection of a beam of light from a mirror mounted at one end of the balance beam onto a photosensitive paper wrapped around a rotating drum. A pen-and-ink recording unit is also available.

A modification of the photographically recording Chevenard thermobalance for electrical recording utilizes a linear-variable differential transformer (62). Constant-load, single-pan projection-type balances can be modified for automatic recording by this technique; the effective single deflection range of these instruments is the range of the projection system, which may be greater than 100 g.

Wendlandt (191) converted a torsion-wire type of balance into an automatic recording thermal balance, maintaining the null position by means of a beam of light falling between two cadmium sulfide photocells. Balance equilibrium was maintained by means of a reversible synchronous motor connected to the torsion wire. The same motor rotated a recording drum in proportion to the change in weight, with the pen being driven at constant speed by another motor.

Groot and Troutner (67) published construction details and operating characteristics of an automatic recording thermal balance made from standard laboratory equipment, using photoelectric null-point detection of unbalance and restoration of force by an electromagnetic device. Garn (60) used a linear-variable differential transformer as the sensing element in his null-point instrument. In this case, the null position was maintained by a motor-driven balance chain whose position was continuously recorded on a strip-chart potentiometer.

5. Digital Readout and Increased Sensitivity

Analytical balances increasingly have options for digital readout of the weight on the pan. In one approach a digital voltmeter is employed to read directly, including the last digit without estimation of vernier positions. Another option is to use a printing mechanism to record the desired weight of the sample on adhesive-backed tape, which then can be attached to notebooks, sample containers, or other records. Such print-out eliminates errors in reading or transcribing figures.

As such improvements become available, the demand increases for *gravimetric titrations,* wherein the weight of titrant dispensed from a weight buret can be indicated directly without any volumetric measurement. When titrations are carried out gravimetrically, errors of drainage, meniscus reading, and temperature change in glassware or solutions are eliminated. Also, since weighing is inherently more precise than volume measurement, the size of samples and the amount of stock solutions can be reduced. The preparation of solutions on a gravimetric basis is independent of temperature and is more rapid and more accurate than when volumetric methods are used. Furthermore, the cleaning of weight burets is easier and less critical, since a slight amount of dirt may cause no appreciable error, whereas a dirty volumetric buret is certain to be troublesome. In addition, tedious calibration of volumetric apparatus can be eliminated.

As for higher sensitivity, analysts can fool themselves in assuming that a more *sensitive* balance directly provides greater *accuracy* in weighing. They may forget that air-buoyancy corrections become important when accuracies of about 1 part in 1000 are desired. For most weighings, the air-buoyancy correction can be neglected, since other errors in most procedures will be greater. But as advances in precision and accuracy occur in readily available balances, the air-buoyancy uncertainty becomes a limiting factor, unless practical means of conducting weighings in a vacuum without serious delays in evacuating the weighing chambers are developed.

Lukaszewski (112), in considering the subject of accuracy in thermogravimetric analysis, lists the following sources of error: (*1*) air-buoyancy and convection effects, (*2*) measurement of temperature, (*3*) nature of the atmosphere in the furnace, (*4*) heating rate, and (*5*) heat of the reaction studied. In his discussion of buoyancy effects, he points out that errors from this source appear as a progressively changing nonlinear function related to the heating rate. Large errors, therefore, can occur in the final loss of weight at the completion of the test; furthermore, the overall shape of the weight-loss curve may cause misinterpretations of reaction kinetics. Spurious gains in weight sometimes observed can result from (*1*) decreasing air buoyancy with increasing temperature, (*2*) increasing convection with increasing temperature, (*3*) effects of heat on the balance mechanism itself, (*4*) thermal turbulence in the hot zone of the furnace, (*5*) random fluctuations in the recording mechanism, (6) furnace induction effects, (7) electrostatic effects, and (8) environment of the thermobalance.

Lukaszewski recommends symmetrical crucibles of minimum volume, together with support assemblies as small and as streamlined as possible and the use of small quantities of sample, with crucible/sample ratios of the order of 20:1. He further recommends that initial trial buoyancy determinations be made with the crucible at each relevant heating rate to be used, as well as comparison of the buoyancy effects to be obtained for the empty crucible versus the crucible filled with inert material.

C. TITRIMETRY

The literature for *instrumented* titrimetric analysis became as voluminous during 1950–1970 as it was for noninstrumented titrimetric techniques in the previous century. Particularly in the United States there had been a distinct reluctance among instrumentation researchers to use analysis instruments which require a supply of reagents. In the words of Stirling and Ho (171), they had visions of "masses of complex and fragile glassware, all of which may shatter at the slightest tremor and cause the immediate demise of the associated expensive electronic equipment, as corrosive fluid shorts its way from terminal board to terminal board." At least, the provision of continuous supplies of reagents is a real maintenance burden; hence the early reluctance to resort to "wet" analyzers.

Of course, the replacement of glass and metal parts with plastics, such as polyolefins and polyfluorocarbons, has essentially eliminated impact hazards and corrosion problems (as well as lubrication difficulties) and has met the need for long-lived, tight-fitting yet nonfreezing joints and valves. All of these, in addition to vast improvements in electronics and in electrode technology (covered in other volumes of this Treatise), have contributed to a resurgence of interest in titrimetric techniques.

The first automatic end-point titrator actually appeared in 1914, and the first automatic titrator with automatic sampling in 1933. What these lacked for complete automation were a sample-handling system, a sample-injection system (analogous to the analyst's pipet), a dilution stage, dispensing systems, an end-point indicator, a readout system, and, finally, a flushing system for cleaning, plus arrangements for possible auxiliary steps such as heating and mixing. All of these are essential in an on-line fully automatic titrator, although not necessarily for laboratory work.

Sample-handling systems often must deal with a process fluid which fouls lines, causing problems usually solved by controlled dilution just after the fluid enters the sampling line. Sample selection is usually combined with dilution, and various systems have been evolved for this. Automatic siphon pipets and solenoid valves are commonplace in laboratory titrators. One technique uses a multiple valve similar to some gas chromatographic sampling valves. Here the sample contained in the valve bore is automatically connected to an automatic siphon pipet to accomplish sampling and accurate volumetric dilution with excellent precision.

Another technique involves a weighted hypodermic plunger, moving between stops and raised by hydrostatic pressure when the diaphragm inlet valve is opened. Simultaneously, an outlet valve in the head closes, and the changeover of the valves allows the plunger to fall, thus delivering the sample.

Another commercial instrument uses a pipet with an overflow and drain system and provides the top level with a pair of electrical contacts to determine when the level is reached. Filling and continuous flushing-out are followed by a standard-timed holding period with the feed shut off, during which the top drain empties, yielding a reproducible meniscus. The diluent pipet in both these instruments uses electrical contacts for level sensing and for operation of the solenoid flow-switching valves. When the sample reaches the reaction vessel, either titration or open-loop colorimetric determinations can be conducted.

The titrimetric dispensing system may take three forms: (1) an automated buret with miniscus reader, (2) an automated hypodermic syringe, or (3) an electrolytic generation system using coulometric principles. The indicating system can be potentiometric, conductimetric, amperometric, thermometric, or photometric. Readout systems vary with the dispensing system used and may give either analog or digital information. For example, either a follow-up meniscus system in a buret or a driven piston may accommodate an encoding disk fitted on the servo-drive, giving digital information directly.

Many end-point-indicating transducers provide a sharp change of slope in the voltage versus volume characteristics at the end point, thereby generally yielding good control signals. On the other hand, recognizing the exact end point is somewhat difficult in certain potentiometric titrations which fail to follow the traditional S-shaped curve. The inflection point corresponding to the end point may be picked up by using a derivative voltage obtained through electronic differentiation of the indicator output. Nearly always this yields more sharply defined wave forms able to operate electronic switches, which then actuate shut-off devices.

Second-derivative shaping may be used with potentiometric titrations, with the end point being taken as the point at which the derivative voltage passes through zero. Third-derivative shaping may offer advantages with spectrophotometric detectors and, in some cases, potentiometric indicators, since it changes polarity more rapidly near the end point.

1. Components and Early Work

As indicated above, the literature of titrimetric instrumentation sometimes emphasizes components put together to form an automatic titrator. An example is the description by Gordon and Campbell (63) of an automatic buret. Their arrangement utilized an interchangeable metal bellows at the bottom of the buret for measuring the pressure of the height of liquid. Deflections of the bellows were measured by a linear-variable differential transformer, yielding continuous graphic recording of changes in volume.

Farquharson (56) developed a precise, automatic, inexpensive buret reader to find automatically the level of a colorless liquid in a glass tube with indication readout on a digital register. The first model had a range of 40 cc with a precision of ±0.02 cc on a single determination (95% confidence level). He further utilized a photocell detector with a cable and drum elevating and measuring mechanical system.

An impulse method whereby stepping motors drove piston burets was claimed by Oehme (143) to have several advantages based on the low inertia of the driving system. The burets delivered single microliters with an accuracy of about 5%. This arrangement was said to have particular advantages for student experiments.

Lingane (108) in 1949 described an automatic potentiometric titration apparatus, applicable to all kinds of reactions and all types of electrode combinations. He introduced the slow approach to the equivalence point, so essential to a highly accurate technique, as a unique feature. His instrument utilized a motor-driven syringe to deliver titrant with greater precision than could be achieved with conventional buret delivery, and was capable of closely varied delivery rates to suit the characteristics of a particular titrimetric reaction. His titration cell was connected to a recording potentiometer which either recorded the entire curve or actuated a mercury switch in the motor circuit to stop delivery of the titrant at the appropriate equivalence point. In subsequent decades Haslamb and Squirrell (73), Irving and Pettit (85), and many others described similar systems involving minor modifications and improvements, but all based on the same general arrangement.

2. Specialized Applications

A turbidimetric titration apparatus was described by Stearne and Urwin (1967) for the automatic recording of transmitted and scattered light. Their instrument was based on the usual principles of light scattering, but included a number of special features. Malmstadt and Hadjiioannou (118) reported a direct automatic derivative spectrophotometric EDTA titrimetric procedure for the determination of calcium and magnesium. Other photometric titrators for particular uses were described by Flaschka and Sawyer (57) for very high precision and by Flaschka and Speights (58) for the inclusion of facilities providing variable path length, zero suppression, and scale expansion. The device described by Greuter (66) made possible the recording of titration curves, with several kinds of end-point detection. He discussed the errors inherent in some methods of doing this.

Narayanan and Sundararajan (138) showed that the precision of an end point in an impedance titration can be improved by compensating the alternating voltage of the titration cell with an external alternating voltage, utilizing a differential amplifier. A sharp jump in the output of the difference amplifier thus occurs at the end point of the titration.

Since conductance end points have advantages for certain kinds of titrations, particularly of low concentrations of carboxylic acids in organic systems, Boardman and Warren (17) developed an automatic conductance titrimeter, specifically for work in nonaqueous media. Another conductometric arrangement was described by Holm-Jensen (77), whereby conductivities of electrolytes were graphically recorded by means of commercially available, strip-chart, dc potentiometric records.

In regard to the high-frequency region, Clayton et al. (36) reported that a disadvantage in such titrimeters is that the sensitivity to concentration changes depends on the geometry of the cell containing the sample to be titrated, as well as on the natural frequency of the instrument. They derived an equation relating the specific conductance of the sample solution to the physical dimensions of the high-frequency cell necessary for maximum sensitivity.

Schonebaum and Breekland (158) described experiments showing that many metals can be determined via EDTA titrations in commercial automatic titrators. Miyake (130) described an automatic recording titrator for conventional applications in acid–base and redox reactions where the titrant was added intermittently and the titration curve recorded stepwise, with potential changes being registered between titrant deliveries. A particular application for sewage analysis was described by Crowther (42) with the objective of shortening the time required for the determination of ammonia and oxidized nitrogen.

Cornwell and Cheng (40) described a specific analytical method utilizing automatic potentiometric EDTA and redox titrations for determining the tellurides of lead and tin. First, the sum of lead and tin was titrated at pH 4.5, through addition of a known excess of EDTA and back-titrating with a standard lead solution. After addition of ammonium fluoride to mask the tin, the EDTA released from the tin–EDTA complex was titrated with standard lead solution. Alternatively, after the determination of total lead and tin, lead was determined by back titration with standard lead on a separate sample aliquot, using tartaric acid in this case to mask the tin. Tellurium was separated as tellurous acid, which was then dissolved in dilute sulfuric acid solution and oxidized by permanganate, the excess being back-titrated automatically with iron(II) solution.

Hallikainen and Pompeo (69) described a continuous-recording electrometric "titrometer" for monitoring the removal of mercaptan sulfur from gasoline. It included two gear-driven pumps for constant-rate delivery, one for the sample and the other for the titrant, into a titration vessel. Their flow rates were continuously adjusted to maintain the combined solution at a steady equivalence value, in the reaction of alcoholic silver nitrate with the mercaptans in the gasoline.

Ridley et al. (154) demonstrated a titration for the dissolved-oxygen content of water by a modification of Winkler's method in an automatic titrator.

A special automatic titrator for chloride was described by Cotlove et al. (41). It was based on silver ions coulometrically generated between a pair of silver electrodes immersed in the solution to be titrated. The silver ions generated precipitated with the chloride ions, and the end point was then indicated electrometrically by a pair of silver electrodes. The actual measured parameter was elapsed time, since the silver ions were generated at a constant rate.

Magnetic titrations using a special titration cell and procedure were described by Heit and Ryan (74).

The particular area of thermometric titrations was thoroughly covered by Jordan in Chapter 91, Vol. 8, of this Treatise. His review of "enthalpy titrations" should encourage additional applications of this technique of end-point detection in cases where the heat-of-reaction levels permit. Priestley et al. (151) also discussed the usefulness and applications of thermometric titrimetric methods, with particular reference to automated routine titrimetric analysis. Christensen et al. (32) described a titrimetric thermometric calorimeter having the precision and accuracy of convenional solution calorimetry. They achieved a low heat leakage and fast equilibration (1–3 sec), enabling quantities of heat as small as 4 cal to be measured with accuracies of 0.1%.

3. Coulometric Techniques

Taylor and Smith (178) developed a coulometric analytical method and a titration coulometer, providing precisions of 1 part in 100,000 for analyzing and evaluating chemical standards. They reported coulometric analysis to be simpler, quicker, and easier to perform than many of the corresponding gravimetric and titrimetric determinations. Titrations reliable to a few thousandths of 1% for both strong and weak acids can be achieved, and the same is true of bases.

The coulometric titration is based on the exact relationship between the amount of electricity used in electrolysis and the amount of chemical reaction produced (Faraday's laws of electrolysis). Results can be based on electrical standards, with no assumptions necessary as to the purity of any primary chemical standards.

Coulometric titration analysis may be classified as either constant-current or controlled-potential (variable-current) coulometry. In the first of these modes an accurately measured current is maintained until the desired electrochemical reaction is completed; then the time required to complete the reaction is multiplied by the value of the current and divided by the Faraday constant. The result is a measure of the amount of reactive material present.

Controlled-potential, or variable-current, coulometric titrations depend on maintaining the potential of the working electrode at a value that will selectively result in the desired reaction. The electrolysis current then decreases continuously as the reaction proceeds and should be zero at completion. The varying current utilized is integrated to find the average value before Faraday's law can be used to determine the amount of reactive component. Extremely small concentrations are adequate in this method, since integrators can measure very small amounts of current.

Shain and Huber (159) described an automatic titrator for use with constant-current potentiometric titrations wherein the termination of titrant flow was based on the shape of the titration curve. They described applications to ferrous, manganous, and thiosulfate and chloride ions and compared them with manual titration values, showing "extremely fine" precision and accuracy.

Takahashi (176) constructed an automatic continuous coulometric titrator and demonstrated its application to acid–base titrations. Smith and Taylor (162) described a highly precise coulometer which permitted time integration of currents totaling 100 coulombs or more with a precision of 1 part in 100,000. In their example the current to be integrated oxidized hydroquinone in an electrolysis cell, producing quinone and acid. The quinone was then reduced by constant-current coulometric titration, with its end point indicated by pH measurement.

Jeffcoat and Akhtar (86) described an automatic instrument suitable for coulometric titration of titrant in an external cell for plant quality-control use. Likewise Christian (33) investigated several coulometric titrations to illustrate the reliabilities of automatic current recording, including reversible–reversible and irreversible–irreversible titration couples, as well as reduction–oxidation and precipitimetric reactions. He gave typical titration curves showing precise recoveries of titrated standards.

Much activity on continuous coulometric titrations has occurred. For example, Takahashi and Sakurai (177) constructed a continuous coulometric titration apparatus for studying relationships between the concentration of the component to be determined and the electrolytic current, as well as transient phenomena. They found that these relationships were influenced by the preset voltage, the sensitivity of the amplifier, the feed rates, the shape and volume of the titration cells, and the chemical reaction rate. Their examples were arsenous and ferrous ions, determined successively through the reaction with electrolytically generated bromine.

Johansson (87) developed a coulometric titrator with operational amplifiers. This instrument gave a potentiometric titration capability of ±0.05 pH unit with glass and calomel electrodes.

Breiter (25) used analog-to-digital equipment to record electrode potential as a function of time during galvanostatic transients of either the current, the ohmic component, or the capacitive component of the ac impedance at the electrode–solution interface during potentiostatic transients. After transfer from paper tape to magnetic tape, the data were processed by computer. The main features of the programming and examples are given in his paper.

Pike and Goode (148) designed a constant-current coulometer suitable for manual or automatic titrations, resulting in an apparatus that provided 11 increments of current in the range 250 μA to 200 mA, giving a standard deviation of better than 30 ppm.

Drew and Fitzgerald (46) described their experiences in external photochemical titration generation. The technique involved preparing the titrant needed by passing a photogenerator through a quartz tube irradiated by a low-pressure mercury arc. The effects of pump and lamp stability, flow rate, and length of photolysis zone were evaluated. The coulometric titrator developed by Hoyt (80) was applied successfully to the determination of water via a Karl Fischer type of reaction. Coulometric *in situ* generation of the moisture-sensitive Karl Fischer reagent was used for accurate determination of the water content of a variety of compounds.

4. Versatile "Universal" Titration Devices

A universal titration console described by Duggan and Stevens (47) was an easily operated device for ordinary pH titrations and for titrations at constant pH. It permitted rapid accumulation of *corrected* titration data, using small sample volumes and reproducible conditions of mixing, buret delivery, and constant temperature. Klaasse (98) described a versatile commercial titrator for measurement and control applications, based on glass-cylinder plunger type of automatic buret, with automatic refilling, and suitable for period sampling, automatic analysis, and full recording. He described applications in the micro, semimicro, and macro ranges. The titrations could be indicated digitally as well as recorded, together with dependent variables of pH, color, amperometric current, and their derivatives.

Anton and Mullen (7) modified a commercial titrator to make it more versatile, giving the operator a choice of liquid or coulometric titrants and potentiometric, amperometric, thermometric, photometric, or conductimetric end-point detection systems. Their photometric mode employed a monochromator for maximum sensitivity, providing adaptability to coulometric, turbidometric, or chemiluminescent titrimetry. An instrument described by Milazzo (128) was an electrochemical automatic titrator having the ability to reveal false end points arising from slowness of the analytical reaction near the true equivalence point.

IX. MEASUREMENT PROBLEMS

A publication of Ruh et al. (155) suggests: "If it had been possible in the past for all instrumentation for laboratory measurements to yield precise information conveniently in terms of absolute units that expressed the physical laws involved, it is probable that many of the problems now confronting us would not have arisen." They found it interesting to explore difficulties traceable to the choice of measurement units used. In some areas of measurement, the metric system is firmly entrenched. In others, the English and/or arbitrary schemes not traceable to any distinct system are used.

For example, the use of a variety of scales in the measurement of temperature is confusing and annoying. In the United States the Fahrenheit scale is commonly employed in engineering circles and by the layman, whereas the sciences utilize the Celsius (or centigrade) scale, and certain specialized fields such as sugar refining and brewing still use a third scale, the Reaumur. Temperature is but one area among many in science where it is important that participants employ a common language.

Likewise, it would be impossible in the space available here to consider all of the problems involved in the use of various units for the measurement of viscosity. The literature of the petroleum industry, to cite one example, provides descriptions of a large number of viscosity systems, and devices for measurement, serving in one respect or another. The evaluation of the resistance of fluids to flow is one type of measurement which seems to have suffered to an unusual degree from personal attempts to maintain the status quo and also from expediencies stemming from war emergencies. For example, ASTM Committee D-2 on Petroleum Products and Lubricants has placed many reports in the record dealing solely with the problem of what temperatures should be established as standard for the measurement of viscosities.

Along with pleas for more extensive use of the *metric* system, Ruh et al. state that problems encountered in measurement operations arise from a variety of causes, including the following:

1. Failure to base instrumentation on fundamental units or principles.
2. Errors in judgment.
3. Reluctance to deal in another's language.
4. Desire to maintain the status quo.
5. Difficulties in instrument maintenance.
6. Lack of convenience in use of existing methods.
7. Complexity of interrelations and conversions.
8. Extension of limits of measurement.
9. Failure to recognize inadequacies in existing methods.
10. Preservation of proprietary interests.
11. Influence of unusual circumstances such as war.
12. Yielding to expediency.

They further suggest the following activities or actions to reduce or eliminate many measurement problems:

1. Develop new tests or methods of measurement based on fundamental principles.

2. Eliminate inferior methods as rapidly as possible.

3. Avoid repetition of past errors.

4. Make maximum use of existing standardization activities.

5. Work toward ultimate international standardization whenever possible.

With regard to teaching, Kolthoff (100) has reminded chemists, *"Instrumental methods of analysis* is a new name for a very old subject . . . *all* measurements are made with an instrument." The complexity which has accompanied the trends toward modern electronic devices creates the problem of designing college courses with proper balance between experimental techniques and theoretical fundamentals. Academic authorities are properly resisting the demand to train graduate analytical chemists who are specialists in spectroscopy, polarography, chromatography, optical methods, or other specific techniques. Yet there is a need for adequate familiarity with *applied* concepts well grounded on *basic* fundamentals found neither in college-trained technicians nor in exclusive theorists. "Antihardwareism" is no more desirable than is "antiintellectualism."

X. THE LITERATURE OF INSTRUMENTS

The literature abounds in rich sources for those concerned with the design, maintenance, and economic aspects of instrumentation. Appropriation of the pertinent parts of these sources for the use of chemists in solving analytical problems is difficult. Unfortunately also, many publications have become unduly repetitious. Teachers, concerned with imparting the fundamentals of instrumental chemical analysis to college students, have published a variety of useful textbooks, and these are valuable refreshers, or even introductions to unfamiliar techniques. Process instruments are described, catalog style, in several volumes. Treatises on physical measurements have appeared and have been revised in up-to-date editions in commendable efforts to keep the information current (see the reference lists at the end of the chapter).

With regard to the current periodical literature, a bewildering selection of journals confronts the chemist who desires to cross disciplines. Utilization of devices developed for biochemical analysis, for example, in other areas requires familiarity with publications not normally encountered by other kinds of chemists. Resort to indexing and abstracting services is one answer when searching time permits, although the time lag can be frustrating. Inspection of the bibliographies in chapters in this book will suggest pertinent scientific journals to monitor.

The U.S. National Bureau of Standards' Office of Basic Instrumentation operates an instrument reference center tailored to the unique requirements of the instrumentation field. Because of the involvement of this field across all branches of science and technology, a huge variety of journals and reports is covered. In addition to a multidimensional classification system, the "peek-a-boo" mechanical punched-card system can be a convenient search aid.

Cross fertilization between disciplines occurs through the technical news periodicals appearing weekly or monthly, as well as through competent technical sales representatives of instrument manufacturers. National meetings of various societies, especially those featuring commercial exhibits, often play a similar role. Regional meetings are sometimes located at academic centers where significant instrumentation research is in progress.

The Instrument Society of America has a standards and practices department which issues documents that can be more informative than textbooks in revealing and emphasizing what is really important in chemical instrumentation. An example is the standard entitled "Dynamic Response Testing of Process Control Instrumentation" (83). This document establishes the basis for dynamic response testing of control equipment with pneumatic output and electric output for closed-loop actuators, for externally activated control valves, and for other final control elements. General recommendations applicable to all dynamic response testing and a glossary defining terms used in the standard are also included.

The desire for speedy publication by various groups has led to a strong trend toward annual symposia and special meetings featuring more or less specialized subjects, followed by hard-cover publication of the papers as presented. The following titles are examples (including also some annual review series):

Advances in Analytical Chemistry and Instrumentation
Advances in Chromatography
Advances in Computers
Advances in Instrumentation
Advances in Magnetic Resonance

Advances in Mass Spectrometry
Advances in X-Ray Analysis
Analysis Instrumentation
Annual Review of NMR Spectroscopy
Applied Spectroscopy Reviews
Biomedical Sciences Instrumentation
Developments in Applied Spectroscopy
Electroanalytical Chemistry
Instrumentation in the Chemical & Petroleum Industries
Instrumentation in the Iron and Steel Industry
Instrumentation in the Pulp and Paper Industry
Instrument Maintenance Management
International Conference on Modern Trends in Activation Analysis
International Conference on Thermal Analysis
International Instruments & Measurements Conferences
International Symposium on Gas Chromatography
Proceedings of the International Congress of Polarography
Proceedings of the International Gas Chromatography Symposium
Progress in Analytical Chemistry
Progress in Nuclear Magnetic Resonance Spectroscopy
Vacuum Microbalance Techniques

All of these as well as some nonsymposia reviews (usually annual), such as Reilley's (G58) "Advances in Analytical Chemistry and Instrumentation," are examples of annually published volumes which may see the contributors' papers in print within 6–12 months after the meeting itself. They have the advantage of being more timely and conveniently grouped (if the symposium was well organized) than the usual journal literature. The most useful portions of such books are the papers on applications, as sources for ideas, especially for meeting new needs to monitor and control processes, whether in development or manufacture.

Research in industrial laboratories, when of general interest and not in conflict with proprietory considerations, is published in the same journals as is academic work. Patents, a relatively untapped source of information, contain many facts of value, especially for teaching, since companies usually postpone open-literature publication of their research until patent-protection questions are cleared. There is a special listing of selected patents at the end of this chapter.

In summary, there is no shortage of written material, and the research analyst or teacher can ill afford to confine himself to a very limited selection of regular reading.

XI. TRENDS

Time and human effort are the most expensive ingredients in many chemical analyses. Hence new instrument designs, which provide increased capacity for useful information at decreased cost per bit of data, are those which find most ready acceptance. Unfortunately, demands for increased complexity or precision usually lead to higher total costs, often out of *apparent* proportion to the gains achieved. The nature of the problem and its ultimate objectives must guide decisions to invest money and effort in instrumentation for a new venture.

Scouting research in instrumentation may have either very general objectives of broad applicability or quite specific ones to meet a given problem involving a particular quality-control or materials-specification situation. Most often the former (broad objectives for general-use instruments) will be the object of interest to instrument manufacturers, while the latter (specific applications) will be the concern of the user. Bringing the two together in ways that meet both kinds of objectives requires intelligent study and evaluation by chemists and engineers alike.

A. MULTIPURPOSE SYSTEMS

Manufacturers desire to build simple multipurpose devices of wide sales potential. One prevailing trend is to design standard components which can be put together in various ways. This approach, together with the provision of accessories for standard instruments, meets most objectives of both groups. For example, a so-called generalized analog instrument synthesizer described by Morrison (132) was capable of testing more than 40 analytical techniques.

Bard's (12) multipurpose electroanalytical instrument incorporated an X–Y recorder, using also a built-in time base. In addition, it included an electromechanical combination amperostat–potentiostat and a variable-speed polarizing unit. The instrument measured electrode potential (rather than cell voltage) and hence did not involve corrections for IR drop, especially with macroelectrodes. Barna et al. (13) described a transistorized modular instrumentation system for handling photomultiplier signals. Many operations were available for work with scintillation counters, for example, coincidence determination, gating, amplitude discrimination, multiplexing, mixing, amplification, and pulse-height analysis. The design emphasized direct interchangeability of modules, minimizing calibration procedures.

Ewing (54) commented on the multipurpose trends evident in chemical laboratory instrumentation, reviewing De Ford's use of Philbrick vacuum-tube operational amplifiers to accomplish a variety of electro-analytical analyses. Unterkofler and Shain (184) developed a system of plug-in panels to obviate the inconvenience and noise-pickup problems associated with patch cords and multiple switching. Each plug-in unit provided the signal generators, feedback networks, range switching, and other capabilities needed to program a series of eight amplifiers to do their jobs for a particular electroanalytical method.

Morrison (133) and independently Ewing and Brayden (55) used the patch-panel concept in what amounted to an analog computer, nick-named by Morrison an "instrument synthesizer." The prewired patch-board approach was also used by Booman (18) to devise an instrument with functions such as pulse, square-wave, linear-scan, linear-scan-and-hold, stair-step, and bidirectional linear-scan wave forms.

B. SPEED AND UNATTENDED OPERATION

Speed is one of the key advantages to be gained in using instruments, and tying in the measuring system with a computer can accelerate an already fast analysis. An example is the Fourier transform infrared spec-trometer system, which combines an interferometer and a computer in an integrated package. The interferometer essentially generates the spec-tral information all at once (i.e., without scanning); hence spectral inter-pretation has been impractical in the past, because of the Fourier analy-sis needed to convert the data into usable plots of intensity versus fre-quency or wavelength. With an appropriate computer, however, this transformation is so fast that spectra can be produced 300 times faster than with a dispersion instrument. Steel (168) has studied the effect of using collimated light on the interferometer performance.

Increased speed can uncover other bottlenecks in some techniques; for example, the faster column flow achievable in both liquid and gas chromatography places higher demands on the detectors. By using coated capillary columns, elution times of only a few seconds are possible, and some workers have replaced conventional detectors by neutron activation transducers. When flow rates are raised to the turbulence level (e.g., 10 times the laminar region), detector response times must be even faster. In liquid chromatography the use of turbulent flow can shorten the analysis time by large factors. Turbulent-flow chromatography calls for longer columns and higher pressure drops than are required under con-ventional laminar conditions.

As discussed elsewhere, gas chromatography can be speeded up also by combining a second analytical instrument with the chromatograph. If an infrared spectrophotometer or a mass spectrometer (especially the time-of-flight version) is combined with the gas chromatograph, the whole job of separation and identification can be shortened. High-speed liquid chromatography without serious loss of resolution was made possible primarily through the use of controlled-porosity siliceous column packings, enabling, for example, separation of the prime nucleotides in only a few minutes.

Fast detection for high-speed liquid chromatography has been achieved in one example through the use of a bipolar pulse instrument allowing a precise conductance measurement at 50-μsec intervals. When two successive constant-voltage charge pulses, having opposite polarities with equal magnitude, are applied, the instantaneous current measured on the cell at the end of the second pulse is a reliable indication of the conductance and is obtained with very little solution heating.

In the laboratory, automation of equipment permits it to operate unattended during off-shifts on routine measurements, allowing a potential threefold productivity increase. This trend is most attractive where instruments of high investment costs are involved. It carries with it, however, a need for reliable sample changing and data storing. Examples include spectrophotometry, mass and emission spectroscopy, gas chromatography, and X-ray diffraction. The availability of small computers at lower costs makes possible the effective handling of large numbers of data. Such systems free the analyst for less routine, more demanding methods-development activities.

As is true of all technology in which automation has occurred, the future of the outmoded machine lies in the scrap heap, while man has yet to become obsolete. The push button has not yet been discovered which brings riches without human effort.

REFERENCES

A. References Cited

1. Aaker, D. A., *Anal. Chem.*, **37**, 1252 (1965).
2. American Society for Testing and Materials, *ASTM Metric Practice Guide*, **E380**, 1966.
3. Anderson, N. G., *Science*, **166**, 317 (1969).
4. Angstrom, K., *Ofersigt Kongl. Vetenskaps-Akad. Förh.*, **1895, 643**.
5. Anon., *Instr. Control Systems*, **32**, 684 (1959).
6. *Ibid.*, **32**, 1357 (1959).
7. Anton, A., and P. W. Mullen, *Talanta*, **8**, 817 (1961).
8. Aronson, M. H., *Instr. Control Systems*, **32**, 1350 (1959).
9. Aslund, N. R. D., and B. T. Cronhjort, *IBM J. Res. Develop.*, **8**, 160 (1964).

10. Baker, W. O., *ASTM Bull.*, **245**, 42 (1960).
11. Bakker, E. J., J. S. Frost, and G. D. Ogilvie, *Anal. Chem.*, **42**, 1117 (1970).
12. Bard, A. J., *Anal. Chim. Acta*, **21**, 365 (1959).
13. Barna, A., J. H. Marshall, et al., *Rev. Sci. Instr.*, **36**, 166 (1965).
14. Bartlett, E. S., and D. N. Williams, *Rev. Sci. Instr.*, **28**, 919 (1957).
15. Beals, L. S., Jr., *Instruments*, **24**, 1290 (1951).
16. Binnington, D. S., and W. F. Geddes, *Ind. Eng. Chem., Anal. Ed.*, **8**, 76 (1936).
17. Boardman W., and B. Warren, *Chem. Ind. (London)*, **1965**, 1634.
18. Booman, G. L., *Anal. Chem.*, **38**, 1141 (1966).
19. Bostock, W., *J. Sci. Instr.*, **29**, 209 (1952).
20. Bottom, V. E., *Phys. Today*, February 1964, p. 24.
21. Bowman, H. A., and L. B. Macurdy, *J. Res. Natl. Bur. Std.*, **63C**, 91 (1959).
22. Boyle, Robert, *Phil. Trans.*, **10**, 329 (1675).
23. Brackett, F. S., *J. Opt. Soc. Am.*, **50**, 1193 (1960).
24. Brefort, J., *Bull. Soc. Chim.*, **1949**, 1542.
25. Breiter, M. W., *J. Electrochem. Soc.*, **113**, 1071, 1966.
26. Brown, E. R., D. E. Smith, and D. D. De Ford, *Anal. Chem.*, **38**, 1130 (1966).
27. Brown, F. W., 1960 Conference on Standards and Electronic Measurements, Boulder, Colo., June 1960.
28. Cahn, L., *Instr. Control Systems*, **35**, 107 (1962).
29. Campbell, C., and S. Gordon, *Anal. Chem.*, **29**, 298 (1959).
30. Chevenard, P., S. Wache, and R. de la Tullaye, *Bull. Soc. Chim.*, **11**, 41 (1944).
31. Childs, C. W., P. S. Hallman, et al., *Talanta*, **16**, 629 (1969).
32. Christensen, J. J., R. M. Izatt, and L. D. Hansen, *Rev. Sci. Instr.*, **36**, 779 (1965).
33. Christian, G. D., *J. Electroanal. Chem.*, **11**, 94 (1966).
34. Churchman, C. W., and P. Ratoosh, eds., *Measurement: Definitions and Theories*, Wiley, New York, 1960.
35. Clark, J. W., *Rev. Sci. Instr.*, **18**, 915 (1947).
36. Clayton, J. C., J. F. Hazel, et al., *Anal. Chim. Acta*, **14**, 269 (1956).
37. Cole, H., *IBM J. Res. Develop.*, **13**, 5 (1969).
38. Condon, E. U., *Science*, **110**, 339 (1949).
39. Coor, T., *J. Chem. Educ.*, **45** (7), A533; (8), A583 (1968).
40. Cornwell, J. C., and K. L. Cheng, *Anal. Chim. Acta*, **42**, 189 (1969).
41. Cotlove, E., H. V. Trantham, and R. L. Bowman, *J. Lab. Clin. Med.*, **51**, 461 (1958).
42. Crowther, J. C., *Chem. Ind. (London)*, **1959**, 327.
43. Davis, R. S., and W. Kipiniak, "Computer Controlled Chromatograph System," paper presented at 18th Annual Symposium, N.J. and N.Y. Sections, Instrument Society of America, Apr. 5, 1966.
44. Dervichian, D. G., *J. Phys. Radium*, **6**, 221 (1935).
45. Dingle, H., Review of book, "Measurements: Definitions and Theories," *Sci. Am.*, **202** (No. 6), 189 (1960); *Brit. J. Phil. Sci.*, **1**, 5 (May 1950).
46. Drew, H. D., and J. M. Fitzgerald, *Anal. Chem.*, **41**, 974 (1969).
47. Duggan, E. L., and V. L. Stevens, *Anal Chem.*, **29**, 1076 (1957).
48. Du Bois, H. E. J. G., and H. Rubens, *Wied. Ann.*, **48**, 236 (1893).
48a. Eisenman, George, D. O. Rudin, and J. U. Casby, *Science*, **126**, 831 (1957).
49. Eisner, L., *J. Chem. Educ.*, **41**, (7) A491; (8) A551; (9) A607 (1964).

50. Elliott, Arthur, and J. H. Dickson, *Laboratory Instruments: Their Design and Application,* 2nd ed., Chemical Publishing Co., New York, 1960.
51. Epstein, S., *Appl. Spectry.,* 14 (1), 7 (1960).
52. Ewald, P., *Ind. Eng. Chem., Anal. Ed.,* 14, 66 (1942).
53. Ewing, G. W., *J. Chem. Educ.,* 42, 32 (1965).
54. *Ibid.,* 46 (10), A717 (1969).
55. Ewing, G. W., and T. H. Brayden, *Anal. Chem.,* 35, 1826 (1963).
56. Farquharson, J., *Rev. Sci. Instr.,* 31, 723 (1960).
57. Flaschka, H., and P. Sawyer, *Talanta,* 8, 521 (1961).
58. Flaschka, H., and R. Speights, *Talanta,* 15, 1467 (1968).
59. French, I. M., *J. Sci. Instr.,* 38, 516 (1961).
60. Garn, P. D., *Anal. Chem.,* 29, 839 (1957).
61. Glover, B. W., R. D. Orwell, and P. J. Adams, *Hilger J.,* X (1), 7 (1967).
62. Gordon, S., and C. Campbell, *Anal. Chem.,* 28, 124 (1956).
63. *Ibid.,* 29, 1706 (1957).
64. *Ibid.,* 32, 271 (1960).
65. Grant, P. M., *IBM J. Res. Develop.,* 13, 15 (1969).
66. Greuter, E., *Z. Anal. Chem.,* 222, 224 (1966).
67. Groot, C., and V. H. Troutner, *Anal. Chem.,* 29, 835 (1957).
68. Hagan, P. J., and J. R. de Laeter, *J. Sci. Instr.,* 43, 662 (1966).
69. Hallikainen, K. E., and D. J. Pompeo, *Instr. Soc. Am. J.,* 8, 22 (1952).
70. Hamilton, W. C., *Science,* 169, 133 (1970).
71. Hancock, H. A., Jr., and I. Lichtenstein, *J. Chromatographic Sci.,* 7, 290 (1969).
72. Hannon, D. M., D. E. Horn, et al., *IBM J. Res. Develop.,* 13, 79 (1969).
73. Haslam, J., and D. C. M. Squirrell, *J. Appl. Chem. (London),* 9, 65 (1959).
74. Heit, M. L., and D. E. Ryan, *Anal. Chim. Acta,* 29, 524 (1963).
75. Hill, R. L., "Transistorized Building Blocks for Data Instrumentation," *Natl. Bur. Std. (U.S.) Tech. Note* 168, U.S. Government Printing Office, Washington, D.C., 1963.
76. Hirsch, R. F., *J. Chem. Educ.,* 44 (12), A1023 (1967); 45 (1), A7 (1968).
77. Holm-Jensen, I., *Anal. Chim. Acta,* 33, 198 (1965).
78. Hooley, J. G., *Can. J. Chem.,* 35, 374 (1957).
79. Howlett, L. E., *Science,* 158, 72 (1967).
80. Hoyt, J. L., *Anal. Chim. Acta,* 44, 369 (1969).
81. Huntoon, R. D., *Science,* 158, 67 (1967).
82. Hyatt, E. P., I. B. Cutler and M. E. Wadsworth, *Am. Ceram. Soc. Bull.,* 35, 180 (1956).
83. Instrument Society of America, *Standard* ISA–S26, "Dynamic Response Testing of Process Control Instrumentation," 1968.
84. International Union of Pure and Applied Chemistry. Commission on Molecular Structure and Spectroscopy, *Multilingual Dictionary of Important Terms in Molecular Spectroscopy,* National Research Council of Canada, Ottawa, 1966.
85. Irving, H., and L. D. Pettit, *Analyst,* 84, 641 (1959).
86. Jeffcoat, K., and M. Akhtar, *Analyst,* 87, 455 (1962).
87. Johansson, G., *Talanta,* 12, 111 (1965).
88. Johnson, B., T. Kuga, et al., *IBM J. Res. Develop.,* 13, 36 (1969).
89. Johnson, D. R., J. W. Cassels, E. G. Brame, and D. W. Westneat, *Anal. Chem.,* 34, 1610 (1962).

90. Jones, G. A., and J. R. Tinklepaugh, *U.S.A.F. Tech. Rept.* **6448,** (November 1950).
91. Kaiser, R., *Z. Anal. Chem.,* **222,** 128 (1966).
92. Keidel, F. A., *Anal. Chem.,* **31,** 2043 (1959).
93. Keinath, G., *Instr. Automation,* **31,** 827 (1958).
94. Kelley, J. B., and H. H. Marold, *Instr. Control Systems,* **32,** 1361 (1959).
95. Kendall, B. R. F., *J. Sci. Instr.,* **43,** 215 (1966).
96. Kennedy, S. W., *Nature,* **210,** 936 (1966).
97. Kieselbach, Richard, *Anal. Chem.,* **32,** 1749 (1960).
98. Klaasse, J. M., "A Versatile Titrator with Measurement and Control Applications," paper presented at Instrument Society of America Conference, *Preprint* **80–59,** September 1959.
99. Klopsteg, P. E., *Science,* **132,** 1913 (1960).
100. Kolthoff, I. M., quoted in editorial by W. J. Murphy, *Anal. Chem.,* **29,** 1 (1956).
101. Kuhlmann, W. H. F., *Der Merchaniker,* **18,** 146 (1910).
102. Learner, R. C .M., *J. Sci. Instr.* (J. Phys. E), Ser. 2, **1,** 589 (1968).
103. Lewin, S. Z., *Anal. Chem.,* **30** (7), 17A (1958).
104. *Ibid.,* **33** (3), 23A (1961).
105. *Ibid.,* **34** (2), 25A; (8), 27A (1962).
106. Lewin, S. Z., *J. Chem. Educ.,* **36** (1), A7; (2) A67 (1959).
107. Lincoln, K. A., *Rev. Sci. Instr.,* **31,** 537 (1960).
108. Lingane, J. J., *Anal. Chem.,* **20,** 285 (1948).
109. Lloyd, E. C., *Rev. Sci. Instr.,* **39,** 1953 (1968).
110. Lohmann, I. W., *J. Sci. Instr.,* **21,** 999 (1950).
111. Longchambon, H., *Bull. Soc. Franc. Mineral. Crist.,* **59,** 145 (1936).
112. Lukaszewski, G. M., *Nature,* **194,** 959 (1962).
113. Macinante, J. A., and J. Waldersee, *J. Sci. Instr.,* **41,** 1 (1964).
114. Macurdy, L. B., *J. Res. Natl. Bur. Std.,* **68C,** 135 (1964).
115. Macurdy, L. B., and H. A. Bowman, *Instr. Automation,* **31,** 1972 (1958).
116. Maddock, A. J., *Research (London),* **14,** 430 (1961).
117. Malmstadt, H. V., and S. R. Crouch, *J. Chem. Educ.,* **43,** 340 (1966).
118. Malmstadt, H. V., and T. P. Hadjiioannou, *Anal. Chim. Acta,* **19,** 563 (1958).
119. Martin, A. I. P., and R. L. M. Synge, *Biochem. J.,* **35,** 1358 (1941).
120. Marzetta, L. A., *Instr. Technol.,* **16,** 51 (1969).
121. McGucken, William, *Nineteenth-Century Spectroscopy,* Johns Hopkins Press, Baltimore, 1969.
122. McNish, A. G., *Electro-Technology,* **71** (5), 114 (1963).
123. McNish, A. G., *Intern. Sci. Technol.,* p. 58, November 1965.
124. McPherson, A. T., *Instr. Automation,* **32** (1), 92 (1959).
125. Melnechuk, T., *Intern. Sci. Technol.,* p. 14, August 1962.
126. Michel, H., *Scientific Instruments in Art and History,* Viking Press, New York, 1968.
127. Middleton, W. E. K., *A History of the Thermometer and Its Uses in Meterology,* Johns Hopkins Press, Baltimore, 1966.
128. Milazzo, G., *J. Electroanal. Chem.,* **7,** 123 (1964).
129. Mitchell, J. Jr., and S. S. Lord, Jr., *Rubber Chem. Technol.,* **34,** 1553 (1961).
130. Miyake, S., *Talanta,* **13,** 1253 (1966).
131. Moll, W. J. H., and H. C. Burger, *Phil. Mag.,* **50,** 621 (1925).

132. Morrison, C. F., *Anal. Chem.*, **35**, 1820 (1963).
133. Morrison, C. F., Jr., *Generalized Instrumentation for Research and Teaching*, Washington State University Press, Pullman, Wash., 1964.
134. Müller, R. H., *Anal. Chem.*, **27** (4), 101A (1968).
135. *Ibid.*, **29**, 1118 (1957).
136. Müller, R. H., and R. L. Garman, *Ind. Eng. Chem., Anal. Ed.*, **10**, 436 (1938).
137. Müller, R. H., and F. D. Lonadier, *Anal. Chem.*, **30**, 891 (1958).
138. Narayanan, U. H., and K. Sundararajan, *J. Electroanal. Chem.*, **6**, 397 (1963).
139. Neal, J. L., F. G. Hurtubise, et al., *ISA Trans.*, **1**, 193 (1962).
140. Nelson, J. E., *J. Chem. Educ.*, **45** (9), A635; (10), A787 (1968).
141. Oden, S., *Bull. Geol. Inst. Univ. Upsala*, **16**, 15 (1918).
142. Oden, S., *Trans. Faraday Soc.*, **17**, 327 (1922).
143. Oehme, F., *Z. Anal. Chem.*, **222**, 244 (1966).
144. Perone, S. P., *J. Chromatographic Sci.*, **7**, 714 (1969).
145. Peters, Horst, and H. G. Wiedemann. *Z. Anorg. Allgem. Chem.*, **298**, 202 (1959).
146. Peterson, A. H., *Instr. Automation*, **28**, 1104 (1955).
147. Petrie, Sir (William Matthew) Flinders, *Numism. Chron.*, 4th ser., **18**, 115 (1918).
148. Pike, J. A., and G. C. Goode, *Anal. Chim. Acta*, **39**, 1 (1967).
149. Pinder, A. R., *Chem. Ind.* (*London*), **1961**, 758, 1180.
150. Poulis, J. A., and J. M. Thomas, *J. Sci. Instr.*, **40**, 95 (1963).
151. Priestley, P. T., W. S. Sebborn, et al., *Analyst*, **88**, 797 (1963).
152. Rabatin, J. G., and R. H. Gale, *Anal. Chem.*, **28**, 1314 (1956).
153. Reilley, C. N., *J. Chem. Educ.*, **39** (11), A853; (12), A933 (1962).
154. Ridley, J. E., D. B. L. Elliott, et al., *Analyst*, **85**, 508 (1960).
155. Ruh, E. L., J. J. Moran, and R. D. Thompson, *ASTM Bull.*, **240**, 31 (1959).
156. Savitsky, A., *Anal. Chem.*, **33** (12), 25A (1961); also "Computer Handling of Spectra. II: Progress," paper presented at Pittsburgh Conference on Analytical Chemistry and Applied Spectroscopy, March 1963.
157. Savitsky, A., and M. J. E. Golay, *Anal. Chem.*, **36**, 1627 (1964).
158. Schonebaum, R. C., and E. Breekland, *Talanta*, **11**, 659 (1964).
159. Shain, I., and C. O. Huber, *Anal. Chem.*, **30**, 1286 (1958).
160. Shirley, J. W., *Ambix*, **4**, 61 (1949).
161. Smith, J. J., *ISA J.*, **12** (7), 47 (1965).
162. Smith, S. W., and J. K. Taylor, *J. Res. Natl. Bur. Std.*, **63C**, 65 (1959).
163. Smythe, L. E., *Talanta*, **15**, 1177 (1968).
164. Soulen, J. R., *Anal. Chem.*, **34**, 136 (1962).
165. Spinrad, R. J., *Science*, **158**, 55 (1967).
166. Spragg, S. P., *Anal. Chim. Acta*, **38**, 137 (1967).
167. Stearne, J. M., and J. R. Urwin, *Makromol. Chem.*, **56**, 76 (1962).
168. Steel, W. H., *J. Opt. Soc. Am.*, **54**, 151 (1964).
169. Stevens, V. L., and E. L. Duggan, *Anal. Chem.*, **29**, 1073 (1957).
170. Stine, K. E., J. E. Stewart, R. A. Weagant, and H. J. Sloane, "Applications of Digital Techniques to Infrared Spectroscopy," paper presented at Pittsburgh Conference on Analytical Chemistry and Applied Spectroscopy, March 1967; Analyzer, *8*, (2), 12 (1967).
171. Stirling, P. H., and H. Ho, *Ind. Eng. Chem.*, **53** (9), 59A (1961).
172. Stock, J. T., *Development of the Chemical Balance*, H. M. Stationery Office, London, 1969.

173. Stolar, G., *Instr. Control Systems,* **34** (1), 85 (1961).
174. Suito, E., and M. Arakawa, *Bull. Inst. Chem. Res., Kyoto Univ.,* **22,** 7 (1950).
175. Tabbutt, F. D., *J. Chem. Educ.,* **39,** 611 (1962).
176. Takahashi, Akir, *Japan Analyst,* **8,** 661 (1959).
177. Takahashi, T., and H. Sakurai, *J. Chem. Soc. Japan, Ind. Chem. Sect.,* **67,** 1802 (1964).
178. Tayor, J. K, and S. W. Smith, *J. Res. Nat. Bur. Std.,* **63A,** 153 (1959); **63C,** 65 (1959).
179. Thomas, J. M., and B. R. Williams, *Quart. Revs.,* **19,** 231 (1965).
180. Thompson, A. E., *J. Chromatog.,* **6,** 539 (1961).
181. Tryhorn, F. G., and W. F. Wyatt, *Trans. Faraday Soc.,* **23,** 238 (1927).
182. U.S. National Bureau of Standards, *Letter Circ.,* **LC1035,** "Units and Systems of Weights and Measures, Their Origin, Development, and Present Status," 1960.
183. United States Navy Electronics Laboratory, "Suggestions for Designers of Electronic Equipment," 1959–60 ed.
184. Unterkofler, W. L., and I. Shain, *Anal. Chem.,* **35,** 1778 (1963).
185. Van Dijck, L. A., and L. M. Simons, *Chem. Weekblad,* **55,** 407 (1959).
186. Verdery, R. B., Jr., *Chemist Analyst,* **38,** 68 (1949).
187. Wallis, E., *Chem. Ind. (London),* **1964,** 1571.
188. Walsh, J. T., R. E. Kramer, and C. Merritt, Jr., *J. Chromatographic Sci.,* **7,** 348 (1969).
189. Webb, R. D., *ISA J.,* **6** (5), 66 (1959).
190. Weber, J. N., *Spectrochim, Acta,* **16,** 1435 (1960).
191. Wendlandt, W. W., *Anal. Chem.,* **30,** 56 (1958).
192. Wheatland, D. P. (assisted by Barbara Carson), *The Apparatus of Science at Harvard, 1765–1800,* Harvard University Press, Cambridge, Mass., 1968.
193. Westneat, D. F., W. E. Keder, and H. W. Safford, *Chemist Analyst,* **44,** 12 (1955).
194. White, C. E., *Instr. Control Systems,* **33,** 1001 (1960).
195. Wildhack, W. A., *Science,* **112,** 515 (1950).
196. Woodson, W. E., *Human Engineering Guide for Equipment Designers,* University of California Press, Los Angeles, 1954.
197. Young, I. G., paper presented at Second Biennial International Gas Chromatography Symposium, Instrument Society of America, East Lansing, Mich., June 1959.
198. Zellmer, N. A., *Electronics,* **39,** 66 (Jan. 24, 1966).
199. Zenchelsky, S. T., and P. R. Segatto, *Anal. Chem.,* **29,** 1856 (1957).

B. General References on Instrumentation and Electronics

G1. Ahrendt, W. R., and C. J. Savant, Jr., *Servomechanism Practice,* 2nd ed., McGraw–Hill, New York, 1960.
G2. Alfrey, G. F., *Physical Electronics,* Van Nostrand, Princeton, N.J., 1964.
G3. Bair, E. J., *Introduction to Chemical Instrumentation,* McGraw–Hill, New York, 1962.
G4. Banner, E. H. W., *Electronic Measuring Instruments,* 2nd ed., Macmillan New York, 1958.
G5. Bedini, S. A., *Early American Scientific Instruments and Their Makers,* Smithsonian Institution Museum of History and Technology, Washington, D.C., 1964.

G6. Beilby, A. L., ed., et al., *Modern Classics in Analytical Chemistry*, American Chemical Society, Washington, D.C., 1970.

G7. Bennett, G. E., G. R. Richards, and E. C. Voss., *Electronics Applied to the Measurement of Physical Quantities*, Ministry of Supply, London, 1952.

G8. Berl, W. C., ed., *Physical Methods in Chemical Analysis*, 2nd ed., Academic, New York, 4 vols., 1960, 1961.

G9. Biffen, F. M., and W. Seaman, *Modern Instruments in Chemical Analysis*, McGraw–Hill, New York, 1956.

G10. Born, Max, and E. Wolf, *Principles of Optics*, 4th ed., Pergamon, New York, 1970.

G11. Burr-Brown Research Corp., *Handbook of Operational Amplifier Applications*, Tucson, Ariz., 1963.

G12. Cannon, C. G., *Electronics for Spectroscopists*, Interscience, New York, 1960.

G13. Chirlian, P. M., *The Analysis and Design of Electron Circuits*, McGraw–Hill, New York, 1965.

G14. Conrady, A. E., *Applied Optics and Optical Design*, Dover, New York, 1957.

G15. Cooke-Yarborough, E. H., *Introduction to Transistor Circuits*, Interscience, New York, 1957.

G16. Davie, O. H., *The Elements of Pulse Techniques*, Reinhold, New York, 1964.

G17. Delahay, C., *Instrumental Analysis*, Macmillan, New York, 1957.

G18. Dixon, J. P., *Modern Methods in Organic Microanalysis*, Van Nostrand, London, 1968.

G19. Donaldson, P. E. K., ed., *Electronic Apparatus for Biological Research*, Academic, New York, 1958.

G20. Eckschlager, Karel, *Errors, Measurement and Results in Chemical Analysis*, Van Nostrand, Princeton, N.J., 1969.

G21. Elliott, A., and J. H. Dickson, *Laboratory Instruments: Their Design and Application*, 2nd ed., Chemical Publishing Co., New York, 1960.

G22. Epstein, Sam, *Scientific Instruments: How To Build and Use Them*, Franklin Publishing Co., Englewood, N.J., 1970.

G23. Ewing, G. W., *Analytical Instrumentation: A Laboratory Guide for Chemical Analysis*, Plenum, New York, 1966.

G24. Ewing, G. W., *Instrumental Methods of Chemical Analysis*, 2nd ed., McGraw–Hill, New York, 1960.

G25. Flaschka, H. A., A. J. Barnard, Jr., and P. E. Sturrock, *Quantitative Analytical Chemistry: Vol. I: Introduction to Principles; Vol. II: Short Introduction to Practice*, Barnes & Noble, New York, 1969.

G26. Fletcher, F. J., ed., *Instrumental Methods*, Van Nostrand, Princeton, N.J., 1966.

G27. Francon, M., *Modern Applications of Physical Optics*, Interscience, New York, 1963.

G28. French, J. A., *Standards and Practices for Instrumentation*, 2nd ed., Instrument Society of America, Pittsburgh, Pa., 1966.

G29. Fribance, A. E., *Industrial Instrumentation Fundamentals*, McGraw–Hill, New York, 1962.

G30. Instrument Society of America, *Standards and Practices for Instrumentation*, 3rd ed., Instrument Society of America, Pittsburgh, Pa., 1970.

G31. Jenkins, F. A., and H. E. White, *Fundamentals of Optics*, 3rd ed., McGraw–Hill, New York, 1957.

G32. Johnson, B. K., *Optics and Optical Instruments*, Dover, New York, 1960 (1947).

G33. Joslyn, M. A., ed., *Physical, Chemical, and Instrumental Methods of Analysis*, 2nd ed., Academic, New York, 1970.

G34. Koenig, H. E., and W. A. Blackwell, *Electromechanical System Theory*, McGraw–Hill, New York, 1961.

G35. Krugers, J., and A. I. M. Keulmans, eds., *Practical Instrumental Analysis*, American Elsevier, New York, 1965.

G36. Langmuir, D. B., and W. D. Hershberger, eds., *Foundations of Future Electronics*, McGraw–Hill, New York, 1961.

G37. Larach, Simon, ed., *Photoelectronic Materials and Devices*, Van Nostrand, Princeton, N.J., 1965.

G38. Lion, K. S., *Instrumentation in Scientific Research: Electrical Input Transducers*, McGraw–Hill, New York, 1959.

G39. Lurch, E. N., *Fundamentals of Electronics*, Wiley, New York, 1960.

G40. Mahood, R. F., *Instrument Maintenance Management*, Vol. 2, Plenum, New York, 1967.

G41. Malmstadt, H. V., and C. G. Enke, *Digital Electronics for Scientists*, Benjamin, New York, 1969.

G41A. Malmstadt, H. V., C. G. Enke, and E. C. Toren, Jr., *Electronics for Scientists*, Benjamin, New York, 1962.

G42. Marton, L., ed., *Advances in Electronics and Electron Physics*, Academic, New York, 1958.

G43. Meinke, W. W. and B. F. Schribner, eds., *Trace Characterization—Chemical and Physical*, U.S. National Bureau of Standards, Washington, D.C., 1967.

G44. Meloan, C. E., and R. W. Kiser, *Problems and Experiments in Instrumental Analysis*, Charles E. Merritt, Inc., Columbus, Ohio, 1963.

G45. Morgenthaler, L. P., *Basic Operational Amplifier Circuits for Analytical Chemical Instrumentation*, McKee–Pedersen Instruments, Danville, Cal., 1967.

G46. Morrison, C. F., *Generalized Instrumentation for Research and Teaching*, Washington State University Press, Pullman, Wash., 1964.

G47. Müller, R. H., E. H. Baum, et al., *Automatic Chemical Analysis*, New York Academy of Sciences, New York, 1960.

G48. Olsen, G. H., *Electronics: A General Introduction for the Non-specialist*, Plenum, New York, 1968.

G49. Pannenborg, A. E., ed., *Encyclopedia of Physics. Vol. XXIII; Electrical Instruments*, Springer–Verlag, Berlin, 1967.

G50. Parr, Geoffrey, and O. H. Davie, *The Cathode-Ray Tube and Its Applications*, 3rd ed., Reinhold, New York, 1959.

G51. Partridge, G. R., *Principles of Electronic Instruments*, Prentice–Hall, Englewood Cliffs, N.J., 1958.

G52. Pearson, E. B., *Technology of Instrumentation*, Van Nostrand, Princeton, N.J., 1958.

G53. Pecsok, R. L., and L. D. Shields, *Modern Methods of Chemical Analysis*, Wiley, New York, 1968.

G54. Philbrick Researches, Inc., *Applications Manual for Computing Amplifiers*, Dedham, Mass., 1966.

G55. Phillips, L. F., *Electronics for Experimenters in Chemistry*, Wiley, New York, 1966.

G56. Poyer, J., J. Herrick, and T. B. Weber, *Biomedical Sciences Instrumentation*, Vol. 3, Plenum, New York, 1966.

G57. Price, L. W., *Electronic Laboratory Techniques*, J. and A. Churchill, London, 1969.

G58. Reilley, C. N., *Advances in Analytical Chemistry and Instrumentation*, Vols. 1–8, Interscience–John Wiley, New York, 1960–70.

G59. Reilley, C. N., and D. T. Sawyer, *Experiments for Instrumental Methods*, McGraw–Hill, New York, 1961.

G60. Sarbacher, R. I., *Encyclopedic Dictionary of Electronics and Nuclear Engineering*, Prentice–Hall, Englewood, N.J., 1959.

G61. Saucedo, R., and E. E. Schiring, *Introduction to Continuous and Digital Control Systems*, Macmillan, New York, 1968.

G62. Schwarz, J. C. P., ed., *Physical Methods in Organic Chemistry*, Holden–Day, San Francisco, 1964.

G63. Smith, J. F., and W. B. Brombacher, "Guide to Instrumentation Literature," *Natl. Bur. Std. (U.S.) Misc. Pub.* **271**, Washington, D.C., 1965.

G64. Strobel, H. A., *Chemical Instrumentation*, Addison–Wesley, Reading, Mass., 1960.

G65. Strouts, C. R. N., H. N. Wilson, and R. T. Parry-Jones, eds., *Chemical Analysis: The Working Tools*, 2nd ed., Vols. 1–3, Oxford University Press, New York, 1962.

G66. Suprynowicz, V., *Introduction to Electronics for Students of Biology, Chemistry, and Medicine*, Addison–Wesley, Reading, Mass., 1966.

G67. Susskind, C., ed., *The Encyclopedia of Electronics*, Reinhold, New York, 1962.

G68. Technicon Symposia, *Automation in Analytical Chemistry*, Vols. 1 and 2, Mediad, Inc., White Plains, N.Y., 1967; *Advances in Automated Analysis*, Vols. 1, 2, and 3, Technicon Corp., Tarrytown, N.Y., 1969.

G69. Thomas, H. E., and C. A. Clarke, *Handbook of Electronic Instruments and Measurement Techniques*, Prentice–Hall, Englewood Cliffs, N.J., 1967.

G70. Tyson, F. C., Jr., *Industrial Instrumentation*, Prentice–Hall, Englewood Cliffs, N.J., 1961.

G71. Von Koch, H., and G. Ljungberg, *Instruments and Measurements*, 2 vols., Academic, New York, 1961.

G72. Warschauer, D. M., *Semiconductors and Transistors*, McGraw–Hill, New York, 1959.

G73. Weissberger, Arnold, ed., *Physical Methods of Organic Chemistry*, Interscience, New York, 1959 et seq.

G74. Whittick, J. S., R. F. Muraca, and L. A. Cavanagh, *Analytical Chemistry Instrumentation: A Survey*, National Aeronautics and Space Administration, Washington, D.C., 1967.

G75. Wiberg, K. B., *Computer Programming for Chemists*, Benjamin, New York, 1965.

G76. Wildhack, W. A., R. C. Powell, and H. L. Mason, eds., "Accuracy in Measurements and Calibrations," *Natl. Bur. Std. (U.S.) Tech. Note* **262**, Washington, D.C., 1965.

G77. Willard, H. H., L. L. Merritt, Jr., and J. A. Dean, *Instrumental Methods of Analysis*, 4th ed., Van Nostrand, Princeton, N.J., 1965.

C. Special References:
Selected Specialized Analytical Texts
Recommended for Their Emphasis on Instruments

1. ELECTROANALYTICAL AND TITRIMETRIC TECHNIQUES

S1. Bark, L. S., and S. M. Bark, *Thermometric Titrimetry*, Pergamon, New York, 1969.

S2. Bates, R. G., *Determination of pH*, Wiley, New York, 1964.

S3. Breyer, B., and H. H. Bauer, *Alternating Current Polarography and Tensammetry*, Interscience–Wiley, New York, 1963.

S4. Durst, R. A., ed., *Ion-Selective Electrodes*, U.S. Government Printing Office, Washington, D.C., 1969.

S5. Eisenman, G., ed., *Glass Electrodes for Hydrogen and Other Cations: Principles and Practice*, Dekker, New York, 1967.

S6. Gyenes, I., *Titration in Non-aqueous Media*, Van Nostrand, Princeton, N.J., 1967.

S7. Ives, D. J., and G. J. Janz, *Reference Electrodes: Theory and Practice*, Academic, New York, 1961.

S8. Lavallée, Marc, O. F. Schanne, and N. C. Hebert, eds., *Glass Microelectrodes*, Wiley, New York, 1969.

S9. Leveson, L. L., *Introduction to Electroanalysis*, Butterworths, Washington, D.C., 1964.

S10. Lingane, J. J., *Electroanalytical Chemistry*, 2nd ed., Interscience, New York, 1958.

S11. Meites, Louis, *Polarographic Techniques*, 2nd ed., Interscience–Wiley, New York, 1965.

S12. Milner, G. W. C., and G. Phillips, *Coulometry in Analytical Chemistry*, Pergamon, New York, 1967.

S13. Murray, R. W., and C. N. Reilley, *Electroanalytical Principles*, Interscience–Wiley, New York, 1964.

S14. Phillips, J. P., *Automatic Titrators*, Academic, New York, 1959.

S15. Stock, J. T., *Amperometric Titrations*, Interscience–Wiley, New York, 1965.

S16. Squirrell, D. C. M., *Automatic Methods in Volumetric Analysis*, Van Nostrand, Princeton, N.J., 1964.

S17. Tyrrell, H. J. V., and A. E. Beezer, *Thermometric Titrimetry*, Barnes & Noble, New York, 1968.

S18. Wilson, C. L., and D. W. Wilson, eds., *Comprehensive Analytical Chemistry*, Vol. IIA: *Electrical Methods*, American Elsevier, New York, 1964.

S19. Zuman, Petr, *Organic Polarographic Analysis*, Pergamon, London; Macmillan, New York, 1964.

2. THERMAL ANALYSIS

S20. Duval, Clement, *Inorganic Thermogravimetric Analysis*, Elsevier, London, 1963.

S21. Slade, P. E., Jr., and L. T. Jenkins, *Thermal Characterization Techniques*, Dekker, New York, 1970.

S22. Smothers, W. J., and Yao Chiang, *Handbook of Differential Thermal Analysis*, Chemical Publishing Co., New York, 1966.

S23. Wendlandt, W. W., *Thermal Methods of Analysis*, Interscience–Wiley, New York, 1964.

S24. Garn, P. D., *Thermoanalytical Methods of Investigation*, Academic, New York, 1965.

3. CHROMATOGRAPHY

S25. Ambrose, Douglas, and B. A. Ambrose, *Gas Chromatography*, Van Nostrand, Princeton, N.J., 1962.

S26. Bayer, Ernst, *Gas Chromatography*, American Elsevier, New York, 1961.

S27. Ettre, L. S., and W. H. McFadden, eds., *Ancillary Techniques of Gas Chromatography*, Interscience–Wiley, New York, 1969.

S28. Ettre, L. S., and Albert Zlatikis, eds., *The Practice of Gas Chromatography*, Interscience–Wiley, New York, 1967.

S29. Johnson, J. F. and R. S. Porter, *Analytical Gel Permeation Chromatography*," Interscience–Wiley, New York, 1968.

S30. Keulemans, A. I. M., *Gas Chromatography*, 2nd ed., Reinhold, New York, 1959.

S31. Knox, J. H., *Gas Chromatography*, Wiley, New York, 1962.

S32. Littlewood, A. B., *Gas Chromatography: Principles, Techniques, and Applications*, Academic, 2nd ed., New York, 1970.

S33. Purnell, Howard, *Gas Chromatography*, Wiley, New York, 1962.

S34. Schupp, O. E., III, *Gas Chromatography*, Interscience–Wiley, New York, 1968.

S35. Stock, R., and C. B. F. Rice, *Chromatographic Methods*, Reinhold, New York, 1963.

S36. Tranchant, Jean, ed., *Practical Manual of Gas Chromatography*, American Elsevier, New York, 1969.

4. MASS SPECTROMETRY

S37. Ahearn, A. J., ed., *Mass Spectrometric Analysis of Solids*, American Elsevier, New York, 1966.

S38. Jayaram, R., *Mass Spectrometry: Theory and Applications*, Plenum, New York, 1966.

S39. Kiser, R. W., *Introduction to Mass Spectrometry and Its Applications*, Prentice–Hall, Englewood Cliffs, N.J., 1965.

S40. McDowell, C. A., ed., *Mass Spectrometry*, McGraw–Hill, New York, 1963.

S41. Reed, R. I., ed., *Mass Spectrometry*, Academic, New York, 1965.

S42. Reed, R. I., *Modern Aspects of Mass Spectrometry*, Plenum, New York, 1968.

S43. Roboz, John, *Introduction to Mass Spectrometry: Instrumentation and Techniques*, Interscience–Wiley, New York, 1968.

5. X-RAY TECHNIQUES

S44. Alexander, L. E., *X-ray Diffraction Methods in Polymer Science*, Interscience–Wiley, New York, 1969.

S45. Bertin, E. P., *Principles and Practice of X-ray Spectrometric Analysis*, Plenum, New York, 1970.

S46. Birks, L. C., *X-ray Spectrochemical Analysis*, 2nd ed., Interscience–Wiley, New York, 1969.

S47. Birks, L. S., *Electron Probe Microanalysis*, Interscience–Wiley, New York, 1963.

S48. Clark, G. L., ed., *The Encyclopedia of X-rays and Gamma Rays*, Reinhold, New York, 1963.

S49. Engström, A., *X-ray Microanalysis in Biology and Medicine*, American Elsevier, New York, 1962.

S50. Jenkins, R., and J. L. de Vries, *Practical X-ray Spectrometry*, Springer–Verlag, Berlin, 1967.

S51. Klug, H. P., and L. E. Alexander, *X-ray Diffraction Procedures*, Wiley, New York, 1954.

S52. Liebhafsky, H. A., H. G. Pfeiffer, E. H. Winslow, and P. D. Zemany, *X-ray Absorption and Emission in Analytical Chemistry*, Wiley, New York, 1960.

S53. Wilson, A. J. C., *Elements of X-ray Crystallography*, Addison–Wesley, Reading, Mass., 1970.

6.. OPTICAL AND SPECTROMETRIC
TECHNIQUES—VISIBLE AND ULTRAVIOLET

S54. American Society for Testing and Materials, *Manual on Recommended Practices in Spectrophotometry*, 3rd ed., Philadelphia, 1969.

S55. Clark, G. L., ed., *The Encyclopedia of Spectroscopy*, Reinhold, New York, 1960.

S56. Dean, J. A., and T. C. Rains, eds., *Flame Emission and Atomic Absorption Spectrometry*, Dekker, New York, 1969.

S57. Edisbury, J. R., *Practical Hints on Absorption Spectrometry*, Hilger & Watts, London, 1966.

S58. Elwell, W. T., and J. A. F. Gidley, *Atomic-Absorption Spectrophotometry*, 2nd ed., Pergamon, New York, 1966.

S59. Herrmann, Roland, and C. T. J. Alkemade, *Chemical Analysis by Flame Photometry*, 2nd ed., Interscience, New York, 1963.

S60. Kortum, G., *Reflectance Spectroscopy: Principles, Methods, Applications*, Springer–Verlag, Berlin, 1969.

S61. Lothian, G. F., *Absorption Spectrophotometry*, 3rd ed., Hilger, London, 1969.

S62. May, Leopold, *Spectroscopic Tricks*, Plenum, New York, 1967.

S63. Meehan, E. J., *Optical Methods of Analysis*, Interscience–Wiley, New York, 1964.

S64. Ramirez-Munoz, Juan, *Atomic-Absorption Spectroscopy and Analysis by Atomic-Absorption Flame Photometry*, American Elsevier, New York, 1968.

S65. Slavin, Walter, *Atomic Absorption Spectroscopy*, Interscience–Wiley, New York, 1968.

S66. Stearns, E. I., *The Practice of Absorption Spectrophotometry*, Wiley, New York, 1969.

S67. Wendlandt, W. W., and H. G. Hecht, *Reflectance Spectroscopy*, Interscience–Wiley, New York, 1966.

S68. Wendlandt, W. W., *Modern Aspects of Reflectance Spectroscopy*, Plenum, New York, 1968.

S69. White, C. E., and R. J. Argauer, *Fluorescence Analysis: A Practical Approach*, Dekker, New York, 1970.

S70. Zander, M., *The Application of Phosphorescence to the Analysis of Organic Compounds*, Academic, New York, 1968.

7. SPECTROMETRIC TECHNIQUES—
INFRARED AND NUCLEAR MAGNETIC RESONANCE

S71. Bovey, F. A., *Nuclear Magnetic Resonance Spectroscopy*, Academic, New York, 1969.

S72. Colthup, N. B., S. E. Wiberley, and L. H. Daly, *Introduction to Infrared and Raman Spectroscopy*, Academic, New York, 1964.

S73. Cross, A. D., *An Introduction to Practical Infra-red Spectroscopy*, 2nd ed., Butterworths, Washington, D.C., 1964.

S74. Hackforth, H. L., *Infrared Radiation*, McGraw–Hill, New York, 1960.

S75. Harrick, N. J., *Internal Reflection Spectroscopy*, Interscience–Wiley, New York, 1967.

S76. Kendall, D. N., *Applied Infrared Spectroscopy*, Reinhold, New York, 1966.

S77. Martin, A. E., *Infrared Instrumentation and Techniques*, American Elsevier, New York, 1966.

S78. Mathieson, D. W., *Nuclear Magnetic Resonance for Organic Chemists*, Academic, New York, 1967.

S79. Potts, W. J., Jr., *Chemical Infrared Spectroscopy*. Vol. I: *Techniques*, Wiley, New York, 1963.

S80. Slichter, C. P., *Principles of Magnetic Resonance*, Harper & Row, New York, 1963.

S81. Smith, R. A., F. E. Jones, and R. P. Chasmar, *The Detection and Measurement of Infrared Radiation*, 2nd ed., Clarendon Press, Oxford, England, 1968.

S82. Stewart, J. E., *Infrared Spectroscopy: Experimental Methods and Techniques*, Dekker, New York, 1970.

S83. Thompson, H. W., ed., *Advances in Spectroscopy*, Interscience, New York, 1959.

S84. White, R. G., *Handbook of Industrial Infrared Analysis*, Plenum, New York, 1964.

See also volumes issued periodically which contain papers presented at Symposia such as those listed in Section X ("The Literature of Instrumentation").

D. Selected Patent References (Abstracts)

1. ELECTROANALYTICAL AND TITRIMETRIC TECHNIQUES

P1. Bergson, G., and P. Ketelsen, U.S. 3,316,166 (1967). Apparatus for electrolytic measurement of oxygen concentration in a gas stream; the stream is passed through a wetter-cell to establish moisture content at an equilibrium condition.

P2. Brems, N. E., U.S. 2,994,590 (1961). Titration device for recording a curve of pH versus volume of added titrating agent.

P3. Clark, L. C., Jr., U.S. 3,330,905 (1968). A polarographic analysis cell that contains an anode and a cathode with an active surface facing a selectively permeable membrane, the cathode being depolarized by the constituent to be measured.

P4. Dowson, A. G., and I. J. Buckland, U.S. 3,050,371 (1962). Process for determining the concentration of dissolved oxygen in a liquid. The dissolved oxygen is transferred to an inert gas which is placed in contact with the cathode of a galvanic cell, whereby a current is generated in proportion to the amount of oxygen separated.

P5. Eiseman, George, D. O. Rudin, and J. U. Casby, U.S. 3,041,252 (1962). Glass electrode for measuring potassium ions, having an active portion made of a compound consisting of potassium oxide, rubidium or cesium oxide, aluminum oxide, and a network-forming oxide, such as SiO_2 or GeO_2.

P6. General Electric Co., Ger. 1,810,459 (1969). Gas analyzer consisting of a solid oxygen-ion electrolyte as an analyzing oxygen electrode, together with a gas titrator.

P7. Harvey, A. C., Can. 647,544 (1962). A detector of the state of the chemical constitution of a liquid, making use of the measurement of an electric current produced on the depolarization of a galvanic couple formed by electrodes of dissimilar metals, particularly to measure residual chlorine in water.

P8. Hersch, P. A., U.S. 3,408,269 (1968). A method for microelemental analysis of an organic sample, including converting the carbon and hydrogen into CO_2 and H_2O, subsequently converting the CO_2 into CO, and converting the CO into iodine vapor. Thereafter, the carrier-gas stream is conveyed to a coulometric cell consisting of a cathode of an inert conductive material and an anode of active carbon, silver, or mercury, said cell generating a current proportional to the rate at which iodine vapor is delivered thereto. The current is integrated for an indication of the amount of hydrogen and carbon in the sample.

P9. Jankowski, C. M., U.S. 2,930,747 (1960). A titrator electrode pair construction for partial immersion in solutions, such as are employed in mercaptan and olefin titrations, containing a one-piece body of poly(tetrafluoroethylene) bar stock.

P10. Liebhafsky, H. A., U.S. 2,882,329 (1959). A thermogalvanic cell consisting of a first electrode made of metal, a second electrode formed of the same metal (e.g., Cu), and a cation/permeable ion-exchange resin membrane electrolyte having as its mobile cation the cation of the electrode metal.

P11. Lindsley, C. H., U.S. 3,073,682 (1963). Process and apparatus for completely automatic titration of samples delivered and tested in batch form.

P12. Miller, J. W., U.S. 3,030,280 (1962). Method and application for determining the end point of a titration wherein the bromine number of an olefin is to be measured.

P13. Molloy, E. W., U.S. 3,406,109 (1968). Sensor cell for polarographic analysis, including an anode and a cathode supported in an electrically insulated body having a convex face with a membrane spaced from the face to form an electrolyte chamber, the membrane being permeable to the constituent being measured.

P14. Poulos, N. A., U.S. 3,096,258 (1963). Electrolytic analysis cell consisting of a silver-wire electrode coated with a paste consisting of a silver salt and an electrolyte solution; a nonconducting absorbent material wetted with the electrode solution, which surrounds the coated silver wire; and a platinum-wire electrode.

P15. Riseman, J. H., and R. A. Wall, U.S. 3,306,837 (1967). Electrode assembly for determining the concentration of ions of one nature in the presence of ions of a second nature; two electrodes, each containing a glass membrane sensitive to one of the ion species, are included.

P16. Robinson, R. H., U.S. 3,313,720 (1967). Apparatus for measuring the amount of dissolved oxygen in an electrically conductive liquid; three long electrodes, preferably made of thallium, lead, and aluminum, are adapted for immersion in the liquid.

P17. Shapiro, J. J., U.S. 3,343,078 (1967); A self-containing Karl Fischer titration device having a meter and a hollow tubular glass electrode assembly with a sealed uranium glass tip with a pair of exposed platinum contacts.

P18. Sharpsteen, J. R., Jr., H. A. Scheraga, and J. S. Butterworth, U.S. 2,962,425 (1960). Electrolytic analyzer whereby material containing at least one constituent such as blood serum on a carrier is analyzed by treating the material and carrier with a solution containing metal ions having a specific affinity for the constituent, so as to leave a film of the solution on the carrier. The procedure involves placing at least two electrodes in contact with the solution, moving at least one electrode in relation to the carrier and in contact with the material, and indicating the voltage generated between the electrodes.

P19. Strickler, A., U.S. 3,308,041 (1967). Continuous titrator, including means to coulometrically generate a titrant; means to mix the liquid sample and titrant; electrolytic means to sense the ion concentration of the product; means to direct the product from the mixer to the sensor; and a controller electrically coupled between the sensor and titrant supply to control the rate at which the titrant is generated.

P20. Union Carbide Corp., Ger. 1,523,082 (1969). Continuous polarograph for determination of concentration of ingredients in a mixture in which mercury from a mercury electrode drops continuously through an electrolytic solution.

P21. Vertes, M. A., and H. G. Oswin, U.S. 3,470,071 (1968). Detector for sensing the presence of noxious gases, composed of an anode, a cathode, an electrolyte, and an external circuit including a warning signal connecting the anode and cathode. The anode and cathode each comprise a liquid-impermeable and gas-permeable polymer membrane coated at one surface with a catalytic layer. The detector is sensitive, simple in construction, and compact, making it useful as a personal detector.

2. THERMAL ANALYSIS

P22. Chiu, Jen, U.S. 3,453,864 (1969). Test cell, for measuring changes in electrical resistivity as a function of temperature of a relatively small quantity of material, composed of two concentric cylindrical electrodes with the sample disposed at least partially between the electrodes.

P23. Wasilewski, J. C., U.S. 3,160,477 (1964). Thermometric titrating apparatus for measuring the heats of reaction during a titration.

3. CHROMATOGRAPHY

P24. Ashmead, H. L., U.S. 3,381,519 (1968). A gas chromatograph system capable of reducing unwanted amplitude variations caused by unknown vagaries in the chromatograph parameters; means for combining the compensating signal and detector output signal to produce a compensated detector output signal is provided.

P25. Clardy, E. K., U.S. 3,468,156 (1969). Method and apparatus for chromatographic analysis of a binary mixture without peak separation.

P26. Ferrari, A., Jr., Can. 660,694 (1963). Method and apparatus for continuous chromatographic analysis.

P27. Hinsvark, O. N., U.S. 3,298,786 (1967). Method for determining the oxygen content of an organic substance by introducing a sample into a closed, heated reaction chamber having an inert-gas atmosphere; the products are reduced, by carbon, to carbon monoxide and hydrogen gas and passed through a chromatographic separating column.

P28. Picker Chromatography, Inc., Ger. 1,805,717 (1969). Column chromatograph for liquids with an injection aperture member incorporating a solvent supply passage that communicates with a solvent inlet passage.

P29. Rogers, L. B., and A. G. Altenau, U.S. 3,312,042 (1967). A gas–solid chromatographic column having a suitable tube packed with devolatilized clathrate crystals formed from a coordination compound of a transition metal, and water or nitrogen compounds.

P30. Skeggs, L. T., U.S. 3,097,927 (1963). Chromatographic analysis apparatus; the eluent from the chromatography column is collected and stored in a length of column, into which the eluent is pumped as it flows from the column.

P31. Sternberg, J. C., U.S. 3,322,500 (1967). Apparatus for chemical analysis that fragments at least part of an organic sample, using an electrical glow discharge device. The fragmentation products are chromatographically separated, detected, and measured.

4. X-ray Techniques

P32. Kuntke, Alfred, and Robert Weigel, Can. 661,068 (1963). X-ray tube having a rotary anode comprising a shaft and a bearing structure for rotatably supporting the anode relative to a stationary support.

P33. N. V. Philips' Gloeilampenfabrieken, Fr. 1,234,443 (1960). A disk-shaped rotating anode for X-ray tubes, comprising a layer of essentially pure tungsten bonded to a layer composed not entirely of this element and having a slower rate of recrystallization because of the presence of at least one oxide, nitride, silicate, carbide, or boride.

5. Optical and Spectrometric Techniques—Visible and Ultraviolet

P34. Ferrari, Andres, and Jack Isreeli, U.S. 3,109,713 (1963). Apparatus for the quantitative colorimetric analysis of liquids, using a colorimeter having a closed flow cell.

P35. Incentive Research and Development A.B., Brit. 1,172,390 (1969). Analyzing apparatus for determining the absorption of ultraviolet radiation by a given amount of the substance in the path of a beam of the radiation; included are a radiation source, a chopper, a cell for the substance, a radiation-responsive cell, and signal circuitry.

P36. Picker Chromatography Inc., Ger. 1,806,177 (1969). Optical absorption photometer having a source of electromagnetic radiation, a device for definition of two parallel, collimated rays, a device for placing material in the path of both rays, and a device which compares the intensity of the radiation passing through the two pieces of material to determine their relative absorptions.

P37. Schneider, G. W., Jr., U.S. 3,030,192 (1962). Colorimetric analyzer for determining the amount of a constituent in a fluid.

P38. Serfass, E. J., U.S. 3,440,016 (1969). Colorimetric analyzer for continuously mixing accurately controlled amounts of liquid sample and reagent so that the flow of reagent will be stopped when the supply is depleted.

P39. Simmons, E. C., U.S. 3,469,789 (1969). An atomizer for introducing a sample into a flame or a similar atom-disassociating means for analysis of the sample constituents.

P40. Smythe, W. J., M. H. Shamos, and G. A. Griffin, U.S. 3,480,369 (1969). Method and apparatus for the colorimetric analysis of a liquid sample flowing through a flow cell as a continuous stream which is segmentized by gas segments.

P41. Technicon Instruments Corp., Brit. 919,759 (1963). Automatic analysis apparatus which includes the closed flow cell of a colorimeter for the colorimetric analysis of a liquid stream.

P42. Tuwiner, S. B., U.S. 3,381,572 (1968). Colorimetric testing device for determining the concentration of a component of a solution to which an indicator has been added.

6. Spectrometric Techniques—Infrared

P43. Bayly, J. G., and W. H. Stevens, U.S. 3,021,427 (1962). Monitoring system for indication of the heavy-water content of air, comprising means for condensing water vapor from the air, and means for estimating the deuterium content of the condensate by measuring the absorption of infrared radiation by the condensate at an absorption peak of the H—O—D molecule.

P44. Franzen, Wolfgang, U.S. 3,077,538 (1963). Method and apparatus for measuring the intensity of infrared and microwave radiation, including a modified superconducting bolometer.

P45. Hartmann and Braun, A. G., Brit. 1,164,775 (1969). Gas analyzer for determining the concentration of carbon monoxide in the exhaust of an internal combustion engine by its absorbing effect on infrared radiation.

P46. Jones, R. W., U.S. 3,470,374 (1969). An infrared detection system using a laser generating a reference radiation level for comparison with a radiation level generated by a target for indicating the temperature of the target. The output of the laser is regulated so that, when a null is obtained between reference radiation and target radiation, the readout on a register controlling the reference radiation source indicates the temperature.

P47. Mausteller, J. W., A. D. Billetdeaux, and R. S. Freilino, U.S. 3,471,698 (1969). Use of infrared techniques to detect a contaminant that is deposited as a film on a reflective surface, where the contaminant has a defined infrared absorption band. Important applications include the detection of various insecticides and other liquid and airborne contaminants that may be deposited on metal or other infrared-reflective surfaces.

P48. Perkin–Elmer Corp., Brit. 1,171,689 (1969). Raman spectrometer produces large exciting radiation flux density by multiple reflection of a laser beam through a sample container.

P49. Ure, R. W., Jr., and E. V. Somers, Can. 662,233 (1963). Infrared detector comprising an infrared-sensitive solid-state material and a cold junction of a thermal electric refrigeration device electrically insulated from, but in direct thermal contact with, the sensitive material.

P50. Warshaw, H. D., U.S. 2,856,540 (1958). Apparatus for detecting infrared radiation from an object in the presence of an extraneous foreign radiation, involving a silicon oxide filter coating on a lens.

P51. Wright, R. H., and P. N. Daykin, Can. 641,696 (1962). Spectrometer adapted for use in the far-infrared spectrum, which uses a so-called zone plate to effect dispersion of the radiation.

7. MISCELLANEOUS DEVICES

P52. Beloit Iron Works, Brit. 887,556 (1962). Process and apparatus for estimating the moisture content of material by generating a microwave signal of a frequency which is absorbed by water.

P53. Blair, D. E., U.S. 3,392,158 (1968). Process for separating a polymer into molecular weight fractions; a substrate, such as an endless solvent-resistant conveyor belt, is coated with the polymer and passed continuously and simultaneously through a series of liquid solvent baths of increasing solvent strength in relation to the polymer.

P54. Burk, M. C., and L. B. Roof, U.S. 3,468,157 (1968). Apparatus for detecting the composition of gas, comprising an acoustical chamber with a pair of spaced transducers, an oscillator for passing a signal to one of the transducers, a variable-frequency oscillator for establishing a second signal, a means to mix the outputs of the oscillators and of the second transducer, and a phase-angle null detector between the mixed signals for adjusting the variable-frequency oscillator.

P55. Collins, H. R., Jr., U.S. 3,480,032 (1969). A temperature-compensated viscosity-measuring device, which produces a first output signal representative of a measurement of the viscosity of a first fluid, combined with a temperature-compensating circuit.

P56. Coulter Electronics Ltd., Ger. 1,806,456 (1969); Particle size analyzer to determine the size above and below which certain fractions are present.

P57. Friedman, Herbert, and T. A. Chubb, U.S. 3,037,387 (1962). Apparatus for determining dew point, which has a hydrogen glow tube with a LiF window producing a light source with a strong emission and an aluminum reflective surface coated with MgF_2.

P58. General Electric Co., Brit. 950,188 (1964). An electrical vapor detector for detecting certain substances or impurities in gases, having an ion source of activated ceramic semiconductor material and an ion collector (nickel–chromium wire) enclosed within a gastight housing.

P59. Iwata, Shigeo, U.S. 3,469,455 (1969). Apparatus for measuring changes in the weight of a sample; the sample container is supported at the outer end of an elongated support bar laterally extending from a sample-loading member included in a weighing mechanism.

P60. Le Clerk, P. G., U.S. 3,065,345 (1962). Determination of the quantity of neutrons in radiation by irradiating a glass body having contiguous parallel strata which have different responses to neutron radiation.

P61. Luft, K. F., U.S. 3,479,860 (1969). An improved gas-flow-measuring system for application in portable apparatus for determining oxygen in gas mixtures by using its paramagnetism.

P62. Maier-Leibnitz, Heinz, Fr. 1,167,052 (1958). Apparatus for determining the composition of a gas or vapor by measuring the slowing down of alpha rays.

P63. Mutter, W. F., and D. C. Watson, U.S. 3,479,880 (1969). Apparatus for delivering liquid samples to a chromatograph in which a probe composed of a pair of tubular needles is stroked down through the top of a vial containing the liquid to be sampled and held in a sampling position on a turntable.

P64. Natelson, Samuel, U.S. 3,453,082 (1969). Instrument for automatically assaying the gas content of liquid samples sequentially conveyed to a reaction chamber which contains a gas absorbent. The pressure change in the reaction chamber is recorded and constitutes a measure of the gas absorbed.

P65. Simon, Wilhelm, U.S. 3,453,866 (1969). An isopiestic molecular weight instrument that includes a vapor tank to contain a solvent-saturated atmosphere, and temperature-measuring elements positioned upright, with the sample deposited above them.

P66. Technion Instruments Corp., Ger. 1,523,050 (1969). Sampling device for automatic analysis in which sample vessels are mounted on a rotary table and guided past filling stations during stepwise motion of the table.

CONTINUOUS AUTOMATIC INSTRUMENTATION FOR PROCESS APPLICATIONS

By C. D. Lewis, *Plastics Department, E. I. du Pont de Nemours & Company, Wilmington, Delaware*

Contents

I. INTRODUCTION

Continuous analysis may be defined as automatic determination of one or more components of a *flowing* stream or other dynamic system in which sampling is performed mechanically. Analyses which require intermittent sampling in a repetitive cycle are considered continuous if they are applied to flowing samples; gas chromatography is such a method. Usually, results are presented on a recorder. Continuous analyzers on flowing process streams are essential components of many modern chemical plants. In these cases, results may be recorded on a control panel or fed into a computer in such a way that automatic control of process variables can be achieved. Continuous analyzers measure composition; there are also many instruments for continuous measurement which reflect composition while not being regarded as strictly analytical. Measurement of color, reflectance, and density, for example, is often essential for process or quality control. These measurements are not considered in detail in this chapter, but many of the principles discussed here can be applied.

Several reviews of commercial continuous analyzers have been published (38,57,227). A monograph on automatic process monitoring appeared in 1960 (226); the general literature is scattered, however, and no critical surveys have been made.

This chapter summarizes historical aspects of the subject, considers its principles, and illustrates methods of measurement which have been applied to continuous analysis.

A. HISTORICAL

Chemical plants have traditionally used measurements of parameters such as pressure, temperature, and flow as means of control; modern plants must have rapid compositional analysis as well. The potential for continuous analysis appeared with the development of devices for detecting rapidly the changing physical properties of gas mixtures undergoing concentration changes. Thermal conductivity measurements using resistance cells were employed in a practical manner as early as World War I to detect hydrogen in air in dirigible hangars (64). The success of this technique has been indicated by the wide use of thermal conductivity cells for analysis of binary gas mixtures.

Practical analyzers were built in the 1920s and 1930s for plant use in continuous measurements for nonselective properties such as electrical conductivity, density, viscosity, and combustibility. These parameters characterized many simple processes quite well and gave more useful information about the system than was provided by pressure and temperature alone. More refined devices for continuous analysis were slow to appear since many syntheses were based on batch processes.

In order to meet the needs for rapid analysis of the complex continuous processes developed during World War II and the succeeding decade, specific and selective laboratory analytical instruments such as spectrophotometers were moved out onto the plant streams. But more rugged analyzers had to be developed. An increasing variety of continuous analyzers now employ elegant techniques, such as pH, gas chromatography, ultraviolet, visible, and infrared spectroscopy, mass spectrometry, and coulometry (69). Unique specificity has been achieved in special cases, such as the Pauling analyzer, which depends on the paramagnetism of oxygen (194). Gas chromatography, in spite of its relatively recent development, has been rapidly accepted because it fulfills many needs for multicomponent analyses.

Continuous analyzers find a wide variety of applications. A very high percentage of the continuous analyzers in the United States are applied in the chemical, petroleum, and related industries, so that "process-stream analyzer" is nearly a synonym for "continuous analyzer." A few plants are already under completely automatic control with such analyzers playing an important part. It should not be overlooked, however, that many applications of analyzers with automatic sampling exist in laboratory and small-scale process units where the kinetics of reactions are being studied or the purity of a reagent or product needs to be monitored. Continuous analyzers for air pollutants are common in many industries and even in certain communities. Furthermore, many medical applications have been reported, ranging from control of anesthetic mixture composition to analysis of blood during a surgical operation—that is, the dynamic system is a human being! The evaluation of the intensity of certain bands of radiation in space by analyzers in earth satellites has received wide publicity (84).

B. GENERAL CONSIDERATIONS

1. Principles

In common with laboratory analytical instruments, continuous analyzers use many of the same basic techniques such as infrared and ultraviolet spectrophotometry. The principles of mass spectrometry and gas chromatography can be carried over more or less directly from the laboratory to continuous analyzers. There are also several techniques which were significantly exploited first for continuous measurements. These include thermal conductivity, electrolysis current for water, galvanic cells for Cl_2 and H_2, and differential refractometry. Absorption or scattering of visible light by smokes and aerosols is far more likely to be used for monitoring than for a batch measurement because it is relative and difficult to calibrate.

Continuous analyzers differ from laboratory instruments in many ways: the former tend to be faster, simpler, more rugged, less selective but more sensitive, smaller, safer, and more reliable. Many of these differences reflect requirements which arise from the need for the analyzers to operate while unattended for long periods of time.

Speed is an axiomatic requirement, since analyzers in industry are judged on the basis of cost per analysis and utility in aiding plant control. Analytical information must be obtained quickly if it is to be useful in modern high-speed plants.

Simplicity implies easy maintenance and uncomplicated design. Gas chromatography has been accepted rapidly for process-stream analyzers in part because the analyzers are composed largely of components which have been long familiar to plant personnel. Other analyzers are simplified in comparison to laboratory instruments at the expense of versatility.

Ruggedness toward shocks, vibration, dust, and the weather is required if the instrument is to withstand the rigors of a plant environment. Fragile and delicate components are avoided.

Selectivity may be compromised with caution if greater *sensitivity* to real changes in composition is gained. In many applications nondispersing infrared analyzers are more sensitive toward a given component than are laboratory spectrophotometers, but interferences are more troublesome in the analyzers. More sensitive signals in the sense of higher voltage levels are frequently encountered in analyzers—for example, condenser microphones for measurements of infrared radiation, and the optical amplifiers used in refractometers.

Size of analyzers is becoming a factor as a result of rising installation costs and the trend to miniaturization of control panels in highly automated plants.

Safety is essential when analyzers are operated in areas where combustion hazards may occur, for example, petroleum refineries and hospital operating rooms. Most commercial analyzers are designed to enclose electrical components in explosion-proof housings.

Reliability is an end product of many factors and has obvious monetary implications in research as well as production applications. Little "down-time" can be tolerated in an instrument which is expected to work around the clock.

Current trends are pushing all these factors to their limits. Power plants working at high steam pressures require sensitive analysis for oxygen in boiler feed water (60). Process control looks to even higher speeds of response, and the sheer volume of instruments in use requires utmost reliability.

The decision on whether to use a batch or continuous system of analysis will often be made in favor of the latter when more rapid or more sensitive analysis will be achieved. Such a decision will also be promoted by the existence of toxic or radioactive hazards in a batch analysis, or of difficult sampling problems with sensitive compounds or trace concentrations which might be lost or altered in individual samples.

2. Component Functions

Continuous analysis involves the same operation as laboratory analysis. A complete instrument for such an analysis possesses in general the following functions in a unit or group of associated units (Fig. 105.1).

a. SAMPLING SYSTEM

Equipment is included for taking a sample at a controlled rate, pressure, and temperature from the main process stream or other source of material. The flow within the analyzer may be intermittent because of the requirements of the measurement.

b. SEPARATION AND TREATMENT OF THE DESIRED CONSTITUENT

Where necessary, provision is made for preparing the initial sample for analysis by further operations such as separation steps and addition

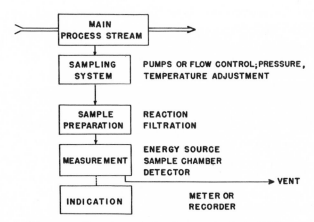

Fig. 105.1. Functions of Continuous Analysis (145).

of reagents. The latter may remove an interference or generate a color or another new property required for the subsequent measurement. Separation may be as simple as a filtration, to remove particulate matter and thus avoid clogging lines, or as complex as the complete operation of a miniature chemical plant.

c. MEASUREMENT

The actual measurement is made on the sample, generally within the confines of a flow cell, by a suitable detector or transducer, preferably one rapid in response. With the exception of radiological and thermometric methods, a source of radiant or electrical energy or physical force is also needed.

d. INDICATION

The results of the analysis are indicated in terms of the concentration of the desired component in the sample by an output signal, usually monitored by a chart recorder or other device which provides a permanent record. Most recorders can be equipped with control functions, so that the analyzer can then be incorporated in a system for automatic process control.

In the modern trend toward modular or "building block" construction to simplify maintenance, parts of features b, c, and d may be represented by interchangeable units, for example, the time programmer in a gas chromatograph, the amplifier in an infrared analyzer, and, for many analyzers, a standard potentiometric recorder. If the analyzer is part of elaborate instrumentation in a complex process, the signal may be read by a data-logging system (Section V.B.I.).

C. SELECTION OF ANALYZER

In the selection of a continuous analyzer to perform a given analysis, there exist two broad approaches: (1) automation of the existing laboratory analysis, a "classical" approach; and (2) reconsideration of the whole analytical problem, or the "fresh look" approach (6). The first approach has the advantage of saving time through familiarity with the effects of interference and the like, but it has the disadvantage that the resulting analyzer may be burdened unnecessarily with the expense or complexity of, for example, a mass spectrometer, titrimeter, separation, or a colorimetric reaction. The alternate approach may result in a method more suited to continuous analysis by virtue of simplicity, but has the disadvantage of time lost in testing interferences and exploring other possible drawbacks.

In either approach, several factors should be considered carefully: in addition to selectivity, sensitivity, and simplicity, applicability to the physical state of the sample, speed of response, and destruction versus preservation of the sample should be studied. Selectivity becomes especially important in situations wherein unsuspected components may appear, for example, in small-scale process units. The sensitivity of a method which initially appears to be borderline may actually prove to be acceptable in service on-stream because continuous sampling tends to improve precision by reducing the magnitude of blank values. Flowing samples in pipelines for trace water analysis, for example, are much easier to protect from contamination by atmospheric moisture than are batch samples. A desire for simplicity may lead to a choice of interference filters instead of a monochromator for wavelength selection in spectrophotometry. On the other hand, separation or chemical reactions may be necessary complicating features to achieve selectivity.

The physical state of the sample or unusual sampling problems may limit the field of choice. Most methods of continuous analysis apply to gaseous samples, and many to liquids, but few as yet apply to solid samples (6). Since most electrical transducers are rapid in response, the speed of response of the total analyzer tends to be limited by the rates of sampling and is hence a problem in design of the sampling system. Methods which tend to destroy or contaminate the sample will not be preferred for samples which are valuable or limited in quantity. Albright (6) offers a guide for rapid screening of analysis techniques applicable to different degrees of discrimination of chemical entities (element or atom, molecules, nonspecifics). This approach may be helpful for the nonchemist, but its greater value may lie in underlining the need for a "fresh look."

The variety of analyzers which may apply to the determination of a given element or compound in a mixture is illustrated in Table 105.I for water, oxygen, and carbon dioxide. The table is intended, not to be historically complete, but rather to include analyzers which have had some importance because of sensitivity or selectivity in given applications. It is apparent that, as in laboratory analysis, no one technique is universally applicable but that one of several techniques may be useful in given circumstances.

It should also be appreciated that most continuous analyzers are initially expensive, and that the installation and startup of an analyzer in a plant will cost several times its purchase price. If, in addition, the analyzer requires continuing maintenance, it will become a burden unless early and spectacular payout in terms of new information about, or better control of, the process is achieved. There can be no substitute for careful consideration in the early stages of selection of method and design of the sampling system with regard to estimated cost. Although maintenance and calibration costs are parts of the overall cost picture, they can be balanced against reasonable optimism toward the value of new data to be gained. It is difficult to place a sufficiently high value on data obtained by an analyzer during a plant startup or process upset.

D. CALIBRATION

While the precision of an analyzer depends on its sensitivity to small real changes in concentration, the accuracy of indication depends primarily on the accuracy of calibration. Calibrations in turn depend on the accuracy with which standard samples are made up. Liquids offer few problems in this respect.

The preparation of standard gas samples requires care, increasing as concentration is decreased or the desired storage pressure is increased. Minor and trace components tend to be lost by adsorption on walls of vessels, fittings, and piping lines (Section II), unless the walls are carefully equilibrated with the composition actually desired. Polar components such as alcohols, acids, and water are especially prone to disappear. A change in ambient temperature may alter the extent of adsorption on the walls and hence change the concentration in the gas.

Mixtures made up under pressure do not spontaneously mix with any

TABLE 105.I
Continuous Analysis

Method	Range	Sample	References
A. For water			
Resistance of hygroscopic salt	6–99% R.H.	Air, unreactive gases	112
Dew point	≥ 1 ppm	Air, oxygen, hydrogen, nitrogen	112
Hair hygrometer	10–100% R.H.	Air	112
Thermal conductivity	$\geq 0.1\%$	Gases and vapors	64
Heat of adsorption	1–5000 ppm	Unreactive gases, hydrocarbon vapors	86
Coulometry	1 ppm–2%	Gases, liquids	136
Near-infrared	≥ 5 ppm	Gases, liquids	107,159
Refractometry	$\geq 0.01\%$	Most organic liquids, aqueous solutions, sulfuric acid	86
Capacitance	$\geq 0.05\%$	Certain liquids, solids	112
Nuclear magnetic resonance	0.1–10%	Solids	5
Neutron absorption	% ranges	Solids	136
Conductometry	0–25%	Sheet materials	112
B. For oxygen			
Thermal conductivity	$\geq 0.1\%$	Gases, vapors	64
Paramagnetism	0.01–100%	Gases, vapors	17,194
Magnetic–thermal	0.01–100%	Gases, vapors	104,213
Catalytic reaction with hydrogen	2–200 ppm	H_2 or unreactive gases	196
Colorimetric reaction with cuprous ion	1–100 ppm	Gases	74
Colorimetric reaction with sodium anthraquinone sulfonate	1–50 ppm	Hydrocarbon vapors	129
Galvanic cell	1–100 ppm	Hydrogen, alkaline or unreactive gases	108
Sonimetry	40 ppm–100%	Gases	140
Polarography	0.005–100%	Aqueous systems	33,47
Reduction–conductivity	ppb	Aqueous systems	61
Reaction with thallium–conductivity	ppm	Aqueous systems	119
C. For carbon dioxide			
Volume change on reaction	$\geq 0.5\%$	Unreactive gases	112
Infrared	≥ 0.2 ppm	Gases, vapors	17
Thermal conductivity	$\geq 0.1\%$	Gases, vapors	64
Density	% ranges	Combustion gases	112
pH	Variable	Aqueous systems	223

haste. Mechanical mixing devices are necessary inside the pressure vessel; otherwise, convection must be set up by heating one end and cooling the other. Good practice requires that successive analyses be made to check on the thoroughness of mixing.

Many of these problems can be overcome by using dynamic gas-mixing systems, such as those described by Troy (251) and Schnelle (219). Since accuracy may suffer through metering and sampling problems in a poor system, rotameters should be of high quality, carefully calibrated, and surfaces should be such as to minimize adsorption of trace components. Such systems are especially suitable where the major component is inexpensive, as in air pollution work; cost of operation will become a factor if the major components are expensive. Safety precautions may be vital if some components are hazardous.

Schnelle's system (Figure 105.2) relies on two types of components: (1) predictable-flow rotameters, whose delivery can be accurately calculated for any gas for which density and viscosity are known under the experimental conditions, and (2) inert materials of construction, such as Pyrex industrial glass tubing, Teflon TFE–fluorocarbon resin, and stainless steel (219). By using a series of accurate dilutions, the concentration of a mixture can be adjusted to any value between 100 ppb and 100%.

A dry/wet-gas-blending technique for calibrating parts-per-million water analyzers has also been described (105). A device for producing from less than 1 to 15 ppm of smog constituents in an air stream has been described by Bryan and Silis (31). The problem of preparing mixtures of toxic gases in air in the range of 1–10,000 ppm has been considered in detail by Saltzman, who used an all-glass flow system (214). Lovelock describes a technique of successive dilutions which is especially useful for establishing the dynamic range of an analyzer (155).

Known mixtures of a gas and a volatile liquid may be prepared by diffusion. As described by Altshuller and Cohen (8), the liquid is held in a narrow vertical tube at constant temperature, and its vapor mixes with a carrier gas moving past the tube opening. Predictable concentrations in the range of 10–10,000 ppm of the vapor in the gas are readily obtained by adjusting tube diameter, temperature, and other variables.

II. THE SAMPLE

Sampling is the most important single feature for complete automation of the analysis, yet is is often a neglected feature in designing for a

Fig. 105.2. Gas-mixing system for calibrating analyzers (219). (Courtesy of Instrument Society of America.)

particular application or installation (50,114). Since many aspects of the sample, notably its cleanliness, are beyond the control of the analyzer, and since much of the sampling system is mechanical of necessity, it is difficult to achieve freedom from maintenance.

The speed of response of an analyzer to a change in concentration of a component of the sample is closely related to the design parameters of its sampling system. Schnelle (220) has pointed out that the accuracy of photometric gas analyzers may be affected by the rate of flow of the sample. This is a consequence of the effect of flow rate and pressure drop in a sampling vent line upon the pressure in the sample cell. An increase in flow will increase pressure drop in the lines and operating pressure in the sample cell. Also, the rate of flow of a hot gas in a cooling device upstream from the analyzer may affect the temperature of the gas in the sample cell if cooling is inefficient. Since gas density is a function of temperature, the sensitivity of the analyzer will then be a function of gas flow rate.

A. COLLECTION OF THE SAMPLE

In modern chemical plant technology, gas and liquid streams are generally under pressure, in order to achieve desired reactions and to move them rapidly, often at elevated temperatures. Sample collecting thus poses problems different from the traditional ones of gas analysis at atmospheric pressure, where sampling often requires partial vacuum techniques. Sample treatment must achieve suitable environmental conditions for many types of analyzers (Section II.B). A sample under modest pressure may require only a simple take-off or siphoning to move it into the analyzer. A sample taken at some point distant from the analyzer must be transported promptly to achieve a desired response rate. Pumping may also be required to achieve a specified flow rate through the analyzer.

A special type of sample gathering used for trace constituents of gases employs paper tape as a concentrating device. The tape is fed from a supply roll to the site of sampling. A time interval of exposure in one location is allowed to guarantee that a certain minimum amount of the component of interest will be collected by filtration for further examination by the analyzer (180).

Many types of sampling pumps have been applied in analyzers. Several varieties which are low in cost use metal fingers or rollers to displace liquid in flexible plastic or rubber tubing (239). Tubing can be chosen to resist most aqueous chemicals, but organic compounds with solvent properties pose a greater problem in achieving useful service life.

Other principal varieties of positive displacement pumps use gears or combinations of bellows or pistons with check valves. Representative types are shown in Fig. 105.3. Gear pumps are available to give flow rates of 5 ml/min and higher. Gear pumps are subject to wear by minute particles in the stream; adequate filtering allows them to be used for sampling in capillary viscometers, however. Bubbles may cause binding. Piston pumps are available at flow rates as low as 30 ml/hr (213). Bellows pumps introduce sample lag at low flow rates, because of the volume of the bellows.

A pump for low flows has been described which acts like a motorized hypodermic needle (134). A hypodermic metering tube is alternately filled with liquid sample, and the trapped sample is ejected to the analyzer by the pressure of an inert gas. This pump is essentially a four-ported valve, and the metering tube is moved by a pneumatic operator to connect appropriate pairs of ports alternately. Flow rate depends on frequency of actuation. The pump delivers a pulsing type of flow, with a little gas mixed with the liquid. This is acceptable for coulometric analyzers for oxygen or water in liquid samples.

Most small positive displacement pumps deliver a pulsing flow. The pulsations can be smoothed out at the expense of increased sample lag by a damping system consisting of a coil and a reservoir in series. Deliveries accurate to ±1% can be achieved with care in the application of small pumps.

Fig. 105.3. Positive displacement sampling pumps. (Courtesy of Research Appliances Co.; Milton Roy Co.; Beckman Instruments, Inc.)

B. TREATMENT OF THE SAMPLE

After being collected, the sample must be adjusted in pressure, temperature, and cleanliness to the needs of the analyzer. This is an area for sound engineering practice, but certain chemical aspects of the problem must be kept in mind.

Pressures of both gas and liquid streams must be reduced to values near atmospheric pressure to prevent overloading of sample cells. Pressure reducers of relatively simple design are generally available for samples up to 5000 psi. For higher pressures, regulators are complex electronic devices. If flow fluctuations can be tolerated, capillary let-down techniques may be used. Rupture disks should be provided to vent excessive pressures safely.

The speed of response of sampling systems for gas analyzers is intimately related to pressure considerations. Since mass is directly related to the pressure of a gas at constant volume, a given sampling system will hold up more material at higher pressures. Thus the gas should be moved through the system at the lowest practical pressure to achieve rapid response of the analyzer to changing concentrations at the inlet to the sampling system (113).

The temperature of the sample generally must be controlled. To achieve constant mass flow, samples require control of both pressure and temperature. Infrared analyzers require a fixed weight of sample in the beam unless compensation techniques are applied. Vapor samples must be held above the dew point. Refractometers require that any change in temperature be very slow; heat exchangers are generally satisfactory here. Analyzers of many types are thermostatted for stability of sample pressure and flow conditions. Thermostatting of small volumes to changes as slight as $0.001°$ can be achieved with small, reliable proportional controllers with a reset function. Thermal conductivity cells require this degree of environment control for precise work.

Sample cleanliness is of utmost importance. Entrained liquids or solids lead to mechanical troubles in the sampling system and tend to give nonrepresentative samples, if they do not cause more serious problems.

Mists and fogs entrained with gas samples will cause trouble just as they do in large-scale processing of such streams. Mist droplets may cause light scattering or window frosting in photometric analyzers. A sample containing a mist is generally unsuitable for handling by automatic sampling devices because it is not representative. Mists change the flow rates of gases in capillaries. Although certain types of filters may remove mists, the filters must then be maintained. It is generally advisable to apply baffling devices or miniature cyclone separators to precipitate mists and draw off the entrained liquid continuously. An effective mist impingement filter has been described (220). Several filtering devices are shown in Fig. 105.4.

A major problem in the preparation of gas and liquid samples for analyzers is the removal of adventitious solids. Pressure regulators and flow-measuring devices generally have small openings which are easily plugged by a few solid particles. Sampling devices and pumps, particularly gear pumps, must be kept clean. Few photometric analyzers can tolerate any accumulation of solids in the cells. The measuring cell of an electrolytic hygrometer can be shorted out by solid particles.

Such solids are generally removed by filters, which are available in a variety of materials, often natural fibers such as the cellulosics. Synthetic fibers such as the acrylics, asbestos, glass wool, and fine metal screens allow higher temperatures to be used and reduce adsorption losses. A recent filter design which is strikingly effective uses sintered metal construction. Very fine pores (down to 2 μ) can be obtained, yet the pressure drop is very low compared to that of fiber-type filters of similar effectiveness (111). Submicron particles require special materials, such as cellulosic membranes, for good retention (168).

It is easy to specify mist- and solid-removal devices for a sampling system but more difficult to anticipate their undesirable effects on the final analysis. Such devices tend to increase volume, lengthen response time, and hold up minor components. Trace components may be removed more or less completely through adsorption or even through chemical reactions such as corrosion on relatively large surfaces of filters and traps. Polar compounds, notably water, alcohols, amines, and acids, are especially prone to be lost in such a fashion. Materials of construction which are least likely to cause such troubles are polytetrafluoroethylene, stainless steel, and, in general, those with the smoothest surfaces. Glass has a polar surface which can be tolerated in some cases if its exposed area is held to a minimum. Copper, brass, and aluminum are relatively reactive.

Similar limitations apply to sealing materials, such as O rings. When elasticity is essential, fluorinated and silicone elastomers offer inertness toward a wide variety of inorganic agents and hydrocarbons. Organics with more powerful solvent properties may tend to be absorbed, particularly ketones, nitriles, amines, and anhydrides at elevated temperatures.

In trace analysis contaminants are often found in a new sampling system. A common contaminant is moisture adsorbed on new piping lines, and a long purging period may be required for the equilibration of a moisture analyzer in such a situation.

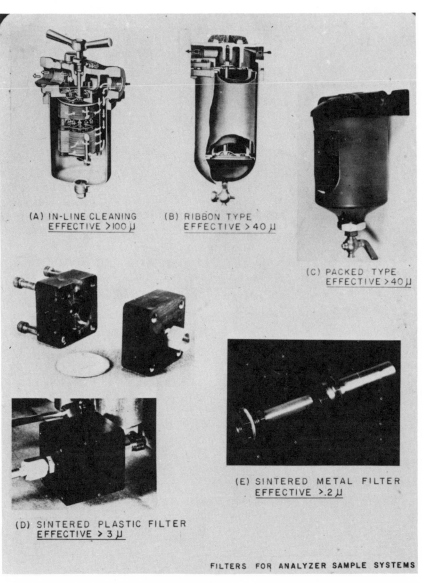

(A) IN-LINE CLEANING
EFFECTIVE >100 μ

(B) RIBBON TYPE
EFFECTIVE >40 μ

(C) PACKED TYPE
EFFECTIVE >40 μ

(D) SINTERED PLASTIC FILTER
EFFECTIVE > 3 μ

(E) SINTERED METAL FILTER
EFFECTIVE >.2 μ

FILTERS FOR ANALYZER SAMPLE SYSTEMS

Fig. 105.4. Sample filters (220). (Courtesy of Instrument Society of America.)

C. METERING THE SAMPLE

Two functions are involved here: (*1*) controlling the volume or mass rate of sample motion for a continuous measurement and (*2*) metering a discrete sample to a device which runs intermittent, repetitive analyses, such as a chromatograph. Controlled flow rate is essential for stability in many analyzers.

As discussed above, pumps gather samples from low-pressure sources and transport them through the analyzer. Metering pumps are desirable, since relatively constant flow or pressure is often required for the stability of the final measuring element. The constant of metering can be checked by occasional measurements of delivery rate at the analyzer exit.

Where sufficient source pressure is available, the flow can often be adequately controlled by combinations of pressure regulators and restrictions, that is, by maintaining constant pressure drop across an orifice, a capillary, or a valve of high quality. Changes in flow rate are then monitored by indicating or recording rotameters. Actual metering of the flow requires calibration of the rotameters or use of precision rotameters.

Special sampling devices have been developed for analyzers which require discrete samples repeated at intervals, such as gas chromatographs. Motor- or solenoid-operated valves are available which trap a fixed volume of sample at known temperature and pressure from a flowing stream. The sample is then swept into the analyzer at a command from a programming device, with the aid of carrier gas under pressure (157). Sampling valves are available for minute liquid samples, as well (17,134,162,167,195).

D. SEQUENCING DISCONTINUOUS SAMPLES

By the application of appropriate sampling and programming devices, it is possible to use a continuous analyzer for analysis of a series of batch samples. It is also possible to achieve semicontinuous analysis of a flowing stream in a cyclic analyzer such as an automatic titrator.

Samples for one colorimetric analyzer are withdrawn from a series of small reservoirs in turn by means of a tilting drawoff tube in conjunction with a timing and indexing mechanism (245). Suction is provided continuously by a pump. After the desired time interval for one sample, the positioning mechanism operates to tilt the drawoff tube while bringing the next sample into register.

Automatic titrators generally have had the sample added and withdrawn by a human operator. Recent developments incorporate sample reservoirs, drains, and valves so that, with the aid of a sequencing mechanism and level-sensing devices, the old sample is drained from the titrating vessel and the new one measured, added, and titrated automatically (35,106). It requires little change to have the sample reservoir flushed continually with a flowing stream. Repeat analyses can then be run to achieve essentially continuous analysis of the stream.

III. SEPARATION AND TREATMENT
OF THE DESIRED CONSTITUENT

Facilities for separation and treatment may introduce considerable complexity into an analyzer. Virtually every continuous analyzer requires some treatment of the sample to adjust its pressure, cleanliness, and temperature to levels accepted by its internal requirements. It may then be necessary to obtain selectivity for the analysis of a component of interest by an actual separation or chemical reaction before the measurement. The first kind of treatment has become almost routine and normally does not introduce undue complexity. The second is likely to require precise control of two or more flow rates. The need for maintenance of pumps and flow controls may tend to offset the gain in specificity, when compared to a less selective but also less complex method of analysis.

A. CONDITIONS FOR REPRODUCIBILITY

Of considerable importance in an analyzer in which separation or reaction is involved is the quality of the mechanical parts. Early analyzers of these types suffered in precision from the low degree of reliability of key items such as pumps for reagent supplies. Small changes in flow rates of reagent and sample had large effects on analyzer readings. The present availability of reliable small pumps with accuracies well within 1% relative, and of pressure- and flow-controlling relays, has done much to achieve stable reaction and separation conditions.

Other mechanical devices which may be encountered include timing mechanisms and tape transport and positioning devices. Time programming of sequenced operations is encountered in many analyzers which employ repetitive analytical cycles. The reliability of timers, switches, and relays is important for reproducible performance.

A reagent may have to be changed or regenerated frequently if its quality is critical. Automatic regeneration is sometimes used with a sensitive or valuable material. If the reagent is a solid adsorbent, its stability or activity may be very important.

B. ATTAINMENT OF SELECTIVITY

1. Chemical Reactions

These will be considered according to the phases involved in the reactions.

a. GAS–GAS REACTIONS

Analyzers in which a gas reacts with a gas with evolution of heat are long established (Sections IV.C.2 and 3). In a recent development, the "aerosol ionization detector," the sample gas is reacted with a vapor to generate minute solid particles or aerosols (241). The particles are easily detected by an ionization chamber.

Examples reported of highly selective applications of the aerosol detector include the following. Hydrogen halides are detected through addition of ammonia; NO_2, SO_2, and other acidic gases are detected through addition of HCl vapor; halogenated hydrocarbons are passed through a bed of hot copper oxide to form an aerosol of copper halide. Tetraethyllead and nickel carbonyl in air are said to be measurable after pyrolysis to the corresponding free metal. A limit of sensitivity of a few parts per billion is reported to be attainable (241).

Gas reaction analyzers lend themselves to novel detection methods because of marked differences in the physical properties of reagents and products. The gas titration analyzer of Weber uses pneumatic detection of the extent of reaction to achieve a relatively simple and reliable analysis for fluorine, an extremely corrosive substance (259). Reaction of fluorine with added sulfur dioxide results in the formation of sulfuryl fluoride:

$$SO_2 + F_2 \rightarrow SO_2F_2$$

Reduction in the number of moles of gas occurs as a result of reaction. When the reaction with flowing gases is conducted at constant inlet pressures and temperature, the resultant product gas flow varies only with fluorine concentration. This variation is reflected in the pressure at the reactor exit, where a pressure measurement indicates the concentration of fluorine in the sample gas.

If the reaction takes place in Monel at 200°C, interference from uranium hexafluoride, hydrogen fluoride, or oxygen appears to be insignificant (259). A precision of $\pm 1\%$ fluorine, above concentrations of 1% of this element, is achieved.

b. Gas–Liquid Reactions

A large variety of continuous analyzers rely for their specificity on the reactions of gas samples with liquid reagents. One of the early analyzers for carbon dioxide in moderate concentration was simply an automatized Orsat apparatus using reaction with caustic and a volumetric measurement (112).

Some analyzers using liquid reagents have long been employed for trace measurements, for example, of oxygen in process gases and of SO_2 in air (74,248). It was not possible in early work to obtain the desired selectivity by any direct measurement then available. The reaction technique has now been extended to develop many of the newer specific analyzers for air pollution and other industrial hygiene applications. Sensitivities to parts per billion can be obtained in certain cases; such sensitivity is essential for meaningful analyses of some pollutants.

Such analyzers generally share a number of critical features, such as efficient scrubbing of the gas sample with the liquid reagent, precise metering of the reagent, and adequate regeneration of used reagent. Fragility of glass parts is often a problem. The final measurement is typically made with a simple detector, often a colorimeter, since adequate specificity is obtained in the reaction step. Many such analyzers tend to be literally automated batch analyses rather than to represent new approaches to analytical problems.

Electrical conductivity has long been a basis for measurement in analyzers employing such reactions. For many years sulfur dioxide in industrial air has been measured in an analyzer, the Thomas Autometer, which depends on oxidation of SO_2 with hydrogen peroxide and measurement of conductivity increase due to the sulfuric acid formed (248). Since the solution is accumulated for an interval in the measuring cell, cyclic operation is employed to empty the filled cell and start a new analysis. The sensitivity is reported to be better than 0.05 ppm SO_2 (88).

A large variety of analyzers employing gas–liquid reactions have been described (Table 105.II), and the literature should be consulted for details. Because of their importance in air pollution work, several of these are commercially available (73,88).

In analyzers employing gas–liquid reactions, the design of the gas-absorbing device is vital for obtaining quantitative results. Three basic absorbing devices have been used in continuous analyzers for air pol-

TABLE 105.II
Reaction Analyzers (Gas–Liquid)

Analyze for	Reaction	Detector	Sensitivity	Remarks	References
		A. Air Samples			
SO_2	$+ H_2O_2 \rightarrow H_2SO_4$	Electrical conductivity	0.05 ppm	Cyclic	248
Organic halides	$\xrightarrow{\text{pyrolysis}}$ HX	Electrical conductivity	Low ppm	Applied to CCl_4, CH_2Cl_2, CH_3Cl	88
Oxidants (O_3)	$+ KI \rightarrow I_2$	Colorimetric	10 ppb	Not specific	148,163,210
"Oxidant precursor"	Expose to UV; $+ KI \rightarrow I_2$	Colorimetric	10 ppb	Not specific	210
O_3	Heat $+ \xrightarrow{\text{catalyst}} O_2$	UV	0.4 ppb	Differential	17
O_3	$+$ Phenolphthalin phenolphthalein $\rightarrow$	Colorimetric	...	...	17
NO_2	Diazotization; aromatic coupling $\rightarrow$ color	Colorimetric	10 ppb	...	249
NO	$+ O_3 \rightarrow NO_2$, then as above	Colorimetric	10 ppb	...	249
F^-	$+$ Zr–Eriochrome R $\rightarrow$ color	Colorimetric	Low ppm	Cyclic	2
Nerve gases	$+$ Indoxyl $\rightarrow$ color	Fluorescence	"Trace"	...	46,267
Boron hydrides	$+$ Triphenyltetrazolium chloride $\rightarrow$ color	Differential reflectance	0.1 ppm	On tape; not specific	142
		B. Gas Samples			
HCN	$+ Ag^+ \rightarrow AgCN + H^+$	Potentiometric	4 ppb	...	3,240
H_2S	$+ Ag^+ \rightarrow Ag_2S + H^+$	Potentiometric	...	...	3
SO_2, H_2S, RSH, RSSR, CS_2	Br_2 oxidation of S compound	Potentiometric	0.1 ppm SO_2	Reagent generated coulometrically; olefins, phenols, NO_2, Cl_2 interfere	32,143
O_2	Oxidation of cuprous ion/ammonia	Colorimetric	1 ppm	Reagent is continuously regenerated	74
O_2	Oxidation of sodium anthraquinone–β–sulfonate $\rightarrow$ color change	Colorimetric	1 ppm	Reagent is continuously regenerated	129
O_2	$+ H_2 \rightarrow H_2O$	Dew point	1 ppm	...	196

lutants. One type consists of a glass column packed with glass helixes. This is used with countercurrent gas–liquid flow in recorders for total oxidant and for nitrogen oxides. Another type is a "wetted-wall" column absorber based on the design of Thomas et al. (248). A third design of absorber uses a vertical coil, 80 ft of 7-mm-i.d. glass tubing (206). The wetted-wall type has the least holdup time. These absorbers work typically with ratios of liquid/air flow of 1:100 to 1:1000 to achieve a useful concentration of the scrubbed components in the liquid phase.

Another scrubber for sampling gases in the parts per billion range may be generally applicable (240). A plastic cylinder is provided with a closely fitting core on the surface of which is cut a helical groove. The air sample flows with the absorbing solution down through the groove, and the gas is absorbed with close to 100% efficiency. By adjusting the flow rates of the sample and liquid, this device is applicable to concentrations of HCN and H_2S from 4 ppb up into the parts-per-million range.

c. Gas–Solid Reactions

Reacting gas samples with solid reagents to achieve specificity is old in the art, and several principles may be mentioned. This technique has the advantage that flow control is not highly critical to achieve a fixed degree of reaction. The very early work on the analysis of carbon monoxide in air utilized catalytic oxidation of the CO, over Hopcalite, followed by thermal detection of the heat of reaction (169).

Selectivity of gas analysis has often been achieved by removal of one component of a mixture. Water interfering in the thermal conductivity analysis of gases is often removed from the sample by drying agents such as activated alumina or silica gel. Carbon dioxide is similarly removed by soda lime. This permits differential analysis for CO_2 by comparing the response of the total stream in a simple detector to that of the stream after removal of the CO_2.

It is feasible in some cases to convert a component of interest selectively to one that can be detected more readily. Thus Rosenbaum et al. (211) oxidized hydrocarbons in air catalytically to CO_2, to allow more sensitive measurement in a nondispersing infrared analyzer. The range was from less than 1 ppm up to at least 300 ppm CO_2.

Certain gases can also be measured by reaction with reagents on rolls of paper tape. Sensitivity is high when the tape is moved slowly to accumulate the component being measured. Hydrogen sulfide reacts with lead acetate impregnated on tape to give dark lead sulfide, which can be measured with a reflectance photometer. An analyzer based on this principle has been available for many years (180); its sensitivity is about 1 ppm H_2S.

Measurement of hydrogen fluoride has been the objective of several analyzers which depend on the quenching of fluorescence of a reagent (36,250). An analyzer built by Stanford Research Institute uses magnesium oxinate reagent impregnated on filter-paper tape. The tape is exposed to the sample in a differential manner; one section is exposed to the raw sample and an adjacent section to sample which has been passed through a tube coated with sodium bicarbonate to remove HF.

The exposed areas of the tape are compared in a differential fluorescence photometer. Sensitivity is reported to be about 0.1 mg HF per cubic meter or 0.25 ppb. The reading is not significantly affected by small concentrations of SO_2 or other common air pollutants.

d. LIQUID–GAS REACTIONS

At least one analyzer employing the reaction of a liquid sample with a gaseous reagent has been reported. An analyzer for dissolved oxygen in boiler feed water uses added nitric oxide to react with the oxygen to form nitrous and nitric acids (61). The change in the electrical conductivity of the water is then a measure of oxygen content in the parts-per-billion range. In order to avoid interference from carbonate, phosphate, and other ions, the incoming sample is treated with an ion-exchange resin to remove both cations and anions.

e. LIQUID–LIQUID REACTIONS

Analyzers employing the reactions of liquid samples with liquid reagents received relatively little publicity until quite recently. Successful analyzers of this type have overcome the difficulty of controlling two liquid flow rates in a precise ratio. In early analyzers flow control was accomplished by gravity flow from constant-head devices. Positive mixing of the streams is best assured by pumping at least one of the streams. The need for reliable pumps has led to considerable development efforts.

One of the early analyzers using liquid reagents involved two successive reactions on the sample in order to achieve specificity. Kieselbach's analyzer has been described as a complete miniature chemical plant for detecting stream pollution by organic matter (137). A sample of waste water from chemical processes is taken for analysis at a constant flow rate and is filtered (Fig. 105.5). Barium hydroxide reagent is added to precipitate carbonates, and another filtration is made. "Wet-combustion" reagent (202) is then added, and the organic materials are oxidized at elevated temperatures. The resulting carbon dioxide is stripped out of the liquid by an oxygen stream and measured by thermal conductivity. Concentration changes of as little as 1 ppm of organic matter, calculated as carbon, can be detected.

Also designed for a specific measurement on a liquid sample is a sulfate-ion analyzer for application to the wet process for phosphoric acid manufacture (49). The analysis is carried out by the precipitation of sulfate as barium sulfate and turbidimetric measurement of the precipitate in a differential photometer.

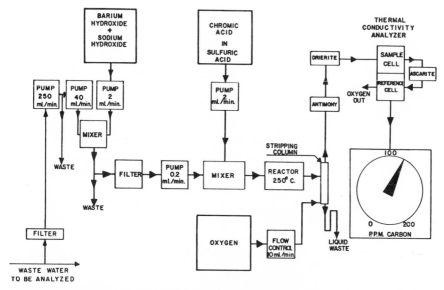

Fig. 105.5. Schematic diagram of analyzer for organics in waste water (137). (Courtesy of *Analytical Chemistry*.)

Analyzers using thermometric measurements following liquid–liquid reaction have been developed for acetic acid in acetic anhydride and for the strength of nitration acid (51,145).

In recent years versatile analyzers have become available which automate every step of liquid–liquid analytical reactions. They are so designed that any of a variety of chemical systems can be applied, both organic and inorganic. There are two general types; one has continuous flow of sample and reagents, as represented by the Autoanalyzer system (245), while the other has a sequence of batchwise manipulations, like a robot chemist (35,106).

The first type has had more rapid acceptance because of its greater versatility. Three analyzers of this type have appeared, and many articles describe continuous applications (53,126,174). Such analyzers utilize features such as pumping, reagent addition, filtering, mixing, and dialysis before measurement. Figure 105.6 illustrates the modular construction, which facilitates rapid changes of conditions, or even of application, when desired. Colorimetric measurements are generally employed, but other methods can also be used.

The heart of such analyzers is the proportioning pump. In one case (245), all liquids are pumped at proportionate rates by the squeezing

Fig. 105.5. Photograph of colorimetric analyzer employing liquid–liquid reactions. (Courtesy of Technicon Instruments Corp.)

action of a single set of rollers on parallel lengths of flexible tubing which are supported by a manifold. Delivery rate is governed by the diameter of the tubing. After pumping the sample, the reagent or portions of it are mixed with the sample and, if desired, are heated in a tempering bath for a chosen time. The reaction mixture can be dialyzed continuously, if necessary. The product is then led through the flow cell of a continuous measuring device and on to a discharge system. The value of the measured variable is continuously recorded or printed out.

Literally hundreds of different continuous applications are theoretically possible, if one judges from the batch analyses which have been performed. Continuous analyzers have actually been assembled for such determinations as glucose in blood, nitrate in fertilizer, sugar in water, phosphate in water, and silica in water (174).

Perhaps the most dramatic application of such a chemical reaction analyzer has been to the continuous monitoring of blood glucose concentration in living human subjects (76). In this case, a continuous flow of anticoagulant is added to the sample stream as it is withdrawn, without introducing anticoagulant into the patient's system.

Another interesting application of such an analyzer lies in the automation of the Kjeldahl nitrogen determination (75). In this case, a continuous digestion takes place in a helical glass tube rotating in a furnace. Ammonia from the conversion of organic nitrogen in the digester is then measured with a colorimetric reagent such as alkaline phenol–sodium hypochlorite. Precision is reported to be $\pm 0.35\%$ N (maximum deviation) at the 44% level.

Jonnard's analyzer is interesting in that the conditions for a particular analysis, such as selection of reagent and of appropriate filter and sensitivity for the colorimeter, can be "dialed in" by electrical selector switches (126).

In some versions of the robot-chemist type of analyzer, a fixed volume of sample is delivered to a titration vessel, and titrant is added from an automatic buret to a colorimetric or potentiometric end point (35,106,207). The volume of titrant delivered is recorded, the titration vessel is emptied, and a new cycle is started. In other versions of the robot chemist, a fixed volume of reagent is added and the extent of color formation is measured (1,225).

Because of its complexity, the robot chemist would seem to be an approach of last resort, yet applications have been made to analyses such as those for free chlorine, chloride, chromium, copper, cyanide, fluoride, formaldehyde, hydrazine, iron, dissolved oxygen, orthophosphates, sulfites, dissolved silica, and total hardness (225). Also, applications have been reported for measuring parts-per-billion concentrations in boiler feed water, such as dissolved oxygen and dissolved silica (225).

2. Physical Techniques of Separation

Instead of utilizing chemical reaction, selectivity in an analysis may be provided by isolating the component of interest through a separation technique. Again, such techniques may be classified according to the phases involved, such as gas–liquid. Included here are the extremely powerful and widely applicable chromatographic techniques. Generally, one does not attempt to isolate the pure component directly in the separation, but rather to obtain it as a binary mixture with an inert carrier gas or liquid. Once a binary mixture is obtained, any of several familiar methods of measurement can usually be applied. Thus, from the standpoint of overall complexity of the final analyzer, the elaboration resulting from the separation technique is at least partially compensated for by the simplicity of the final measurement.

a. Gas–Liquid Separations

When it is difficult to analyze directly for a gas dissolved in a liquid, because of low concentration or interference from the liquid, it is often possible to apply a stripping technique to obtain a vapor phase which is enriched in the component of interest. This approach has been reported by Friel (85) to be useful in the measurement of oxygen in liquids in fractional parts-per-million concentrations. Stripping the sample with a countercurrent flow of pure nitrogen provides a gas sample useful for an oxygen analyzer working on the galvanic cell principle. Maley (159) has described a very similar technique for the analysis of parts per million of methanol in cyclohexane, using a nondispersing infrared analyzer for the final measurement. Cole et al. (52) have applied stripping to hydrocarbon liquids to determine trace amounts of water, using an electrolytic analyzer. In a chemical analyzer system, carbon dioxide is evolved from acidified blood plasma with the aid of CO_2-free air and is then determined colorimetrically with an alkaline phenolphthalein reagent (229).

Operation of a gas–liquid stripping system imposes the requirement of controlling accurately both a gas flow and a small liquid flow. It is not difficult, by adjustment of relative flow rates, to obtain a gas stream with 10 ppm or more of the component of interest for each part per million in the original liquid sample.

b. Liquid–Liquid Extraction

Extraction is a common method of achieving selective separation in laboratory work, but it is difficult to automate extraction because of possible uncertainty in locating the interface between the liquids being contacted. One of the reported uses of continuous liquid–liquid extraction occurs in an analyzer for uranium ion in aqueous streams (186). The extractor is essentially a vertical glass tube for countercurrent contact of the immiscible fluids. The lighter solvent phase is introduced at the bottom of the column and then passes up through the heavier medium. Efficient contact is assured by rotating an axial rod at 2500–3000 rpm in the extraction zone.

c. Dialysis

Dialysis is widely used in laboratory biochemical work to achieve separations of difficult materials such as suspended solids. This technique has little appeal for general analytical separation because of slow rates. Effective dialysis at reasonable rates is achieved in a chemical analyzer, however, by using continuous concurrent flow along a long spiral path. This geometry is achieved by having grooves which are mirror images of each other cut in the surfaces of two plates. A membrane is then sandwiched between the two surfaces, with the grooves facing each other to provide parallel paths on both sides of the membrane.

Applications of continuous dialysis have been made to (*1*) determination of Terramycin in pharmaceutical preparations (98); (*2*) analysis for nitrate in fertilizers (34); and (*3*) determination of ammonia in biological materials (151).

d. Gas Chromatography

Gas chromatography (GC) is generally applicable to the rapid, multicomponent analysis of a wide variety of gases and volatile liquids. It is very popular for process-stream analyzers because of its versatility and selectivity. Sensitivity is high, and less than 0.1% of many compounds can be detected.

This popularity has been gained in only a few years from the first published description of the GC technique for laboratory analysis. Three years elapsed between the paper of James and Martin in 1952 (122) and the first report of a successful plant analyzer (27). By 1959, such analyzers were reported to be selling better than any other type. It now seems certain that they will be used very widely for direct process control in the petroleum and chemical industries (261).

Gas chromatography is a separation process which converts a sample mixture into a series of two-component mixtures, using as one component a flowing inert carrier gas. The separation takes place on a column into which the sample is injected, in the stream of carrier gas. The two-component mixtures are then analyzed in succession by simple detectors for binary gas mixtures.

Several features are necessary to design a plant analyzer, starting from the essentials of the laboratory instrument. In addition to the separating column, detector, and recorder, the plant analyzer possesses a sampling system and automatic attenuator for large signals. Since a number of components are analyzed in each sample, the analysis cannot be truly continuous, and a programmer is required to govern the steps in the repetitive cycle. A complete GC analyzer is shown in Fig. 105.7.

The heart of the sampling system is a valve which takes a discrete sample of the process stream and transfers it to the column at the proper time. In order to minimize time lag, there should be rapid flow in the sampling system up to the sampling valve. Valves for sampling gases

Fig. 105.7. Gas chromatographic analyzer with explosion-proof housing for hazardous areas (157). (Courtesy of Perkin–Elmer Corp.)

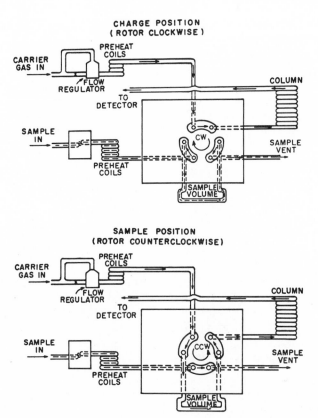

Fig. 105.8. Schematic diagram of gas-sampling valve (157). (Courtesy of Perkin-Elmer Corp.)

and vapors of volatile liquids are well developed and can be obtained in both linear and rotary configurations.

The rotary valve can be regarded as the pneumatic analog of a six-pole single-throw switch (Fig. 105.8). The rotary configuration possesses an advantage for high-temperature work and for samples containing powerful organic solvents in that polytetrafluoroethylene can be used as the sealing material. The linear valves thus far designed have required elastomeric seals which are less tolerant of severe temperature and chemical environments. Generally, such valves are used for volumes of the order of 0.25–5 ml but a valve for a high-speed analyzer can provide a sample as small as 0.05 ml (127).

For the sampling of volatile liquid process streams, it has been common practice to use a vaporizer and a gas-sampling valve, but liquid-sampling valves are now available. Volatilization becomes unsatisfactory with liquids which contain tars or dissolved solids, or which are unstable or prone to polymerize. Several devices can be used for direct injection of minute liquid samples (17,134,162,167,195). In all these liquid-sampling valves, the minute volume of sample is trapped, after appropriate flushing, in a capillary or groove of metal which is then moved into a position to be emptied into the carrier-gas stream; one such valve is diagrammed in standby position in Fig. 105.9. Vaporization of the minute sample is accomplished in a small preheated section at the column inlet. Tars or solids formed upon varporization will accumulate in the preheater rather than in the sampling device.

Columns for the chromatographic separations are of two general types. Fixed gases and low-molecular-weight hydrocarbons are separated on solid adsorbents such as silica gel, alumina, and molecular sieves. Other volatile materials are generally separated by partition liquids which are coated on an inert solid support. The partition liquid must possess low volatility and good thermal stability to minimize change of its analytical characteristics as the column ages in use. Phenomena of separation on the column are considered in detail elsewhere in this Treatise.

Detectors for the analysis of the separated component/carrier-gas mixtures are based most commonly on the thermal conductivity principle, although many types of analyzers have served as detectors in laboratory gas chromatography. Thermal conductivity detectors have enjoyed wide acceptance in process applications in the United States. Both hot-filament and thermistor types have been successful. A high-speed thermal conductivity cell has been developed (40). Catalytic combustion detectors are rugged and provide adequate sensitivity for some hydrocarbon analyses (63,97). Thermal detectors utilizing heat of adsorption have also been used for hydrocarbons (70).

Needs for sensitivity of detection greater than that offered by standard thermal conductivity cells have led to improvement of the latter and a trend toward ionization detectors (4). Kieselbach (138) has described shielded thermistors and auxiliary electronics which have been used for analyses of parts-per-million concentrations. In his system, the base-line thermistor output is stable even at 50 μV full-scale recorder sensitivity.

Flame ionization detectors have been introduced for plant use, with appropriate explosion-proofing. Flames, radioactive sources, and high voltage in ionization detectors add potential hazards to plant instruments. Ionization detectors have been reported to be in use in plant analyzers in Europe (162). They are most effective with small-diameter or capillary columns because of their rapid response.

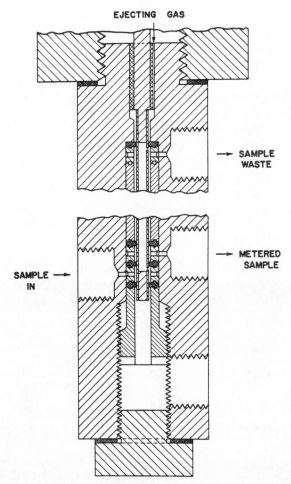

Fig. 105.9. Diagram of liquid-sampling valve (134). (Courtesy of *Analytical Chemistry.*)

Detector signals are large for major components and small for minor components. Attenuators corresponding to each peak of interest are, therefore, preset to keep the corresponding signals on scale on the recorder. They are controlled by the programmer, to be switched into the circuit when the appropriate peak appears in the cycle.

Recorders are standard potentiometric chart recorders. In handling microvolt-level signals in trace analyses, a signal amplifier is required (21,138). Trace analysis can be achieved without amplification in favorable cases by using a large sample and a fast column and detector so that tall, narrow peaks are obtained (4). In recording the output of ionization detectors, an impedance-changing amplifier such as an electrometer is generally employed (155).

Although the recorder follows the peaks for the individual components if desired, an abbreviated presentation of the chromatogram is used routinely. This is a "bar-graph" presentation, which records only a bar whose length is the same as the height of the peak (Fig. 105.10). Components of no interest can be ignored in programming so that the recorder does not respond to them.

Memory devices and peak-sensing devices allow a recorder to hold a continuous reading of a single peak between analytical cycles. As a result trend recorders can be used, and control signals proportional to the concentration of any given component can be provided (78,81).

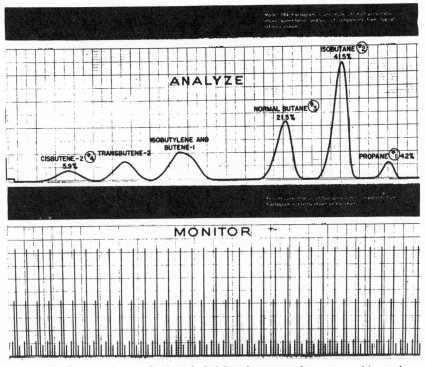

Fig. 105.10. Bar graph record of peak heights from gas chromatographic analyzer (157). (Courtesy of Perkin–Elmer Corp.)

Programmers are timing devices which call into play the sequential steps of the analysis: sampling, base-line adjustment, attenuation, back-flushing, and stream switching. Cam timers are commonly used, while the new tape-controlled programmers offer ease in changing programs (80,169). One type of programmer has a large timing dial with peripheral contacts which allow a timed function to be changed with considerably more ease than is possible with typical cams (48).

Special features which can be obtained in plant GC analyzers include high-speed operation, multiple sample streams, use of multiple columns, and back flushing. High-speed operation is obtained through the use of small-diameter columns and small-volume detectors designed to have high speed of response. Analytical cycles within 1 min have been reported (97).

Multiple streams of similar composition can be analyzed, after directing each stream in turn through the sampling valve. Even a different concentration level of a given component in each of several streams can be handled by proper programming of corresponding sets of attenuators.

Analysis of mixtures containing components of widely different elution times can be accelerated by sequencing the action of two or more columns (15). The use of short and long columns in series with appropriate valving allows both early and late components to be eluted and directed through the detector in reasonable overall times.

"Heavy ends," components which move only very slowly on a given column and which are not pertinent to the measurement, can be removed by back flushing. The direction of flow through the column is reversed after the components of interest have been eluted and detected (15). This is important to prevent poisoning of the column by late-eluting materials. Back flushing is included as a preventive maintenance feature in every chromatograph used by one chemical company.

Criteria by which GC analyzers may be judged include resolution, sensitivity, short-term instability (noise), long-term instability (drift), minimum range, effect of voltage supply change, effect of ambient temperature changes, ease of maintenance, and ease of operation. Most of these criteria may also be applied to other types of analyzers.

1. Resolution is a measure of the extent of separation between peaks for components which tend to elute close together. It is affected by sample size and system geometry (dead volume) as well as by column variables.

2. Sensitivity is a measure of the smallest concentration of a component which can be reliably detected. It depends on resolution, sample size, elution time of the component, and noise.

3. Noise, or *short-time instability,* consists of fluctuations in the recorded signal (when no sample is present) which arise from instabilities due to flow, temperature, and electrical effects in the detection-recording system.

4. Drift, or *long-term instability,* is a gradual shift in the base-line signal. It may be due to temperature changes or loss of liquid phase from the column.

5. The *minimum* useful *range* of an analyzer depends largely on its sensitivity. There is little point in having a minimum range of less than 100 times the sensitivity. Narrower ranges merely amplify the noise level.

6. Effects of voltage supply variations are likely to be manifested in signal variations, if proper stabilizing circuits are not included in the analyzer.

7. Ambient temperature changes will affect the elution times of components and the base-line signal of thermal conductivity detectors, unless the analyzer is well thermostatted.

8. Ease of maintenance is related to the overall complexity and the accessibility of all working parts of the analyzer to testing and replacement. Maintenance needs are vital factors in determining operating costs.

9. Ease of operation may be judged from the procedure necessary when operating variables have to be adjusted to maintain the desired conditions or to set up for a new analysis. Access for adjustment of the programmer, thermostat, and flow controls, and the ease with which columns and sample sizes are changed, are important factors.

The gas chromatograph is so versatile and popular that the total variety of applications may only be guessed at. Most of the applications to date have been to gas and volatile liquid mixtures in the petroleum and petrochemical industries (157). Acceptance has been slower in the organic and inorganic chemical industries. Such applications tend to be more difficult in terms of the operating temperatures required to achieve separation in reasonable analysis times, or the availability of columns to perform the separations.

Table 105.III lists typical applications of gas chromatographic analyzers (157). The table includes the major process unit, the specific applications, and the components which are usually measured.

The many applications reported of such analyzers in automatic control loops (261) include the following: propane in natural gasoline from

TABLE 105.III
Applications of Gas Chromatographic Analyzers (157)

Process and units	Analysis
Natural gasoline plant	
Absorber	Methane, ethane, CO_2, propane
Deethanizer	Methane, ethane, propane
Depropanizer	Ethane, propane, propylene, isobutane
Debutanizer	Propane, isobutane, n-butane, isopentane
Butane Splitter	Isobutane, n-butane, isopentane, n-pentane
Butane isomerization	
Isomerization product	Propane, isobutane, n-butane
Butane splitter	
overhead	Propane, isobutane, n-butane
bottoms	Isobutane, n-butane, isopentane
Alkylation	
Feed	Propane, isobutane, n-butane, butylenes
Debutanizer	Propane, isobutane, n-butane, isopentane
Depropanizer	
overhead	Ethane, propane, propylene, isobutane
bottoms	Propane, isobutane, n-butane, isopentane
Deisobutanizer	
overhead	Propane, isobutane, n-butane, isopentane
bottoms	Propane, isobutane, n-butane, isopentane
Ethylene	
Absorber	H_2, CO, CO_2, methane, ethane, ethylene
Deethanizer	CO, CO_2, methane, ethane, ethylene
Ethane splitter	Ethane, ethylene, propane
Depropanizer	Ethane, ethylene, propane, isobutane
Polymerization	
Reactor feed	Methane, ethane, propane, propylene, isobutane
Splitter	Ethane, propane, propylene, isobutane
Depropanizer	Ethane, propane, propylene, isobutane
Inorganic chemistry	
Ammonia synthesis	CO, A, N_2, H_2, methane, ammonia
Sulfur recovery	O_2, N_2, H_2S, SO_2
Organic chemistry	
Alcohol plants	Water, methanol, ethanol, propanol, acetone
Benzene, toluene, xylene plants	Hexane, benzene, toluene, xylenes
Freon production	Mixed Freons

a depropanizer column (79), isopentane in a debutanizer column (78), butene-1 + isobutylene from a furfural absorber in a butadiene plant (78), $H_2S + SO_2$ in tail gas from the oxidation of H_2S to free sulfur (252), and control of a catalytic reactor in a pilot plant from a product/reactant ratio (260).

Present trends in research indicate that process gas chromatographs will be extended in application to odor and flavor measurement, pyrolysis analysis of polymers, analysis of other solid and nonvolatile samples, and parts-per-million measurements of impurities in gases and liquids. Even parts-per-billion analysis can be achieved through introduction of a preliminary concentration step.

The present limitation of response time to that of the repetitive cycle seems certain to be minimized by the advent of high-speed analyzers. Several components can now be measured in 1 min through the use of small samples (127), columns (65,152), and detectors (79,255,260). Further increases in speed, by a factor of 5 or 10, appear likely.

e. LIQUID–LIQUID AND ION-EXCHANGE CHROMATOGRAPHY

In these forms of chromatography, the moving phase is a liquid. The fixed phase, which is the active part of a column, may consist either of a liquid immiscible with the moving phase and immobilized by an inert solid, or of a reversibly reactive solid such as an ion-exchange resin. There have been no reports as yet of process applications of these techniques, but they have become very valuable in the laboratory when automated to a high degree. The sample is placed on the column manually, but subsequent operations of elution, detection, and measurement of components are automatic in a number of cases, such as the analysis of amino acids (233,264). Process applications are likely when rates of elution, presently measured in hours, are accelerated.

Among the interesting detectors for analysis of binary mixtures in chromatography of this general type are differential refractometers with cells of very small volume (19) and vapor-pressure detectors. Sternberg and Carson (238) have described an application of a thermistor detector which measures the relative concentration of a nonvolatile solute in a volatile solvent by the phenomenon of vapor-pressure lowering (Section IV.C.2.c). In liquid gradient-elution chromatography using solid absorbents, a thermistor can serve as a qualitative continuous detector of component zones. The thermistor, placed in the absorbent bed, detects thermal effects associated with adsorption and desorption of the liquid components. Blumer (25) has applied this technique to elution analysis and to total-displacement analysis of hydrocarbon mixtures. Improved sensitivity for concentration peaks, with freedom from thermal drift, is obtained by differentiating the thermistor output signal.

Monitoring of chromatographic eluents for radioactive components can be performed best with special plastic or glass flow cells designed to give efficient counting with Geiger tubes (26). Even weak beta emitters can be handled in cells with thin windows of polytetrafluoroethylene (24). A very sensitive scintillation flow detector for continuously monitoring alpha or beta activity in liquids or gas streams is available (44).

f. Liquid–Solid Separations

The most common example of this type of separation is filtration. This is often employed as a means of keeping clean liquid samples supplied to analyzers (Section II.B) and in such cases involves a separation of adventitious solids. In the organics analyzer of Kieselbach, two filtrations are used (137). The first employs a standard fixed filter for removal of relatively coarse suspended matter in the water sample. The second filtration uses paper tape for a continuous chemical separation, the removal of carbonates which have been precipitated to prevent interference in the subsequent determination of CO_2. The tape is pulled slowly over a suction port in a smooth cylinder. The sample to be filtered floods the underside of the paper, dropping off freely to prevent plugging.

IV. METHODS OF MEASUREMENT

A. ELECTRICAL METHODS

The electrical properties of molecules have been applied to the measurement of concentration in a variety of continuous analyzers. Such methods are seldom very specific (an exception is an electrolytic analyzer for parts per million of water), but they are generally sensitive to low concentrations of active species. Measuring circuits tend to be relatively simple and readily adapted to durable, rugged analyzers.

The electrical methods include those in which a change of concentration can be related to a change (1) in the voltage developed at immersed electrodes (potentiometry and pH), (2) in the current flowing in the system (conductance, polarography, coulometry), or (3) in the capacitance of the system (capacitance and dielectric constant). The greatest variety of methods is associated with type 2, while type 3 can be applied to the widest variety of samples, representing the gaseous, liquid, and solid states.

1. Potentiometry

a. pH

Hydrogen-ion concentration in aqueous systems is most commonly measured as its negative logarithm, or pH. The glass electrode responds to pH linearly over a wide range of concentration. The potential developed between a glass electrode and a stable reference, typically the calomel electrode, is the common measure of pH.

Critical points for on-stream applications of pH include the fragility of the glass electrode, its susceptibility to contamination and to interference by other cations, and the need for signal amplification. Physically rugged electrode designs suitable for use in process vessels and pipelines have been available for many years (Fig. 105.11) so that fragility is not a severe problem. The ease of contamination of the glass membrane means that frequent maintenance of electrodes may be required with

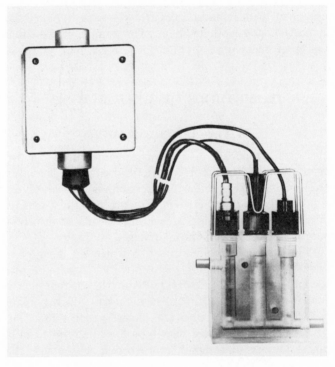

Fig. 105.11. Flow cells for pH analyzer.

streams containing foreign matter. Principal sources of interferences are the alkali metal cations, notably sodium.

The temperature of the sample is another common variable with marked effect on the signal obtained at constant concentration of hydrogen ions. This is generally regulated by electrical compensation in the signal circuit. Operation above 100°C is not possible at present. Since the electrode potential is in the millivolt range and no appreciable current can be drawn, a sensitive electrometer amplifier is required (input impedance *ca.* 10^{11} Ω). Input circuits and leads are thus subject to effects of stray capacitance and leakage currents if not carefully designed.

Applications of pH techniques are made widely to measure the acidity and alkalinity of process streams. Among the numerous applications of analyzers employing pH measurements are boiler feed water in steam-generating plants, water used in cooling towers, neutralized acid sludge wastes in petroleum plants, fermentation media for antibiotic production, and cane juice in sugar refining (45). An immersion probe employing a pH-sensitive electrode has been developed for measuring dissolved carbon dioxide in blood (223). A similar system is available for CO_2 in respired air (17). Sensitivity to small variations in the CO_2 content of polluted air has been claimed for a pH measurement in an aerated suspension of calcium carbonate (150).

There are common applications of this measurement to control the additions of reagents to maintain a desired pH. Since rate phenomena are involved in achieving complete reaction, it is imperative that the pH be measured at the point where control is desired. Dependence of pH on stream flow rates may be observed.

b. Sodium Ion

Since pH measurements are subject to interference from sodium ion, it is possible to measure sodium ion concentration at constant pH. This has been achieved recently with the aid of special sodium aluminate glasses, which give the glass electrode specificity for sodium ion in the presence of potassium and lithium ions (144).

c. Chloride Ion

When suitable electrodes are employed, concentration can be determined potentiometrically in a number of other oxidation–reduction systems. Chloride ion can be measured in terms of the potential of a silver/silver chloride electrode relative to a calomel reference electrode (125). The useful range of this electrode is 0.1–10,000 ppm chloride.

Since the voltage output is related to the logarithm of the chloride ion concentration, this may be considered a pCl measurement.

d. RESIDUAL CHLORINE

To control the chlorination of drinking water supplies, a potentiometric measurement of free chlorine is used. With a roughened gold electrode as little as 1 ppm of chlorine can be detected (256). In the chlorination of pulp stock, the concentration of sodium hypochlorite can be measured by the oxidation–reduction potential of a platinum/silver electrode pair (224). These electrodes have been made particularly rugged by sheathing of the respective metals on stainless steel rods. Since pulp stock is viscous, it is diluted before the measurement. In this application, dilution does not change the oxidation–reduction potential, which is a measure of the following ratio: log (oxidized form/reduced form). Oxidation–reduction potentials have also been applied to the measurement of toxic Cr^{+6} in aqueous wastes (124) and to the cupric/cuprous ion ratio in the dyeing of acrylic fibers.

2. Conductance

Measurement of electrolytic conductivity is a nonselective technique which has long seen wide application because of its sensitivity and simplicity. Very low concentrations of electrolyte can be measured in aqueous solutions. Although high concentrations can also be determined, anomalous effects are more likely to occur, such as maxima in conductivity–concentration relationships. Little selectivity can be achieved in multicomponent systems unless all components other than the one of interest can be held constant, that is, by use of differential techniques.

Conductivity is measured in cells with rigid, flat, parallel electrodes of a metal (generally platinum) which is inert to the chemical system. For the relative measurements of greatest interest in process work, ruggedness and ease of cleaning are primary design features for the conductivity cells (Fig. 105.12). Geometry is governed by the electrode area that is necessary to achieve the desired sensitivity. Cell resistances in the range of $10–10^5$ Ω are typical. Glass is commonly used for mounting and for shielding electrodes in corrosive liquids.

Since conductivity is the reciprocal of resistance, measurements are universally made with resistance bridges. Alternating-current excitation of the bridge is used to minimize polarization of the electrodes. A bridge may be part of the self-balancing circuit of a recording potentiometer, or the unbalance voltage in the bridge when conductivity changes may

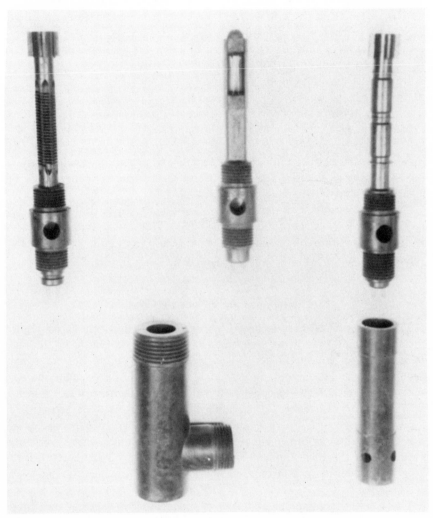

Fig. 105.12. Flow cells for conductivity analyzers.

be amplified and rectified to drive any of a variety of dc monitoring devices.

Long-established applications for conductometric analyzers, in addition to the Thomas Autometer (Section III.B.1.b), include measurements of acidic gases and of chlorine in air as solutions in deionized water

(96). In another analyzer, parts per million of O_2 are measured indirectly as NO_2 after reaction with NO and absorption in water to form dilute HNO_3 (61). Dissolved electrolytes can be detected at parts-per-million concentrations in purified water derived from distillation, deionization, or steam condensation for boiler feed.

By means of differential techniques, sensitivity can be improved so as to control the quality of water used to wash textiles, or of municipal drinking water to which fluorides have been added.

Maintenance of the conductivity cell can be reduced if an electrodeless cell is used. Such a cell operates by employing a loop containing the process stream as the coupling between the primary and secondary windings of a transformer (96). Applications are restricted to materials of high conductivity, 10^4 μmhos and higher.

Another useful cell design of recent origin employs three electrodes to make effectively two cells in one (96). This is advantageous if it is desired to eliminate the effect of capacitance from polarization of the cells, since this effect is canceled out in certain bridge circuits.

3. Capacitance and Dielectric Constant

The dielectric properties of compounds are used as the basis for a relatively small number of continuous analyzers for both liquid and solid samples. Measurements on gases are also possible but have not been reported. The actual measurement is that of the capacitance between electrodes with fixed geometry. The basic instrumentation is an ac bridge for measuring capacitance. This circuit is well established for the detection of liquid levels through the change in capacitance of a probe as it is approached or contacted by a liquid–vapor interface (112). Since the measurement is nonselective, only binary mixtures, or a polar component in a mixture of nonpolar substances, can be analyzed accurately.

The sensitivity for polar materials is such that water can be measured down to part-per-million concentrations in hydrocarbons. Other applications include measurements of the propane/oil ratio in petroleum refinery dewaxing operations (246); the aromatic content of the feed to solvent extraction units and to catalytic reformers (37,172); the "basic sediment and water" in crude oil (116); the methyl ethyl ketone in benzene–toluene mixtures; and the water in organic solvents and in granular materials such as starch, coal, meal, cereals, and polymers (23). The moisture in paper and textile fabrics can also be measured (112).

Even isomeric compounds can sometimes be distinguished, as in the measurement of the ratio of toluene 2,4-diisocyanate to toluene 2,6-diisocyanate in polyurethane manufacture (236).

4. Polarography

It is well known that many ions and compounds can be reduced at a dropping mercury electrode which has a sufficient applied potential in relation to a reference electrode. Similarly, oxidation can be carried out at a polarized platinum electrode. The potential necessary to achieve the reduction (or oxidation) is characteristic of the component of interest. When a supporting electrolyte is used in the laboratory, the applied voltage is scanned, and a plot of diffusion current through the test solution versus voltage is obtained. This shows breaks or waves for all reducible components.

For process work, it may be sufficient to apply a fixed potential to reduce a given component in a flowing stream. In this case, recorded current can be related to variations in concentration of the component of interest. Because the dropping mercury electrode is not very convenient to use in a process instrument, solid metal electrodes are sometimes employed. If the dropping mercury electrode is used, mercury flow conditions are critical in order to avoid false readings from flow and vibration effects of the moving sample on the mercury drops (253).

Applications of polarographic techniques include measurements of dissolved oxygen (253) in water and of residual chlorine in drinking water. Commercial units are available (80). The reduction of oxygen at the dropping mercury electrode is well known. Oxygen can also be reduced at a platinum electrode which is held at -0.8 to -1.1 V in relation to a silver/silver oxide reference electrode. Measured current is related to oxygen concentration. This electrode pair allows the construction of a physically rugged probe for dissolved oxygen measurements. To prevent contamination of the electrodes by impurities in liquid samples, a plastic film which is permeable to oxygen is used to isolate the measuring cell from the liquid system (33,47). Oxygen then diffuses from the liquid sample through the plastic film and into the probe for measurement. Analyses for SO_2, NO_2, Cl_2, and Br_2 are also said to be possible through application of similar principles (218).

Applications of the polarographic oxygen analyzer have been made to fresh and sea water, biological fluids, suspensions of algae, and similar systems (33). The sensitivity has been described in terms of concentration by volume; changes in 0.01 ml/O_2 per liter of sea water can be detected.

5. Ionization Methods

Several highly sensitive but relatively nonselective measurements on gas mixtures depend on the conduction properties of ionized gases. It has long been known that such properties depend markedly on gas composition.

The need for sensitive detectors for gas chromatography has recently led to the development of a variety of such "ionization detectors" to measure the concentration of eluted components as they emerge in the carrier-gas stream. In general, only a binary mixture is involved so that little selectivity is needed. These detectors are the subject of an excellent review by Lovelock (155), who points out that their potentialities for continuous analysis, apart from gas chromatography, have not yet been explored.

These detectors vary notably in the means used to achieve ionization; all employ dc polarizing voltage to measure conduction of the ionized gas. Electrical means of ionization include low-pressure glow discharge (203), radio-frequency corona discharge (130,139), and electron impact by thermionic electrons. Ionization by radiant energy is considered in Section IV.B.9.

6. Mass Spectrometry

Mass spectrometry combines electrical, separation, and, often, magnetic methods to achieve rapid multicomponent analysis. Electrical methods are used to produce ions which can then be separated by magnetic or electrical fields. The initial ionization usually results in a fragmentation pattern which is characteristic of the molecule of interest and hence gives qualitative information. The currents resulting from collection of the various individual ions yield quantitative data. In process-stream analyzers, the method is generally applicable to mixtures such as hydrocarbon gases which are in the vapor state at ambient temperatures. Calculations of some intricacy are required, and equipment of general utility is expensive. These considerations have limited the application of mass spectrometry in process-stream analysis (13).

In any mass spectrometer, ions are produced by bombardment of the sample with electrons from a hot filament. The ions are then accelerated in an electric field. This is done under high vacuum to ensure efficient subsequent collection of the ions that are formed. Various instrument designs differ in the methods of separating the ions of the parent molecule and its fragments. Positive ions are accelerated to high velocity in an electric field and are then collected by an electrometer in the order of increasing mass/charge ratio, giving a series of peaks—a mass spectrum.

The traditional mass spectrometer employs a strong magnetic field to separate or resolve the ions by focusing them into narrow beams representing the various mass/charge ratios. A process-stream analyzer of this type is rapid in response but generally is limited to a maximum mass/charge ratio of about 80, or to compounds of molecular weight below 81. Six preselected peaks can be scanned in less than 1 min.

Another magnetic mass spectrometer for process work uses crossed electrostatic and magnetic fields to achieve resolution over a wider mass range through cycloidal focusing of the ion beams (209). Generally the electrostatic field is varied to produce the spectrum. The upper limit of the cycloidal instrument for good resolution is a mass of about 150. A photograph of this analyzer is shown in Fig. 105.13.

A third type of mass spectrometer uses radio-frequency accelerating voltages to resolve different masses in the absence of a magnetic field (67). Since the transit time of a particle from source to collector is a fraction of its mass/charge ratio, any desired mass number may be brought into register by adjusting the frequency. The range of this type of analyzer is about 12–100 mass/charge units, and a peak measurement has a time constant of about 2 sec.

A fourth type of mass spectrometer takes advantage of the relative velocities, or "time of flight," of ions in the absence of electric fields. The ions initially are formed in bunches by a pulsed electron beam. The ion bunches are accelerated into a field-free drift tube, where they separate into groups having various velocities depending on the mass/charge ratios. These groups register in succession at the collector electrode. The range extends to mass/charge ratios of about 200.

7. Coulometry

Spontaneous electrolysis in a galvanic cell can be used to measure specifically traces of a variety of substances: oxygen, (109,133), hydrogen (82), cyanide ion (14,212,247), and fluoride ion (115). Cyanide is measured in a cell with a silver anode, a platinum cathode, and $0.1M$ sodium hydroxide as electrolyte (14). Initial current in the cell is proportional to initial cyanide concentration over the range 0.1–1 ppm, with 70 μA being obtained for 1 ppm CN^- in the cell. Very little or no interference has been noted in tests with added nitrite, nitrate, fluoride,

Fig. 105.13. Cycloidal-focusing mass spectrometer analyzer.

chloride, sulfite, sulfate, or phosphite ions or ammonia. Hypochlorite gives moderate interference. Such a cell can be used to monitor cyanide concentration in the air (247) if the air stream is continuously bubbled through the electrolyte. A cyanide analyzer has been adapted to the detection and estimation of air-borne protein in concentrations of a few micrograms per liter, following pyrolysis of the protein aerosol to form cyanide.

Fluoride ion can be measured in a cell comprising an aluminum anode, a platinum cathode, and either $0.2M$ acetic or $0.017M$ benzoic acid as electrolyte (14). No interference is produced by added nitrate, sulfate, and cyanide. Some interference is caused by phosphate and sulfide, however, and chloride is a serious source of interference. Sensitivity is 4.5 $\mu A/\mu g$ F$^-$ in the cell. Application of this cell to the monitoring of soluble atmospheric fluoride has been studied (115).

Although the principles of galvanic cell analysis for oxygen have been known for some time, application to trace analysis has been made only recently. Oxygen will depolarize a carbon/zinc cell of the Fery type and allow a proportional current to flow (171,184). This cell has been used as the basis of commercial oxygen analyzers with ranges as narrow as 0–1A full scale or as wide as 0–25% (121). Electrode life is relatively short. The electrolyte is acidic; moistened solid ammonium chloride has been used (121,171). Carbon monoxide and up to 24% carbon dioxide do not interfere.

A galvanic cell with much greater sensitivity for oxygen was developed by Hersch (108). This device reduces oxygen at a silver cathode in a cell, which uses a lead or cadmium/cadmium hydroxide anode and potassium hydroxide as electrolyte. The current obtained is proportional to oxygen concentration at parts-per-million levels. The electrode reactions are:

$$\text{Cathode: } O_2 + 4e + 2H_2O \xrightarrow{\text{Ag}} 4OH^-$$
$$\text{Anode: } \quad 2Cd + 4OH^- \rightarrow 2Cd(OH)_2 + 4e$$
$$\text{Overall: } 2Cd + 2H_2O + O_2 \rightarrow 2Cd(OH)_2$$

Since the cell current is several microamperes for each part per million of oxygen, sensitivity of 1 ppm by volume full scale can be achieved on a good recording potentiometer. Calibration is facilitated if an electrolytic oxygen generator is placed ahead of the galvanic cell. Acidic substances will interfere if not removed from the gas stream.

In one inexpensive design, an anode of lead foil is surrounded by a sheet of porous poly(vinyl chloride), impregnated with KOH solution and wrapped loosely with silver gauze (109) (Fig. 105.14). Such an oxygen analyzer can be a useful aid to bench-scale research. It is inexpensive to construct and easy to manipulate. A low-resistance microammeter suffices for the analytical readings.

Keidel (133) has described an industrial analyzer, fabricated of stainless steel, in which essentially quantitative coulometric performance is achieved. A porous silver frit is used as cathode, together with a cadmium anode (Fig. 105.15).

The oxygen galvanic cell has no blank current when the system is free of reducible substances. Some organic vapors may interfere if they are not removed from the sample (109). Current output is not linear above 100 ppm, but the range can be extended in linear fashion by removing a predetermined portion of the oxygen in the sample.

The galvanic cell is applicable for determining oxygen in gases such as nitrogen, hydrogen, hydrocarbons, ammonia, and carbon monoxide. Dissolved oxygen in certain liquids has also been measured after interposition of a transfer step to provide a gaseous sample (68,85).

A similar analytical cell measures traces of hydrogen at the parts-per-million level (82). In this case, hydrogen is oxidized at a platinized platinum anode. The cathode is a calomel half-cell, and the electrolyte is potassium chloride in hydrochloric acid. The electrode reactions in this cell are:

$$\text{Anode:} \qquad \text{H}_2 + 2\text{H}_2\text{O} \xrightarrow{\text{Pt}} 2\text{H}_3\text{O}^+ + 2e$$
$$\text{Cathode:} \ \text{Hg}_2\text{Cl}_2 + 2\text{H}_3\text{O}^+ + 2e \rightarrow 2\text{Hg} + 2\text{HCl} + 2\text{H}_2\text{O}$$
$$\text{Overall:} \qquad \text{H}_2 + \text{Hg}_2\text{Cl}_2 \rightarrow 2\text{Hg} + 2\text{HCl}$$

This galvanic cell method has been applied to the measurement of hydrogen in nitrogen and hydrocarbon gases. If oxygen is present in

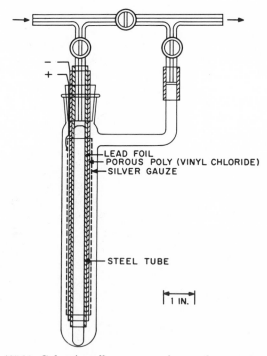

Fig. 105.14. Galvanic cell oxygen analyzer—glass construction.

Fig. 105.15. Galvanic cell oxygen analyzer—stainess steel construction.

the sample, it must first be removed in order to prevent interference. Scrubbing with sodium anthraquinone β-sulfonate solution is very effective for this separation.

These instruments for oxygen and hydrogen determination might be termed "coulometric analyzers." Another device in this class is the electrolytic water vapor analyzer described by Keidel (132) for measuring parts-per-million concentrations of moisture in a wide variety of gases. This analyzer depends on the current which flows when water is absorbed

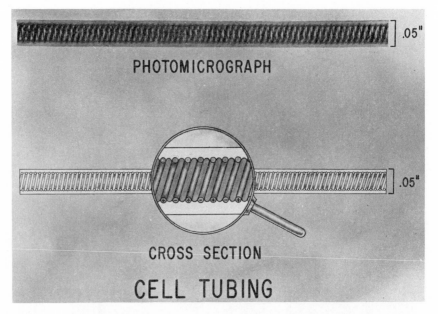

Fig. 105.16. Detail of cell for electrolytic water vapor analyzer.

and then electrolyzed in a special cell, illustrated in Fig. 105.16. A dc potential is applied to two platinum wire electrodes which are coiled inside a tube of Teflon TFE-fluorocarbon resin. Water vapor in a gas flowing through the tubular cell is absorbed from the stream by phosphoric anhydride. Electrolysis to oxygen and hydrogen occurs under the influence of the dc potential, and the electrolysis current is proportional to the number of moles of water absorbed per unit time. The construction of intertwined wires in a thin tube provides a large electrode area to give high efficiency in both electrolysis and absorption.

Applications have been made to the measurement of water vapor in air, nitrogen, hydrogen, Freon fluorinated hydrocarbon refrigerants, hydrochloric acid, sulfur dioxide, and phosgene. Butenes polymerize in the cell (105).

The electrolytic analyzer is readily modified for the measurement of water in hydrocarbon solvents. This is possible by using a stripping technique to separate water from a liquid sample continuously in a small column (52) with a dry inert gas as carrier (Fig. 105.17). Sensitivity of the order of 0.1 ppm water can be achieved.

One coulometric analyzer has been widely applied for measurement

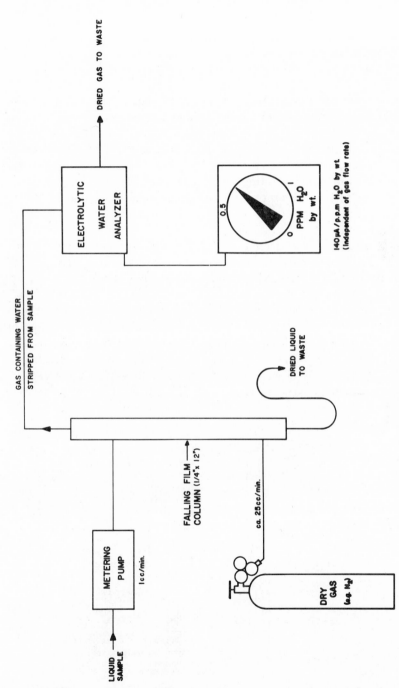

Fig. 105.17. Schematic diagram of analyzer for water in liquids.

6345

of oxidizable sulfur compounds in gases. The Titrilog is capable of measuring hydrogen sulfide, sulfur dioxide, mercaptans, thiophene, and organic sulfides and disulfides in gases with a maximum sensitivity of 0.1 ppm (32,143).

The measurement of the sulfur compounds is accomplished by titration with bromine (Fig. 105.18). The bromine is generated electrolytically from acidic potassium bromide solution in which the sulfur compounds are absorbed from the gas streams. The bromine concentration is sensed by electrodes whose potential is measured by a feedback amplifier that controls the bromine-generating current. The rate of generation is equivalent to the rate of absorption of the sulfur compound, and hence the current is proportional to the concentration of sulfur compounds in the sample stream.

The titration is not selective among the various types of sulfur compounds. It can be made selective (*1*) for sulfides plus mercaptans, by removing H_2S from the sample, or (*2*) for sulfides, by removing mercaptans plus H_2S.

Continuous coulometric analysis has also been used for water, in a modified Karl Fischer method (16), and for chloride ion in gases by automatic titration with silver ion generated in the titration cell (58). As silver ion is consumed by chloride, the change in concentration is sensed by a pair of electrodes and amplified to adjust a servosystem which controls the generating current. This current is recorded as a

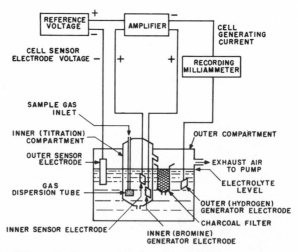

Fig. 105.18. Schematic diagram of coulometric analyzer for sulfur compounds.

measure of chloride ion; as little as 0.1 μg can be detected. The analysis has been applied to the detection of chlorinated pesticides separated from mixtures by gas chromatography (59).

B. RADIANT ENERGY METHODS

Analyzers which employ radiant energy in measurement are among the most common types. Many sensitive and specific analyses are achieved without mechanical contact between the transducer and the sample. Energy in many parts of the electromagnetic spectrum can be used—infrared, visible, ultraviolet, alpha, beta, X, and gamma radiation. Ultraviolet analyzers were first introduced in 1930 (62).

In radiant energy measurements, the energy absorbed is the quantity most frequently measured. Emission, refraction, and optical polarization effects are also employed. Selectivity in analyzing mixtures of compounds of related structures is achieved by employing monochromatic energy or by selective absorption in the sample and detector. By choosing the region of the spectrum, it is possible to analyze for broad types of structures, for example, infrared for organic functional groups, ultraviolet for many unsaturated groups, and X-rays for metals. Gamma rays are applied to mass properties such as density.

Many reviews of radiant energy analyzers have been written from many viewpoints (69,131,187,188,190,191,192,217,254).

1. Infrared Absorption

The most popular and successful methods are those employing infrared absorption. Infrared spectrophotometers are generally automatic in their scanning operations in the laboratory but are too fragile for many process-stream applications. There are reports, nevertheless, that they were used for some critical stream-monitoring applications in synthetic rubber plants during World War II. Repetitive multicomponent analysis by means of programmed wavelength selection was discussed in 1949 (117,265).

Renzetti and Rogers (206) adapted a double-beam spectrophotometer for measurement of hydrocarbons in air at 3.45 μ to fractional part-per-million concentrations on a continuous-flow basis (206). A 10-meter, multiple-reflection sample cell was used with the sample compressed to 150 psi, and the absorbance sensitivity of the instrument was increased fivefold by modification of the optical wedge of the null-balance system. Cells of extremely long path are available for spectrophotometers for air pollution research but not for analyzers (237). Positive identification of air pollutants is possible because of the ability to trap a sample in the cell and to scan its spectrum whenever desired.

For applications of monochromatic infrared absorption in operating plants, where real ruggedness is essential, small, totally enclosed spectrometers have been designed. These include the single-beam (Pfund) monochromator of Herscher and Wright (110) and the double-beam instrument of Savitzky et al. (216). Plant experience with one such spectrometer has been reported in analysis for the ortho isomer in the mixed ethyltoluenes (110). The scanning ability of laboratory instruments is not provided by these analyzers.

Dispersive infrared instruments are to be preferred to nondispersive instruments for special cases, for operation with liquid samples or with very long optical paths, or at wavelengths longer than about 10 μ. Problems with selectivity and energy in such cases are more easily handled by a monochromator. With highly corrosive samples, the external location of the sample cell in laboratory spectrophotometers offers a practical advantage—a leak in the cell will not be likely to destroy the instrument before it is discovered.

The best-known radiant energy method for selective chemical analysis of gases in plants is nondispersing infrared absorption. There are well over 1000 such instruments in use in the United States, more than half of them in the chemical industry (69). Their ruggedness and reliability are further attested to by increasing reports of their acceptance as integral components of automatic process-control loops. Of the many reviews of nondispersing infrared analyzers, one critique from the standpoint of the plant customer is that of Wall (254).

The nondispersive instrument is basically a filter photometer, or "colorimeter" in the infrared region. Although the filter generally employs the vapors of the compound of interest, or a material of similar absorption characteristics, interference filters are becoming of interest. There are both negative- and positive-filter types, and both appear to stem from the experiments of Pfund, carried out about 1939 (200). He used a stream of hot gas as a selective source of radiation, in one case, and also a gas-filled thermal detector whose absorption properties made it selective in response. The theoretical aspects have been discussed by Koppius (141).

In the negative-filter type, which is largely of historical interest, a high concentration of vapor of the component of interest is placed in a sealed cell located in half of a split infrared beam from a single source (Fig. 105.19). A detector such as a differential thermopile (193)

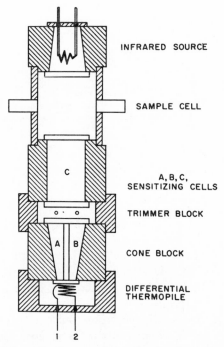

INFRARED SOURCE

SAMPLE CELL

A, B, C,
SENSITIZING CELLS

TRIMMER BLOCK

CONE BLOCK

DIFFERENTIAL
THERMOPILE

Fig. 105.19. Schematic diagram of negative-filtering infrared analyzer.

or bolometer will then respond to changes in absorption caused by variations in concentration of this compound in a sample cell which is placed
in both beams. Interferences from compounds in the sample with absorption bands overlapping those of the analyte can often be minimized
by a suitable filter cell containing the interfering component and placed
in both beams (254). Although reasonably rugged, negative-filter instruments are not particularly sensitive and have been largely superseded
by positive-filtering analyzers.

The positive-filter type of infrared analyzer uses double beams, a
sample cell in one beam, and a selective detector which responds only
to the absorption bands possessed by the component of interest. This
is achieved by placing vapors of such a compound in a pneumatic condenser microphone, as originally proposed by Luft in 1943 (156). The
electrical response of the detector then changes with concentration in
the sample cell (Fig. 105.20).

Alternating-current operation by use of a beam chopper leads to rela-

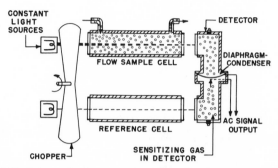

Fig. 105.20. Schematic diagram of positive-filtering infrared analyzer.

tive independence of the signal to variations in ambient temperature and source intensity. Apparent source fluctuations arising from changes in concentration of noninterfering absorbers in the sample are also minimal in effect. Additional stability has been provided in some cases by null-balance operation with servo control of an optical wedge or of source voltage in one beam (169).

The response of the positive detector only to energy in the bands possessed by the filter material contributes to good signal/noise ratios, and many applications to analyses at parts-per-million concentrations have been reported since 1951 (178,200,257). Although this is particularly true for strongly absorbing analytes such as carbon monoxide, carbon dioxide, and acetylene, it has also been achieved with aliphatic hydrocarbons in air pollution studies (149). Signal/noise ratio can sometimes be further improved by restricting the wavelengths detected through the use of filters or choppers which are selectively absorbing. Littman and Denton (149) used a polyethylene chopper to achieve selectivity for aliphatic hydrocarbon vapors in the atmosphere at concentrations below 1 ppm (149).

Samples for infrared analyzers are usually gases. It has been possible, however, to monitor water or low-boiling alcohols dissolved in hydrocarbons after first stripping the liquid sample with an inert gas to give a vapor phase enriched with the water or alcohol (159).

An interesting compensation technique is employed in the differential positive detector of Liston et al. (146). In this case, interferences are greatly minimized by sensitizing a second detector only to the interfering material and adjusting for equal response of each detector for the interference. Now, taking the ratio of outputs of the two detectors cancels out the interference. Examples include (1) eliminating the effect of water vapor upon methanol vapor analysis, and (2) making gas analyses substantially independent of sample pressure and temperature.

The latest avenue of promising development for nondispersive infrared absorption leads in the direction of filter photometry (232) using interference filters. These filters make possible high selectivity in compact optical systems approaching the simplicity of colorimeters. Characteristics of such filters in quantity production include bandwidths as low as 0.5% of the peak wavelength, measured at 50% of the peak transmission, and transmission up to 70% of that of the uncoated substrate. Such filters are available from 1.0 to 20 μ.

Applications of filter photometers include measurement of hydrocarbons in exhaust gases at 3.43 μ (185) and of water vapor in the upper atmosphere (183). The latter application employs a compact instrument noteworthy for its use of lead sulfide photoconductive detectors and of a vibrating chopper, rather than a rotating vane (Fig. 105.21). Lead sulfide detectors give high signal/noise ratio with high-frequency response (230). Even higher sensitivity can be achieved by accepting a long time constant (hence, narrower bandwidth in the amplifier and lower noise level), which corresponds to typical response lags on the sampling of flowing streams. The long wavelength of the lead sulfide cutoff at about 3.5 μ tends to limit applications at present. Newer photoconductive detectors, made of such materials as lead selenide, lead telluride, and indium antimonide, promise progressively longer cutoff out to $ca.$ 8 μ (230).

Filter photometry is the basis of a "universal" analyzer (232). Interference filters are used to cover any desired part of the broad spectral range from 0.24 to 2.7 μ, that is, from the ultraviolet through the visible and near-infrared into part of the infrared.

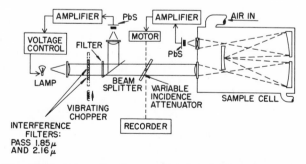

Fig. 105.21. Schematic diagram of infrared filter photometer.

2. Ultraviolet and Visible Absorption

The instrumentation in the ultraviolet and visible regions of the spectrum is much more sensitive than that normally employed in the infrared, because much more source energy is usually available, but it appears to be less popular for selective analysis (89). The simplicity and reliability achieved in ultraviolet–visible instruments are reflected in the large numbers of detectors for smoke, turbidity, and "off-color" in use in industry (89). Selective chemical analysis, such as is required for meaningful air pollution studies, depends in many cases on a chemical reaction as a first step, followed by color measurement. This contrasts with the direct selective measurements obtained in many infrared analyzers.

Ultraviolet analyzers were patented as early as 1930 (62). Dispersing instruments have been rarely applied in process-stream work, although laboratory spectrophotometers are reported to have been used for continuous monitoring of styrene in butadiene streams during World War II. Recording spectrophotometers became available shortly thereafter (89). Renzetti (205) describes a repetitive-scanning spectroradiometer for monitoring ozone at parts-per-million concentrations in air, based on earlier work by Stair (235). Mercury lines in the 265–354 mμ region are scanned using a 300-ft outdoor path, with automatic attenuation of the intense 313-mμ line. Ozone concentration is calculated from the chart record of the absorbance at 265 mμ after correction for background absorbance at 313 mμ.

Ozone has been measured in air by comparing the absorption of the sample at 254 mμ with that of a treated sample from which ozone has been removed by catalytic decomposition (88).

In place of monochromators, it is much more common in continuous analysis to obtain selectivity by use of line sources, together with narrow-bandpass filters to isolate individual spectral lines. An early instrument used mercury lines at 254, 313, or 365 mμ to detect contaminants such as mercury vapor, aromatic hydrocarbons, SO_2, CS_2, chlorine, and metal carbonyls in the air (100). This employed simple double-beam filter optics with a sample cell in one beam, the other beam serving as a fixed reference, and phototube detectors. Troy (251) has described an improved portable version of this analyzer and listed its sensitivity to various toxic gases and vapors when used with and without certain filters to select wavelength (Table 105.IV).

One photometric analyzer for stream monitoring possesses the stabilizing features of a null-balance optical system and automatic zeroing (Fig.

TABLE 105.IV
Minimum Detectable Concentrations of Gases and Vapors
Using Portable Ultraviolet Photometer (251)

Substance	Maximum allowable concentration, 8-hr exposure, ppm by vol.	Minimum detectable concentration, ppm		
		No filter[a]	Filter[b] A	Filter[c] B	Filter B with external meter
Acetone	500	12	310	70	23
Benzene	35	0.8	d	9	3
Bromine	1	5	0.6	9	3
Carbon disulfide	20	45	54	24	7.5
Chlorine	1	25	7	7	2.3
Chlorobenzene	75	0.8	d	10	3
Cyclohexanone	100	25	210	70	23
Mercury	0.01	0.00003	d	0.0003	0.0001
Nickel carbonyl	1	0.015	0.7	0.15	0.045
Nitrobenzene	1	0.015	d	0.18	0.06
Nitrogen dioxide	25	2	0.3	1.7	0.55
Ozone	0.1	0.025	d	0.3	0.09
Phosgene	1	5.5	d	60	20
Sulfur dioxide	10	2.2	26	8	2.6
Tetrachloroethylene	100	1.0	d	12	4
Toluene	100	0.7	d	8	2.5

[a] No filter: mercury source has 81% of its energy at 254 mμ.
[b] Filter A: passes largely 334, 365, 405, 436, and 546 mμ.
[c] Filter B: passes some 254 mμ but largely 287, 313, and 334 mμ.
[d] Not detectable at concentration levels found in atmosphere.

105.22). Light sources such as tungsten can be used (90). Sensitivity with a minimum range of about 0.005 absorbance unit full scale can be obtained. This is 40 times better than the value achievable with many laboratory instruments. Applications of this analyzer have been made to measurements of aromatic compounds, ketones, chlorine, phosgene, SO$_2$, NO$_2$, and similar absorbers in the ultraviolet. In the visible region of the spectrum, analysis can be made for very low levels of color and turbidities in liquids to less than 1 ppm. Sample cells as thin as a few thousandths of an inch have been used.

Some photometric analyzers are single-beam instruments and obtain base-line stability by careful regulation of power for the source and electronics. Phototubes and photomultiplier tubes are commonly used

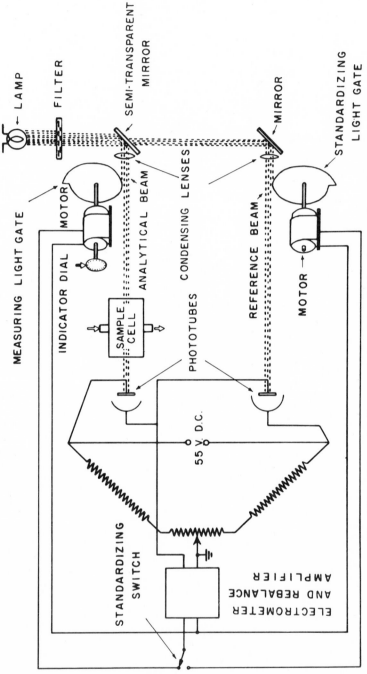

Fig. 105.22. Schematic diagram of ultraviolet–visible photometric analyzer.

6354

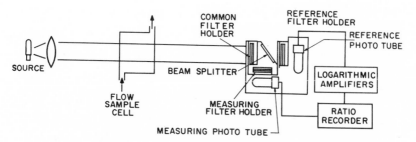

Fig. 105.23. Schematic diagram of ratio photometric analyzer, (90a).

as detectors in general-purpose instruments, while photocells have served adequately in some special analyzers (206).

The problem of drifts arising from sample cell clouding, variable haze in the sample, and like effects can be met by periodic zeroing. Another solution for window fouling is to use a ratio detector and two sample cells, one in each beam but one unusually thin, as compensation (89).

It is not always easy to eliminate interferences from overlapping absorption bands in the ultraviolet–visible photometric analyzers. The bands are usually quite broad, with only a narrow portion generally being used for the measurement. Consequently, negative-filtering techniques (such as those used in nondispersing infrared work) cannot be applied. The solution, in general, is to use a ratio technique and to compare the absorptions at two different wavelengths. This may be done in double-beam instruments with the aid of a sample cell and appropriate filters in each beam (89). Another approach is the use of two single-beam analyzers, each set at a different wavelength, recording the ratio of their outputs (39). A recent double-beam ratio analyzer uses simple logarithmic amplifying circuits in a difference mode so that output is linear in absorbance and hence proportional to sample concentration (Fig. 105.23).

When a component of interest shows little absorption or nonselective absorption in the ultraviolet–visible region, it is often possible to convert it by chemical reaction to a colored or fluorescent substance. This then can be measured with high sensitivity by colorimetric techniques (Section III.B.1.6).

3. Methods Using High-Energy Radiation

Uses of alpha, beta, and gamma radiation in measurement depend on the highly penetrating nature of these rays. Systems which utilize penetrating radiation do not have to be in intimate physical contact with the specimen or sample stream in order to obtain good measurements; this is an advantage from such standpoints as cost of construction, necessity for confining pressure or a corrosive material, or protecting a sensitive material from atmospheric contamination.

There are many applications of high-energy radiation to density and thickness gaging, based on the close correlation between the stopping power of a material and its mass. Applications to specific analyses, however, are relatively few (231). Continuous measurements employing alpha radiation have been rare because of its weak penetrating power; an application to gas analysis has been reported (231).

a. BETA-RADIATION

Beta radiation is moderately penetrating and is therefore used for thickness measurements on thin metal sheets, paper, rubber, and plastics as well as certain analyses of liquids.

A "hydrogen analyzer" (99) utilizes the absorption of beta particles by liquid samples. Density can be measured to ± 0.0002 g/ml; or, if the density of the sample is known independently, chemically bonded hydrogen in organic liquids can be recorded to the nearest 0.05%. The temperature of the sample is closely controlled in order to hold density variations from this source to a minimum. A ^{90}Sr source is used.

A technique related to beta-ray absorption utilizes the determination of beta-ray back-scattering intensity for analysis of the carbon/hydrogen ratio in hydrocarbons (173). When beta particles strike matter, some are "reflected" back in the direction of the source and can be measured. The intensity of back-scattered electrons depends on the atomic number of the sample and is subject only to a second-order effect from density or sample temperature. A laboratory method has been described by Gray et al. (95), who point out the ease with which this technique can be extended to continuous monitoring.

The beta-ray back-scattering analyzer requires a source, a detector, and radiation-measuring devices, in addition to the sample cell. The source (^{90}Sr and ^{90}Y) is placed between the sample and the ionization chamber detector, mounted, and shielded so that only scattered radiation can reach the chamber. The sample cell has a thin (0.001-in.) mica window. The ionization chamber has a thin stainless steel window and is filled with argon. The current developed in the ionization chamber is amplified and recorded as a measure of back-scattering intensity. The readings are relative, so that the analyzer must be calibrated with standard samples (95).

The carbon/hydrogen ratio has been measured by the beta-ray back-scattering technique for hydrocarbons with ratios in the range 5.2–15.7, with accuracies of better than 0.5% relative. Mixtures of hydrocarbons, such as p-xylene–isooctane, have been analyzed.

b. Gamma-Radiation

From well-known absorption relationships, it is deduced that the density of material confined in a fixed thickness can readily be measured with penetrating radiation. With gamma rays, this can be done directly on the process pipeline; in many cases, no special sample cell is needed (231). Furthermore, if the material is a binary mixture or behaves like one, the density will be a measure of its concentration.

Density gages based on gamma rays find application in measurements such as (1) densities of oil products to control blending operations; (2) density of flotation media used to classify coal for various uses; (3) percentage conversion of styrene–butadiene rubber latex (42); (4) concentration of sulfuric acid; and (5) concentrations of slurries, as of catalyst in oil.

Gamma sources include radium, ^{60}Co, and ^{137}Cs. Detectors are ionization chambers, generally high voltage.

c. X-Radiation

(1) X-Ray Absorption

Like alpha, beta, and gamma radiation, X-rays penetrate matter in inverse relation to the density of the medium. Under proper conditions, however, X-ray absorption analysis can be used to measure atomic concentrations in a variety of substrates. Selectivity can be achieved only by using a monochromator or a monochromatic source, such as a radioactive isotope which emits X-rays by K-capture. Such selectivity takes advantage of the absorption edge effect, that is, the phenomenon that the wavelength of the X-ray emitted by an element is strongly absorbed by the elements slightly below it in atomic number but is only moderately absorbed by itself and the elements above it. A specific analyzer for vanadium can be constructed using two sources, ^{54}Mn and ^{55}Fe. Vanadium has minimum absorption for the X-rays from ^{54}Mn, whose absorption edge is at a higher atomic number, but has maximum absorption for the X-rays from ^{55}Fe, whose absorption edge is just below that of vanadium. The difference in absorption of the two wavelengths by the same sample can be related to the vanadium concentration. An analyzer has also been built for lead in air (12). All the elements present should be known, since a strong absorber of either wavelength will interfere.

Beerbower (18) has modified a laboratory X-ray instrument for process use, for the determination of 0.05–2.7% sulfur in hydrocarbons. He describes a cell construction which combines the ion source, sample space, and Geiger tube in close proximity. The cell is formed from sheets of Mylar polyester film and Lucite acrylic resin. Other pertinent sample considerations include adequate flow rate, to keep the sample-cell time constant less than the ratemeter constant, and control of pressure and temperature, to maintain constant density at constant concentration.

Other reported applications include measurement of the concentration of a calcium soap in a lubricating oil and monitoring of ammonium chloride in the wash water from a scrubber (18).

(2) X-Ray Fluorescence (X-Ray Emission Spectrography)

X-ray fluorescence (or X-ray emission spectrography) may be applied to the determination of elements, whether in compounds or free, above atomic number 19 (83). It is thus most useful for metal analysis and is of great practical interest because of its speed and ease of sampling; in many cases this technique can replace the tedious wet chemical analysis of metals and alloys. As the name implies, secondary X-rays are excited when the sample is irradiated by a source of X-rays. Since the wavelengths of the secondary X-rays are characteristic of the element undergoing excitation, isolation of a particular wavelength makes possible both qualitative and quantitative analysis. A diffraction crystal monchromator is used to select wavelength, and a Geiger counter to detect and measure the radiation of interest.

X-ray fluorescence stream analyzers are available commercially (10). One model consists of a spectrometer, a high-voltage supply for the X-ray tube, and an electronics console containing ratio and calibration circuits. The spectrometer contains two curved crystal monochromators, one for analysis and the other for a compensating reference signal.

X-ray fluorescence analyzers are being used to measure such variables as (1) plating weights of tin and zinc on steel plate, (2) various elements in ore concentrates, and (3) copper and bromine concentrations in solutions (22,92). Probable future process applications are indicated by laboratory procedures for stray metal in oil, the control of additives blending with lubricating oils by analyzing for barium, calcium, and zinc, and analysis of bromine and lead in gasoline.

Sample forms suitable for measurement include liquids, slurries, powders, and solids. A smooth surface on a metal sample is desirable (93). No contact with the surface is required, and an exposure as short as 1 or 2 sec to the beam will give a useful signal from a given area of the sample. The analysis obtained is substantially that of the surface, with an effective depth typically between 100 and several thousand microns. Such an analysis is satisfactory for tin plated on steel, and the total fluorescence from the tin is related to the thickness of the coating. When several elements are present at a surface, the fluorescence emitted at the various characteristic wavelengths is proportional to the amounts of the respective elements. Multichannel X-ray fluorescence measurements thus permit the determination of relative concentrations or, if a total intensity measurement is also made, of absolute concentrations of the various elements.

4. Emission Spectroscopy

Inorganic components of many materials are analyzed by automatic methods involving arc or flame excitation of emission spectra.

a. ARC EMISSION

Arc emission is very widely used for multicomponent metal analysis in the steel industry and, increasingly, in light metal industries. Although sampling is usually batchwise, a high degree of automation of the spectrometers makes possible complete analysis with meter readouts in a few minutes per sample. This is accomplished by positioning multiple photoelectric detectors at the outer circumference of the spectrograph to monitor individual predetermined wavelengths emitted by the elements of interest. Automatic correction circuits for background intensity have been provided in such instruments to simplify interpretation of the analytical results (103). Many excellent reviews of such emission methods have appeared (30,101,102,135,182).

A portable automatic emission spectrograph for recording low levels of beryllium oxide in air is commercially available (175). This analyzer samples the air through a paper tape for an interval of 1 min, integrates light emitted at 3131.1 Å after electric spark excitation, and cycles repetitively. Detection utilizes a photomultiplier, with integration of the signal for 30 sec by an integrating amplifier. This analyzer is sensitive to as little as 0.1–0.5 μg of BeO per cubic meter of air. Such a principle of collection and detection is novel and should be applicable to the determination of other metals and their compounds in air.

Fig. 105.24. Flame photometer attachment for colorimetric analyzer.

b. FLAME EMISSION

Flame photometry by its nature requires a flowing sample, and it is not surprising that this technique has been used on a continuous-flow basis. Opler and Miller (181) used a recording flame photometer to monitor the sodium ion concentration in elution from an ion-exchange column.

An automatic flame photometer which can be applied to continuous monitoring of many types of samples has been described by Isreeli et al. (120) as an attachment to a chemical analyzer system. The sample is precisely diluted with a solution of an internal standard and delivered to the burner at a fixed flow rate by a proportioning pump. A dialysis step is included to provide further dilution and separation where desired, as in the case of biological fluids. A burner with an integrating chimney lined with magnesium oxide serves to illuminate the cadmium sulfide photodetectors (Fig. 105.24). The analytical detector views the emission from an internal standard of lithium nitrate. Measurement of the analytical/reference ratio is recorded. The sensitivity is 0.8 meq/liter for sodium and 0.1 meq/liter for potassium.

Another continuous flame photometer developed for measurements on boiler feed water is reported to have a sensitivity of 0.1 ppb of sodium (60,158). It is also applicable to boron.

It is well known that the optical character of a flame can be affected by organic components in the feed. The emissivity detector for gas chromatography takes advantage of this phenomenon, using a photocell to measure emission (94). With coal gas as the combustible agent, sensitivity comparable to that of thermal conductivity detectors is obtainable. Emissivity varies with the nature of the compound, and therefore qualitative applications are possible. Comparisons of relative response have been reported for aromatic, paraffinic, and naphthenic hydrocarbons in thermal conductivity and in emissivity detectors (94).

5. Light Scattering

The ability of suspended particles to scatter light makes possible the detection of particles in gases with high sensitivity. Light scattered in the forward direction has been used as the basis of a continuous photometer for aerosols developed by Sinclair (71). A dark-field optical system makes particles in the sample stream appear brightly lit and register on a photomultiplier, whose output is amplified and read out (201). Very low particle concentrations can be detected in this way. Actual particle counts cannot be made, however, because of variations in particle shapes and sizes which affect the amount of light scattered by each particle.

6. Raman Spectrometry

No reports of application of Raman spectrometry to continuous analysis have been noted, but such use seems to be only a matter of time, now that laboratory instruments are commercially available. The need for analytical applications to aqueous systems and to sulfur compounds, for example, which are difficult or impossible for infrared analysis, may lend impetus to such a development. Problems are apparent in the design of flow cells which will conserve scattered radiation, in quantitative interpretations, and in the complexity and fragility of present apparatus.

7. Polarimetry

An automatic recording polarimeter has been described (215). Rotation of the plane of polarization of a plane-polarized light beam by the sample is measured by a photometric system whose output is used to position a disk polarizer to maintain optical balance. The position of the polarizer is recorded as a measure of the rotatory power of the sample. The analyzer has a precision within $\pm 0.5°$ specific rotation in application to a plant stream of molten resin at 180°C. The simplicity of the electro-optical system results in a rugged and trouble-free analyzer.

8. Refractometry

Another method utilizing radiant energy, differential refraction, does not require spectral dispersion or filtering in automatic instruments but does employ photometric detection. The technique is capable of high sensitivity for analyses of binary and pseudobinary mixtures, although it is less selective than many of the spectroscopic methods.

A rugged, automatic refractometer with an optical balance-detection system has been described (91). The temperature effect, to which refractive index is so sensitive, is minimized by surrounding a sealed cell containing a reference material with a coaxial cell through which the liquid sample flows. The windows of the cells are positioned at angles such that the refractions due to sample and reference subtract. Any difference between the two materials greater than an initial zero setting results in a net refraction of the light beam passing through the cell assembly. This refraction is detected by a phototube bridge after the focused beam is split by a fixed dividing target. The bridge is rebalanced by a servo-driven glass plate. The position of the balancing glass plate is transmitted by standard techniques to a suitable indicator or recorder to give a measure of the extent of refraction. A schematic diagram of such a refractometer is shown in Fig. 105.25.

An analyzer of this type can achieve a sensitivity of 0.000004 refractive index unit, equivalent, for example, to 40 ppm of water in glacial acetic acid (91). Very compact, rugged assemblies have been built for industrial use. Similar instruments have been reported to have reliability adequate to allow their use to control distillation columns in petroleum refineries (262). Many practical analyses have been performed routinely by refractometry, including measurements of the concentration of sugar solutions and of solutions of high-molecular-weight organic com-

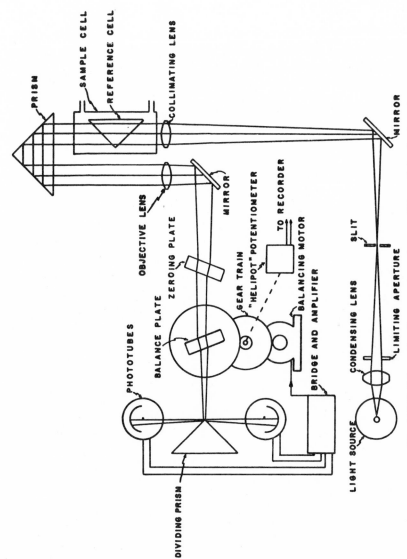

Fig. 105.25. Schematic diagram of continuous differential refractometer.

PRISM

SAMPLE CELL

REFERENCE CELL

COLLIMATING LENS

OBJECTIVE LENS

MIRROR

MIRROR

ZEROING PLATE

BALANCE PLATE

PHOTOTUBES

GEAR TRAIN

"HELIPOT" POTENTIOMETER

TO RECORDER

BALANCING MOTOR

BRIDGE AND AMPLIFIER

DIVIDING PRISM

LIGHT SOURCE

CONDENSING LENS

SLIT

LIMITING APERTURE

pounds, as well as the composition of mixed Freon fluorinated hydro-carbons for aerosol propellants (91). Application of differential refrac-tometry has recently been made to liquid chromatogram effluents, with the aid of very small sample cells (19).

In many cases, the reading of a refractometer is completely insensitive to stray colors, tar, and dirt in the sample stream. This gives it a distinc-tion among continuous analyzers in that it is nearly free of drift in zero point and in sensitivity. Only infrequent checking of the initial calibration is needed.

9. Ionization Methods

Several new, sensitive detectors for continuous gas analysis, all de-pendent on ionization by radiant energy and the properties of ionized gases, have been developed (155). Examples are the cross-section ioniza-tion detector, argon detector, electron capture ionization detector, elec-tron mobility detector, photoionization detector, and aerosol ionization detector. The argon instrument has seen extensive application and devel-opment in gas chromatography.

Most of these detectors operate at atmospheric pressure, where conduc-tion does not occur easily except with the aid of a potent source of charged particles, that is, ionizing energy. Since very small currents flow, these detectors generally require electrometer amplifiers for readout. Such methods usually operate best when detecting a small concentration of one gas in a second or major component, for example, component peaks in the effluent gas stream of gas chromatography. It is thus desir-able in general that the major component be nonconducting and that the minor one give rise to charged particles, under the experimental conditions. If the major component does contribute appreciable back-ground conduction, the minor component should exert a marked influence on this background.

a. Cross-Section Ionization Detector

This detector depends on the absorption of ionizing radiation according to the atomic cross sections of constituent elements (155). It can be used with any carrier gas and for the measurement of all gases and vapors. Its sole disadvantage is a relatively poor sensitivity. Sensitivity is high, however, when the carrier gas or major component is a light gas such as helium or hydrogen.

b. Argon Detector

The argon detection is basically a cross-section ionization instrument with argon as the carrier gas and a source of beta rays (154). It is said to depend for its operation on the excitation of argon to its metastable state, by bombardment with electrons from the source, and the ionization of introduced organic molecules by energy transferred to them from the metastable argon atoms. When the organic material enters the ionization chamber of the detector, there ensues an increase in current flow, following the equation:

$$I = \frac{CA(x + y) + Bx}{CA/1 - ae^{b(V-1)}} + B$$

where C = concentration of organic material; A, B, a, b = constants; V = applied potential; x = primary electron concentration; and y = initial metastable atom concentration.

Although this suggests the possibility of approaching infinite current at some finite value of C, current-limiting techniques can be applied. Either of two types of output–concentration relationships can be obtained through proper design: (1) with internal space charge, a linear relationship over a dynamic range of 10^5 or more, or (2) with a constant energizing current, a log–linear relation of output potential with concentration.

Sensitivities down to parts per billion are possible, and response speeds in the millisecond range have been achieved. Currents obtained are such that electrometer amplifiers are not essential, and high-impedance recorders have been used. The upper limit of useful range is less than 0.1% concentration.

Ionization will be achieved with any test substance, organic or inorganic, whose ionization potential is equal to or less than the stored energy of metastable argon (11.7 eV). The response in general is related to the mass of the test molecule. Calibration tends to be required for molecular weights below 100 and for compounds whose ionization potential approaches 11.7 eV, since the probability of ionization then decreases. Thus the rare gases and other fixed gases, hydrogen halides, and some of the small hydrocarbons are not detected. Water vapor and air interfere with the performance of the argon detector, even though they are not detected. Compounds containing halogen or nitro groups require special care if errors due to recombination effects are to be avoided. A variety of sources, notably radium, ^{90}Sr, and tritium, have been employed.

c. ELECTRON CAPTURE IONIZATION DETECTOR

Very sensitive analyses of binary mixtures having a halogen or oxygen compound as one component are possible with this detector, which takes advantage of ion recombination phenomena (155). An ion chamber containing a free-electron gas is maintained at a potential just sufficient for the collection of all the free electrons produced. The introduction of a substance which captures electrons causes a decrease in current flow related to the concentration of the substance. Selective sensitivity to classes of organic compounds can be achieved by choice of carrier gas and applied potential. Use of this type of detector for identifying functional groups in gas chromatographic effluents has been suggested.

d. Electron Mobility Detector

The electron mobility detector is a sensitive device for detecting the permanent gases, water vapor, CO, and CO_2 (155). It depends on a reduction in the excitation of the carrier gas by substances which absorb some of the energy of the electrons. It is necessary to achieve heavy background ionization by using a high potential and a carrier gas (argon or helium) which is known to be or is deliberately contaminated. Then the reduction in current flow in the presence of the test substance can be observed.

e. Photoionization Detector

Ultraviolet light from a glow discharge can be used as ionizing radiation for a detector (153). The glow discharge can be maintained in helium, argon, nitrogen, or hydrogen, preferably at a pressure below 100 mm. The test gas is led into the same chamber with the discharge, and ionization is detected by conductivity between a pair of electrodes. The detector does not respond to a permanent gases or water vapor, but most other substances can be detected (153).

f. Aerosol Ionization Detector

The aerosol ionization detector is an ionization gage which is selective for particles so readily ionized that they greatly reduce the ion current. As indicated earlier, an appropriate reagent or treatment is used to form an aerosol from the component of interest, for example, NH_4Cl from NH_3 and HCl (169,241). The ion chamber is of straightforward design and includes a weak source of alpha particles, such as radium foil. Since only the component of interest is detected, in the ideal case, sensitivity is very high—as much as 1 ppb referred to the original sample. Only solid particles are ionized and detected; hence dusts and other aerosols are potential interferences which must be removed from the sample stream before the analytical reagent is added.

C. THERMAL METHODS

1. Thermal Conductivity

The historical basis for many early continuous gas analyzers, thermal conductivity continues to serve a wide group of analytical needs both in continuous plant analyzers and, as the detector, in most applications of gas chromatography. Thermal conductivity is elegantly simple in its essentials, requiring only low-voltage direct current for energization and Wheatstone bridge circuits for output to standard recording equipment (64). The cells are rugged, having undergone evolution since the first practical use (monitoring hydrogen in air in dirigibles in World War I) and during industrial application since 1920. Thermal conductivity analyzers are reliable and relatively low in cost, properties which provide incentive to consider them for many applications in which extreme selectivity is not required.

Absolute measurements of thermal conductivity are very difficult; therefore process applications always use differential techniques, generally with standard gases as reference materials. Since thermal conductivity is a single-valued function of gas composition and temperature only, and since temperature is always held constant for stability, only binary gas mixtures are readily analyzed for components. Mixtures of greater complexity are usually analyzed after preliminary separations, unless it is only desired to detect *any* change in composition.

Typical apparatus for analysis in the percentage range consists of a dc power supply, a thermal conductivity cell in a suitable thermostat, a recorder or other readout device, and, in special cases, a signal pre-amplifier. When higher sensitivity is needed, greater stability of voltage and temperature must be achieved to prevent noise and drift. More massive cells are required to prevent local temperature gradients. Careful control of gas pressure and flow becomes necessary. When it is desired to measure concentrations in parts-per-million ranges, special constructions of thermal elements and of the circuits are devised to achieve the utmost stability of base-line signal (138).

Typical cells for thermal conductivity work consist of metal blocks drilled for passage of gas streams, with chambers for the thermal elements. For the latter, both heated filaments and thermistors are commonly employed. Filaments are generally of platinum, tungsten, or Kovar, depending on service requirements; platinum has the widest temperature range, with a temperature coefficient of resistance about $+0.4\%/°C$. Thermistors are semiconductors prepared by fusing mixtures of transition metal oxides. They have negative temperature coefficients of resistance, typically $-4\%/°C$. Filament cells are preferred for high-temperature use, above *ca.* 225°C. Thermistors lose sensitivity at such temperatures but offer higher output signals than filaments at lower temperatures, especially below 100°C. Signal/noise ratios, however, are roughly equivalent for filaments and thermistors in well-designed cells.

Thermistors offer some advantage in geometry because of their small size.

Thermal conductivity cells typically house the detecting and reference thermal elements in close proximity in order to minimize thermal gradients between them. A useful cell design for gas chromatography work is shown in Fig. 105.26. The reference element is in a sealed or otherwise constant atmosphere of one component. Since the thermal elements are flow- and pressure-sensitive, cell and system design should minimize variations in these factors. Flow sensitivity can be reduced by using diffusion of the gas stream in its access to the thermal element, but this tends to increase response time. Kieselbach's shielded thermistor is exceptional in response (138).

Signal sensitivity is markedly affected by cell geometry; thus chamber volume should be small. In high-speed gas chromatography analyzers, which use small-diameter columns and small samples, miniature thermal conductivity cells are employed. This type of cell retains good sensitivity in spite of the use of small samples. Sensitivity in GC applications depends on measuring the peak component concentration in a few seconds before it can be diluted by mixing with the lower concentrations which precede and follow it in the stream. If the chamber of the thermal conductivity cell has about the same diameter as the chromatographic column, such mixing will be minimized.

The basic Wheatstone bridge circuit is generally employed to measure the resistance of the thermal element. The bridge is balanced at zero concentration or some other reference value. In automatic analyzers, in order to make measurements of varying concentration in the sample, the resulting unbalance voltage is read directly on a device such as

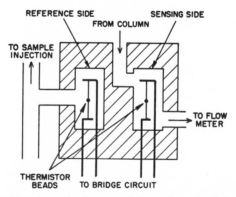

Fig. 105.26. Schematic diagram of thermal conductivity cell.

a recorder. If high sensitivity demands greater freedom of the thermal elements from the effects of variations in bridge voltage and ambient temperature, compensating devices may be included in the bridge, as Littlewood (147) and Kieselbach (138) have done.

Applications of thermal conductivity analyzers are widespread, and the method is most effective when differences in thermal conductivity between the components of a binary gas mixture are greatest. Hence gases with very high thermal conductivity, such as hydrogen and helium, are readily analyzed for in mixtures. Likewise, impurities in hydrogen and helium are easily measured; these gases are therefore popular as carrier gases for gas chromatography, when thermal conductivity detectors are used. Applications of thermal conductivity analyzers include measurement of the hydrogen/nitrogen ratio in ammonia synthesis gas, sulfur dioxide in the gases of the contact process for sulfuric acid, impurities in gases, such as argon separated from liquid air, and hydrogen dissolved in water, after stripping with oxygen to obtain a gaseous mixture.

Mixtures more complex than binary ones can be analyzed by thermal conductivity (1) if only one component varies, in relation to the others; (2) if all the components but the one of interest have the same or similar thermal conductivity; or (3) if one component can be selectively removed to allow a differential analysis of the initial and treated mixtures. Thus ammonia can be measured through its selective removal from hydrogen–nitrogen mixtures, and the same is true of carbon dioxide in flue gases.

Minter and Burdy (170) describe a thermal conductivity cell which uses both convection and conduction chambers to obtain compensation for the thermal conductivity of a third component of the sample. This cell was applied to the determination of ethylene in a mixture of air and carbon dioxide.

2. Thermometry

Analyzers based on thermometry utilize the measurement of temperature in three similar situations.

a. COMBUSTIBILITY

Gas combustibility measurements have long been important both for detecting atmospheric hazards and for providing a guide to the performance of a fuel–air mixture in power generation and other applications of industrial heating. The analysis takes advantage of the heat which is generated when a flowing combustible mixture and an excess of air contact a noble metal catalyst filament such as platinized platinum. It is convenient to measure the temperature rise of the filament by a closely associated thermocouple or by use of the filament itself as a resistor in a Wheatsone bridge circuit. The temperature change of the filament affects its resistance and hence the voltage output of the bridge. A filament exposed to a stagnant sample of a typical mixture provides compensation for other physical properties of the gas, such as thermal conductivity and specific heat (112).

Combustibility analysis is nonspecific among flammable compounds and finds chemical application chiefly in monitoring for combustion hazards of vapors in the atmosphere. Compact, portable forms of such analyzers facilitate leak testing in process areas. These analyzers are usually calibrated in terms of per cent of the lower explosive limit (63,169).

Combustion analysis applies most specifically to oxygen when hydrogen is added to the sample. In the presence of hydrogen, heat is generated only when oxygen is present and is thus proportional to oxygen concentration.

Catalytic combustion cells are also used as detectors for measuring combustible components in process gas chromatography (97,169).

b. HEAT OF REACTION

Several interesting analyzers have been based on the measurement of the heat of reaction generated in a flowing gas stream. Analysis for carbon monoxide in air by such means was established during World War I. Hopcalite is used to catalyze the reaction of CO with oxygen, and the temperature rise across the reactor is measured as an indication of concentration. Thermopiles located at appropriate points in the gas stream provide useful signals from the heat effect. Sensitivity is of the order of 0–500 ppm CO full scale (112).

An analyzer for parts per million of water in gases utilizes the heat liberated in the adsorption of water on a desiccant (86). Two tubes of desiccant are used and are regenerated alternately in a cyclic operation. Analyzers for oxygen and for hydrogen in gases use combusion with added hydrogen or air over a noble metal catalyst to provide a heat effect which is measured in similar fashion (72).

As little as 0.1 ppm ozone in dry air can be measured by the heat of decomposition of ozone over Hopcalite (41). Thermistors, one in the sample stream and one in the treated air, measure the heat evolved in ozone decomposition as a differential signal. This analyzer is designed for atmospheric studies of ozone.

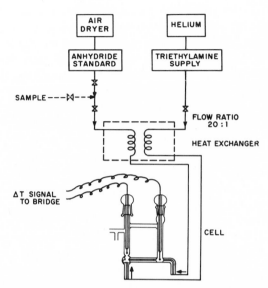

Fig. 105.27. Schematic diagram of thermometric analyzer.

Thermometric analysis has had many laboratory applications to the batch analysis of liquids, but only two examples are known of continuous analyzers for liquids. One analyzer resulted from automation of a laboratory method which was originally developed to measure acetic acid in acetic anhydride (164). In this analyzer, a small proportion of triethylamine is added to the acetic anhydride sample stream in order to react exothermally with any acetic acid which is present (Fig. 105.27). The resulting temperature rise is indicated by the difference in resistance between an analytical thermistor in the mixture and a reference thermistor placed in the incoming sample. Careful design of the glass measuring cell and control of the liquid flow allow measurement of 0.02%) acetic acid in acetic anhydride.

The principle of automatic analysis here should be capable of extension to many other reactions, both acid–base and reduction–oxidation. A similar analyzer takes advantage of the heat of reaction of added benzene to measure the strength of nitration acid continuously (51).

c. HEAT OF VAPORIZATION

Relative measurements of the vapor pressure of volatile liquids can be made with thermistors. Sternberg's thermistor detector senses the heat effect in a thin film of evaporating liquid (238). With a moving liquid stream, an equilibrium temperature is reached which will change only if the composition of the liquid changes. A differential technique is used to provide a reference for the analytical measurement. A reference thermistor contacts a pure solvent, while the analytical thermistor contacts the mixture or solution of interest. The outputs of the two thermistors are balanced in a bridge circuit. The sensitivity of the detector is *ca.* 10^{-8} mole of solute in the eluent from a liquid–liquid chromatographic column.

The same heat of vaporization effect has been applied by Ehrmantraut (166) and by Wilson et al. (263), in laboratory devices for measuring molecular weights of nonvolatile solutes at known concentrations in volatile solvents. These applications suggest the possibility of continuous analysis of polymer solutions.

3. Flame Ionization Detector

The flame ionization detector is an analyzer for binary mixtures which depends on ionization in a flame, such as hydrogen burning in air. The sample gas is passed through the flame, and electrodes for ion measurement are placed in or near the flame. Although the mechanism of ion formation is obscure, sensitive and stable detectors are easily made (167). The flame ionization detector shows a wide range of concentrations over which response is linear; it is useful from about 1 ppb up to about 1% of a given component in the sample gas. It is a selective detector in that air and water vapor in the sample gas have negligible effects. Disadvantages lie in the effects of geometry, flow rate, and gas composition on the *relative* responses of different components. This requires calibration for each component.

The flame ionization detector responds to all organic compounds except formic acid. The response is greatest with hydrocarbons and diminishes with increasing substitution of elements such as oxygen, hydrogen and halogens. It does not respond to CO, CO_2, CS_2, or to elements or inorganic compounds outside groups I and II of the periodic table.

A useful design of flame ionization detector is that of Desty et al., shown in Fig. 105.28 (66,155). Important features of this detector include the heavy metal jet, to minimize the temperature of the jet and hence its thermionic emission, and the porous metal diffuser, to provide laminar flow of air within the chamber. Dust particles increase noise level and should be removed by suitable filters from the gases supplied.

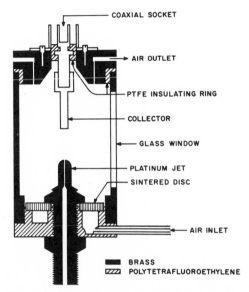

Fig. 105.28. Schematic diagram of flame ionization detector.

The flame ionization detector provides the basis for highly sensitive monitors for hydrocarbon pollution of the atmosphere. Since the flame does not register the normal components of the atmosphere, it may be used for detecting concentrations of hydrocarbons in air of 1 ppm or less. Apparatus for such analysis of air has been described by Andreatch and Feinland (9).

Flame ionization detectors have achieved great popularity in laboratory gas chromatography, particularly for measuring the small amounts of material eluted from capillary and small-diameter packed columns. These detectors are so sensitive that, in gas chromatography, the volatility of liquid phases on the column may give rise to appreciable background signals. Acceptance of this type of detector in process-stream analyzers is likely, in spite of the potential hazards of hydrogen and the flame.

Properties of flames other than ion formation change with the composition of the gas mixture burned. Successful detectors have been based on both optical emission (Section IV.B.4.b) and thermal properties. The "hydrogen flame detector" (221) uses thermocouples to measure the temperature of the flame as a function of gas composition.

4. Freezing Point

Although freezing point is widely used as a criterion of purity of organic compounds, only one apparatus has been described for completely automatic measurement of this property (110a). In this analyzer, a liquid sample is drawn from a plant stream on a repetitive cycle and is fed to a jacketed cell in which it is agitated and cooled at about 1°/min. A resistance thermometer is used for the temperature measurement, and the cooling curve is recorded. As the temperature approaches the crystallizing point, a cycle timer is started. This timer allows time for the freezing to be well established and then controls the remaining operations in the cycle: the melting and purging of the sample, cooling of the cell, and admission of a fresh sample. A complete cycle takes 30 min. The analyzer records over the range 0–10°C with a precision of about 0.1°C.

An interesting approach is represented by the recording cryoscopic apparatus of Simons (228), which lacks mechanization of sampling. The temperature of the sample solution is sensed by a thermistor, and super-cooling is controlled to minimize interference from this source. If the temperature of the sample drops 0.4°C below the freezing point known for the pure solvent, nitrogen gas precooled by liquid nitrogen is admitted briefly by a solenoid valve to cool the outer wall of the freezing cell and induce crystallization.

5. Dew-Point Temperature

The moisture content of a gas can be determined by measuring its dew-point or saturation temperature. Apparatus is available to make this measurement continuously on a flowing gas stream down to a dew point of −68°C, corresponding to 0.1 ppm (87). The dew point is measured by an optical system which observes the condensation of moisture on a cooled mirror to whose surface a thermocouple is attached. The mirror is cooled constantly by a refrigeration system which can maintain the mirror at −68°C. If samples with higher dew points are encountered, the mirror is warmed by a heater which is controlled by an amplifier–thyratron combination operating from the photocell signal. The mirror is then held at the dew-point temperature by the balance between cooling and the heat input called for by the control system.

The dew-point recorder can be applied to gases having dew points in the range from ambient temperature to −68°C. Moisture can be measured in such gases as air, oxygen, hydrogen, nitrogen, and furnace atmospheres. For moisture contents below 0.1%, the electrolytic moisture analyzer (132) will perform such analyses for considerably less initial cost.

6. Boiling Point

Initial boiling point and end point in distillation are empirical temperature criteria of considerable importance to the quality of certain petroleum products such as gasoline. Both can be determined rapidly and continuously by simulated distillation apparatus (99,243,266). Good control of liquid level and temperature are essential to reproducible operation.

D. MAGNETIC METHODS

Methods which depend for selectivity on the presence of a magnetic field include the measurement of paramagnetic materials, nuclear magnetic resonance, and electron paramagnetic resonance. (Mass spectrometry has been considered as an electrical method; see Section IV.A.6.)

1. Paramagnetic Oxygen Determination

The paramagnetic properties of oxygen have been widely applied in continuous analysis since Pauling first devised his highly selective oxygen meter (194). In this analyzer, the gas sample flows through a chamber to which a permanent magnetic field is applied. Any oxygen molecules present are selectively deflected into the field and cause the rotation of a quartz dumbbell which is suspended in the chamber. The position of the dumbbell is indicated on a scale by a lamp-and-mirror arrangement. Small portable analyzers of this type have a sensitivity of about 300 ppm oxygen.

Greater stability and sensitivity (10 ppm) are provided in recording versions of the paramagnetic oxygen analyzer (17). In one type, the dumbbell is kept in a null position by the action of an electrostatic field which balances the deflection torque (Fig. 105.29). The current necessary to generate the electrostatic field at balance is recorded as proportional to oxygen concentration.

Some related oxygen recorders use the "magnetic wind" or thermal–magnetic principle. In such analyzers, an electrically heated wire is located in a magnetic field in a cell through which the sample gas passes (104,213). Deflection of oxygen molecules in the field sets up convection which cools the hot wire. Measurement of the resistance of the filament relative to a reference filament outside the magnetic

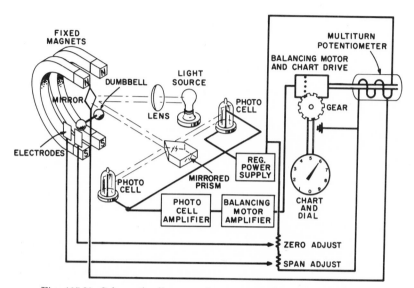

Fig. 105.29. Schematic diagram of paramagnetic oxygen analyzer.

field then gives an indication of oxygen concentration. Such analyzers with hot filaments are more rugged but usually less sensitive than those with quartz dumbbell detectors.

Another oxygen analyzer utilizing paramagnetism works on a "magnetic pressure" principle which gives complete freedom from thermal conductivity effects (258). The sample passes through a ring about which a permanent magnet is rotated. As the magnet rotates, an alternating pressure due to oxygen is produced in the ring and generates an ac voltage proportional to oxygen concentration in a condenser microphone detector which is connected pneumatically to the ring.

The magnetic methods are specific for oxygen because it is the only common strongly paramagnetic gas. Oxides of nitrogen and of chlorine (NO, NO_2, ClO_2, ClO_3) are weakly paramagnetic and interfere only if present in high concentration in the sample. All other gases are weakly diamagnetic and do not interfere. In analyzers which depend on the thermal–magnetic principle, interference may be encountered from variations in the thermal conductivity of the sample. High concentrations of hydrogen or methane may interfere in this way.

Common applications of the magnetic methods for oxygen analysis include determinations in flue gas and furnace atmospheres, and quality control of gases separated from liquid air.

2. Nuclear Magnetic Resonance

Nuclear magnetic resonance (NMR) is a relatively new technique which has already been used in process-stream analysis because of its applicability to solid organic materials. The theory of NMR is quite complex and has been well summarized for practical purposes by Aikman et al. (5). Very briefly, the NMR technique measures the resonance absorption of radio-frequency energy of fixed frequency by atomic nuclei as a function of the strength of an intense magnetic field in which the sample is placed. Only certain nuclei respond; of these, hydrogen, deuterium, ^{6}Li, ^{13}C, ^{14}N, ^{17}O, ^{19}Fl, and ^{35}Cl are typical. The various nuclei absorb at widely different field strengths. In liquid chemical compounds, small "chemical shifts" occur in the absorption of responding nuclei as a function of chemical and spatial structure. Thus many organic structures and water can be qualitatively identified and quantitatively determined from the chemical shifts of their protons. In solids, only broad absorptions can generally be seen, but the presence of adsorbed liquids, such as water, is readily observed from their sharp hydrogen resonance bands.

The equipment for NMR spectroscopy typically includes a large permanent magnet with a set of coils to allow the field strength to be swept over a limited range, an rf oscillator operating in the megacycle range to energize the sample coil, a low-frequency oscillator to modulate the weaker rf signals, an rf receiver, detectors, and a dc recorder. Absorption of rf energy by the sample nuclei (held in or passing through the sample coil) at resonance is thus detected and recorded as curves of intensity of absorption versus field strength.

The best-known application of NMR in continuous analysis has been the measurement of moisture content of solids such as cornstarch (5,7). The NMR signal indicates the quantity of protons of the two types occurring in the moist starch—the ones in the carbohydrate structure and the ones in the "liquid" adsorbed water. The starch protons absorb broadly, but the water protons give a narrow band superimposed on the broad band. Use of low-frequency modulation of the signal and appropriate filtering leads to the recording of a derivative signal peak arising largely from the narrow band absorption of the water. The peak amplitude correlates with the moisture content of the starch. Because the peak is obtained from a swept signal, repetitive or cyclic, rather than truly continuous, analysis is necessary. The solid sample can move continuously, because the sweep rate can be fairly rapid, of the order of a few seconds per sweep. Moisture contents in the range of 4–15% are readily measured.

An analyzer for liquid samples of a petroleum product has been described by Nelson et al. (176). Since it is desired to measure two absorption peaks, in this case in a high-resolution NMR spectrometer, the peaks are individually integrated electronically. In order to allow time for the nuclei to align with the field, the sample is moved quite slowly through the field and sample coil. Subsequent developments in stabilizing the magnetic field now make it possible to utilize peak heights directly, rather than areas, in such a measurement.

One of the intriguing features of NMR for future exploitation is its possibility of high-speed readout. Oscilloscopes can be used for scanning the spectrum. If sample response in the field can also be improved, very rapid analyses should be possible.

3. Electron Paramagnetic Resonance

The electron paramagnetic resonance technique has not yet been applied extensively in continuous analysis, but its specific laboratory application to free-radical detection in gases, liquids, and solids has aroused so much interest that process applications in the near future seem likely. An analyzer for oxygen in char has been patented (234).

Electron paramagnetic resonance (EPR) resembles nuclear magnetic resonance in theory and apparatus. The most important difference lies in the frequency range of resonance of free or unpaired electrons, that is, the microwave region of the spectrum, compared to the megacycle region for NMR. Generally, broad bands are observed in EPR, although different nuclei in a compound of interest introduce fine structure. Quantitative results may be obtained by integrating the band area. Paramagnetic atoms such as oxygen and iron will interfere with the detection of free radicals.

It would seem that EPR could be applied to automatic analysis for free radicals in solids such as polymers through adaptation of the microwave cavity sample holder to moving samples and readout of the appropriate integrated absorption bands. As with NMR, very rapid analyses should be possible. Transition metals are measured in catalysts by a recently developed EPR analyzer (71).

E. OTHER METHODS

Many other physical properties have been used for continuous analysis of materials. These include density, vapor pressure, and sonimetry, for gases; and density, viscosity (28,77), and surface tension, for liquids.

TABLE 105.V
Miscellaneous Methods of Continuous Analysis

Technique	Principle	Applications	Reference
Gas density	Gravimetry	Up to 2.5 sp. gr.	17
	Kinetic energy (Ranarex)	Up to 3.0 sp. gr.; CO_2 in flue gas	204
	Δp across orifice at constant flow	UF_6 streams	198
Liquid density	Mechanized balance and hydrometers	Up to 3.5 sp. gr.	143a
	Δp between dip tubes at different levels in a tank	General for large vessels	112
	γ-Ray absorption	Particularly for "in-pipeline" measurements	231
Vapor pressure	Absolute pressure at constant temperature	Petroleum products	99
Viscosity	Mechanical falling ball		77
	Mechanized piston		179
	Δp across orifice at constant flow		99
	Drag on rotating spindle	Pulp, paper stock, foods	29
	Damping of vibrating reed	Polymers; in pipelines	20
	Power input to mixer		66a
	Float suspension		77

These methods are essentially nonselective and hence generally applicable only to binary mixtures. Selectivity useful for analytical purposes has been achieved only in special cases wherein the desired component has an unusual value of the property being measured. An example is the high density of carbon dioxide in combustion gases. A summary of those methods, which see little application to selective analysis, is given in Table 105.V.

The utility of sensitive methods despite poor selectivity is demonstrated by the significant use of gas density detectors in gas chromatography. In all these methods, temperature variation must be controlled or corrected for, in order to achieve analytical accuracy.

1. Gas Density Balance

Small volumes of sample are analyzed and high sensitivity is achieved by gas density balance detectors, which have been designed for use in gas chromatography. Such devices depend on a dynamic balance between

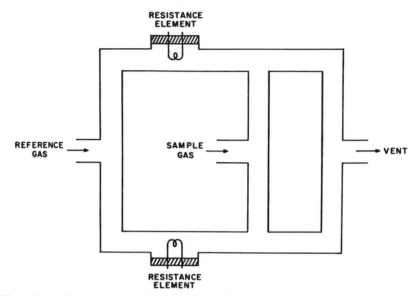

Fig. 105.30. Diagram of gas density balance detector (161). (Courtesy of Gow–Mac Instrument Co.)

flowing streams of an inert reference gas and of the gas to be analyzed. The streams flow in narrow channels in a block, connected in such a way that a change in density of the sample alters the relative rates of flow of the reference gas over two heated resistors (Fig. 105.30). This changes the relative temperatures of the resistors, measured either by thermocouples (160) or by the use of thermistors as the heated resistors in a bridge circuit (161). Amplification of thermocouple output is necessary. Indication is said to be directly proportional to the molecular weight of the sample, so that the balance is easily calibrated in terms of weight per cent concentration.

Such detectors have the advantage that the sample never contacts the heated elements, thus preventing degradation of the sample and fouling of the elements. A disadvantage lies in the necessity for good thermostatting, because of ambient temperature effects on the heated elements.

2. Surface Tension

In the Film Tensiometer, for continuous measurement of surface tension, a thin film of the liquid moves continuously between two vertical rods (197). One rod is fixed in position, and the other is free to move against a fixed tension. The force on the movable rod developed by the film is measured by a motion transducer and recorded. Compensation by a reference film at the same temperature and flow rate corrects for the effects of temperature variations. Sensitivity is within 0.5 dyne/cm. The analyzer can be applied to aqueous or organic solutions or emulsions.

3. Sonimetry

Measurement of the speed of sound in a gas mixture can serve as an analytical method for binary mixtures, since the velocity depends on the density and specific heat of the gas. Analytical uses of this technique have been known in the laboratory since 1927. It is a sensitive method, since the velocity of sound can be measured with high precision.

If sonic measurements on two pure gases are made in the same apparatus, the difference in sonic velocities can be expressed by the phase relationship of sound waves in the two media (140):

$$\Delta\phi_{ab} = Lf\left(\frac{1}{V_a} - \frac{1}{V_b}\right)$$

where $\Delta\phi_{ab}$ = phase change (wavelengths); L = path length of the sound (cm); V_a, V_b = velocity of sound in the respective pure gases (em/sec); and f = frequency (Hz). Phase change, and hence sensitivity, are thus seen to increase with both path length and frequency. The velocity of sound in a pure gas is related to its molecular weight by

$$V = \sqrt{\gamma RT/M}$$

where γ = ratio of specific heats of the gas, R = gas constant, T = absolute temperature, and M = molecular weight. Sensitivity increases with components of lower molecular weight. The composition of a mixture of two gases can be calculated from measurements of sonic velocity, as follows:

$$X_a = \frac{1/V_m + 1/V_b}{1/V_a + 1/V_b} \times \frac{\Delta\phi_{mb}}{\Delta\phi_{ab}}$$

where X_a = mole per cent of component a, V_m = velocity of sound in the mixture, and $\Delta\phi_{mb}$ = phase change in wavelengths between the mixture and pure b (a function of L, f, V_m and V_b, similar to $\Delta\phi_{ab}$, above).

The apparatus for measurement of sonic velocities comprises a tube containing a sound transmitter and a receiver fixed in position, a thermostat for the tube, a sonic oscillator, an amplifier for the receiver, and

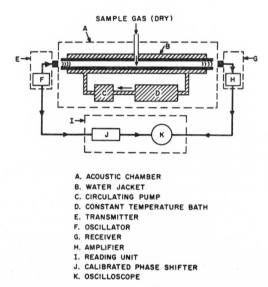

A. ACOUSTIC CHAMBER
B. WATER JACKET
C. CIRCULATING PUMP
D. CONSTANT TEMPERATURE BATH
E. TRANSMITTER
F. OSCILLATOR
G. RECEIVER
H. AMPLIFIER
I. READING UNIT
J. CALIBRATED PHASE SHIFTER
K. OSCILLOSCOPE

Fig. 105.31. Schematic diagram of sonic gas analyzer (140). (Courtesy of Instrument Society of America.)

a readout device (Fig. 105.31). Piezoelectric crystals are used for sound transmission and reception. A typical tube for high-sensitivity sonimetry is 8 mm o.d. by 1 meter long. For readout, an oscilloscope may be used to display the phase difference between transmitted and received signals. For monitoring, a phase shift meter and an associated recorder are employed. Readings are independent of pressure variations. If phase shifts greater than one wavelength are encountered, the number of whole wavelengths in the difference must be monitored by frequency-counting techniques.

A sonic analyzer has been applied to the measurement of respired air (140). Changes of 36 ppm in oxygen content have been detected after removal of carbon dioxide and water from the sample. Tenfold better sensitivity is possible for hydrogen in binary mixtures.

A gas chromatography detector based on sonic analysis has been described by Noble (177). This detector combines two sound tubes or cells for a differential measurement, one fed with pure carrier gas (helium) as a reference and the other fed with chromatograph effluent. The large volume of sample needed to provide a sensitive measurement of phase shift in a long sound tube exceeds the requirements of other common detectors for gas chromatography.

V. PRESENTATION AND CALCULATION OF DATA

This section will consider the problems of interpreting and filing a large number of data collected by continuous analyzers and by laboratory instruments such as spectrophotometers. The former are more likely to be used for single analyses, in contrast to multicomponent determinations by the latter. In either case, the data must first be assimilated by recording in some permanent fashion, with provision for notifying the operator promptly of significant events such as process upsets. Routine or interpretive calculations will have to be performed, so that the data must next be transcribed into a form suitable for the calculating device. The results of the calculations must be recorded or communicated to other interested parties. Further computations may then be performed, for example, accounting and process yields. Finally, the summary information should be stored for future reference in such a way that it can readily be retrieved.

No single scheme or "system" for performing such chores will meet the detailed requirements of all automatic instruments. Much of the general problem has been solved for process instruments by data-logging systems in pilot plants and full-scale plants. Systems have now appeared to handle the needs of specific laboratory instruments. Components can be purchased for the assembly of individual systems, so that some saving can be made in the considerable capital cost of complete systems. The most important lack is that of hardware or manageable computer programs to convert the raw data (e.g., spectral curves) into useful results such as concentrations of components. Savitzky (215a) has discussed these problems and suggested solutions for certain applications, such as infrared spectra and titration curves.

Principles and components of information systems for analytical instruments are discussed briefly in the following sections, together with some representative systems for the handling of data from mass spectrometers and infrared spectrophotometers.

A. ANALOG AND DIGITAL REPRESENTATION

The first problem to be faced is that of matching the output data from the analytical instrument to the input capacity of the system. Practically all analytical instruments at present yield their data in *analog* form, that is, each datum point is represented by a position on a continuous scale of magnitude. The scale may be a slide-wire, meter, rotating shaft, current, pressure, or voltage. The final "logged" or printed data, however, should be in *digital* form, that is, each datum point represented by one or more numerical digits (selected by discrete mechanical or electrical contacts, before printout).

Digital records are more compact than analog records. The digital form is also more popular for further calculations which may be performed on the data, for example, component concentrations and yields. Although analog computers are available for such calculations and offer simplicity when few numerical operations are to be performed, digital computation equipment is preferred in general for speed and accuracy in complex tasks.

B. DATA ASSIMILATION AND REDUCTION

Two general types of problems are encountered in handling data from an automatic instrument. One is that of fitting the concentration readings from an analyzer into a data-logging system; this is often encountered in process applications. The other problem is designing or specifying a system to convert the raw data from a complex laboratory spectrophotometer into concentrations of components.

1. Data-Logging Systems

A data-logging system accepts analog data from a multitude of process instruments in turn and prints out process information such as temperatures, pressures, flows, levels, and concentrations. Provisions may be included for detecting and warning the operator of process upsets. Two basic forms of data-logging systems are common, with functions as indicated below:

System A	System B
Switch channel of input data	Switch channel and computer program
Amplify signal	Amplify signal
Linearize signal	Digitize signal
Digitize linearized signal	Compute linearized datum, flows, and
Display signal	yields (storage)
Detect off-limit channel	Display
Punch card or tape for separate computer	Print final output
Print final output	Control functions
Control functions	

System A includes circuits to linearize data, for example, thermocouple voltages to temperatures, for every input channel. It depends on an external computer to calculate flows and yields for accounting purposes at some later time.

System B includes a computer which can perform all such calculations internally, store the results, and recall them from storage whenever desired. The present trend is toward System B. It represents, in general, less hardware and investment than System A, yet it possesses the versatility of the computer. A modern process installation is likely to require a computer, which is easier to integrate with System B.

The data-logging system may be summed up in terms of its important functions of data collection, detection of off-limit variables, and digestion of data. Typically, 50 or more variables are measured and logged in a cycle. The electrical output of the typical continuous analyzer is readily accepted by the system input, and concentrations and yields of products can be calculated with the aid of the analytical data.

2. Laboratory Systems

Data-reduction systems for complex laboratory instruments such as infrared spectrophotometers and mass spectrometers are less complex than data-logging systems for processes but present greater difficulty in computation. Such systems, in general, only digitize and log the data from two channels from each instrument, that is, the quantitative and qualitative axes of the spectrum. Calculation of component information from the logged spectral data requires a good computer and a well-designed program. Inversion of a matrix for solving a complex mass spectrum requires a computer with large storage capacity, such as an IBM 1800.

Such a data system may comprise simply a two-channel "digital readout" device which produces logged output data for later communication with an external general-purpose scientific computer. The complexity of the logging system will depend on such factors as (1) what type(s) of analytical instrument it is applicable to, (2) whether voltages or angles are to be measured, (3) what speed of operation is required, and (4) whether any numerical operations are to be performed on the data before printout.

Presently available data-reduction systems for the laboratory are capable of handling only one kind of instrument. Problems are best illustrated by examples.

For laboratory mass spectrometers one type of digitizing equipment converts the mass peak and the accelerating voltage to printed digits (56). The amplified electrometer signal is read at every peak value by an electronic voltmeter. At each peak reading, the accelerating voltage is balanced by a potentiometer whose shaft position is read by a mechanical digitizer. Both readings are printed simultaneously in decimal digits, and paper tape may be punched at the same time with coded readings. The accelerating voltage is logged as the corresponding mass/charge ratio. As many as two peaks per second may be read out.

The automatic computation of the mass spectrum is complicated by a need for some human intervention. All likely components of the sample must be assigned on the basis of prior knowledge, and a computer program must either be written or be available. The computer then can calculate quantitative results. Once the program is written for a given type of sample, many similar samples may be handled completely automatically. The computer program must include the mass patterns of all likely components of the sample or refer to patterns in the computer's memory.

Two systems for recording digital output from infrared spectrophotometers have been described. The International Telephone & Telegraph Spectroanalyzer (119a) consists of a digital readout system for a commercial spectrophotometer and a computer specifically designed for routine multicomponent analysis (Fig. 105.32). Although the recorder is calibrated linearly in transmittance, the digital readout of intensity is transmitted in absorbance units through the use of a specially designed shaft-position encoder. Digital indication of wavelength is obtained from a wavelength counter, and both readings are punched onto paper tape. With samples whose components are qualitatively known, the sample tape is then run through the computer together with prerecorded tapes carrying coefficients corresponding to the various components. These coefficients have been obtained from the solution of simultaneous differential equations on a general-purpose computer. The computation in the Spectroanalyzer finally reads out directly the concentration of each component of the sample. It is important that no unknown absorbing components be present in the sample. The Spectroanalyzer system was designed originally for routine analysis of steroid mixtures of four or five components, but much wider applications are possible.

A digital readout system designed by the Perkin–Elmer Corporation (199) consists of transmittance and wave number encoders, together with electronic components designed to punch paper tape in a code suitable for subsequent calculation by a general-purpose computer. Both variables are digitized by shaft-position encoders. The wave number interval at which recordings are to be made can be selected as 0.1, 1.0, and 10 cm^{-1}.

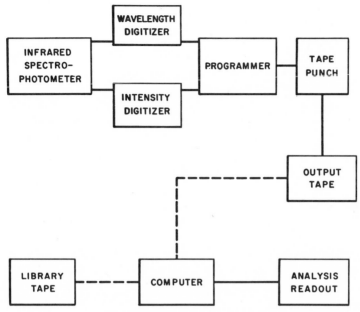

Fig. 105.32. Schematic diagram of IT & T Spectroanalyzer. (Courtesy of International Telephone & Telegraph Corp.)

A digital readout device is available for the General Electric recording spectrophotometer. Intensity and wavelength readings are recorded on punch cards, which can serve as inputs to a computer to obtain integrated color values in the tristimulus system.

Digital recording of data from an emission spectrograph requires a different approach, since there is no scanning of wavelengths. In the system developed by Scribner at the U.S. National Bureau of Standards for the analysis of alloys, photomultiplier detectors measure the intensity of each emission line desired (222). Each detector output charges a condenser whose voltage charge is proportional to intensity. When readout is required, each condenser discharges through a circuit which gives a pulse rate in proportion to the initial charge. The pulse rate is measured by counters, one for each condenser, and the count is transferred to punch cards when a fixed counting interval has elapsed. The entire analysis and readout of 12 lines takes less than 1 min.

One digital readout system has been described thus far for process gas chromatography. The total-analysis digital system developed by Phillips Petroleum Company integrates the area of each peak, applies a sensitivity factor to each component area, and presents the area as a percentage of the total sample (128). The normalization to percentage basis is accomplished by an automatic gain control circuit which keeps the summed area of all peaks at 100 units. Final results are typewritten. As many as 15 components per sample can be analyzed. The sensitivity factors must be programmed into the system in advance, and hence all the components must be qualitatively known.

A rudimentary readout system for gas chromatography (199) integrates each peak as it elutes and prints out the accumulated interval at the start of each successive peak on paper tape. To find the area of each peak after the first, it is necessary to subtract successive readings.

3. Components of Systems

a. Digitizers

Digitizers accept analog information and convert it to digital form, that is, a series of contacts or voltages which represent a number by the presence or absence of information. The common types of digitizers are shaft-position encoders and electronic voltmeters.

Shaft-position encoders are suited to the conversion of information available in the position of a slide-wire (as in a recording potentiometer), or the angle of rotation of a prism, goniometer, or similar rotary device. The encoder includes a positioning device which moves with the rotating shaft, and a geometrical pattern (in a plane at right angles to the shaft) of contact or optical areas which determine the output signals at every angular position. Mechanical encoders have electrical contact areas in concentric rings, so that the positioning device will make one or more contacts, up to a maximum of the number of rings, in many discrete positions around the circumference of the encoder. If there are ten rings, 2^{10}, or 1024, such angular positions are possible. Greater angular resolution can be provided by larger encoders or by gearing encoders in series. Mechanical encoders are the least expensive digitizing devices.

Photoelectric encoders use light and dark areas in a geometric pattern, with lights and photocells placed on opposite sides of the pattern, to give high or low output in the photocell corresponding to each area. Magnetic encoders use a pattern of areas or coils which give either of two levels of induced voltages in the various output bits.

Electronic voltmeters convert input voltages to a series of contact positions or output voltages to represent digits. Most electronic voltmeters act as self-balancing potentiometers, comparing the input voltage with internally generated standard voltages until balance is reached.

In one type of voltmeter, the standard voltages are generated in stepwise fashion in a switching network of resistors or a series of contacts on a potentiometric slide-wire. The readout represents the switch or contact positions at balance. Such voltmeters balance in 1 or 2 sec.

Some electronic voltmeters generate the comparison voltage in a sweep which is linear with time. The time equivalent to the voltage at balance is represented by a series of pulses. The pulses are counted electronically, and the count is then read out as digital contacts or voltages. Voltmeters of this type are capable of making as many as 100 measurements per second. One application of such voltmeters is found in readouts for mass spectrometers.

The use of digitizers introduces the subject of codes, that is, the language in which digital machines communicate with each other and with their human operators. The latter require the decimal numbers, 0, 1 . . . , 9, in the final readings, but equivalent mechanical or electrical systems (e.g., 10-position switches) would be much too complex for every step of a large-scale computer's operations. Hence binary numbers are used in computers and accessory equipment. The table shows one code of binary equivalents of the decimal digits:

Decimal number	Binary number
0	0000
1	0001
2	0010
3	0011
4	0100
5	0101
6	0110
7	0111
8	1000
9	1001

Since 0 and 1 can be represented by the open and closed positions of a relay or by the absence or presence of voltage at a terminal, the resulting simplicity of machine components is evident.

The code shown above is "pure" binary, and each digit is a "bit." This code is rarely used for numbers greater than 9 because it progresses in powers of 2, in contrast to our predilection for decimals. Another drawback is the change of several binary digits between many adjacent decimal numbers, for example, four binary digits change in going from 7 to 8. This could lead to chaotic ambiguity in a mechanical encoder if the contact brushes were halfway between angular positions 7 and 8 at the time of readout.

More practical codes have been devised. The cyclic progressive or Gray code is widely used in shaft-position encoders because only the least significant digit is affected in ambiguous readings. The binary-coded decimal form is used in computers for ease of conversion to decimal digits in readout. Since numbers of several decimal digits are in general use, computers work in terms of "words" of 17–40 binary digits. The extra bits are used for giving various machine instructions, error checking, and the like.

b. DATA CONVERTERS

Between stages of a data system, information must be transmitted in terms that the receiving stage can utilize. Requirements will influence the choice of analog or digital mode for the data to be transmitted. In a large petroleum refinery or chemical plant, data transmission over distances of hundreds or a few thousands of feet may be necessary from instruments in the field to the logging device. Transmission of analog signals such as dc voltages can be satisfactory over such distances if reasonable care is taken to shield the lines against pickup of stray voltages. Signals in the low millivolt range may require amplification in the field to minimize noise pickup in transmission.

One of the factors controlling digital communication is the code, as discussed above. Another is speed; until recently, tape punches and typewriters could accept only up to 20 characters per second. Still another is the choice between serial (one at a time) and parallel (simultaneous) handling of the code bits. Shaft-position digitizers have typically 12 bits (three binary-coded decimal digits) available for parallel readout by a computer or printer, but a tape punch or typewriter can accept only 4 bits at a time. Hence an intermediate device called a programmer must serialize the reading before presentation to the transcribing device in the smaller groups of 4 bits. The programmer can include buffer storage and programming. Buffer storage is used in many places in data systems so that one device will not unduly load another. The buffer is generally composed of thyratrons or relays. These hold a reading until the tape punch or typewriter can record all of it; then the storage is cleared to accept a new reading from the digitizer. The programmer uses stepping switches to advance the bits to be read.

Data for further machine operations can also be recorded on punch cards or magnetic tapes; both can accept parallel inputs. Punch cards record decimal numbers, requiring later conversion to digital code in computation. Magnetic tapes require ac input but can record information at high speed. They are well suited for recording rapidly changing variables for later playback to computers operating at any input rate desired.

c. MEMORY

Several methods of storing data for later retrieval are available, and various factors govern their selection. Printed matter, punched paper tape, and punch cards are all relatively inexpensive. Printed matter is inaccessible for machine retrieval because methods for retrieval of such information are only now being developed. Paper tape and punch cards are retrievable by machine at a relatively slow rate, and physical storage and handling presents a considerable problem if much information is to be stored. Punch cards can be searched on relatively small equipment of the business machine type, however; such equipment is usually available to the average laboratory. Large amounts of infrared and X-ray data on standard materials are now available on punch cards.

Magnetic tape is widely accepted for storage and retrieval. It offers the advantages of relatively low cost of the tape, high density of recorded information, and comparatively fast retrieval when desired.

The most readily accessible memories, and also the most expensive, are found in the digital computers themselves, where compactness is essential. Magnetic drums and core storage offer high densities of information and very rapid access to all the storage. Because of expense, such devices can usually not be justified for storage of process data and should be used only for brief computer programs and subroutines, where access time of the order of milliseconds is essential. Lengthy programs and compilations of data essential to the computer are stored on auxiliary magnetic tape. Advanced methods of computer storage such as etched films and cryotrons will eventually offer very high densities and rapid access.

d. PROGRAMS

"Programming" is literally the technique of giving orders to the system. The term is applied to a variety of devices in automatic instruments; the equipment employed depends primarily on the speed of operation required. The operation of cyclic analyzers such as gas chromatographs is generally accomplished by "program timers" with multiple electrical contacts controlled by cams, at intervals from 1 sec to several minutes. In converting parallel readout of digitizers to serial readout, "programmers" with stepping switches moving at about 25 steps per second are employed. "Programming" is now generally applied to the operation of high-speed computers; "programs" consist of orders precisely stated in machine language and operating at high speed from punched cards, punched tape, or magnetic tape. A program for a digital computer can refer to directions for routine operations which are already stored in the computer's memory.

Since programming a digital computer requires considerable mathematical skill, "compiler" programs have been devised to allow the computer to write its own detailed program from simple instructions. Compilers (or "interpretive routines") such as ALGOL and FORTRAN accept instructions in special algebraic notations which are called common languages; these notations are easily learned by scientists and engineers. The compilers are adaptable to many computers. When the instructions for a particular problem are given to a computer in a common language, the compiler (generally stored on magnetic tape) is also made part of the instructions. This directs the computer to interpret the common language properly and to prepare an "object program," which then actually conducts the solution of the problem.

For routine calculations, part of the programming of a digital computer can be handled by temporary interconnection of various components of the computer in a particular pattern. This operation is termed "patch boarding" because it is facilitated by having the connections brought out to removable boards, which have sockets for plugs and resemble a telephone switchboard.

C. TYPES OF COMPUTERS

The computers considered to this point have been digital machines. These are used in process work for the correction of raw data, for accounting-type calculations, and in some instances for actual control of the process. Digital computers find ready application here because they are highly accurate and versatile, with considerable experience in both business and scientific calculations.

Large analog computers have not been used at all as yet for data logging. This is primarily a result of relatively low accuracy, typically 0.1%, with 0.01% available in the best cases. Analog computers are becoming increasingly popular, however, for the calculation of process dynamics. Since these calculations tend to involve the response characteristics of process instruments such as analyzers, process applications may be expected.

Analog computers are ideally suited to the solution of many types of differential equations, since the terms in the equations are set up as voltage inputs to "operational" amplifiers. The operations which can be performed include multiplication, squaring, division, integration, and differentiation. The voltage inputs are set up on coefficient potentiometers. The operation is selected by choice of feedback circuit, RC network in the amplifier output, and the like. Voltage outputs are read on dials or recording devices. It is thus possible to achieve an *analog* for the behavior of every pressure, temperature, level, flow, and concentration in a process, through trial and error solutions. The time scale can be compressed or expanded.

Analog methods are not unknown to process instrumentation, since differentiation and integration have long been used in the rate and reset functions of individual process controllers. Recording of mass flow from differential pressure indicators also involves analog integration.

D. FINAL OUTPUTS

Data systems used in process applications can provide their final output data in a variety of modes useful to the process operators. The permanent records are generally typewritten, with off-limit readings indicated in a different color. Critical variables which go off-limit generally activate visible or audible alarms. Important variables, particularly those capable of rapid change, are sometimes displayed graphically as a function of time on cathode-ray tubes. Several such variables can be displayed simultaneously by time-sharing of the sweep.

When a computer is incorporated in a data-logging system, more sophisticated displays of process data are possible, and actual control of the overall process by the computer becomes feasible. Since the computer has memory, it can present a record of one variable as a function of another on an X–Y function plotter. Because the computer has access to all the process data, it can digest large numbers of data and predict trends. Control of analytical instruments by a computer in a data system may be illustrated by the gas chromatograph. The computer can (*1*) direct the gas chromatograph to sample a particular stream at a given time, (*2*) sense and remember the height of each component peak, (*3*) take ratios of particular peak heights, and (*4*) reset a process controller in accord with the desired trend for the component ratio.

If the process operator can supply equations which describe the overall process, the computer can test these equations. Once valid equations are found, the computer can take over control of the process by resetting the controllers for individual variables in such a way as to maintain the desired trends. Recently computers have been placed in "closed-loop" control of plants producing chemicals such as ammonia and vinyl chloride (43). Increasing automation makes it likely that this trend will accelerate.

VI. LITERATURE SOURCES

The literature of continuous analysis is scattered, and only one recent book on the subject is available. Sources, like the instruments themselves, represent many fields of science and engineering. Journals which carry pertinent articles include the following:

1. *Review of Scientific Instruments*
2. *Journal of the Optical Society of America*
3. *Analytical Chemistry*
4. *Nature (British)*
5. *Control Engineering*
6. *Instruments and Automation*
7. *Journal of the Instrument Society of America*
8. *Instruments and Control Systems*

Details of circuits, optics, and chemistry generally are found in the first four sources listed. Such papers tend to have little information on the maintenance needs and "life expectancy" of the analyzers described. Patterson (187–192) includes continuous analyzers in his reviews.

The other four American journals in the above list (there are British equivalents) provide general background but give few details of construction, although schematics of circuits and diagrams of mechanical features are sometimes included. Sampling is often pointed out as critical, but few solutions to sampling problems appear except in papers devoted exclusively to them (113,114,220). These articles are aimed at instrument engineers, chemical engineers, and technicians. They are likely to emphasize ruggedness of the analyzers and other points affecting or reflecting maintenance experience. References are often notable for their absence. Reviews of the field appear in these journals, but they discuss only commercial instruments and are seldom critical. Individual analyzers for air pollutants have been described in papers, and there has been one detailed review of this portion of the field in *Archives of Industrial Health* (88).

Columns on instrumentation by invited authorities appear in the advertising sections of *Analytical Chemistry* and *Industrial and Engineering Chemistry*. These articles are often critical as well as topical. It is unfortunate that some useful information therein may be lost because it is general library practice to remove advertising sections before binding individual issues of journals together.

The only American book devoted exclusively to continuous analysis is Siggia's (227). It describes only commercial instruments and has few literature references. An older British book by Jones (123) is similarly restricted but provides background on principles and sampling. A variety of industrial techniques in continuous sampling and measurement is discussed in a symposium (232a), and a book on gas sampling (250a) has been published. Brief discussions of many analyzers are found in a handbook of process instruments (55). Conn and Avery (54) discuss infrared analyzers in some depth, and several analyzers are described in the ICI treatise (242).

Several pamphlets devoted largely or exclusively to continuous analysis have been published by technical societies and journals (38,57,174,226). Two of these are based on symposia on automatic chemical analysis (38,174). Another represents a symposium on automatic process monitoring (226). Symposia on analysis instrumentation are sponsored at intervals by the Instrument Society of America, and most of the papers are later published in its *Proceedings*. Typically, these are concerned with commercial instruments. The American Chemical Society (118), the American Society for Testing Materials, (11), and air pollution control groups have held similar symposia. Atomic Energy Commission reports are available on a few analyzers (141a,179a).

Additional books and pamphlets will be found in the supplemental bibliography.

REFERENCES

1. Adams, D. F., and R. K. Koppe, *Anal. Chem.*, **31**, 1249 (1959).
2. Adams, D. F., R. K. Koppe, and N. E. Matzek, *Anal. Chem.*, **33**, 117 (1961).
3. Adams, E. M., "Instruments for Sampling and Analyzing Organic Vapors in Air," in *Encyclopedia of Instrumentation for Industrial Hygiene*, C. D. Yaffe, D. H. Byers, and A. D. Hosey, Eds., University of Michigan, Ann Arbor, 1956, pp. 41–48.
4. Adams, V. H., and D. J. Fraade, *IRE Trans. Ind. Electronics* **PGIE–11**, 1–9 (December 1959).
5. Aikman, A. R., R. S. Codrington, and F. F. Kirchner, *Control Eng.*, **4** (6), 105 (June 1957).
6. Albright, C. M., Jr., *Ind. Eng. Chem.*, **52**, 484 (1960).
7. Alexander, F. M., *ISA Proc.*, **14**, *Paper* **22–59** (1959).
8. Altshuller, A. P., and I. R. Cohen, *Anal. Chem.*, **32**, 802 (1960).
9. Andreatch, A. J., and R. Feinland, *Anal. Chem.*, **32**, 1021 (1960).
10. Applied Research Laboratories, Glendale, Calif.

11. American Society for Testing Materials, "Symposium on Instrumentation in Atmospheric Analysis," *ASTM Spec. Tech. Publ.*, **250**, Philadelphia, Pa., 1959.
12. Aughey, H., *J. Opt. Soc. Am.*, **39**, 292 (1949); Koppius, O. G., *J. Opt. Soc. Am.*, **39**, 294 (1949).
13. Bailey, C. W., and R. F. Howard, *ISA Proc.*, **14**, *Paper* **126–59** (1959).
14. Baker, B. B., and J. D. Morrison, *Anal. Chem.*, **27**, 1306 (1955).
15. Baker, W. J., and T. L. Zinn, *Control Eng.*, **8** (1), 77 (January 1960).
16. Barendrecht, E., *Nature*, 1181, Apr. 25, 1959.
17. Beckman Instruments, Inc., Fullerton, Calif.
18. Beerbower, A., in *Analysis Instrumentation*, **III**, Control *Eng. Reprint* **492**, 1959.
19. Bell, L. W., Jr., *ISA Conf. Preprint* **40SL61**, 1961.
20. Bendix Aviation Corp., Cincinnati, Ohio.
21. Bennett, C. E., S. Dal Nogare, L. W. Safranski, and C. D. Lewis, *Anal. Chem.*, **30**, 898 (1958).
22. Bernhard, A. E., *ISA Conf. Preprint* **18SL61**, 1961.
23. *Bituminous Coal Res.*, **18** (1), 6 (Spring 1958).
24. Blaedel, W. J., and E. D. Olsen, *Anal. Chem.*, **32**, 789 (1960).
25. Blumer, M., *Anal. Chem.*, **32**, 772 (1960); Claxton, G., *J. Chromatog.*, **2**, 136 (1959).
26. Boyd, G. E., L. S. Myers, Jr., and A. W. Adamson, *J. Am. Chem. Soc.*, **69**, 2849 (1947).
27. Bradford, B. W., D. Harvey, and D. E. Chalkley, *J. Inst. Petrol.*, **41**, 80 (March 1955).
28. Brookfield, D. W., and R. A. Minard, *Ann. N.Y. Acad. Sci.*, **91**, Art. 4: 838 (1961).
29. Brookfield Engineering Co., Stoughton, Mass.
30. Brown, H. C., *J. Metals*, **6**, *AIME Trans.*, **200**, 349 (1954).
31. Bryan, F. A., and V. Silis, *Am. Ind. Hyg. Assoc. J.*, **21**, No. 5, p. 423 (October 1960).
32. Bryden, R., and A. Dean, *Ann. N.Y. Acad. Sci.*, **87**, Art. 2: 813 (1960).
33. Carritt, D. E., and J. W. Kanwisher, *Anal. Chem.*, **31**, 5 (1959).
34. Catanzaro, E. W., *Ann. N.Y. Acad. Sci.*, **87**, Art. 2: 808 (1960).
35. Central Scientific Co., Chicago 13, Ill.
36. Chaikin, S. W., T. D. Parks, and C. I. Glassbrook, U.S. Pat. 2,741,544 (Apr. 10, 1956); Chaikin, S. W., C. I. Glassbrook, and T. D. Parks, 123rd Meeting American Chemical Society, Los Angeles, Calif., 1953.
37. Chamberlain, N. F., B. W. Thomas, J. B. Beaugh, and P. B. Land, *Ind. Eng. Chem.*, **48**, 1990 (1956).
38. *Chem. Eng., Proc. Control*, reprint, June 1957.
39. *Chem. Eng. News*, **34**, 1510 (1956).
40. *Chem. Eng. News*, **37** (35), 52 (Aug. 31, 1959).
41. *Chem. Eng. News*, **38** (27), 55 (July 4, 1960).
42. *Chem. Eng. News*, **38** (30), 50 (July 25, 1960).
43. *Chem. Eng. News*, **38** (45), 60 (Nov. 7, 1960).
44. *Chem. Eng. News*, **39** (16), 63 (Apr. 17, 1961).
45. Cherry, R. H., *Ann. N.Y. Acad. Sci.*, **91**, Art. 4: 850 (1961).
46. Cherry, R. H., G. M. Foley, C. O. Badgett, R. D. Eanes, and H. R. Smith, *Anal. Chem.*, **30**, 1239 (1958).

47. Clark, L. C., Jr., U.S. Pat. 2,913,386 (Nov. 17, 1959).
48. Claudy, H. N., C. C. Helms, P. R. Scholly, and D. R. Bresky, *Ann. N.Y. Acad. Sci.,* **72** (13), 779 (Mar. 20, 1959).
49. Claudy, H. N., F. W. Karasek, B. O. Ayers, and J. G. Skinner, *Anal. Chem.,* **31,** 1255 (1959).
50. Clift, T. L., *Ann N.Y. Acad. Sci.,* **91,** Art. 4: 825 (1961).
51. Cohn, G., *Ann. N.Y. Acad. Sci.,* **91,** Art. 4: 873 (1961).
52. Cole, L. G., M. Czuha, R. W. Mosley, and D. T. Sawyer, *Anal. Chem.,* **31,** 2048 (1959).
53. Coleman Instruments, Inc., Maywood, Ill.
54. Conn, G. K. T., and D. G. Avery, *Infrared Methods,* Academic, New York, 1960, p. 174.
55. Considine, D. M., *Process Instruments and Controls Handbook,* McGraw-Hill, New York, 1957.
56. Consolidated Electrodynamics Corp., Pasadena, Calif.
57. *Control Engineering, Analysis Instrumentation Series, Reprint* **493,** 1959.
58. Coulson, D. M., and L. A. Cavanagh, *Anal. Chem.,* **32,** 1245 (1960).
59. Coulson, D. M., L. A. Cavanagh, J. E. de Vries, and B. Walther, *J. Agr. Food Chem.,* **8,** 399 (1960).
60. Crandall, W. A., *Ann. N.Y. Acad. Sci.,* **87**, Art. 2: 911 (1960).
61. Czuha, M., Jr., and L. C. Thayer, 1st International Congress of ISA, Sept. 20, 1954.
62. Darrah, W. A., U.S. Pat. 1,746,525 (Feb. 11, 1930) (reissued); Glasser, L. G., *J. Electrochem. Soc.,* **97,** 201C (1950).
63. Davis Instrument Division of Davis Emergency Equipment Co., Inc., Newark 4, N.J.
64. Daynes, H. A., *Gas Analysis by Measurement of Thermal Conductivity,* Cambridge University Press, Cambridge, England, 1933.
65. De Ford, D. D., and B. O. Ayres, *Anal. Chem.,* **32,** 1711 (1960).
66. Desty, D. H., C. J. Geach, and A. Goldup, in *Gas Chromatography 1960,* R. P. W. Scott, ed., Butterworths, London, 1960, p. 46.
66a. De Zurik Corp., Sartell, Minn.
67. Donner, W., *Electronics,* **28** (11), 137, (November 1955); *ISA J.,* **3,** 89 (1956).
68. Dowson, A. G., and I. J. Buckland, *Nature,* **177,** 712 (1956).
69. Dudenbostel, B. F., *Ind. Eng. Chem.,* **48** (1), 81A (1956); *Chem. Week,* **78,** 138, 140, 142 (1956).
70. Dudenbostel, B. F., and W. Priestley, Jr., *Ind. Eng. Chem.,* **48** (9), 55A (1956).
71. Elion Instruments, Bristol, Pa.
72. Engelhard Industries, Inc., Newark 2, N.J.
73. Faith, W. L., *Air Pollution Control,* Wiley, New York, 1959.
74. Fellows, C. G., *ISA Paper* **52-3-2,** 1952.
75. Ferrari, A., *Ann. N.Y. Acad. Sci.,* **87,** Art. 2: 792 (1960).
76. Ferrari, A., G. Kessler, F. M. Russo-Alesi, and J. M. Kelly, *An Instrumental Approach to Chemical Pharmacology in Vivo,* Technicon Controls Corp., Chauncey, N.Y.
77. Fischer and Porter, Hatboro, Pa.
78. Fourroux, M. M., *Proc. 37th Ann. Conv. Natl. Gas Assoc. Am., 1958,* pp. 16–18; *J. Inst. Petrol.,* **45** (422), 29A (February 1959).

79. Fourroux, M. M., F. W. Karasek, and R. E. Wightman, *Oil Gas J.,* **58** (12), 96 (Mar. 21, 1960).

80. The Foxboro Co., Foxboro, Mass.

81. Fraade, D. J., and E. E. Escher, 136th National Meeting, paper presented at American Chemical Society, Atlantic City, N.J. (September 1959).

82. Frey, H. J., and R. C. Voter, U.S. Pat. 3,003,932 (Oct. 10, 1961); to E. I. du Pont de Nemours & Company.

83. Friedman, H., and L. S. Buks, *Rev. Sci. Instr.,* **19,** 323 (1948).

84. Friedman, H., and co-workers, cited by R. Tousey, *J. Opt. Soc. Am.,* **47,** 21 (1957).

85. Friel, D. D., *Ind. Eng. Chem.,* **52,** 494 (1960).

86. *Gas Age,* **117** (6), 20 (1956).

87. General Electric Co., Schenectady, N.Y.

88. Giever, P. M., and W. A. Cook, *A.M.A. Arch. Ind. Health,* **233,** (March 1960); Byran, R. J., and J. C. Romanovsky, *Instr. Automation,* **29,** 2432 (1956).

89. Glasser, L. G., *Control Eng.,* **4,** 87 (1957).

90. Glasser, L. G., *J. Opt. Soc. Am.,* **45,** 556 (1955).

90a. Glasser, L. G., R. J. Kanzler, and D. J. Troy, *ISA Proc.,* **4,** 11 (May 1958).

91. Glasser, L. G., and D. J. Troy, *ISA Paper* **56–1–1,** ISA Meeting, New York, Sept. 17, 1956.

92. Goodwin, P. S., in *Analysis Instrumentation,* **III,** *Control Engineering Rprint* **492,** 1959.

93. Goodwin, P. S., J. L. Jones, and M. F. Hasler, paper presented at Pittsburgh Conference of Analytical Chemistry and Applied Spectroscopy, March 1956.

94. Grant, D. W., in *Gas Chromatography, 1958,* D. H. Desty, ed., Academic, New York, 1958, p. 153.

95. Gray, P. R., D. H. Clarey, and W. H. Beamer, *Anal. Chem.,* **31,** 2065 (1959).

96. Green, R. M., *ISA Proc.,* **14,** *Paper* **130–59,** 1959.

97. Greenbrier Instrument Co., Ronceverte, W. Va.

98. Grenfell, T. C., D. J. McLaughlin, and J. M. Kelly, *Ann. N.Y. Acad. Sci.,* **87,** Art. 2: 857 (1960).

99. Hallikainen Instruments, Berkeley, Calif.

100. Hanson, V. F., *Ind. Eng. Chem., Anal. Ed.,* **13,** 119 (1941).

101. Hasler, M. F., *J. Opt. Soc. Am.,* **41,** 870 (1951).

102. Hasler, M. F., *Spectrochim. Acta,* **6,** 69 (1953).

103. Hasler, M. F., E. Davidson, H. Orr, and W. H. Barry, *Mikrochim. Acta,* **1955,** 596.

104. The Hays Corp., Michigan City, Ind.

105. Halter, R. C., and L. M. Johnson, paper read at American Petroleum Institute, Division of Refining, Houston, Tex., May 8, 1961.

106. Harrison, W. J., paper read at Pittsburgh Conference of Analytical Chemistry and Applied Spectroscopy, March 1960.

107. Hecht, G. J., J. A. Edinburgh, and V. N. Smith, *ISA J.,* **7** (2), 40 (1960).

108. Hersch, P., *Nature,* **169,** 792 (1952); Brit. Pat. 707,323, to Mond Nickel Co., Ltd. (Apr. 14, 1954).

109. Hersch, P. A., *Anal. Chem.,* **32,** 1030 (1960).

110. Herscher, L. W., and N. Wright, *J. Opt. Soc. Am.,* **43,** 980 (1953).

110a. Hodgson, E., in *Automatic Measurement of Quality in Process Plants,* Academic, New York, 1958, p. 29.

111. Hoke, Inc., Cresskill, N.J.
112. Holzbock, **W. G.**, *Instruments for Measurements and Control,* 2nd ed., Reinhold, New York, 1962.
113. Houser, E. A., *Control Eng.,* **7** (8), 129 (August 1960).
114. Houser, E. A., *ISA J.,* **7** (9), 95 (September 1960) ; Percy, L. E., *Instr. Control Systems,* **33,** 1755 (1960).
115. Howard, O. H., and C. W. Weber, *Arch. Ind. Health,* **19,** 355 (1959).
116. Howe, W. H., *Inst. Radio Engrs., Trans. Ind. Electronics,* **PGIE–5,** 56 (April 1958).
117. Hudson, R. L., and K. W. Hering, *Control Eng.,* **1,** 87 (1954).
118. *Ind. Eng. Chem.,* **52,** 481 (1960).
119. Industrial Instruments Engineering Corp., 89 Commerce Rd., Cedar Grove, N.J.
119a. International Telephone and Telegraph Co., Nutley, N.J.
120. Isreeli, J., M. Pelavin, and G. Kessler, *Ann. N.Y. Acad. Sci.,* **87,** Art. 2: 636 (1960).
121. Jacobson, M. G., *Anal. Chem.,* **25,** 586 (1953).
122. James, A. T., and A. J. P. Martin, *Biochem. J.,* **50,** 679 (1952).
123. Jones, E. B., *Instrument Technology.* Vol. II: *Analysis Instruments,* Butterworths, London, 1956.
124. Jones, R. H., and R. J. Joyce, Symposium on Water Quality Measurement and Instrumentation, U.S. Public Health Service, Cincinnati, Ohio, 1960.
125. Jones, R. H., and T. J. Kehoe, *Ind. Eng. Chem.,* **51,** 731 (1959).
126. Jonnard, R., *Ann. N.Y. Acad. Sci.,* **87,** Art. 2: 669 (1960).
127. Karasek, F. W., and B. O. Ayers, *ISA J.,* **7** (3), 70 (March 1960).
128. Karasek, F. W., M. C. Burk, and B. O. Ayers, *Anal. Chem.,* **33,** 1543 (1961).
129. Karasek, F. W., R. J. Loyd, D. E. Lupfer, and E. A. Houser, *Anal. Chem.,* **28,** 233 (1956).
130. Karmen, A., and R. L. Bowman, *Ann. N.Y. Acad. Sci.,* **72** (13), 714 (1959).
131. Kay, K., *Anal. Chem.,* **29,** 589 (1957).
132. Keidel, F. A., *Anal. Chem.,* **31,** 2043 (1959).
133. Keidel, F. A., *Ind. Eng. Chem.,* **52,** 490 (1960).
134. Keidel, F. A., and C. D. Lewis, *Anal. Chem.,* **33,** 1456 (1961).
135. Kemp, J. W., *Anal. Chem.,* **28,** 1838 (1956).
136. Kertzman, J., *Ann. N.Y. Acad. Sci.,* **91,** Art. 4: 901 (1961).
137. Kieselbach, R., *Anal. Chem.,* **26,** 1312 (1954).
138. *Ibid.,* **32,** 1749 (1960).
139. Klaasse, J. M., and W. Hampton, *ISA Conf. Reprint* **6–H60,** 1960.
140. Kniazuk, M., and R. Prediger, paper read at Tenth Annual Instrument Conference and Exhibit, Instrument Society of America, Los Angeles, Calif., Sept. 12–16, 1955.
141. Koppius, O. G., *Anal. Chem.,* **23,** 554 (1951).
141a. Kritz, W. R., U.S. at Energy Comm. *AEC Res. Develop. Rept.* **DP–356** (April 1959).
142. Kuhns, L. J., R. H. Forsyth, and J. F. Masi, *Anal. Chem.,* **28,** 1750 (1956).
143. Landsberg, H., and E. E. Escher, *Ind. Eng. Chem.,* **46,** 1422 (1954) ; Wallace, C. F., U.S. Pats. 2,350,378 (1944) and 2,482,078 (1949).
143a. Leeds and Northrup Co., Philadelphia, Pa.
144. Leonard, J. E., *Analyzer,* **1** (4), 9 (1960).

145. Lewis, C. D., *J. Chem. Educ.*, **35**, 87 (1958).
145a. Lewis, C. D., paper read at 132nd Meeting, American Chemical Society, New York, September 1957.
146. Liston, M. D., A. J. Andreatch, and C. Beebe, *ISA Paper* **56–11–3**, Instrument Society of America Conference, New York, September 1956. New York, N.Y., September 1957.
147. Littlewood, A. B., *J. Sci. Instr.*, **37**, 185 (1960).
148. Littman, F. E., and R. W. Benoliel, *Anal. Chem.*, **25**, 1480 (1953).
149. Littman, F. G., and J. Q. Denton, *Anal. Chem.*, **28**, 945 (1956).
150. Lodge, J. P., Jr., *Chem. Eng. News*, **39** (38), 83 (Sept. 18, 1961).
151. Logsdon, E. E., *Ann. N.Y. Acad. Sci.*, **87**, Art. 2: 801 (1960).
152. Loyd, R. J., B. O. Ayers, and F. W. Karasek, *Anal. Chem.*, **32**, 698 (1960).
153. Lovelock, J. G., *Nature*, **188**, 401 (1960).
154. Lovelock, J. G., in *Gas Chromatography*, R. P. W. Scott, ed., Butterworths. London, 1960.
155. Lovelock, J. E., *Anal. Chem.*, **33**, 162 (1961).
156. Luft, K. F., *Z. Physik* **24**, 97 (1943).
157. Maier, H. J., and H. N. Claudy, *Ann. N.Y. Acad. Sci.*, **87**, Art. 2: 864 (1960).
158. Maley, L. E., *ISA Conf. Preprint* **19SL61**, 1961.
159. Maley, L. E., *ISA J.*, **4**, 312 (1957).
160. Martin, A. J. P., and A. T. James, *Biochem. J.*, **63**, 138 (1956).
161. Martin, R. L., *Anal. Chem.*, **32**, 336 (1960); Gow–Mac Instrument Co., Madison, N.J.
162. Mawson, J., *Ind. Eng. Chem.*, **52** (2), 85A (February 1960).
163. McCabe, L. C., *Ind. Eng. Chem.*, **48** (2), 91A (February 1956).
164. McClure, J. H., T. M. Roder, and R. H. Kinsey, *Anal. Chem.*, **27** 1599 (1955).
165. McWilliam, I. G., and R. A. Dewar, in *Gas Chromatography 1958*, D. H. Desty, ed., Academic, New York, 1958, p. 142.
166. Mechrolab, Inc., Mountain View, Calif.
167. Microtek Instruments, Inc., Baton Rouge, La.
168. Millipore Filter Corp., Bedford, Mass.
169. Mine Safety Appliances Co., Pittsburgh, Pa.
170. Minter, C. C., and L. M. J. Burdy, *Anal. Chem.*, **23**, 143 (1951).
171. Moiseev, A. S., and N. M. Brickman, *J. Appl. Chem. (USSR)*, **12**, 620 (1939).
172. Moore, F. E., *Oil Gas J.*, **56** (36), 96 (Sept. 8, 1958).
173. Muller, R. H., *Anal. Chem.*, **29**, 969 (1957); Muller, D. G., *Anal. Chem.*, **29**, 975 (1957).
174. Muller, R. H., ed., *Ann. N.Y. Acad. Sci.*, **87**, Art. 2: 609–951 (1960).
175. National Spectrographic Laboratories, Inc., Cleveland 3, Ohio.
176. Nelson, F. A., C. A. Reilly, and W. E. Savage, *Ind. Eng. Chem.*, **52**, 487 (1960).
177. Noble, F. W., 13th Annual Conference on Electrical Techniques in Medicine and Biology, Washington, D.C., Oct. 31, 1960, *Digest of Tech. Papers*, pp. 12–13.
178. Noebels, H. J., *Chem. Eng. Progr.*, **52**, 228 (1956).
179. Norcross Corp., Newton 58, Mass.
179a. Oak Ridge Natl. Lab. Rept., **ORNL–3015**, U.S. Department of Commerce Washington, D.C. p. 143.
180. Offutt, E. B., and L. V. Sorg, *Anal. Chem.*, **27**, 429 (1955).
181. Opler, A., and J. H. Miller, *J. Opt. Soc. Am.*, **42**, 779, 784 (1952).

182. Orsag, J., *Spectrochim. Acta,* **6,** 80 (1953).
183. Ostergren, R. H., and K. G. Halvorsen, *ISA Proc.* **11,** 56-3-1 (1956); Wood, R. C., L. W. Foskett, and N. B. Boster, *ISA Proc.,* **9,** 54-36-1 (1954).
184. Paris, A., *Ind. Chim. (Paris)*, **31,** 253 (1933).
185. Parsons, J. L., M. J. Irland, and F. R. Bryan, *J. Opt. Soc. Am.,* **46,** 164 (1956).
186. Patient, D. A., *Ann. N.Y. Acad. Sci.,* **87,** Art. 2: 830 (1960).
187. Patterson, G. D., Jr., *Anal. Chem.,* **27,** 574 (1955).
188. *Ibid.,* **29** (1957).
189. *Ibid.,* **31,** 646 (1959).
190. Patterson, G. D., Jr., and M. G. Mellon, *Anal. Chem.,* **22,** 136 (1950).
191. *Ibid.,* **23,** 101 (1951).
192. *Ibid.,* **24,** 131 (1952).
193. Patterson, W. A., *Chem. Eng.,* **59,** 132 (1952).
194. Pauling, L., U.S. Pat. 2,416,344 to California Institute Research Foundation (Feb. 25, 1947).
195. Penther, C. J., and J. W. Hickling, American Petroleum Institute, Division of Refining, 26th Midyear Meeting, Houston, Tex., May 8, 1961.
196. Pepkowitz, L. P., *Anal. Chem.,* **27,** 245 (1955).
197. Peterson, E. C., and A. R. Martin, in *Chem. Eng. News,* **38** (38), 79, (1960).
198. Weber, C. W., and W. S. Pappas, cited in *Chem. Eng. News,* **36** (37), 72 (1958).
199. Perkin–Elmer Corp., Norwalk, Conn.
200. Pfund, A. F., and C. L. Gemmill, *Bull. Johns Hopkins Hosp.,* **67,** 61 (1940).
201. Phoenix-Precision Instrument Co., Philadelphia, Pa.
202. Pickhardt, W. P., A. N. Oemler, and J. Mitchell, Jr., *Anal. Chem.,* **27,** 1784 (1955).
203. Pitkethly, R. C., *Anal. Chem.,* **30,** 1309 (1958).
204. Pfaudler Permutit Co., New York, N.Y.
205. Renzetti, N. A., *Anal. Chem.,* **29,** 869 (1957).
206. Renzetti, H. A., and L. H. Rogers, *ISA Proc.,* **11,** 56-27-3 (1956); Pilston, R. G., and J. U. White, *J. Opt. Soc. Am.,* **44,** 572 (1954).
207. Research Specialties Co., Richmond, Calif.
208. Ridgefield Instrument Group, Schlumberger Corp., Ridgefield, Conn.
209. Robinson, C. F., and L. G., Hall, *Rev. Sci. Instr.,* **27,** 504 (1956); Robinson, C. F., *Rev. Sci. Instr.,* **27,** 509 (1956).
210. Rogers, L. H., *J. Air Poll. Control. Assoc.,* **8,** 124 (August 1958).
211. Rosenbaum, E. J., R. W. Adams, and H. H. King, Jr., *Anal. Chem.,* **31,** 1006 (1959).
212. Roth, H. H., 99th paper read at Annual Meeting, Electrochemical Society, Washington, D.C., Apr. 10, 1951.
213. Milton Roy Co., Philadelphia 18, Pa.
214. Saltzman, B. E., *Anal. Chem.,* **33,** 1100 (1961).
215. Saltzman, R. S., J. F. Arbogast, and R. H. Osborn, *Anal. Chem.,* **27,** 1446 (1955).
215a. Savitzky, A., *Anal. Chem.,* **33** (13), 25A (1961).
216. Savitzky, A., and D. R. Bresky, *Ind. Eng. Chem,* **46,** 1382 (1954).
217. Savitzky, A., E. H. Woodhull, and A. P. Weber, *Ind. Eng. Chem.,* **48,** 1047 (1956).

218. Sawyer, D. T., and R. S. George, paper read at American Chemical Society, 133rd National Meeting, San Francisco, Calif., April 1958.
219. Schnelle, P. D., *ISA J.*, **4**, 128 (1957).
220. Schnelle, P. D., *ISA Proc.*, **10**, 55-18-1 (1955).
221. Scott, R. P. W., in *Vapor Phase Chromatography*, D. H. Desty, ed., Butterworths, London, 1957, p. 131; Primavesi, G. R., G. F. Oldham, and R. J. Thompson, in *Gas Chromatography 1958*, D. H. Desty, ed., Academic, New York, 1958, p. 165.
222. Scribner, B. F., and R. E. Michaelis, cited in *Chem. Eng. News,* **36** (16), 72 (Apr. 21, 1958).
223. Severinghaus, J. W., and A. F. Bradley, *Electrodes for Blood Oxygen and Carbon Dioxide Determination,* Anaesthesia Research Laboratory, National Heart Institute, National Institutes of Health, Bethesda, Md., 1960.
224. Seymour, G. W., *Proc. Tech. Assoc. Pulp Paper Ind.*, **426** (1957).
225. Sheen, R. T., and E. J. Serfass, *Ann. N.Y. Acad. Sci.*, **87** (2), 844 (1960).
226. Siggia, S., ed., *Ann. N.Y. Acad. Sci.*, **91,** Art. 4: 819 (1961).
227. Siggia, S., *Continuous Analysis of Chemical Process Streams,* Wiley, New York, 1959.
228. Simons, E. L., *Anal. Chem.*, **30**, 979 (1959).
229. Skeggs, L. T., Jr., *Ann. N.Y. Acad. Sci.*, **87**, Art. 2: 650 (1960).
230. Smith, R. A., F. E. Jones, and R. P. Chasmar, *The Detection and Measurement of Infrared Radiation,* Oxford University Press, Fair Law, N.J., 1957, p. 368.
231. Smith, V. N., *Analysis Instrumentation,* **I,** *Control Eng. Reprint* **490,** 1958.
232. Smith, V. N., L. G. Glasser, and E. H. Woodhull, *ISA J.*, **4**, 26 (1957).
232a. Society of Instrument Technology, *Automatic Measurement of Quality in Process Plants,* Academic, New York, 1958.
233. Spackman, D. H., W. H. Stein, and S. Moore, *Anal. Chem.*, **30**, 1190 (1958); Moore, S., D. H. Spackman, and W. H. Stein, *Federation Proc.*, **17,** 1107 (1958).
234. Spry, W. J., Jr., and L. S. Singer, U.S. Pat. 2,957,128 to Union Carbide Corp. (Oct. 18, 1960).
235. Stair, R., *Proc. 3rd Natl. Air Pollution Symp.* Stanford Research Institute, Los Angeles, Calif., 1955, p. 48; Regener, V. H., "Atmospheric Ozone in the Los Angeles Region, University of New Mexico, Albuquerque, N. Mex.., *Sci. Rept.* **3,** Contract AF 19 (122)—381, July 22, 1954.
236. Steingiser, S., W. C. Darr, and E. E. Hardy, *Anal. Chem.*, **31,** 1261 (1959).
237. Stephens, E. R., P. L., Hanst, R. C. Doerr, and W. E. Scott, American Chemical Society, 128th National Meeting, Minneapolis, Minn., September 1955.
238. Sternberg, J. C., and L. M. Carson, Pittsburgh Conference on Analytical Chemistry and Applied Spectroscopy, Feb. 29, 1960, *Paper* **71.**
239. Stirling, P. H., and H. Ho, *Ind. Eng. Chem.,* **53** (2), 51A (February 1961).
240. Strange, J. P., *Anal. Chem.*, **29**, 1878 (1957).
241. Strange, J. P., *Ann. N.Y. Acad. Sci.*, **91** (4), 916 (1961).
242. Strouts, C. R. N., J. H. Gilfillan, and H. N. Wilson, *Analytical Chemistry,* Clarendon Press, Oxford, England, 1955, Vol. I, p. 849.
243. Technical Oil Tool Corp., Los Angeles 38, Calif., *Bull.* **1259–11** (March 1960).
244. Technical Oil Tool Corp., Los Angeles 38, Calif., *Bull.* **5900–11** (March 1960).
245. Technicon Controls Corp., Chauncey, N.Y.; Skeggs, L. T., U.S. Pat. 2,797,149 to Technicon International, Ltd., (June 25, 1957); Whitehead, E. C., and A.

Ferrari, Jr., U.S. Pat. 2,899,280 to Technicon International, Ltd. (Aug. 11, 1959).

246. Thomas, B. W., and J. B. Beaugh, *Petrol. Refiner,* **35** (10), 133 (1956).

247. Thomas, C. O., and B. B. Baker, *Anal. Chem.,* **31**, 1391 (1959).

248. Thomas, M. D., J. O. Ivie, and T. C. Fitt, *J. Ind. Eng. Chem.,* **18**, 383 (1946).

249. Thomas, M. D., J. A. MacLeod, R. C. Robbins, R. C. Goettelman, R. W. Eldridge, and L. H. Rogers, *Anal. Chem.,* **28**, 1810 (1956).

250. Thomas, M. D., G. A. St. John, and S. W. Chaikin, "Symposium on Instrumentation in Atmospheric Analysis," *ASTM Spec. Tech. Publ.* **250**, Philadelphia, Pa., 1959.

250a. Tiné, G., *Gas Sampling and Chemical Analysis in Combustion Processes,* Pergamon, New York, 1961.

251. Troy, D. J., *Anal. Chem.,* **27**, 1217 (1955).

252. Turner, G. S., *ISA Proc.,* **6**, C4-1 (1960).

253. Tyler, C. P., U.S. Pat. 2,993,846 to Esso Research & Engineering Co. (July 25, 1961).

254. Wall, R., *ISA J,* **4**, 267 (1957).

255. Wall, R. F., *Ind. Eng. Chem.,* **51** (8), 53A (August 1959).

256. Wallace and Tiernan, Inc., Belleville 9, N.J.

257. Waters, J. L., and N. W. Hartz, *Instruments,* **25**, 622 (1952).

258. Waters Associates, Framingham, Mass.

259. Weber, C. W., *Anal. Chem.,* **32**, 387 (1960).

260. Wherry, T. C., *Chem. Eng. Progr.,* **56** (9), 49 (September 1960).

261. Wherry, T. C., *ISA J.,* **7**, No. 3, 48 (March 1960).

262. Wherry, T. C., *Oil Gas J.,* **54**, 125 (1956).

263. Wilson, A., L. Bini, and R. Hofstadter, *Anal. Chem.,* **33**, 135 (1961).

264. Woods, K. R., and R. L. Engle, Jr., *Ann. N.Y. Acad. Sci.,* **87**, Art. 2: 764 (1960).

265. Wright, N., and L. W. Herscher, *J. Opt. Soc. Am.,* **36**, 195 (1946).

265a. Yaffe, C. D., D. H. Byers, and A. D. Hosey, eds., *Encyclopedia of Instrumentation for Industrial Hygiene,* University of Michigan, Ann Arbor, 1956.

266. Yanak, J. D., and A. M. Calabrese, in *Analysis Instrumentation,* **III,** *Control Eng. Reprint* **492,** 1959.

267. Young, J. C., J. R. Parsons, and H. E. Reeber, *Anal. Chem.,* **30**, 1236 (1958).

SUPPLEMENTAL BIBLIOGRAPHY

1. Athans, M., and P. Falb, *Optimal Control,* McGral-Hill, New York, 1966.

2. Buckley, P. S., *Techniques of Process Control,* Wiley–Interscience, New York, 1964.

3. Carroll, G. C., *Industrial Process Measuring Instruments,* McGraw-Hill, New York, 1962.

4. Ceaglske, N. H., *Automatic Process Control for Chemical Engineers,* Wiley–Interscience, New York, 1956.

5. Considine, D. M., and S. D. Ross, eds., *Handbook of Applied Instrumentation,* McGraw-Hill, New York, 1964.

6. Dorf, R. C., *Modern Control Systems,* Addison-Wesley, Reading, Mass., 1967.

7. Holzbock, W. G., *Automatic Control: Principles and Practice,* Reinhold, New York, 1958.
8. LaJoy, M. H., *Process Control Analysis,* Instruments Publishing Co., Pittsburgh, 1956.
9. Liptak, B. G., ed., *Instrument Engineers Handbook,* Vols. 1–2, Chilton, Philadelphia, 1969–70.
10. Matthews, W. H., ed., *Instrumentation in the Chemical and Petroleum Industries,* Vol. 4, Plenum, New York, 1968.
11. Perlmutter, D. D., *Introduction to Chemical Control Systems,* Wiley–Interscience, New York, 1965.
12. Robinson, G. H., ed., *Instrumentation in the Chemical and Petroleum Industries.* Vol. 5: *The Challenge of Computers in the Chemical and Petroleum Industries,* Plenum, New York, 1969.
13. Robinson, G. H., ed., *Instrumentation in the Chemical and Petroleum Industries,* Vol. 1, Plenum, New York, 1964.
14. Sage, A. P., *Optimum Systems Control,* Prentice–Hall, Englewood Cliffs, N.J., 1968.
15. Sanders, C. W., ed., *Instrumentation in the Chemical and Petroleum Industries,* Vol. 2, Plenum, New York, 1966.
16. Smith, J. F., and W. G. Brombacher, "Guide to Instrumentation Literature," *Natl. Bur. Std. (U.S.) Miss. Publ.* **271,** Washington, D.C., 1965.
17. Tyson, F. C., Jr., *Industrial Instrumentation,* Prentice–Hall, Englewood Cliffs, N.J., 1961.
18. Ward, J. T., ed., *Instrumentation in the Chemical and Petroleum Industries,* Vol. 3, Plenum, New York, 1967.

SECTION G: Preparation for Analytical Research and Utilization

Chapter 106

DESIGN OF EXPERIMENTS IN ANALYTICAL CHEMISTRY INVESTIGATION

By Fred H. Tingey, *Aerojet Nuclear Company, Idaho Falls, Idaho*

Contents

I. INTRODUCTION

The particular area of applied statistics dealing with the selection and arrangement of experiments in time and space to utilize most efficiently the total effort that can be allocated to a given study is known as experimental design. The idea of a designed experiment is not new to most investigators, since some preliminary planning must go into even the most elementary study if any real degree of success is to be realized. However, the idea of a statistically designed experiment is new to some, and consequently, when introduced to the subject, these individuals tend to look upon the statistical method as an alternative to whatever method they would ordinarily use. It is to be emphasized that, whenever data to be obtained in the course of an investigation are subject to error, the statistical approach provides the only sound and logical means of treatment. There is no question of an equally satisfactory alternative. In this chapter we shall discuss some of these methods as they apply to typical studies in analytical chemistry. Only a relatively brief treatment can be given here. The reader who desires additional information is referred to the many excellent texts on experimental design (6,7,9).

II. DEFINITION AND NOTATION

In order to gain some familiarity with the notation and language convenient to the subject of statistical experimental design, we will consider as an example a study in which several different procedures (possibly resulting from eliminating steps in existing procedures) of preparing sample filaments for mass spectrometer assay are under consideration and it is desired to know whether a change in procedure introduces significant differences into the final measurement. In addition to the intentional changes in procedure, the technician preparing the filament for scanning is admitted as a possible source of variation, as is the preparation step and the actual scanning of a particular filament. Thus the sources of variation which we shall call *factors* are Procedures, Preparations, Technicians, and Analyses.

In order to further identify a particular set of conditions imposed by particular values for the factors and resulting in a given observation, we associate with each factor the various values and refer to them as *levels*. Thus, if five different procedures are under investigation, each procedure constitutes a particular level of the factor Procedures, as does each technician when related to the factor Technicians, and so forth. In this terminology we can identify any given observation in terms of the factors and in particular the combination of factor levels which were involved in obtaining the observation. These various possible combinations of factor levels are called *treatments,* that is, analysis 1, on preparation 2, by technician 1, using procedure 4, say, constitutes a particular treatment from the set of all possible treatments. Application of all treatments a single time is called a *replication* of the experiment. If only a portion of the total experiments required under a particular design is in fact accomplished in the course of the investigation, we say that a certain *fractional replication* of the entire experiment has been made.

It is convenient and expeditious to adopt a symbolism for identifying a given observation or yield resulting from the application of a given treatment to a given test material or *experimental unit*. We will let X denote the observation and use letters $h, i, j, \ldots$ as subscripts to denote particular levels of the factors which resulted in X. Thus the symbol X_{hijk} as related to the factors of the example, where the subscript h is identified with the factor Procedures, i with Preparations, j with Technicians, and k with Analyses, denotes the kth analysis of the ith preparation by technician j using procedure h.

Of fundamental concern in any given experiment are the effects on the yield of varying the various factors. This simply means that, if we identify by symbols, say $a_h, b_i, c_j, \ldots$ the contribution to the corresponding X unique to the hth level of factor A, ith level of factor B, jth level of factor $C, \ldots$, respectively, we are interested in determining whether one can validly infer from the data at hand whether the a_h's are all equal, and/or the b_i's are all equal, and/or the c_j's are all equal, and so forth. In actuality, there is no loss in generality in assuming that the effects are deviations from respective population means or averages. The *population* is defined as that set of effects to which an inference resulting from the analysis of the data is to apply. It is often the case that only a part of a given population of effects for a given factor is involved in a particular experiment.

In many instances, and particularly in analytical chemistry investigations, certain combinations of factor levels sometime yield observations or results which cannot be explained simply in terms of the effects a_h, b_i, c_j, . . . To allow for this contingency we shall adopt combinations of symbols such as $(ab)_{hi}$, $(ac)_{hj}$, $(bc)_{ij}$, $(abc)_{hij}$, . . . to denote the combined effects and use the word *interaction* in referring to them.

One additional consideration of fundamental importance has to do with differentiating between *crossed* and *nested* effects and will be described in terms of the example. First consider the Procedures factor. It seems reasonable to assume that, regardless of the levels of the other factors involved in a particular treatment, a unique effect, a_h, identifiable with the particular procedure A_h, would be present in all observations which involved that procedure. A similar analysis could be made with regard to the Technician factor. Contrast these with the effect to be associated with the Preparation factor. It is to be noted that the preparation identification has meaning only as it is used to denote a particular preparation associated with a given procedure as applied by a given technician. To assume that the effect of preparation 1, say, as represented by b_1, is the same over all procedures and technicians is unreasonable, since there is nothing unique or systematic to what we chose to call preparation 1 (assuming of course that there is no learning effect as the technician makes several preparations using a given procedure). Comparisons between means of observations corresponding solely to a division of the data by preparation number would tend to be alike regardless of variations between replicate preparations corresponding to each technician–procedure combination. Similarly, the analysis or mass assay of a given filament as identified by number has meaning only as it relates to the procedure, technician, and preparation involved.

Such effects are said to be nested, and to emphasize this fact it is conventional in denoting the associated effects to place in parentheses the subscripts corresponding to the nesting. Thus under this notation the symbols to be identified with the preparations effect in the example would be $(abc)_{j(hi)}$ and with the analysis effect $(abcd)_{k(hij)}$. The latter term includes all extraneous variation in addition to that identified with the particular replication (k) of the given treatment. For this reason it is called the *residual random error* and is simply noted by $e_{k(hij)}$. All effects which are not nested are said to be crossed. In terms of basic assumptions, interaction effects between crossed factors sum to zero (have expected value zero) when summed over the corresponding populations of values as defined by any index, whereas nested effects necessarily sum to zero (have expected value zero) only when summed with regard to indices not in parentheses.

III. MODELS

Before one can devise an adequate statistical design in a proposed area of investigation, it is necessary to formulate the manner by which the individual measurements are assumed to be affected by the factors either intentionally or unintentionally introduced into the study. This process of setting down in a rather formal manner an expression relating the observations or results to component factor effects in the study is called *model building*. The appropriate design and subsequent statistical analysis of the data obtained under the design are a consequence of the assumed model. Thus the model-building step is of prime importance. The construction of a model calls heavily upon the experience and knowledge of the experimenter as to how the various factors might exhibit their effects. To illustrate these ideas, let us consider a typical troubleshooting study common to most analytical laboratories. Under investigation is a given analytical technique for determining the concentration of a certain constituent in solution. The technique is to be applied over a range of concentrations and requires use of a particular type of apparatus of which there are several in the laboratory. In addition, it is believed that the particular analyst applying the technique could also exercise an effect upon any given result. If we denote by X_{hijk} the observation corresponding to the kth determination by analyst h on concentration i using equipment j, then in terms of the notation defined above, if only the main effects of the factors are believed to be of consequence, a model of the form

$$X_{hijk} = m + a_h + b_i + c_j + e_{k(hij)}$$

would be adequate to describe the results; however, if interactions are also admitted, then additional terms corresponding to these assumed interactions must be included. For this example, if all interactions are to be included the model would become

$$X_{hijk} = m + a_h + b_i + c_j + (ab)_{hi} + (ac)_{hj} + (bc)_{ij} + (abc)_{hij} + e_{k(hij)}$$

The model in this form corresponds to a three-factor experiment in which all factors are assumed to be crossed. In a particular study it is not necessary that all interactions be included. Only those believed by the experimenter to be of possible significance in view of his experience and the physical nature of the problem at hand should be reflected in the actual model.

A second example is given to illustrate the nested concept. Suppose that a study is to be conducted to determine the errors associated with the lot–batch–sample analysis chain typical of chemical studies. If we denote by X_{hijk} the kth analysis on the jth sample drawn from the ith batch from the hth lot, then, recognizing the nested nature of the effects, the model would be

$$X_{hijk} = m + a_h + (ab)_{i(h)} + (abc)_{j(hi)} + e_{k(hij)}$$

where the A factor corresponds to lots, the B factor to batches, and the C factor to samples. The model in this form corresponds to a three-factor experiment in which all the factors are said to be nested.

A third example is provided to illustrate the model when both crossed and nested factors occur in the same experiment. Consider a study which involves the primary steps of sampling batches, making preparations on each sample, and then analyzing on different instruments in replicate each preparation. Identifying the symbol A with samples, B with preparations within samples, and C with instruments, the model for the kth determination on instrument j of preparation i from sample h as denoted by X_{hijk} would be

$$X_{hijk} = m + a_h + c_j + (ab)_{i(h)} + e_{k(hij)}$$

IV. EXPERIMENTAL DESIGN

A. PRELIMINARY CONSIDERATIONS

Fundamentally, given the model, the experimenter must collect data of such a type and in such a manner that the effects assumed in the model can be estimated and inferences made on the basis of the sampling distributions associated with the estimates as to whether certain sets of effects identified by factors or by combinations of factors could "reasonably" be different from zero. In approaching this problem it is important to note from the model that the measured results of the application of a particular treatment are affected not only by the action of the treatments but also by residual random error. As we shall see, this error variation serves as a yardstick for determining the significance of effects in the subsequent analysis of the data obtained under a given design.

Thus, to the extent that the magnitude of the error can be influenced by the design, every effort should be made to keep the error as small as possible. Methods for reducing experimental error and thus increasing the sensitivity of the experiment fall into three main categories:

1. Refining the experimental technique to include:
   a. Uniformity in applying the treatments.
   b. Control over external influences so that every treatment operates under as nearly the same conditions as possible.
   c. Choice of the correct characteristics to be measured in the phenomenon under study.
2. Increasing the size of the experiment by:
   a. Taking more replications.
   b. Including additional treatments.
3. Handling the individual experiments, as identified by treatments, in such a way that the effects of the variability are reduced. This would include:
   a. Determination of size and shape (when feasible) of the experimental unit to obtain maximum accuracy of measurement.
   b. Taking auxiliary measurements that might be related to the phenomenon under investigation.
   c. Arranging the experimental units in time and space so that (1) the effects are estimated with maximum precision within the total effort to be allocated to the experiment, (2) the estimates of the effects are not confused one with another or with the experimental variation, and (3) the tests to be constructed relative to the significance of certain effects or comparisons are independent of one another. It is in this area that the word "design" as applied to the course of conducting an experimental investigation originated.

B. SIMPLE COMPARATIVE EXPERIMENTS

Experiments conducted to compare the effects of only two different treatments are called simple comparative experiments. Models of such experiments, however, must often take into account environmental factors which can and often do influence the results. To illustrate this we shall consider two examples. The first is a study directed toward determining whether two analysts' results differ, on the average, as each applies the same procedure in analyzing for a given component in a stock solution. The factor under investigation therefore is Analyst, and the levels are identified by the two individuals involved in the experiment. A mathematical model for the results might be

$$X_{hi} = m + a_h + e_{i(h)}, \qquad h = 1, 2$$

In assuming this model the experimenter must be aware it is implicit to the model that (1) apart from experimental error, only the analysts are admitted as exhibiting an effect upon the result, (2) all other effects, to include environmental factors and the replication effect, are included in the residual error term, and these residual error effects are effectively independent.

Measurements obtained under any of the following conditions for this example could violate the model and/or assumptions and thus invalidate the resulting analysis:

1. Analyst A uses one instrument in all his determinations; analyst B, another. Such a procedure would confound or confuse the analyst effect with an instrument effect not provided for in the model.

2. Each analyst obtains all his results as required by the experiment essentially simultaneously, that is, the analyses are made in a group. Under this arrangement it is doubtful that the residual random error effects within a group would be independent since environmental effects, setup effects, and the like would be common to all measurements in the group.

3. Analyst A uses material from one batch of solution; analyst B, from another batch. The analyst effect would be confounded with a batch effect.

A second example which will be used to demonstrate the basic fundamentals of good experimental design for experiments of this sort is as follows. Suppose that we wish to determine whether two different methods of preparing specimens for emission spectrometry analysis affect the analysis for boron in aluminum stock material. It is anticipated that in the stock material an impurity which might possibly affect the boron analysis may not be uniformly distributed and consequently a "test piece" effect must be considered. The model for the experiment would be

$$X_{hij} = m + a_h + b_i + e_{j(hi)}$$

where effects a_h are to be identified with the method of preparation, and the b_i's with the test piece. The experimental procedure consists of:

1. Selecting n pieces of material from a stock sheet and dividing the pieces into two parts.

2. Preparing one piece from each pair for analysis by one method; the other piece, by the other method. The choice as to which piece is to be prepared by which procedure is made in a random manner.

3. Analyzing the prepared specimens in replicate in a random order.

The design of this experiment is characterized by (*1*) blocking the experimental units (test pieces) into pairs as nearly homogeneous in impurity level as possible so that the two pieces making up a pair can be assumed to have the same b_i effect, (*2*) assigning treatments in a random manner to a unit from each pair, (*3*) analyzing each specimen in a random order.

By blocking the experimental unit into pairs we provide a mechanism by which any difference between blocks in terms of true impurity content can be removed from the residual error term, thus making our experiment more sensitive. We note from the model and the associated design that differences between sums of replicates under each preparation within blocks provide estimates for the difference in preparation effects, regardless of any inherent variation in impurity level from block to block.

By assigning the method of preparation to each pair in a block in a random manner and by analyzing the prepared specimens in a random order, we are ensuring that in the absence of a preparation effect no difference between pairs would be detected and that the residual error terms are effectively independent between pairs within a block and between blocks.

C. FACTORIAL EXPERIMENTS

The model for the second example above corresponds to what is called a randomized block experiment involving only two treatments (different preparations); however, it can be considered in the framework of a *factorial* experiment involving two factors, namely, preparations and blocks. In this particular example, however, one should note that the model does not admit an interaction effect between blocks and preparations. Factorial design is characterized by at least one experiment being conducted at each combination of factor levels, whatever the number of factors and the number of associated levels may be, and as such provides the experimenter with observations over a relatively broad range of values as contrasted with the one-at-a-time approach commonly used. By virtue of the fact that the total number of experiments required in a factorial design for even few factors and corresponding levels can be large, this type of experimentation can be expensive. If, however, all main effects and interactions are admitted in the model, the only design that will facilitate the estimation of the main effects of every factor independently of one another, enable the dependence of one factor upon the levels of another (interactions) to be determined, estimate the effects with uniformly maximum precision within the total effort to be expended, and provide an estimate of residual random error by which the significance of the effects can be determined, is the factorial design.

1. 2^n Factorial Design

For any number of factors in a given experiment the factorial design resulting in the fewest number of experiments would involve just two levels of each factor. Such designs, known as 2^n factorial designs, where n is the number of factors, are extremely valuable for exploratory-type experimentation when little experience is available as to the magnitude and nature of the interactions. They also play a fundamental role in investigations directed toward the establishment of optimum procedures or conditions, in the sense of determining conditions for the experimental variables under the control of the experimenter which maximize or minimize some property of the phenomenon under study. Furthermore, certain basic fundamental principles associated with factorial designs in general are easily demonstrated by these simple designs.

When only two levels of each factor are involved in the design, it is convenient to denote the two levels corresponding to the A factor by (1) and a, those for the B factor by (1) and b, and so forth. The symbol (1) refers to the lower level of the factor, the normal condition, or the absence of a given condition, while a refers to the higher level, the change from normal, or the presence of a condition, and so on. It is to be recognized that some duplicity of notation with that previously used to define effects is introduced in this connection, inasmuch as lower-case letters will be used in the 2^n type of factorial experiment to denote both treatments and effects. However, subscripts on the effects will serve to differentiate. Using this notation, then, for the two-factor experiment involving the factors A and B and the levels (1), a and (1), b, respectively, the four combinations of factor levels (treatments) are conveniently identified as in Table 106.I. The extension of this system of notation to more than two factors is immediate.

a. CONFOUNDING IN 2^n DESIGN

When several factors are involved in an experiment, it may not be possible to carry out all treatments under uniform conditions. Thus the treatments must be arranged into blocks for the purpose of execution.

TABLE 106.I
Treatments in a 2^2 Factorial Experiment

Symbol for treatment combinations	Level of factors[a]	
	A	B
(1)	−	−
a	+	−
b	−	+
ab	+	+

[a] + indicates the factor at higher level; −, the factor at lower level.

The consequence of this action is to introduce an additional factor into the experiment, that is, "blocks," which will tend to confuse or *confound* the estimation of certain effects as identified with the factorial experiment. Principles of good experimental design require that the grouping into blocks be made in a very particular manner so that the experimenter can exercise control over the extent and nature of the confounding. To illustrate the principle of confounding, let us consider a factorial experiment involving four factors, each at two levels. If only a single replication of this experiment were accomplished, a total of sixteen individual experiments would be required. Let us suppose, however, that it is not physically possible to maintain uniform experimental conditions over the time required to conduct all sixteen experiments. Such might be the case in a study in which only eight experiments could physically be completed over the period of one working day and it is believed that setup conditions, humidity, and other parameters which vary from day to day might have an effect on the results. The problem therefore is to determine which experiments are to be accomplished on which days.

To demonstrate the consequence of various blockings it is necessary to extend Table 106.I to that for four factors. This extension is accomplished in Table 106.II. An examination of this table suggests a system for determining the blocking. For example, if every treatment for which a plus sign appears in the column headed by A were accomplished on one day and the remaining eight on another day, any differences in the observations or results attributable to day-to-day differences would be reflected in the differences between the means of the observations which had factor A at high level and the means of those which had factor A at the low level. Since, as we shall see, this difference essentially

TABLE 106.II
Treatments in a 2^4 Factorial Experiment

Symbols for treatment combinations	Factors[a]			
	A	B	C	D
(1)	−	−	−	−
a	+	−	−	−
b	−	+	−	−
ab	+	+	−	−
c	−	−	+	−
ac	+	−	+	−
bc	−	+	+	−
abc	+	+	+	−
d	−	−	−	+
ad	+	−	−	+
bd	−	+	−	+
abd	+	+	−	+
cd	−	−	+	+
acd	+	−	+	+
bcd	−	+	+	+
abcd	+	+	+	+

[a] + indicates the factor at higher level; −, the factor at lower level.

defines the estimate of the factor A effect, a confounding of days with factor A results. This is highly undesirable. Blocks resulting from applying the same criterion to the columns headed by B, C, and D would result in a similar type of confounding. To choose a more suitable confounding we must extend Table 106.II to cover interactions or combined effects. This is simply done by multiplying in turn corresponding signs for any combination of columns in Table 106.II, using the rule of signs in the multiplication (the coefficient 1 being omitted in every case) and heading up the resulting columns by the symbols identifying the columns involved in the combination. Table 106.III is such an extension and includes for future reference a column headed by T to correspond to the sum of all treatments.

As was indicated above (and is intuitively appealing), each main effect apart from constant multiplier is simply the linear sum of all treatment results where the signs corresponding to treatments in the sum are as given in Table 106.II or Table 106.III for the effect as symbolically denoted by the column heading. The coefficients, being 1 in every case, are omitted. This procedure for estimating effects applies also to the interactions as symbolically represented in the remaining columns of Table 106.III. The columns in essence define the computational procedure to estimate the effects and thus are labeled accordingly.

We will now reconsider the problem of confounding in the light of Table 106.III. Since in fact the columns define the signs applied to the yields or observations corresponding to the treatments in the linear sums which estimate the effects, we can confound any given   effect, in particular a given interaction, with blocks (day-to-day differences) simply by placing all treatments with a plus sign for that effect in one block and the remaining treatments (those with a minus sign) in the other block. Since the experimenter is free to choose the effect to confound (assuming that the experiment is to be conducted in only two blocks), he naturally chooses the least important, which is usually the highest-order interaction, namely, $ABCD$ in this example. To accomplish this, treatments designated as (1), ab, ac, bc, ad, bd, cd, and $abcd$ would be run on one day and the remaining treatments on another. It is easily verified from Table 106.III that all linear sums corresponding to the effects other than the interaction $ABCD$ are essentially clear of any day-to-day differences, since for each sum half the results corresponding to the treatments run on a given day enter in that sum with a plus sign and the other half with a minus sign, thus canceling any day-to-day effect that might exist.

TABLE 106.III
Treatments and Effects in a 2^4 Factorial Experiment

Symbol for treatment combinations	Effect															
	T	A	B	C	D	AB	AC	AD	BC	BD	CD	ABC	ABD	ACD	BCD	ABCD
(1)	+	−	−	−	−	+	+	+	+	+	+	−	−	−	−	+
a	+	+	−	−	−	−	−	−	+	+	+	+	+	+	−	−
b	+	−	+	−	−	−	+	+	−	−	+	+	+	+	−	+
ab	+	+	+	−	−	+	−	−	−	−	+	−	−	+	+	+
c	+	−	−	+	−	+	−	+	−	+	−	+	−	+	+	−
ac	+	+	−	+	−	−	+	−	−	+	−	+	−	+	+	+
bc	+	−	+	+	−	−	−	+	+	−	−	−	+	+	−	+
abc	+	+	+	+	−	+	+	−	+	−	−	+	−	−	−	−
d	+	−	−	−	+	+	+	−	+	−	−	−	+	+	+	−
ad	+	+	−	−	+	−	−	+	+	−	−	−	+	−	+	+
bd	+	−	+	−	+	−	+	−	−	+	−	+	−	+	−	+
abd	+	+	+	−	+	+	−	+	−	+	−	−	+	−	−	−
cd	+	−	−	+	+	+	−	−	−	−	+	+	+	−	−	+
acd	+	+	−	+	+	−	+	+	−	−	+	−	−	+	−	−
bcd	+	−	+	+	+	−	−	−	+	+	+	−	−	−	+	−
abcd	+	+	+	+	+	+	+	+	+	+	+	+	+	+	+	+

(1) General Principles of Confounding 2^n Design in Two Blocks

Confounding any given effect in just two blocks, as we have seen, amounts to dividing the treatments into two groups such that the comparison between the groups corresponds with the effect it is desired to confound. This is easily accomplished by referring to a table such as Table 106.III and placing in one group all treatments with a plus entry and in the other those with a minus entry corresponding to the effect to be confounded.

An analogous procedure which does not require the somewhat cumbersome completion of a plus and minus table is derived by noting from Table 106.III that in terms of the treatment symbols the linear sum corresponding to the A effect is symbolically $a + ab + ac + abc + ad + abd + acd + abcd - (1) - b - c - bc - d - bd - cd - bcd$. Now, if we treat the symbols (1), a, b, c, and d algebraically, the above expression is simply the algebraic expansion of $(a-1)(b+1)(c+1)(d+1)$. In a similar manner the linear sum corresponding to the B effect is algebraically represented by $(a+1)(b-1)(c+1)(d+1)$, the C effect by $(a+1)(b+1)(c-1)(d+1)$, and the D effect by $(a+1)(b+1)(c+1)(d-1)$. The manner of formation of the terms for each effect is apparent. For interaction effects, it can be demonstrated from Table 106.III that the linear sum corresponding to an interaction effect in its algebraic form is as for the main effects, that is, a product of $(a \pm 1)(b \pm 1)(c \pm 1) \ldots$, where the sign is plus for a given factor if the letter in that factor does *not* correspond to a letter in the interaction being evaluated and the sign is minus if it does. Thus, for the interaction ABC, say, the linear sum of results corresponding to treatments could be obtained without recourse to a table such as Table 106.III by algebraically expanding the expression $(a-1)(b-1)(c-1)(d+1)$. Consequently, depending on the effect it is desired to confound with blocks, the algebraic representation of that effect can be expanded and the blocking made to correspond to the treatments having plus signs and those having minus signs in the expansion.

(2) Higher-Order Confounding in 2^n Factorial Design

Suppose now that we want to partition the treatments into more than two blocks. That is, in our example, say, something less than eight observations could be performed over a given day's time and as a consequence the sixteen observations must be accomplished in more than two sets. Since more than two blocks are to be involved, more than a single effect must be confounded with blocks. For a 2^n factorial design the number of blocks must be a power of 2. It can be shown that, for 2^p blocks, $2^p - 1$ effects must be confounded with blocks. Of these effects, p can be chosen at will with the remaining $2^p - p - 1$ effects as represented by the products of the p chosen effects in all possible ways being confounded as a consequence.

The operation of multiplying effects together introduces a slight complication in notation. We note that for any two pairs of effects, when their symbolical representations as given in the column headings of Table 106.III are multiplied algebraically, terms of the form $A^\alpha B^\beta C^\gamma D^\delta$ result, where the exponents, as represented by the Greek letters, are either 0, 1, or 2. The powers of 0 and 1 present no complication to our system of denoting effects since algebraically any symbol raised to the zeroth power is 1 and the exponent 1 simply need not be indicated, that is, $A^0 = 1$ and $A^1 = A$. However, symbols raised to the power of 2 are not interpretable as such in our system of notation. To determine what is meant by the generalized interaction obtained by multiplying the interaction ABC by BCD, say, as algebraically noted by $(ABC)(BCD)$ $= AB^2C^2D$, we refer again to Table 106.III. We note from this table that, if we multiply in turn the corresponding signs, assuming coefficients of 1 in every case, under the columns headed by ABC and BCD, and compare the resulting column with that headed by the symbol AD, the corresponding entries are identical. A similar observation could be made for any generalized interaction resulting from multiplying together algebraically any pair of effect symbols, where the rule of formation is that any symbol squared in the product form is to be equated to unity. This is equivalent to striking out of the product any letter common to the two effects being multiplied together.

We will now apply these ideas to the previous example. Suppose that we wish to block the sixteen treatments into four blocks of four. Three effects must be confounded with the four blocks as a consequence. Since $4 = 2^2$, and thus $p = 2$, we note that two of the three effects to be confounded can be chosen as we please with the third being the generalized interaction. Suppose that we choose to confound the effects AD and ABC, and as a consequence confound $(AD)(ABC) = BCD$, with blocks. For obvious reasons, the higher-order interactions are usually sacrificed to the block confounding. For four factors in four blocks it is apparent that the above is essentially the best that can be done.

Once we have decided upon the effects to be confounded, we must then determine the treatments to be included in each set or block. There are several ways to do this. Probably the simplest is by recourse to a table analogous to Table 106.III for the experiment at hand. For

TABLE 106.IV

Confounding a 2^4 Factorial in Four Blocks of Four

Treatment symbol	Chosen confounded effect	
	AD	ABC
(1)	+	−
a	−	+
b	+	+
ab	−	−
c	+	+
ac	−	−
bc	+	−
abc	−	+
d	−	−
ad	+	+
bd	−	+
abd	+	−
cd	−	+
acd	+	−
bcd	−	−
$abcd$	+	+

convenience we list the columns corresponding to the effects chosen to be confounded. For our example these would be as given in Table 106.IV.

We note that four different forms or pairs of signs corresponding to the AD effect and the ABC effect, respectively, exist. These are − −, − +, + −, + +. By classifying the treatments according to these pairs, that is, all with − − in one group, − + in another, and so on, we partition the sixteen treatments into four groups of four, namely,

$$− − : [ab, ac, d, bcd]$$
$$− + : [a, abc, bd, cd]$$
$$+ − : [(1), bc, abd, acd]$$
$$+ + : [b, c, ad, abcd]$$

If we let these groups define the treatments for the four blocks, we note that except for the three intentionally confounded effects (i.e., AD, ABC, BCD) each linear sum of yields corresponding to the estimation of any given effect involves all sixteen treatments, and from Table 106.III it can be observed that four treatments in each block have signs such that two treatments have a plus sign associated with the omitted coefficient of 1 and two have a minus sign. As a consequence any block differences cancel out or are effectively removed from the resulting linear sum.

The basis ideas used in confounding in factorial designs of the 3^n and $3^n 2^n$ types (10,24) are similar to those given above but are increasingly more complicated as a result of introducing more levels of the factors. In view of the complications introduced and the limited scope of such designs, especially in analytical chemistry investigations, they should be avoided.

b. Fractional Factorial Design in 2^n Factorial

For a complete mathematical model to include all the main effects and interactions associated with an n-factor experiment, a total of $2^n - 1$ (2^n if the mean, m, is counted) effects must be included. It becomes increasingly more difficult to interpret interactions as the order, that is, the number of factors involved in the interaction, increases. Furthermore, in most cases higher-order interactions can reasonably be assumed not to exist. Consequently, one is reluctant to expend the effort required under a complete factorial design to estimate effects which, a priori, are "insignificant." The fractional factorial design provides one solution to this problem. In addition it provides the experimenter with a sequential approach to his investigation since additional experiments required to complete a full factorial design can be performed if desired.

Depending on the model assumed for the observations or the results generated in the course of an investigation, certain economies of effort over the complete factorial design can be effected. The procedures for constructing fractional factorial designs are closely related to those used in confounding effects with blocks, since for a given case a fractional factorial design can be obtained by selecting a block in an appropriate confounded design. These designs have found greatest use as applied to 2^n factorial designs and thus will be discussed in the framework of the preceding section.

The basic objective in constructing a fractional factorial design in a particular case is to select from the set of all treatments required under the complete factorial a subset which, when applied under the conditions of the model, will enable the main effects of every factor to be independently estimated and will also provide independent estimates of certain selected interactions for which information is desired. These designs, like the complete factorial design, can be constructed so as to provide an independent estimate of the experimental error variation and thereby provide a means by which the significance of the estimated effects can be assessed.

TABLE 106.V
$\frac{1}{2}$ Replicate of a 2^4 Design

Treatment	T	A	B	C	D	AB	AC	AD	BC	BD	CD	ABC	ABD	ACD	BCD	ABCD
(1)	+	−	−	−	−	+	+	+	+	+	+	−	−	−	−	+
ab	+	+	+	−	−	+	−	−	−	−	+	−	−	+	+	+
ac	+	+	−	+	−	−	+	−	−	+	−	−	+	−	+	+
bc	+	−	+	+	−	−	−	+	+	−	−	−	+	+	−	+
ad	+	+	−	−	+	−	−	+	+	−	−	+	−	−	+	+
bd	+	−	+	−	+	−	+	−	−	+	−	+	−	+	−	+
cd	+	−	−	+	+	+	−	−	−	−	+	+	+	−	−	+
abcd	+	+	+	+	+	+	+	+	+	+	+	+	+	+	+	+

One way to construct fractional factorial designs follows immediately from the procedure used in confounding in the preceding section. Consider the example in which a 2^4 experiment was to be conducted in two blocks of eight treatments per block. By confounding the interaction $ABCD$ with blocks, we determined that all other effects could be estimated free and clear of any block if the treatments (1), ab, ac, bc, ad, bd, cd, and $abcd$ were placed in one block and the remaining treatments in the other. Let us examine the consequence of doing only the experiments called for in the first block. To make this appraisal we will abstract for Table 106.V that portion of Table 106.III which corresponds to only those treatments of the first block.

We note from this table that pairs of columns are identical, the pairing being as follows:

1. T, $ABCD$;   5. D, ABC;
2. A, BCD;   6. AB, CD;
3. B, ACD;   7. AC, BD;
4. C, ABD;   8. AD, BC.

Consequently, the linear sum of yields corresponding to the estimation of effect A, say, will be identical to that for effect BCD, for B identical to that for ACD, and so on through the remaining pairs. We say, therefore, as a result, the pairs of effects as defined by the identical columns in Table 106.V are aliased with each other and the estimation procedure cannot differentiate between them but can provide only the combined effect. This would be of little consequence, however, in this example if one could validly assume that all three-factor interactions were negligible and that factor C, say, did not interact with anything. For under these assumptions, the A, B, C, D, AB, BD, AD effects are in reality, apart from a constant multiplier, those estimated in the seven independent linear sums resulting from applying the signs of the different columns of Table 106.V to the results of the eight treatments. One can easily demonstrate by going through the same routine as above that, if the treatments of the second block, namely *a, b, c, abc, d, abd, acd, bcd*, had been accomplished instead of the first set, the same effects would be aliased one with another; however, the linear sum corresponding to a given pair would, apart from a constant multiplier, estimate the difference between the effects rather than the sum, as is the case for the eight treatments selected. The assumptions required to make the design useful are the same in both cases. Consequently, either set could be chosen for the $\frac{1}{2}$ replicate.

The fact that the column headed T is identical to the effect we chose to confound with blocks is of special significance and can be shown to be the case for higher-order confounding in which more than a single effect is involved. This fact is utilized in devising a general procedure for determining which sets of effects are aliased one with another when a certain fractional replication of a 2^n factorial is accomplished, the chosen treatments corresponding to those in a given block under an appropriate confounded design. The rule results from the statements previously made in relation to confounding a 2^n factorial design in 2^p blocks, and as a consequence $2^p - 1$ effects are confounded with blocks. We recall that for such an arrangement p effects can be chosen as the experimenter so desires, the remaining $2^p - p - 1$ effects, as represented by the products of the p chosen effects in all possible ways, being confounded as a consequence. Now, if we consider the set of p chosen effects and the $2^p - p - 1$ additional effects resulting from multiplying together the chosen effects in all possible ways, and add to the group the symbol I to act like unity in the algebraic multiplication of effects, we have a set of symbols which we shall call the defining contrast subgroup. By multiplying in turn every symbol in the defining contrast group by an effect symbol not in the group, we obtain another set of effects which contains the multiplier. Various sets of effects can be generated in this manner until all effects belong to some set. The sets so obtained define those effects aliased with each other under the given fractional factorial design. For the example just discussed this determination is summarized in Table 106.VI.

In order to fully appreciate the ramifications of this technique of design let us apply it to other examples. First we consider the extension of the confounding of the sixteen observations to four blocks of four and select for accomplishment the set of four experiments which falls

TABLE 106.VI
Aliasing in a $\frac{1}{2} \cdot 2^4$ Fractional Factorial Design

	Defining contrast subgroup	
Effect	I	$ABCD$
A	$AI = A$	$A^2BCD = BCD$
B	$BI = B$	$AB^2CD = ACD$
C	$CI = C$	$ABC^2D = ABD$
D	$DI = D$	$ABCD^2 = ABC$
AB	$ABI = AB$	$A^2B^2CD = CD$
AC	$ACI = AC$	$A^2BC^2D = BD$
BC	$BCI = BC$	$AB^2C^2D = AD$

in one of the blocks. This constitutes a $\frac{1}{4}$ replicate of the 2^4 factorial experiment. We recall that the confounding of effects AD and ABC with blocks resulted in the effect BCD also being confounded. Consequently the defining contrast subgroup, when the treatments of any one of the blocks are taken as a fractional replication, consists of the symbols I, AD, ABC, BCD. Suppose that we select the set consisting of the treatments (1), bc, abd, acd as the particular four treatments to be run. In this case an abstract of Table 106.III gives Table 106.VII.

We note from Table 106.VII that the sixteen columns group into four groups of four as identified by the columns in a group being identical or made to be identical by applying the multiplier (-1) to every sign in a column. The groups are as follows:

$$T, \; AD, \; -ABC, \; -BCD$$
$$A, \; D, \; -BC, \; -ABCD$$
$$B, \; ABD, \; -AC, \; -CD$$
$$C, \; ACD, \; -AB, \; -BD$$

TABLE 106.VII
Effects in a $\frac{1}{4} \cdot 2^4$ Fractional Factorial Design

Treatment	T	A	B	C	D	AB	AC	AD	BC	BD	CD	ABC	ABD	ACD	BCD	ABCD
(1)	+	−	−	−	−	+	+	+	+	+	+	−	−	−	−	+
bc	+	−	+	+	−	−	−	+	+	−	−	−	+	+	−	+
abd	+	+	+	−	+	+	−	+	−	+	−	−	+	−	−	−
acd	+	+	−	+	+	−	+	+	−	−	+	−	−	+	−	−

Choosing a different block (one of the other sets of four treatments corresponding to the chosen confounding) would have resulted in the same grouping, only the signs preceding some of the effects in the sets of four would have been different. We observe that the four observations in this $\frac{1}{4}$ replicate of the 2^4 experiment, even under the assumption of nonsignificant interaction effects, present a serious problem inasmuch as main effect A is aliased with, among other things, main effect D. Consequently, this design would not be acceptable and is given here primarily for illustrative purposes. We note that the four sets of aliased effects, apart from signs, could easily have been obtained from the defining contrast subgroup in the manner previously defined by multiplying the elements of the subgroup in turn by A, B, and C.

The procedure of confounding in more than four blocks and analogously obtaining less than a $\frac{1}{4}$ replicate of a 2^n factorial experiment can be extended from the general pattern indicated for the four-block example, that is, for 2^p blocks one sets down the p effects which can be chosen at will. The columns of signs corresponding to these p effects, as obtained from a table such as Table 106.III, are then examined for like p'tples of signs. For $p = 2$, we find all possible combinations of pairs of signs to be $- -, - +, + -, + +$; for $p = 3$, the different triples are $- - -$, $- - +$, $- + +$, $- + -$, $+ - -$, $+ - +$, $+ + -$, $+ + +$; and so on. In general, for any given p then there will be 2^p different p'tples, and consequently the treatment can be arranged accordingly into 2^p different groups.

For a larger number of factors and/or a high number of effects to be confounded, the above system of partitioning of the treatments into blocks or sets corresponding to a given confounding and/or fractional replication can be cumbersome. One can arrive at the same blocking, for a given set of chosen effects which generate the defining contrast subgroup, by listing in one block called the *principal block* the treatments which have an even number (including zero as an even number) of letters in common with the p generators. The other $2^p - 1$ blocks can be obtained by multiplying, under the rule of combination as for the effects, the treatments in the principal block by any treatment not in the block. Thus, for example, if we wish to construct a $\frac{1}{8}$ replicate of a 2^6 factorial experiment, choosing the effects $ABCD$, $CDEF$, and ACE as the three generators or effects to be confounded over the 2^3 blocks, only those treatments as given in Table 106.VIII have an even number of letters in common with these three effects. The extent and the nature of the aliasing of effects resulting for this fractional replicate or any one of the seven other sets obtained from this set in the manner described above are given in Table 106.IX.

TABLE 106.VIII

Treatments for a $\frac{1}{8}$ Replicate of a 2^6 Design with Defining Contrast
Subgroup Generators of *ABCD*, *CDEF*, *ACE*

	Number of letters in common with treatment		
		Effect	
Treatment	*ABCD*	*CDEF*	*ACE*
(1)	0	0	0
abcd	4	2	2
abef	2	2	2
acf	2	2	2
ade	2	2	2
bce	2	2	2
bdf	2	2	0
cdef	2	4	2

It is to be noted from Table 106.VIII that the algebraic product, under the special rule of multiplication we adopted in defining the generalized interaction of two effects, of any two treatment symbols in the principal block gives a treatment in the principal block. This will in fact always be the case and can be used to determine additional treatments to be added to the block as such treatments are found to satisfy the even-number criterion.

We have considered factorial designs and fractional factorial designs

TABLE 106.IX

Aliasing of Effects in a $\frac{1}{8} \cdot 2^6$ Factorial, Defining Contrast
Subgroup Generators: *ABCD*, *CDEF*, *ACE*

I	*ABCD*	*CDEF*	*ACE*	*ABEF*	*BDE*	*ADF*	*BCF*
A	*BCD*	[a]	*CE*	*BEF*	[a]	*DF*	[a]
B	*ACD*	[a]	[a]	*AEF*	*DE*	[a]	*CF*
C	*ABD*	*DEF*	*AE*	[a]	[a]	[a]	*BF*
D	*ABC*	*CEF*	[a]	[a]	*BE*	*AF*	[a]
E	[a]	*CDF*	*AC*	*ABF*	*BD*	[a]	[a]
F	[a]	*CDE*	[a]	*ABE*	[a]	*AD*	*BC*
AB	*CD*	[a]	*BCE*	*EF*	*ADE*	*BDF*	*ACF*

[a] More than three-factor interaction.

involving various numbers of observations. In particular, for a total of eight observations we have seen that such designs can be used:

1. To estimate the effects of three factors and all interactions between them. This constitutes a complete factorial design for three factors.

2. To estimate the effects of four factors and the two-factor interactions between three of them, all other interactions being assumed zero. This constitutes a $\frac{1}{2}$ replication of a 2^4 factorial.

3. To estimate the effects of six factors and one two-factor interaction, all other interactions assumed to be zero or negligble. This constitutes a $\frac{1}{8}$ replication of a 2^6 factorial.

In addition we could determine that eight observations can be used also:

1. To estimate the effects of five factors and an interaction of one factor with each of two others, all other interactions assumed zero or negligible. This constitutes a $\frac{1}{4}$ replication of a 2^5 factorial.

2. To estimate the effects of seven factors, all interactions assumed zero or negligible. This constitutes a $\frac{1}{16}$ replicate of a 2^7 factorial.

Similar summaries can be derived for designs of this type requiring other than eight observations.

c. Determination of Optimum Conditions

Factorial designs of the type 2^n and fractional 2^n are used extensively in the steepest ascent technique of Box and Wilson (5) for determining so-called optimum conditions. This is a fundamental problem in methods development work, and therefore this procedure has found considerable application in the field of analytical chemistry. We shall discuss only the basic fundamentals here; the reader is referred to the original paper or subsequent discussions and extensions (4,11) for a more comprehensive treatment of the method. The basic steps involved in this procedure are as follows:

1. The experimenter conducts an experiment of the 2^n factorial or fractional 2^n type in the space of the base variables.

2. A tangent plane is then fitted to the response surface as described by the results obtained for each treatment.

3. The path of steepest ascent or descent on the tangent plane in terms of the space of the base variables is then determined, and additional experiments are conducted along this path until it appears, in terms of the results obtained, that a region of near maximum or near minimum has been found.

4. Experiments are then conducted in the region of the "apparent" maximum or minimum, and a function of the base variables is fitted to the results obtained. The form and the nature of the fitted function then are used to determine the true nature of the apparent maximum (minimum).

The procedures used under step *4* above are beyond the scope of this chapter. In many cases, however, the results obtained in the application of the first three steps are sufficient to define for all practical purposes the conditions sought, and thus this discussion will be limited to these steps. For a four-factor experiment, say, we shall represent the response obtained by varying the factors A, B, C, D in a manner corresponding to values of x_1, x_2, x_3, x_4, respectively, for the factors, by the function

$$\phi(x_1, x_2, x_3, x_4).$$

The form of this function in any particular case is unknown; however, as a first approximation we shall use the equation of a hyperplane:

$$\phi(x_1, x_2, x_3, x_4) = \beta_0 x_0 + \beta_1 x_1 + \beta_2 x_2 + \beta_3 x_3 + \beta_4 x_4$$

where $x_0 \equiv 1$.

On the basis of data obtained from conducting experiments over the space of the base variables the coefficients (β's) the planar equation are to be estimated. Since only a plane is assumed, only two levels of each variable need be involved in the experiment. Consequently we need consider for the base experiment only 2^n and fractional 2^n designs. From the results of such designs, estimates for the slope coefficients are obtained by applying the signs with coefficient 1 corresponding to the column headings denoting main effects A, B, C, and D and the total in a table such as Table 106.V to the results corresponding to the treatments, summing the numbers so obtained, and dividing the result by the number of treatments (in reality the sum of squares of coefficients). If fractional designs are used, the particular design is chosen which does not confound main effects with other effects believed to be of significance. The reasons for this is apparent in what follows. Thus, if results $y_1, y_2, \ldots, y_8$ were obtained corresponding to the experiments required by the design of Table 106.V, a table of the values of variables x_1, x_2, x_3, x_4 corresponding to factors A, B, C, D, respectively, and the associated response could be summarized in the form of Table 106.X. The estimate, b, for the corresponding β is obtained from the relationship

$$b = \frac{\Sigma yx}{\Sigma x^2}$$

TABLE 106.X

Values of the Independent Variables Corresponding to $\frac{1}{2} \cdot 2^4$
Factorial Experiment and the Responses Obtained

Response	Treatment	x_0	x_1	x_2	x_3	x_4
y_1	(1)	+1	−1	−1	−1	−1
y_2	ab	+1	+1	+1	−1	−1
y_3	ac	+1	+1	−1	+1	−1
y_4	bc	+1	−1	+1	+1	−1
y_5	ad	+1	+1	−1	−1	+1
y_6	bd	+1	−1	+1	−1	+1
y_7	cd	+1	−1	−1	+1	+1
y_8	abcd	+1	+1	+1	+1	+1

where the sum is over all eight responses, and the columns of x's used in the calculation corresponds to the particular slope coefficient being estimated. Thus

$$b_0 = \frac{1 \cdot y_1 + 1 \cdot y_2 + 1 \cdot y_3 + 1 \cdot y_4 + 1 \cdot y_5 + 1 \cdot y_6 + 1 \cdot y_7 + 1 \cdot y_8}{8}$$

$$b_1 = \frac{(1 \cdot y_2 + 1 \cdot y_3 + 1 \cdot y_5 + 1 \cdot y_8) - (1 \cdot y_1 + 1 \cdot y_4 + 1 \cdot y_6 + 1 \cdot y_7)}{8}$$

and so forth.

It is to be noted that by expanding Table 106.X in the usual manner to include product columns the slope estimates so obtained are aliased in precisely the same manner as the corresponding effects were aliased under the given fractional replication design. That is to say, b_1 really estimates $\beta_1 + \beta_{234}$, where β_{234} is the slope coefficient corresponding to the term $x_2 x_3 x_4$ if the planar equation were extended to include all product terms. In the same manner b_2 estimates $\beta_2 + \beta_{134}$ and so on, the nature of the aliasing being obtained for each estimate by reference to the theory of fractional replication as applied to the estimate b_i. In addition, since columns corresponding to headings x_1^2, x_2^2, x_3^2, and x_4^2, obtained by squaring the entries of columns headed by x_1, x_2, x_3, and x_4, respectively, in Table 106.X, are identical to the column headed by x_0, b_0 really estimates $\beta_0 + \beta_{11} + \beta_{22} + \beta_{33} + \beta_{44}$, where β_{jj} denotes the coefficient of x_j^2 if squared terms were also included in an extension of the planar equation. In any given application it is well to keep in mind the nature of the possible biases that could result from an invalid assumption that the plane is an adequate representation in a given case.

When estimates have been obtained for the slope coefficients in the

TABLE 106.XI

Calculation of Path of Steepest Ascent in Four-Factor Experiment

	Variable			
	x_1	x_2	x_3	x_4
1 Base level	B_A	B_B	B_C	B_D
2 Unit	U_A	U_B	U_C	U_D
3 Estimated slope b	b_1	b_2	b_3	b_4
4 Unit × b	$U_A b_1$	$U_B b_2$	$U_C b_3$	$U_D b_4$
5 Change in level per k change in x_1	k	$\dfrac{k}{U_A b_1} U_B b_2$	$\dfrac{k}{U_A b_1} U_C b_3$	$\dfrac{k}{U_A b_1} U_D b_4$
6 Possible trials on the path of steepest ascent	$B_A + jk$	$B_B + j\dfrac{k}{U_A b_1} U_B b_2$	$B_C + j\dfrac{k}{U_A b_1} U_C b_3$	$B_D + j\dfrac{k}{U_A b_1} U_D b_4$

where j is any positive number

planar equation, the path in the base variables of steepest ascent (descent) of the fitted plane can be determined. In defining the procedure to be used, the change in level for the variable corresponding to the change from 0 to 1 in the design space will be called the *unit* and designated for a factor by the symbol U with a letter subscript corresponding to the factor. The value corresponding to the 0 level will be called the *base* and designated by B with the appropriate subscript. Under this notation the calculation is summarized in Table 106.XI for a four-factor experiment. Extension to other numbers of factors is immediate. The calculation of Table 106.XI is based on the fact that, in order for the experiments to proceed along the indicated path of steepest ascent, the factors must be varied in proportion to the estimated slopes as related to the units of the design. This is accomplished in line *4*, which says that for every $U_A b_1$ change in x_1 from its base, x_2 should be changed by $U_B b_2$ amount, x_3 by $U_C b_3$, and x_4 by $U_D b_4$. Consequently, for an arbitrarily chosen convenient increment of k in the variable x_1, say, line *5* gives the corresponding changes for the other variables. The desired experiments, line *6*, as defined by particular values for the base variables, along the path of steepest ascent are obtained by adding to the base (line *1*) multiples of the indicated increments (line *5*). Trials are then conducted along this path.

2. Fractional Replication in Factorial Designs Involving More Than Two Levels of Each Factor

As has been noted, the problem of aliasing and consequently obtaining a suitable fractional replication design when more than two levels of each factor are involved in a factorial-type experiment becomes increasingly difficult as the number of levels of each factor increases.

Except for the case in which only the main effects of the factors are believed to be of consequence and therefore admitted in the mathematical model for the observations, such designs have limited application. A class of designs which can be derived under the assumption of "no interactions" is known as orthogonal squares. For n factors each at t levels (t^n factorial), where $n \leq t$, the orthogonal square design that will facilitate the estimation and corresponding test of significance of the main effects of the factors under the assumption that all interactions are negligible or zero is a $1/t^{n-2}$ replicate. This represents a considerable economy of effort over the full factorial design; however, in analytical chemistry investigations the existence of certain interactions is the rule rather than the exception, and this fact tends to limit the use of these designs in this area.

a. Latin Square Designs

For $n = 3$, the mathematical model for an observation under the no-interaction assumption is

$$X_{hij} = m + a_h + b_i + c_j + e_{j(hi)}$$

where X_{hij} denotes the observation corresponding to the hth level of factor A, ith level of factor B, and jth level of factor C, all factors having the same number, t, of levels. The $1/t$ replicate corresponding to this model, which facilitates estimation of main effects free of aliasing with each other, is known as a Latin square. The manner of construction of the design in a given case could be derived from an extension of the theory of 2^n designs. However, the end result essentially selects, from the set of all possible factor–level combinations corresponding to the full factorial, of which there are t^n, the set that can be arranged in a square array of rows and columns such that the rows correspond to the levels of one factor, the columns corresponds to the levels of another factor, and the third factor is introduced into the body of the array in such a manner that all the levels of the third factor are represented in each row and each column, with each level occurring once and once only in each row and each column. Such a design for $t = 4$ is given in Table 106.XII.

In actuality, for any given number of levels, t, there exist many Latin square arrays which satisfy the basic criterion for such a design. For the above example one could interchange the row subscripts, and/or the column subscripts, and/or the subscripts in the body of the table

TABLE 106.XII

4 × 4 Latin Square Design

Level of factor A	Level of factor B			
	B_1	B_2	B_3	B_4
A_1	C_1	C_2	C_3	C_4
A_2	C_2	C_3	C_4	C_1
A_3	C_3	C_4	C_1	C_2
A_4	C_4	C_1	C_2	C_3

to obtain a different set of treatments still satisfying the Latin square design requirements. It can be shown that there are four basic sets of a 4 × 4 Latin square and 576 different squares in all. The particular square and hence set of 16 treatments to be run in any given case should be selected by a random means. This can be practicably done by setting down a square array of symbols and subscripts, as in Table 106.XII, that satisfies the Latin square criterion and then randomly assigning the particular levels and the factors in the experiment at hand to the symbols and the levels in the array.

b. GRAECO–LATIN SQUARES

For the number of factors equal to 4 ($n = 4$) under the no-interaction assumption, the mathematical model is

$$X_{hij} = m + a_h + b_i + c_j + d_k + e_{k(hij)}$$

where the notation is as before except that the k subscript is used to identify the levels of the additional factor. The $1/t^2$ fractional replication of the t^4 factorial corresponding to this model can be obtained from the Latin square design corresponding to only three factors. This is done by constructing another Latin square of the same size as the first, using the letter D as the designator for the factor appearing in the body of the table, which possesses the properties that, when the two squares are superimposed, each level of the C factor from the first square occurs once and only once in the same cell as each level of the D factor from the second square; and over the t^2 experiments defined by the superimposed array each level of C is in combination with each level of D once and only once. A typical construction is given in Table 106.XIII.

Such pairs of Latin squares having this property are said to be orthogonal. The superimposed resulting square is called a Graeco–Latin square. The name of this design and also of the Latin square design stems from the fact that usually the levels of the C variable are denoted by the Latin letters A, B, C, D, and the levels of the D variable by the Greek letters α, β, γ, δ. With the row and column designations omitted, a Graeco–Latin square of side 4 in this notation corresponding to Table 106.XIII is given in Table 106.XIV.

By examining Table 106.XIII for the Graeco–Latin square it is evident that additional Graeco–Latin squares could be obtained from this square by interchanging rows, columns, or levels of factors C and/or D in

TABLE 106.XIII
4 × 4 Latin Square

Level of factor A	Level of factor B			
	B_1	B_2	B_3	B_4
A_1	C_1	C_2	C_3	C_4
A_2	C_2	C_1	C_4	C_3
A_3	C_3	C_4	C_1	C_2
A_4	C_4	C_3	C_2	C_1

4 × 4 Latin Square Orthogonal to Above Square

Level of factor A	Level of factor B			
	B_1	B_2	B_3	B_4
A_1	D_1	D_2	D_3	D_4
A_2	D_3	D_4	D_1	D_2
A_3	D_4	D_3	D_2	D_1
A_4	D_2	D_1	D_4	D_3

4 × 4 Graeco–Latin Square
Resulting from Superimposing the Above Latin Squares

Level of factor A	Level of factor B			
	B_1	B_2	B_3	B_4
A_1	$C_1 D_1$	$C_2 D_2$	$C_3 D_3$	$C_4 D_4$
A_2	$C_2 D_3$	$C_1 D_4$	$C_4 D_1$	$C_3 D_2$
A_3	$C_3 D_4$	$C_4 D_3$	$C_1 D_2$	$C_2 D_1$
A_4	$C_4 D_2$	$C_3 D_1$	$C_2 D_4$	$C_1 D_3$

TABLE 106.XIV
4 × 4 Graeco–Latin Square

αA	βB	γC	δD
γB	δA	αD	βC
δC	γD	βA	αB
βD	αC	δB	γA

the body of the table. In application one should select in a random manner the particular set of t^2 treatments from all the sets for which the orthogonal property obtains. This can be done effectively in much the same manner as that described in the case of choosing a particular Latin square from those available.

c. ORTHOGONAL SQUARES

With reference to any given Graeco–Latin square design of side t, it is sometimes possible to introduce additional factors into the design, under the assumption of no interaction among any of the factors, super-imposing on the Graeco–Latin square design additional Latin squares, each involving a newly introduced factor, where these squares are orthogonal to each other and to the two Latin squares used in the construction of the Graeco–Latin square. Such a square for the 4 × 4 example of Table 106.XIII, using the letter F to denote the new factor, is given in Table 106.XV. We note from this table that every level of C occurs once in each column, once in each row, once in conjunction with every level of D, and once in conjunction with every level of F. Analogous statements can be made for the levels of D and the levels of F. As in the case of the Graeco–Latin square, orthogonality is not affected by the interchange of rows, columns, C levels, D levels, or F

TABLE 106.XV
Complete Orthogonal Square of Side 4

Level of factor A	Level of factor B			
	B_1	B_2	B_3	B_4
A_1	$C_1D_1F_1$	$C_2D_2F_2$	$C_3D_3F_3$	$C_4D_4F_4$
A_2	$C_2D_3F_4$	$C_1D_4F_3$	$C_4D_1F_2$	$C_3D_2F_1$
A_3	$C_3D_4F_2$	$C_4D_3F_1$	$C_1D_2F_4$	$C_2D_1F_3$
A_4	$C_4D_2F_3$	$C_3D_1F_4$	$C_2D_4F_1$	$C_1D_3F_2$

levels; thus in any given case many orthogonal sets are obtainable from the basic set. Clearly some limitation is imposed on the number of factors that one can effectively study in such a design. For a square of size t on a side one can show that the number of factors, n, must not exceed t, and not always, depending on t, can orthogonal squares be constructed. In fact, for $t = 4$, 5, 6, and 7, it can be shown that there are 3, 4, 0, and 6, respectively, orthogonal squares corresponding to the given t (15).

The mathematical model for an observation obtained under the complete orthogonal set design of Table 106.XV is

$$X_{hijkp} = m + a_h + b_i + c_j + d_k + f_p + e_{p(hijk)}$$

3. Other Fractional-Type Designs

In a previous section we discussed an experiment directed toward comparing two different methods for preparing specimens for emission spectrometry analysis. The design used was characterized by blocking the experimental material into homogeneous sets or test pieces, assigning treatments in each set in a random manner, and finally analyzing the results of each treatment in a random order. For our example only two test pieces in each block were required since only two treatments (methods of preparing specimens) were involved. Providing all treatments are represented in each block, this particular design is known as a complete randomized block and has an associated mathematical model corresponding to a two-factor factorial design in which one of the factors is blocks and the block by other factor interaction is assumed to be negligible.

a. BALANCED INCOMPLETE BLOCKS

It is apparent that the application of a complete randomized block design to a given experiment for which the randomized block model obtains may be physically impossible if the number of treatments (levels of the single factor under investigation) is so large that all treatments cannot be applied in each block, that is, homogeneous sets of sufficient number of experimental units for the application of all the treatments cannot be obtained. Thus the number of treatments per block must be something less than the full set. A randomized block design in which not all treatments are equally represented in every block is said to be an incomplete randomized block design. If every pair of treatments occurs together in a block the same number of times, the incomplete block design is said to be balanced. Balance in incomplete block designs is especially desirable from the statistical analysis point of view and in situations where all treatment comparisons are equally important. For n treatments, k per block balance may obviously be obtained by taking the treatments in all possible combinations; the number of blocks required would be $n!/k!(n-k)!$. This is not necessarily the only manner, however, by which balanced designs may be constructed. For experiments involving higher numbers of treatments it is very often possible

TABLE 106.XVI
Symmetrically Balanced Incomplete Block Designs

Seven Treatments in Seven Blocks of Three

Treatment	1	2	3	4	5	6	7
A_1	x	x	x				
A_2		x	x	x			
A_3	x			x		x	
A_4		x				x	x
A_5	x				x		x
A_6		x			x	x	
A_7		x		x			x

Seven Treatments in Seven Blocks of Four

Treatment	1	2	3	4	5	6	7
A_1		x	x	x			x
A_2	x	x		x		x	
A_3	x		x			x	x
A_4		x	x		x	x	
A_5				x	x	x	x
A_6	x		x	x	x		
A_7	x	x			x		x

Six Treatments in Six Blocks of Five

Treatment	1	2	3	4	5	6
A_1	x	x	x	x	x	
A_2	x	x	x	x		x
A_3	x	x	x		x	x
A_4	x	x		x	x	x
A_5	x		x	x	x	x
A_6		x	x	x	x	x

TABLE 106.XVI (*Continued*)

Five Treatments in Five Blocks of Four

Treatment	1	2	3	4	5
A_1	x	x	x	x	
A_2	x	x	x		x
A_3	x	x		x	x
A_4	x		x	x	x
A_5		x	x	x	x

Four Treatments in Four Blocks of Three

Treatment	1	2	3	4
A_1	x	x	x	
A_2	x	x		x
A_3	x		x	x
A_4		x	x	x

Three Treatments in Three Blocks of Two

Treatment	1	2	3
A_1	x		x
A_2	x	x	
A_3		x	x

to construct balanced incomplete designs involving considerably fewer blocks than would be necessary if one took all combinations of treatments corresponding to the block size.

Table 106.XVI gives several balanced incomplete block designs. We note, for example, that in the design corresponding to seven treatments in seven blocks of three, the number of blocks is considerably fewer than would be required by taking all combinations of the seven treatments in sets of three, that is, $7!/3!\,4! = 35$.

For additional balanced incomplete block designs the reader is referred to the several excellent textbooks on the subject (8,12,14).

b. Symmetrically Balanced Incomplete Blocks

An examination of Table 106.XVI indicates that the designs are balanced not only with respect to blocks but also with respect to treatments, that is, the balance is not destroyed when treatments and blocks are interchanged. When such is the case, the designs are said to be symmetrically balanced and hence have an advantage over the balanced incomplete block designs inasmuch as a block effect corrected for differences between treatments can be estimated as well as the treatment effects corrected for difference between blocks. Thus the effects of two factors, each at the same number of levels, can be investigated under such a design with something less than the number of treatments required under a corresponding full factorial design, providing of course that the interaction between the two factors can be assumed to be negligible.

A necessary and sufficient condition for balanced incomplete block designs to be symmetrically balanced is that the number of blocks equal the number of treatments. Symmetrically balanced incomplete block designs are available in most statistical texts on experimental designs.

V. TESTING HYPOTHESES

After data have been obtained under a given experimental design, the problem becomes one of computing certain functions of the data, which are called statistics, whose sampling distributions (probability distributions) can be derived from the basic assumptions placed on the model. By recourse to a tabulation of such distributions and calculations of corresponding numerical values obtained from the experimental data, inferences with known risk of error can be made. In making these inferences we have occasion to talk about the null hypothesis, designated as H_0. The null hypothesis assumes a specific value for a certain parameter or set of parameters of the model. The assumed value is usually, but not necessarily, zero. For the model $X_{hi} = m + a_h + e_{i(h)}$ a null hypothesis might be that all a_h's are identically zero, that is,

$$H_0: \quad a_h \equiv 0 \qquad \text{for all } h$$

This is equivalent to assuming that the levels of factor A are identical with respect to their effect on the observations.

Since the primary objective of an experiment under this model is to determine the validity of such an assumption, we are concerned with a mechanism for making a "test" of a given hypothesis. The rationale one uses in statistically testing a given null hypothesis, H_0, is as follows:

From a calculated value for a given function of the experimental data, and the corresponding probability distribution of the associated statistic obtained under the null hypothesis assumption on the parameter(s) under test, the probability of obtaining the results observed or others more unfavorable to the null hypothesis assumption than those observed is determined. If the probability is high, no doubt is cast on the validity of the null hypothesis. If the probability is low, it is concluded that either:

a. an improbable event has occurred, or
b. the null hypothesis is false, that is, the assumption placed on the parameters of the model by the null hypothesis is not valid.

Since it is logical to choose a probable rather than an improbable explanation, we reject the null hypothesis if the probability is low enough. The choice of what constitutes "low probability" is to some extent arbitrary. The actual value used in any given study depends to a large extent on the risk the experimenter is willing to assume with regard to making a wrong decision. Whatever the choice may be, the value should be set before obtaining the data and actually performing the test.

Existence of a statistically significant effect is justification for assuming that the null hypothesis is false. However, if no established significance can be associated with the result obtained, the experimenter is not justified in assuming that the null hypothesis is true. All that can be said is that the results obtained do not rule out the null hypothesis as a reasonable assumption on the model. Thus, if the data are very variable and few observations are obtained, differences in effects which the experimenter might consider to be important would probably not be detected. The experimenter must be concerned not only with the error of decision resulting from rejecting the null hypothesis when it is true (significance level of the test), but also with the ability of the test to discriminate with regard to differences of a given magnitude (power of the test). Thus in designing an experiment consideration must be given to

a. the decision error resulting from rejecting the null hypothesis if in fact it is true,
b. the decision error resulting from not rejecting the null hypothesis when in fact differences sufficiently large to be of practical importance exist.

Control over the first type of error, α, is exercised by setting the significance level of the test since the significance level in fact defines this error. Control over the second type of decision error, β, for a given α and magnitude, δ, of difference desired to detect is exercised through setting the total experiment size (number of observations). This is accomplished through replication and/or number of factors. Often the economics of the testing is the factor controlling the total experiment size (number). In this case less attention can be paid, in designing an experiment, to the control of type II error than to the control of type I. For any given experiment size, however, the experimenter should make himself aware of the ability of the design and subsequent test to discriminate between differences of a given magnitude. A satisfactory discussion of how this is done for a given class of designs is beyond the scope of this chapter. Many articles and textbooks (1,3,16), however, treat the subject in various degrees, and the reader is encouraged to familiarize himself with their contents if details are desired. In our treatment we shall be concerned primarily with control over type I error.

VI. ANALYSIS OF VARIANCE

The purpose of this section is to develop for the designs discussed in previous sections the particular functions of the experimental data which are to be computed in order to effect statistical tests concerning assumptions made on the parameters of the mathematical model for the observations.

A. FACTORIAL EXPERIMENTS IN GENERAL

There are many different types of analyses which one could perform on data obtained under the general factorial experimental design. The particular analysis to be used must be keyed to the assumed model for the observation. Since both finite and infinite populations for the levels of factors can be encountered in an actual case, the factors may be crossed or nested, and various interactions may be assumed to be zero, a considerable number of different cases can result, each having its own unique analysis. It would be difficult, if not prohibitive, in a chapter of this sort to detail the analyses for each conceivable type of factorial model, even if relatively few factors were involved; thus we shall proceed to develop the analysis for a particular model, namely, for the case in which all populations to which the inferences are to apply are assumed to be infinite, all factors are assumed to be crossed, and all main effects and interactions are admitted in the model. We shall then show how, through the application of certain routines and rules, the analysis for this model can be altered in a fairly simple manner to correspond to that required for any given factorial model. In so doing the actual computations required may be more than would be the case if the analysis were performed directly; however, the end result is the same and only a single routine need be learned, thus giving some justification to this approach.

The arithmetic detail required in the analysis can best be explained by an example. We shall use a three-factor experiment in which all factor–level combinations are replicated (repeated) n times. Extension of the method to factorial experiments involving more or less than three factors can easily be accomplished by extending and adapting the rules to be defined in the three-factor case to the example at hand. As has been previously noted, the complete three-factor factorial model is given by

$$X_{hijk} = m + a_h + b_i + c_j + (ab)_{hi} + (ac)_{hj} + (bc)_{ij} + (abc)_{hij} + e_{k(hij)}$$

We shall add to the existing notation symbols n_a, n_b, n_c, . . . , n to denote the number of levels of factor A, factor B, factor C, factor. . . , and the number of replicates for each combination, respectively. Also we shall have occasion later to use the symbols N_a, N_b, N_c, . . . , to denote the corresponding population sizes (the number of levels corresponding to factors A, B, and C, . . . , respectively, in the populations to which the inferences resulting from the analysis are to apply). The statistical analysis (17) for a factorial model consists basically of resolving the total variations among all the observations obtained under the model into identifiable components, the number of components corresponding to the number of sets of effects appearing in the model. For a complete three-factor factorial experiment the number of components, as can be seen from the model, is eight when replications are obtained. It may seem strange that the analysis used to determine whether there are fundamental differences among the effects constituting a set is approached through variance analysis. It is to be noted, however, that an assumption that all effects are identical within a set is equivalent to assuming the variation between them to be essentially zero.

For the complete three-factor model the symbols and notations used are summarized in Table 106.XVII. From an examination of this table one can see that the individual observations are first totaled under all possible classifications identified by combinations of factor–level subscripts. The rule for the totals is simply that a dot (.) in a particular position in the subscript for a given total indicates that the sum is over the levels of the corresponding factor. The rule for formulation of a given starred (*) sum of squares is: (1) each squared total involved is divided by the number of observations which make up the total and is summed over the remaining factor levels, (2) the number so obtained

TABLE 106.XVII

Computations for Three-Factor Factorial Experiment—Complete Model

---

Factors: $\qquad\qquad\quad$ $A, B, C.$

Levels: $\qquad\qquad\quad$ $A_1, A_2, \ldots, A_h, \ldots, A_{n_a};$
$\qquad\qquad\qquad\quad$ $B_1, B_2, \ldots, B_i, \ldots, B_{n_b};$
$\qquad\qquad\qquad\quad$ $C_1, C_2, \ldots, Cj, \ldots, C_{n_c}.$

Model: $\quad X_{hijk} = m + a_h + b_i + c_j + (ab)_{hi} + (ac)_{hj} + (bc)_{ij} + (abc)_{hij} + e_{k(hij)}$

To be computed:

$$\sum_k X_{hijk} = T_{hij.} \quad = \text{total of the } n \text{ observations corresponding to the } h\text{th level of } A,$$

$\qquad\qquad\qquad\qquad$ ith level of B, and jth level of C.

$$\sum_{j,k} X_{hijk} = T_{hi..} \quad = \text{total of the } n_c n \text{ observations corresponding to the } h\text{th level}$$

$\qquad\qquad\qquad\qquad$ of A and the ith level of B.

$$\sum_{i,k} X_{hijk} = T_{h.j.} \quad = \text{total of the } n_b n \text{ observations corresponding to the } h\text{th level of}$$

$\qquad\qquad\qquad\qquad$ A and the jth level of C.

$$\sum_{h,k} X_{hijk} = T_{.ij.} \quad = \text{total of the } n_a n \text{ observations corresponding to the } i\text{th level of}$$

$\qquad\qquad\qquad\qquad$ B and the jth level of C.

$$\sum_{i,j,k} X_{hijk} = T_{h...} \quad = \text{total of the } n_b n_c n \text{ observations corresponding to the } h\text{th level}$$

$\qquad\qquad\qquad\qquad$ of A.

$$\sum_{h,j,k} X_{hijk} = T_{.i..} \quad = \text{total of the } n_a n_c n \text{ observations corresponding to the } i\text{th level}$$

$\qquad\qquad\qquad\qquad$ of B.

$$\sum_{h,i,k} X_{hijk} = T_{..j.} \quad = \text{total of the } n_a n_b n \text{ observations corresponding to the } j\text{th level}$$

$\qquad\qquad\qquad\qquad$ of C.

$$\sum_{h,i,j,k} X_{hijk} = T_{....} \quad = \text{total of all } n_a n_b n_c n \text{ observations.}$$

$$S^* \quad = \sum_{h,i,j,k} X_{hijk}^2 - \frac{T_{....}^2}{n_a n_b n_c n}$$

$$S_{abc}^* \quad = \sum_{h,i,j} \frac{T_{hij.}^2}{n} - \frac{T_{....}^2}{n_a n_b n_c n}$$

$$S_{ab}^* \quad = \sum_{h,i} \frac{T_{hi..}^2}{n_c n} - \frac{T_{....}^2}{n_a n_b n_c n}$$

TABLE 106.XVII(*Continued*)

$$S_{ac}^* = \sum_{h,j} \frac{T_{h.j.}^2}{n_b n} - \frac{T_{....}^2}{n_a n_b n_c n}$$

$$S_{bc}^* = \sum_{i,j} \frac{T_{.ij.}^2}{n_a n} - \frac{T_{....}^2}{n_a n_b n_c n}$$

$$S_a^* = \sum_h \frac{T_{h...}^2}{n_b n_c n} - \frac{T_{....}^2}{n_a n_b n_c n}$$

$$S_b^* = \sum_i \frac{T_{.i..}^2}{n_a n_c n} - \frac{T_{....}^2}{n_a n_b n_c n}$$

$$S_c^* = \sum_j \frac{T_{..j.}^2}{n_a n_b n} - \frac{T_{....}^2}{n_a n_b n_c n}$$

is then differenced with the squared total of all observations divided by the total number. The subscripts correspond to the factors over whose levels the squared totals are being summed.

Table 106.XVIII defines the manner in which the total variation as defined by the first expression in this table is resolved into the eight quantities (component sums of squares) given in the subsequent expressions. The rule in Table 106.XVIII for forming the component sums of squares except for the error component, designated with subscript e and obtained by subtraction, is as follows: For given letter subscripts

TABLE 106.XVIII
Variation Analysis in Three-Factor Complete Model

Total variation:

$$S^* = \sum_{h,i,j,k} X_{hijk}^2 - \frac{T_{....}^2}{n_a n_b n_c n} = \text{sum of component sums of squares:}$$

$$S_a = S_a^*$$
$$S_b = S_b^*$$
$$S_c = S_c^*$$
$$S_{ab} = S_{ab}^* - S_a - S_b$$
$$S_{ac} = S_{ac}^* - S_a - S_c$$
$$S_{bc} = S_{bc}^* - S_b - S_c$$
$$S_{abc} = S_{abc}^* - S_{ab} - S_{ac} - S_{bc} - S_a - S_b - S_c$$
$$S_e = S^* - S_a - S_b - S_c - S_{ab} - S_{ac} - S_{bc} - S_{abc}$$

on the left-hand side of a particular equation, the right-hand side is formed by subtracting from the corresponding starred quantity, as defined in Table 106.XVII, all component sums of squares which involve the subscripts on the left, one at a time, two at a time, . . . , $s-1$ at a time, where s is the number of subscripts. The generalization of the definitions and computational procedure to any number of factors is immediate.

Now, depending on certain assumptions made on the model, one can determine for each of these components, when divided by an appropriate number called the degrees of freedom, an expected or average value termed the expected mean square. This expected value provides the key to the manner by which tests of the statistical hypotheses (assumptions on the parameters of the model) are to be made. It is expedient for our purpose to consider the case in which the effects in the model can be considered as random samples from infinite populations. We define by the symbols $V(a)$, $V(b)$, $V(c)$, $V(ab)$, $V(ac)$, $V(bc)$, $V(abc)$, and $V(e)$ the variances (18) or average squared deviations from the population mean of these populations.

With this notation Table 106.XIX summarizes the computations and corresponding expected value for each component for the model under consideration. The model, assumptions, and notations can be easily generalized to an investigation involving any number of factors. From an inspection of Table 106.XIX it can be seen that the number of degrees of freedom for a given component with the exception of the error is obtained by multiplying together the number of levels reduced by one for all factors corresponding to the subscripts of the component. The

TABLE 106.XIX

Expected Mean Squares in Crossed Three-Factor Factorial Infinite Population Model

Component sum of squares	Degrees of freedom	Expected mean square
S_a	$n_a - 1$	$V(e) + nV(abc) + n_cnV(ab) + n_bnV(ac) + n_bn_cnV(a)$
S_b	$n_b - 1$	$V(e) + nV(abc) + n_anV(bc) + n_cnV(ab) + n_an_cnV(b)$
S_c	$n_c - 1$	$V(e) + nV(abc) + n_anV(bc) + n_bnV(ac) + n_an_bnV(c)$
S_{ab}	$(n_a - 1)(n_b - 1)$	$V(e) + nV(abc) + n_cnV(ab)$
S_{ac}	$(n_a - 1)(n_c - 1)$	$V(e) + nV(abc) + n_bnV(ac)$
S_{bc}	$(n_b - 1)(n_c - 1)$	$V(e) + nV(abc) + n_anV(bc)$
S_{abc}	$(n_a - 1)(n_b - 1)(n_c - 1)$	$V(e) + nV(abc)$
S_e	$n_an_bn_c(n - 1)$	$V(e)$
$S*$	$n_an_bn_cn - 1$	

The mean square (M.S.) is formed from the ratio of the component sum of squares to its corresponding degrees of freedom:

number of degrees of freedom for the error component can be found by subtraction (i.e., the total of all components must add to the total number of observations minus one), or is simply the number of replicates, minus one, multiplied by the number of levels of all factors. Under the assumption that all populations are independently normally or Gaussian distributed as characterized by the familiar bell-shaped curve, a statistical test as to whether a given population variance component might be zero is effected as a result of the fact that the ratio of any two mean squares from this table for which the expected values are the same, if the hypothesis to be tested is true, has a completely specified probability distribution with parameters corresponding to the degrees of freedom for the two mean squares. This distribution is known as the F distribution, and a tabulation as given in Table 106.XX has arguments corresponding to the values for the degrees of freedom and the commonly used values for the type I (α) error.

It is conventional, in forming ratios of mean squares in the process of testing hypotheses as to zero value for certain population variances, to use as the numerator for a given pair the mean square which has the larger expected value in the event that H_0 is false. Under this convention, in testing for the significance of an observed ratio we need only determine probabilities of observing F values *larger* than the value actually observed from evaluating the ratio. Values for the ratio of less than 1 therefore automatically result in a "cannot reject H_0" inference. Thus to test the hypothesis that there is no three-factor interaction effect in this model, that is, H_0: $V(abc) = 0$, we would form the ratio

$$\frac{\text{M.S. } (abc)}{\text{M.S. } (e)}$$

and compare the number obtained therefrom to a critical value obtained from the table of F values corresponding to $(n_a - 1)(n_b - 1)(n_c - 1)$ and $n_a n_b n_c (n - 1)$ degrees of freedom. Similarly the ratios

$$\frac{\text{M.S. } (bc)}{\text{M.S. } (abc)}, \quad \frac{\text{M.S. } (ac)}{\text{M.S. } (abc)}, \quad \frac{\text{M.S. } (ab)}{\text{M.S. } (abc)}$$

can be used to test in turn the null hypotheses, H_0, that $V(bc) = 0$, $V(ac) = 0$, and $V(ab) = 0$, respectively, regardless of the outcome of previous tests. We note, however, from Table 106.XIX that, unless certain of the population interaction variances test to be zero, simple ratios of single-component mean squares having the same expected value under the hypotheses of zero variance corresponding to a main effect cannot

TABLE 106.XX

Percentage Points for the F Distribution[a]

ν_1 = degrees of freedom for numerator; ν_2 = degrees of freedom for denominator

ν_2	α	ν_1=1	2	3	4	5	6	7	8	9	10	12	15	20	30	60	120	∞
1	0.10	39.9	49.5	53.6	55.8	57.2	58.2	58.9	59.4	59.9	60.2	60.7	61.2	61.7	62.3	62.8	63.1	63.3
	0.05	161	200	216	225	230	234	237	239	241	242	244	246	248	250	252	253	254
	0.01	4,050	5,000	5,400	5,620	5,760	5,860	5,930	5,980	6,020	6,060	6,110	6,160	6,210	6,260	6,310	6,340	6,370
2	0.10	8.53	9.00	9.16	9.24	9.29	9.33	9.35	9.37	9.38	9.39	9.41	9.42	9.44	9.46	9.47	9.48	9.49
	0.05	18.5	19.0	19.2	19.2	19.3	19.3	19.4	19.4	19.4	19.4	19.4	19.4	19.5	19.5	19.5	19.5	19.5
	0.01	98.5	99.0	99.2	99.2	99.3	99.3	99.4	99.4	99.4	99.4	99.4	99.4	99.4	99.5	99.5	99.5	99.5
3	0.10	5.54	5.46	5.39	5.34	5.31	5.28	5.27	5.25	5.24	5.23	5.22	5.20	5.18	5.17	5.15	5.14	5.13
	0.05	10.1	9.55	9.28	9.12	9.01	8.94	8.89	8.85	8.81	8.79	8.74	8.70	8.66	8.62	8.57	8.55	8.53
	0.01	34.1	30.8	29.5	28.7	28.2	27.9	27.7	27.5	27.3	27.2	27.1	26.9	26.7	26.5	26.3	26.2	26.1
4	0.10	4.54	4.32	4.19	4.11	4.05	4.01	3.98	3.95	3.93	3.92	3.90	3.87	3.84	3.82	3.79	3.78	3.76
	0.05	7.71	6.94	6.59	6.39	6.26	6.16	6.09	6.04	6.00	5.96	5.91	5.86	5.80	5.75	5.69	5.66	5.63
	0.01	21.2	18.0	16.7	16.0	15.5	15.2	15.0	14.8	14.7	14.5	14.4	14.2	14.0	13.8	13.7	13.6	13.5
5	0.10	4.06	3.78	3.62	3.52	3.45	3.40	3.37	3.34	3.32	3.30	3.27	3.24	3.21	3.17	3.14	3.12	3.11
	0.05	6.61	5.79	5.41	5.19	5.05	4.95	4.88	4.82	4.77	4.74	4.68	4.62	4.56	4.50	4.43	4.40	4.37
	0.01	16.3	13.3	12.1	11.4	11.0	10.7	10.5	10.3	10.2	10.1	9.89	9.72	9.55	9.38	9.20	9.11	9.02
6	0.10	3.78	3.46	3.29	3.18	3.11	3.05	3.01	2.98	2.96	2.94	2.90	2.87	2.84	2.80	2.76	2.74	2.72
	0.05	5.99	5.14	4.76	4.53	4.39	4.28	4.21	4.15	4.10	4.06	4.00	3.94	3.87	3.81	3.74	3.70	3.67
	0.01	13.7	10.9	9.78	9.15	8.75	8.47	8.26	8.10	7.98	7.87	7.72	7.56	7.40	7.23	7.06	6.97	6.88
7	0.10	3.59	3.26	3.07	2.96	2.88	2.83	2.78	2.75	2.72	2.70	2.67	2.63	2.59	2.56	2.51	2.49	2.47
	0.05	5.59	4.74	4.35	4.12	3.97	3.87	3.79	3.73	3.68	3.64	3.57	3.51	3.44	3.38	3.30	3.27	3.23
	0.01	12.2	9.55	8.45	7.85	7.46	7.19	6.99	6.84	6.72	6.62	6.47	6.31	6.16	5.99	5.82	5.74	5.65
8	0.10	3.46	3.11	2.92	2.81	2.73	2.67	2.62	2.59	2.56	2.54	2.50	2.46	2.42	2.38	2.34	2.31	2.29
	0.05	5.32	4.46	4.07	3.84	3.69	3.58	3.50	3.44	3.39	3.35	3.28	3.22	3.15	3.08	3.01	2.97	2.93
	0.01	11.3	8.65	7.59	7.01	6.63	6.37	6.18	6.03	5.91	5.81	5.67	5.52	5.36	5.20	5.03	4.95	4.86

v_2	P	1	2	3	4	5	6	7	8	9	10	12	15	20	30	60	120	∞
9	0.10	3.36	3.01	2.81	2.69	2.61	2.55	2.51	2.47	2.44	2.42	2.38	2.34	2.30	2.25	2.21	2.18	2.16
	0.05	5.12	4.26	3.86	3.63	3.48	3.37	3.29	3.23	3.18	3.14	3.07	3.01	2.94	2.86	2.79	2.75	2.71
	0.01	10.6	8.02	6.99	6.42	6.06	5.80	5.61	5.47	5.35	5.26	5.11	4.96	4.81	4.65	4.48	4.40	4.31
10	0.10	3.29	2.92	2.73	2.61	2.52	2.46	2.41	2.38	2.35	2.32	2.28	2.24	2.20	2.15	2.11	2.08	2.06
	0.05	4.96	4.10	3.71	3.48	3.33	3.22	3.14	3.07	3.02	2.98	2.91	2.84	2.77	2.70	2.62	2.58	2.54
	0.01	10.0	7.56	6.55	5.99	5.64	5.39	5.20	5.06	4.94	4.85	4.71	4.56	4.41	4.25	4.08	4.00	3.91
12	0.10	3.18	2.81	2.61	2.48	2.39	2.33	2.28	2.24	2.21	2.19	2.15	2.10	2.06	2.01	1.96	1.93	1.90
	0.05	4.75	3.89	3.49	3.26	3.11	3.00	2.91	2.85	2.80	2.75	2.69	2.62	2.54	2.47	2.38	2.34	2.30
	0.01	9.33	6.93	5.95	5.41	5.06	4.82	4.64	4.50	4.39	4.30	4.16	4.01	3.86	3.70	3.54	3.45	3.36
15	0.10	3.07	2.70	2.49	2.36	2.27	2.21	2.16	2.12	2.09	2.06	2.02	1.97	1.92	1.87	1.82	1.79	1.76
	0.05	4.54	3.68	3.29	3.06	2.90	2.79	2.71	2.64	2.59	2.54	2.48	2.40	2.33	2.25	2.16	2.11	2.07
	0.01	8.68	6.36	5.42	4.89	4.56	4.32	4.14	4.00	3.89	3.80	3.67	3.52	3.37	3.21	3.05	2.96	2.87
20	0.10	2.97	2.59	2.38	2.25	2.16	2.09	2.04	2.00	1.96	1.94	1.89	1.84	1.79	1.74	1.68	1.64	1.61
	0.05	4.35	3.49	3.10	2.87	2.71	2.60	2.51	2.45	2.39	2.35	2.28	2.20	2.12	2.04	1.95	1.90	1.84
	0.01	8.10	5.85	4.94	4.43	4.10	3.87	3.70	3.56	3.46	3.37	3.23	3.09	2.94	2.78	2.61	2.52	2.42
30	0.10	2.88	2.49	2.28	2.14	2.05	1.98	1.93	1.88	1.85	1.82	1.77	1.72	1.67	1.61	1.54	1.50	1.46
	0.05	4.17	3.32	2.92	2.69	2.53	2.42	2.33	2.27	2.21	2.16	2.09	2.01	1.93	1.84	1.74	1.68	1.62
	0.01	7.56	5.39	4.51	4.02	3.70	3.47	3.30	3.17	3.07	2.98	2.84	2.70	2.55	2.39	2.21	2.11	2.01
60	0.10	2.79	2.39	2.18	2.04	1.95	1.87	1.82	1.77	1.74	1.71	1.66	1.60	1.54	1.48	1.40	1.35	1.29
	0.05	4.00	3.15	2.76	2.53	2.37	2.25	2.17	2.10	2.04	1.99	1.92	1.84	1.75	1.65	1.53	1.47	1.39
	0.01	7.08	4.98	4.13	3.65	3.34	3.12	2.95	2.82	2.72	2.63	2.50	2.35	2.20	2.03	1.84	1.73	1.60
120	0.10	2.75	2.35	2.13	1.99	1.90	1.82	1.77	1.72	1.68	1.65	1.60	1.54	1.48	1.41	1.32	1.26	1.19
	0.05	3.92	3.07	2.68	2.45	2.29	2.18	2.09	2.02	1.96	1.91	1.83	1.75	1.66	1.55	1.43	1.35	1.25
	0.01	6.85	4.79	3.95	3.48	3.17	2.96	2.79	2.66	2.56	2.47	2.34	2.19	2.03	1.86	1.66	1.53	1.38
∞	0.10	2.71	2.30	2.08	1.94	1.85	1.77	1.72	1.67	1.63	1.60	1.55	1.49	1.42	1.34	1.24	1.17	1.00
	0.05	3.84	3.00	2.60	2.37	2.21	2.10	2.01	1.94	1.88	1.83	1.75	1.67	1.57	1.46	1.32	1.22	1.00
	0.01	6.63	4.61	3.78	3.32	3.02	2.80	2.64	2.51	2.41	2.32	2.18	2.04	1.88	1.70	1.47	1.32	1.00

[a] Abridged from "Tables of Percentage Points of the Inverted Beta Distribution," *Biometrika*, **33** (1943), and published here with the permission of the authors, Maxine Merrington and Catherine M. Thompson, and the editor of *Biometrika*.

be found. However, we can form linear sums of mean squares such that the sum of the corresponding expected values equals the expected value of a given component under the hypothesis to be tested. For example, we note that under the hypothesis H_0: $V(c) = 0$, M.S. (bc) + M.S. (ac) − M.S. (abc) has the expected value $V(e) + nV(abc) + n_b nV(ac) + n_a nV(bc)$, as does M.S. (c) if in fact $V(c)$ is zero. Similar linear sums can be derived for comparison with $S_b/(n_b - 1)$ and $S_a/(n_a - 1)$.)

Technically speaking, the ratios so formed do not have the F distribution; however, an approximate test is effected by treating the ratio in a given case as if it were F, adjusting the degrees of freedom to be identified with the linear sum by the formula (19,21)

$$f' = \frac{[\text{M.S. } (bc) + \text{M.S. } (ac) - \text{M.S. } (abc)]^2}{\dfrac{[\text{M.S. } (bc)]^2}{(n_b - 1)(n_c - 1)} + \dfrac{[\text{M.S. } (ac)]^2}{(n_a - 1)(n_c - 1)} + \dfrac{[\text{M.S. } (abc)]^2}{(n_a - 1)(n_b - 1)(n_c - 1)}}$$

The general formula for the adjusted degrees of freedom associated with a mean square expressed as a linear sum of other mean squares in the form

$$\text{M.S.} = \sum_i A_i (\text{M.S.})_i$$

is simply

$$f' = \frac{(\text{M.S.})^2}{\sum_i A_i^2 (\text{M.S.})_i^2 / \gamma_i}$$

where γ_i represents the degrees of freedom corresponding to $(\text{M.S.})_i$.

The usual case is that, in the process of testing the interactions by simple ratios, certain of the interactions test to be indistinguishable from zero; as a consequence, when zero is substituted for these interactions in the expected values corresponding to the mean squares to be used in testing the main effects, simple ratios can be found to facilitate the test. This brings us to a discussion of "pooling" mean squares. Whenever, as a result of statistically testing and identifying certain variances as being indistinguishable from zero, one can, upon substituting zero for these variances in the expected values for other components, find several mean squares having the same expected value, a more discriminating test can be effected (smaller type II error) if the mean squares having the common expected value are averaged in a weighted manner, where the weights for each mean square involved are the corresponding degrees of freedom.

For example, if in a given case as a result of forming the ratio

$$\frac{\text{M.S. } (abc)}{\text{M.S. } (e)}$$

we infer that H_0: $V(abc) = 0$ cannot be rejected, we can substitute $V(abc) = 0$ in the expected value for the mean square corresponding to S_{abc} and note as a consequence that the expected values for the mean squares corresponding to S_e and S_{abc} are the same, that is, $V(e)$. Consequently, they can be pooled in the manner described above, resulting in the mean square

$$\frac{n_a n_b n_c (n - 1)\text{M.S. } (e) + (n_a - 1)(n_b - 1)(n_c - 1)\text{M.S. } (abc)}{n_a n_b n_c (n - 1) + (n_a - 1)(n_b - 1)(n_c - 1)}$$

whose expected value is $V(e)$. We note that the operation of pooling consists simply of adding together the appropriate sums of squares and dividing by the sum of the corresponding degrees of freedom.

Now let us consider other three-factor factorial models in order to introduce into the computational scheme the consequences of certain factors in the experiment being nested within other factors, and certain populations of effects being finite. Suppose that, for a given experiment, the experimenter, on the basis of his experience with the phenomenon under investigation, believes the A factor to be crossed, the B factor to be nested within the A factor, and the C factor to be crossed. The model would then be

$$X_{hijk} = m + a_h + c_j + (ab)_{i(h)} + e_{k(hij)}$$

the data consisting of n observations at each factor–level combination. We shall further complicate the model by assuming that each of the populations of a_h effects, c_j effects, and $(ab)_{i(h)}$ effects is finite. Under this assumption it is implicit to the analysis that the particular levels of A, B, and C in the experiment be randomly chosen from the corresponding populations.

An examination of the reduced model, as compared to the corresponding complete model, indicates that certain effects are not admitted; this is equivalent to admitting the effects but assuming the variance between them to be zero. For the above model the variances identified as identically zero by virtue of the model are $V(b)$, $V(ac)$, $V(bc)$, $V(abc)$. This relationship provides a key to the manner in which we can use the method of analysis defined for the complete model to obtain that corresponding to any reduced model and, in particular, the one given above. Suppose that in Table 106.XIX, ignoring for the moment that all populations were assumed infinite in that summary, we substitute in the "Expected Mean Square" column zero for those variances for which no corresponding effects are admitted in the model of the example, that is, $V(b)$, $V(ac)$, $V(bc)$, $V(abc)$. We note as a result that certain mean squares have the same expected value after the substitution; namely, the mean squares corresponding to S_{ac}, S_{bc}, S_{abc}, and S_e, all have expected value $V(e)$, and the mean squares corresponding to S_b and S_{ab} have expected value $V(e) + nV(ab)$. As a consequence they can be pooled in accordance with the procedure previously discussed to arrive at the components pertinent to the given model and as summarized in Table 106.XXI. The pooled mean square and corresponding sum of squares are identified in a particular case by the symbol $\tilde{S}$ with a subscript consistent with the corresponding component effect in the reduced model.

Next we shall consider the expected values of the pooled mean squares obtained in the manner above, since these are needed in order to effect correct statistical tests. They could be read directly from Table 106.XIX if we were concerned always with three factors and infinite populations. Since such is not the case, however, it is necessary to develop for any given model and corresponding mean squares a procedure for determining the expected values of the mean squares. These could be obtained by substituting the equation of the model into each functional form for each mean square and under the basic assumptions on the model, obtained

TABLE 106.XXI

Sum of Squares Corresponding to the Model

$X_{hijk} = m + a_h + c_j + (ab)_{i(h)} + e_{k(hij)}$

Pooled sum of squares	Corresponding sum of squares under complete model	Degrees of freedom
$\tilde{S}_a$	S_a	$n_a - 1$
$\tilde{S}_c$	S_c	$n_c - 1$
$\tilde{S}_{ab}$	$S_b + S_{ab}$	$n_a(n_b - 1)$
$\tilde{S}_e$	$S_{ac} + S_{bc} + S_{abc} + S_e$	$n_a n_b n_c n - n_a n_b - n_c + 1$

through distribution theory the corresponding expected or average values. This method is not for the occasional practitioner. Consequently, Bennett and Franklin (2) have summarized the end result of such a procedure in a set of rules which when applied enable one to write down with comparative ease the expected value of any given mean square obtained under a given factorial model in the manner described above. The procedure requires the construction of a table in which the symbolic representations of the effects of the reduced model are listed in the left-hand margin and define rows, and the indices of summation are listed along the top and define columns. The rules for filling in the tables are as follows:

1. Whenever the column identification index does not appear as a subscript in the row effect identification, write the number of levels in the experiment of the factor corresponding to that index.
2. Whenever the column index corresponds to a letter in parentheses in the row subscript, write unity.
3. Whenever the column index corresponds to a letter in the row subscript but not in parentheses, write one minus the sampling fraction for the population corresponding to the column index. The sampling fraction for a particular factor is simply the ratio of the number of levels of that factor in the given experiment to the number of levels in the corresponding population to which the inferences are to be applied; for example, for factor A, the sampling fraction is n_a/N_a, for B it is n_b/N_b, and so forth.

Each expected mean square is a linear sum of variances. For the pooled error component mean square the sum consists simply of the error variance. For the other pooled component mean squares the variances present in the linear sum for a particular component are the error variance plus all those nonzero variances which have *at least* the same letters as the component in their symbolic representation. Thus the expected mean square associated with $S(abc)$ could involve $V(e)$ and $V(abc)$; $S(ab)$ could involve $V(e)$, $V(ab)$, and $V(abc)$; $S(a)$ could involve $V(e)$, $V(a)$, $V(ab)$, $V(ac)$, $V(abc)$, etc., depending on which are nonzero under the given model. The coefficient of a particular variance in the expected mean square for a given component other than the error is the product of all the numbers in the corresponding row excluding the numbers in the columns associated with the particular component, that is, column h is excluded for component $S(a)$, columns h and i for $S(ab)$, columns h, i, and j for $S(abc)$, and so forth. Application of the

TABLE 106.XXII

Variance Component Calculation for Three-Factor Study:
A Factor Crossed, B Factor Nested within A Factor, C Factor Crossed

Pooled sum of squares	Degrees of freedom	Effect	Index				Expected mean square (finite populations)
			h	i	j	k	
$\tilde{S}_a$	$n_a - 1$	a_h	f_a	n_b	n_c	n	$V(e) + f_b n_c n V(ab)$ $+ n_b n_c n V(a)$
$\tilde{S}_c$	$n_c - 1$	c_j	n_a	n_b	f_c	n	$V(e) + n_a n_b n V(c)$
$\tilde{S}_{ab}$	$n_a(n_b - 1)$	$(ab)_{i(h)}$	1	f_b	n_c	n	$V(e) + n_c n V(ab)$
$\tilde{S}_e$	$n_a n_b n_c n - n_a n_b$ $- n_c + 1$	$e_{k(hij)}$	1	1	1	1	$V(e)$

rules and the computation of expected values for the model under consideration are given in Table 106.XXII, where

$$ f_a = 1 - (n_a/N_a), \ f_b = 1 - (n_b/N_b), \ f_c = 1 - (n_c/N_c). $$

In summary, then, for any factorial-type model the analysis consists of the following steps:

First: Compute the component sums of squares as if all factors were crossed (general factorial model) and relate the associated mean squares to expected mean squares, assuming all populations are infinite.

Second: Corresponding to the particular investigation, identify the zero variance components in the general model and make this substitution in the expected mean squares of the general model to determine which components are to be pooled.

Third: Write the reduced mathematical model corresponding to the pooled system with nesting identified by parentheses in the subscripts.

Fourth: With reference to the reduced model and application of certain rules enumerated above, construct a table to facilitate the expected mean square construction.

Fifth: For each pooled component mean square construct the corresponding expected mean square by applying the rules of composition to determine the individual terms and the associated coefficients. For all populations assumed to be infinite substitute unity for the corresponding f factor whenever it appears in the general linear sums, and zero for the populations assumed to be totally sampled during the course of the investigation.

Table 106.XXIII summarizes the application of these steps to another factorial-type model.

TABLE 106.XXIII
Variance Component Calculation for Three-Factor Study: A Factor Crossed, B Factor Crossed, C Factor Nested

Sum of squares	Degrees of freedom	Complete model $X_{hijk} = m + a_h + b_i + c_j + (ab)_{hi} + (ac)_{hj} + (bc)_{ij} + (abc)_{hij} + e_{k(hij)}$ Expected mean square (infinite populations)	Reduced model $X_{hijk} = m + a_h + b_i + (ab)_{hi} + (abc)_{j(hi)} + e_{k(hi)}$ Expected mean square (infinite populations)
S_a	$n_a - 1$	$V(e) + nV(abc) + n_cnV(ab) + n_bnV(ac) + n_bn_cnV(a)$	$V(e) + nV(abc) + n_cnV(ab) + n_bn_cnV(a)$
S_b	$n_b - 1$	$V(e) + nV(abc) + n_anV(bc) + n_cnV(ab) + n_an_cnV(b)$	$V(e) + nV(abc) + n_cnV(ab) + n_an_cnV(b)$
S_c	$n_c - 1$	$V(e) + nV(abc) + n_bnV(ac) + n_anV(bc) + n_an_bnV(c)$	$V(e) + nV(abc)$
S_{ab}	$(n_a - 1)(n_b - 1)$	$V(e) + nV(abc) + n_cnV(ab)$	$V(e) + nV(abc) + n_cnV(ab)$
S_{ac}	$(n_a - 1)(n_c - 1)$	$V(e) + nV(abc) + n_bnV(ac)$	$V(e) + nV(abc)$
S_{bc}	$(n_b - 1)(n_c - 1)$	$V(e) + nV(abc) + n_anV(bc)$	$V(e) + nV(abc)$
S_{abc}	$(n_a - 1)(n_b - 1)(n_c - 1)$	$V(e) + nV(abc)$	$V(e) + nV(abc)$
S_e	$n_an_bn_c(n - 1)$	$V(e)$	$V(e)$

Pooled sum of squares	Degrees of freedom	Effect	Index				Expected mean square for reduced model
			h	i	j	k	
$\tilde{S}_a$	$n_a - 1$	a_h	f_a	n_b	n_c	n	$V(e) + f_cnV(abc) + f_bn_cnV(ab) + n_bn_cnV(a)$
$\tilde{S}_b$	$n_b - 1$	b_i	n_a	f_b	n_c	n	$V(e) + f_cnV(abc) + f_an_cnV(ab) + n_an_cnV(b)$
$\tilde{S}_{ab}$	$(n_a - 1)(n_b - 1)$	$(ab)_{hi}$	f_a	f_b	n_c	n	$V(e) + f_cnV(abc) + n_cnV(ab)$
$\tilde{S}_{abc}$	$n_an_b(n_c - 1)$	$(abc)_{j(hi)}$	1	1	f_c	n	$V(e) + nV(abc)$
$\tilde{S}_e$	$n_an_bn_a(n - 1)$	$e_{k(hi)}$	1	1	1	1	$V(e)$

B. SIMPLE COMPARATIVE EXPERIMENTS

As we previously noted, the mathematical model for a simple comparative experiment is given by

$$X_{hij} = m + a_h + b_i + e_{j(hi)},$$
$$h = 1, 2; \ i = 1, 2, \ldots, n_b; \ j = 1, 2, \ldots, n$$

where the a_h effects are identified with the factor of interest, and the b_i effects with a nuisance factor usually associated with the environmental conditions under which the experiment is conducted, that is, with likeness of test conditions. Since the model corresponds to a two-factor factorial experiment, with no interaction, the data obtained therefrom can be analyzed by application of the procedures of Section VI.A. However, for the case when $n = 1$, that is, when there is only a single observation for each factor–level combination, an equivalent and somewhat simpler analysis can be made. This is accomplished by defining

$$D_i = X_{1i} - X_{2i}$$

computing

$$\bar{D} = \sum_{i=1}^{n_b} \frac{D_i}{n_b}$$

$$s_D^2 = \sum_{i=1}^{n_b} \frac{(D_i - \bar{D})^2}{n_b - 1}$$

and forming the ratio

$$t_{n_b-1} = \frac{\bar{D}}{s_D} \sqrt{n_b}$$

Under the given model and the assumptions that $a_1 = a_2$ and that the error terms are normally and independently distributed, the ratio t_{n_b-1} has the distribution (Student's t) for which certain percentage points are given in Table 106.XXIV. A test of the hypothesis $H_0: \ a_1 = a_2$ is effected then by comparing the computed value of t_{n_b-1} in a given case to the corresponding tabled values for the distribution and making an inference on H_0 in the manner of Section V, recognizing of course that the larger the values of t_{n_b-1} are, both positively and negatively, the more the data tend to disagree with H_0.

There exist several alternative procedures for testing the null hypothesis $H_0: \ a_1 = a_2$ in simple comparative experiments where the observations are paired. These procedures are more rapid than the classical

TABLE 106.XXIV
Percentage Points of the t Distribution[a]
ν = degrees of freedom

$\nu \backslash \alpha$	0.50	0.25	0.10	0.05	0.025	0.01	0.005
1	1.00000	2.4142	6.3138	12.706	25.452	63.657	127.32
2	0.81650	1.6036	2.9200	4.3027	6.2053	9.9248	14.089
3	0.76489	1.4226	2.3534	3.1825	4.1765	5.8409	7.4533
4	0.74070	1.3444	2.1318	2.7764	3.4954	4.6041	5.5976
5	0.72669	1.3009	2.0150	2.5706	3.1634	4.0321	4.7733
6	0.71756	1.2733	1.9432	2.4469	2.9687	3.7074	4.3168
7	0.71114	1.2543	1.8946	2.3646	2.8412	3.4995	4.0293
8	0.70639	1.2403	1.8595	2.3060	2.7515	3.3554	3.8325
9	0.70272	1.2297	1.8331	2.2622	2.6850	3.2498	3.6897
10	0.69981	1.2213	1.8125	2.2281	2.6338	3.1693	3.5814
11	0.69745	1.2145	1.7959	2.2010	2.5931	3.1058	3.4966
12	0.69548	1.2089	1.7823	2.1788	2.5600	3.0545	3.4284
13	0.69384	1.2041	1.7709	2.1604	2.5326	3.0123	3.3725
14	0.69242	1.2001	1.7613	2.1448	2.5096	2.9768	3.3257
15	0.69120	1.1967	1.7530	2.1315	2.4899	2.9467	3.2860
16	0.69013	1.1937	1.7459	2.1199	2.4729	2.9208	3.2520
17	0.68919	1.1910	1.7396	2.1098	2.4581	2.8982	3.2225
18	0.68837	1.1887	1.7341	2.1009	2.4450	2.8784	3.1966
19	0.68763	1.1866	1.7291	2.0930	2.4334	2.8609	3.1737
20	0.68696	1.1848	1.7247	2.0860	2.4231	2.8453	3.1534
21	0.68635	1.1831	1.7207	2.0796	2.4138	2.8314	3.1352
22	0.68580	1.1816	1.7171	2.0739	2.4055	2.8188	3.1188
23	0.68531	1.1802	1.7139	2.0687	2.3979	2.8073	3.1040
24	0.68485	1.1789	1.7109	2.0639	2.3910	2.7969	3.0905
25	0.68443	1.1777	1.7081	2.0595	2.3846	2.7874	3.0782
26	0.68405	1.1766	1.7056	2.0555	2.3788	2.7787	3.0669
27	0.68370	1.1757	1.7033	2.0518	2.3734	2.7707	3.0565
28	0.68335	1.1748	1.7011	2.0484	2.3685	2.7633	3.0469
29	0.68304	1.1739	1.6991	2.0452	2.3638	2.7564	3.0380
30	0.68276	1.1731	1.6973	2.0423	2.3596	2.7500	3.0298
40	0.68066	1.1673	1.6839	2.0211	2.3289	2.7045	2.9712
60	0.67862	1.1616	1.6707	2.0003	2.2991	2.6603	2.9146
120	0.67656	1.1559	1.6577	1.9799	2.2699	2.6174	2.8599
∞	0.67449	1.1503	1.6449	1.9600	2.2414	2.5758	2.8070

[a] Computed by Maxine Merrington from "Tables of Percentage Points of the Incomplete Beta Function," *Biometrika*, **32**, 168-181 (1941), by Catherine M. Thompson and reproduced by permission of Catherine M. Thompson and the editors of Biometrika.

methods described above and do not require the assumption of normality with regard to the error term of the model. They result, however, in a larger type II error (β) of decision for any given number of pairs than do the classical tests. The most useful of these procedures is Wilcoxon's signed rank test (22). If one can assume that the underlying distributions are symmetric then a test on the equality of the means is effected as follows: Obtain the algebraic difference between the members

TABLE 106.XXV

Percentage Points for Signed Ranks[a]

The probability is approximately α that the Wilcoxon sum W, obtained from n_b ranks and assigned positive or negative signs at random with equal probability, will be less than the values tabulated.

Number of pairs, n_b	Significance level			
	$\alpha = 0.10$	$\alpha = 0.05$	$\alpha = 0.02$	$\alpha = 0.01$
5	0.6	. . .	. . .	. . .
6	2.1	0.6	. . .	. . .
7	3.7	2.1	0.3	. . .
8	5.8	3.7	1.6	0.3
9	8.1	5.7	3.1	1.6
10	10.8	8.1	5.1	3.1
11	13.9	10.8	7.2	5.1
12	17.5	13.8	9.8	7.2
13	21.4	17.2	12.7	9.8
14	25.7	21.1	15.9	12.7
15	30.4	25.3	19.6	15.9
16	35.6	29.9	23.6	19.5
17	41.2	34.9	28.0	23.4
18	47.2	40.3	32.7	27.7
19	53.6	46.1	37.8	32.4
20	60.4	52.3	43.4	37.5
21	67.6	58.9	49.3	42.9
22	75.3	66.0	55.6	48.7
23	83.9	73.4	62.3	54.9
24	91.9	81.3	69.4	61.5
25	100.9	89.5	76.9	68.5

[a] Reprinted from Table 5.7 in C. A. Bennett and N. L. Franklin, *Statistical Analysis in Chemistry and the Chemical Industry*, Wiley, 1954, by permission of the authors and publisher.

of each pair (say, treatment–nontreatment). Assign consecutive rank numbers to the differences, taking into consideration the sign of the differences. Tied values are given a mean rank. Then assign the same algebraic sign to the rank number as occurs in the corresponding difference. Add the rank totals of like algebraic sign and denote the smaller numerical total by W. Compare this total with the tabulated figures given in Table 106.XXV for the appropriate significant level (α) and number of pairs (n_b). If the smaller rank total is less than the tabulated figure, reject with risk α the null hypothesis of no difference between treatments.

C. 2^n FACTORIAL

The analysis of 2^n factorial-type experiments can be accomplished under the procedures of Section VI.A; however, this is often cumbersome and does not make use of the facts that the effect of a given factor in such designs is simply the linear sum of results corresponding to the treatments and that a very simple relationship exists between effects and mean squares. With reference to Table 106.III, a given effect is obtained by applying the signs for the column headed by the effect symbol to the results of the treatments and summing. In the event that there is no replication, that is, only a single observation has been obtained for each treatment, the sum so obtained is divided by 2^{n-1} or simply half the number of treatments so as to produce an averaging of treatments of each sign. If there is a replication and the number of results for each treatment is represented by r, the linear sum obtained in the manner described above (except that the replicates under each treatment are totaled and the signs are applied to the totals) is divided by $2^{n-1}r$ to obtain the corresponding effect. When all the effects have been obtained in this manner, the mean square for each component is obtained by the relationship

$$\text{Mean square} = 2^{n-2}r(\text{effect})$$

where n = number of factors, r = number of replicates, effect = appropriate linear sum of treatment totals (r replicates per treatment) divided by $2^{n-1}r$.

When there is replication, an additional component sum of squares and consequently mean squares is supplied by the variation between replicates for each treatment. This is best obtained as a difference between the sum of squares associated with the 2^{n-1} effects and the total sum of squares for all $2^n r$ individual observations. The degrees of freedom to be associated with this sum of squares are $(r-1)2^n$. For $r=2$

one can simply obtain the error component by computing the sum of squares of the differences between corresponding pairs of replicates.

To effect a test of significance in the manner of Section V and Section VI.A for any given effect, the method of Section VI.A for determining expected values of mean squares corresponding to a given mathematical model for the observations can be used. Appropriate ratios of mean squares are formed accordingly.

The linear sums required in estimating the effects and the mean squares in a 2^n type of factorial design are easily obtained through a systematic tabular method originally devised by Yates (23). This procedure requires first writing down in column form the treatments, or treatment totals when the treatments are replicated, in so-called standard order. The standard order consists of listing first the treatment symbolically represented by (1), that is, the low level of all factors, and then introducing in succession, into the listing, letters of the alphabet, in the conventional order, and listing in sequence, after each new introduction and before the next one, all the treatments obtained by multiplying the treatments in sequence, under the rule of multiplication defined for effects in Section IVC.1.a.(2), with the exception of (1) previously listed, by the newly introduced letter. The standard order, then, for a 2^3 factorial experiment is (1), a, b, ab, c, ac, bc, abc. The next step is to add in sequence by pairs the results from the treatments as listed in standard order and to list the sums in sequence in column 1, say, thus completing the top half of the column. The remaining half of column 1 consists of the differences, listed in sequence between the same pairs, where the first item of the pair is subtracted from the second

TABLE 106.XXVI

Yates's Method of Determining Effects in a 2^3 Factorial Experiment

Ef- fect	Treat- ment	(1)	(2)	(3)
T	(1)	$(1) + a$	$(1) + a + b + ab$	$(1) + a + b + ab + c + ac + bc + abc$
A	a	$b + ab$	$c + ae + bc + abc$	$a - (1) + ab - b + ac - c + abc - bc$
B	b	$c + ac$	$a - (1) + ab - b$	$b + ab - (1) - a + bc + abc - c - ac$
AB	ab	$bc + abc$	$ac - c + abc - bc$	$ab - b - a + (1) + abc - bc - ac + c$
C	c	$a - (1)$	$b + ab - (1) - a$	$c + ac + bc + abc - (1) - a - b - ab$
AC	ac	$ab - b$	$bc + abc - c - ac$	$ac - c + abc - bc - a + (1) - ab + b$
BC	bc	$ac - c$	$ab - b - a + (1)$	$bc + abc - c - ac - b - ab + (1) + a$
ABC	abc	$abc - bc$	$abc - bc - ac + c$	$abc - bc - ac + c - ab + b + a - (1)$

in every case. This process of pairing, adding, and subtracting is applied to the entries of column 1, in exactly the same manner as was done on the treatment results, to obtain column 2, and is continued through a number of columns equal to the number of factors (n). If this is done, the entries in the last column in sequence are the linear sums corresponding to the effects listed in standard order, where the first entry corresponds to the total of all results and the total serves as the identity in the symbolic multiplication of effects. A symbolic representation of this computation for a 2^3 factorial experiment is given in Table 106.XXVI.

D. CONFOUNDED 2^n FACTORIAL DESIGN

The statistical analysis of a 2^n factorial design in which certain of the effects (interactions) are confounded with blocks and there is no replication is no different from that given in Section VI.C, with the exception that the effects confounded with blocks are not meaningful in themselves. If the experiment is replicated r times over $b \times r$ blocks, some complication in the analysis results. The first step in the procedure for determining the sum of squares for each identifiable effect and/or component in this case is to apply Yates's method to the treatment totals to obtain the sum of squares for the unconfounded effects. Next, using the block totals, the variation between blocks is computed by squaring each block total, summing the squared totals divided by the number of observations in each block over all the blocks, and subtracting from the sum the square of the total of all observations over all blocks divided by the total number of all observations. Finally, the variation between all observations is determined by squaring each observation, summing over all observations, and subtracting from this sum the squared total of all observations divided by the total number of observations.

When these quantities have been obtained, the analysis is accomplished by listing the unconfounded sums of squares, each with its single degree of freedom, and then listing the sums of squares identified with the block-to-block variation with associated degrees of freedom equal to one less than the total number of blocks. From these components and the total sum of squares with degrees of freedom one less than the total number of observations, the appropriate error sum of squares and the associated degrees of freedom can be obtained by subtraction, since for any analysis the sum of the individual component sums of squares as well as the degrees of freedom must add up to the corresponding totals

TABLE 106.XXVII
Analysis of Variance for Confounded 2^n Factorial Design
Replicated r Times over $b \times r$ Blocks

Source of variation	Degrees of freedom	How obtained
The $2^n - b$ unconfounded effects, each with a single degree of freedom	$2^n - b$	By Yates's method applied to treatment totals
Between blocks	$br - 1$	By direct computation involving block totals and totals of all observations
Residual error	$(r - 1)(2^n - b)$	By difference
Total	$r2^n - 1$	By direct computation involving individual observations and total over all observations

over the entire experiment. This computation is summarized in Table 106.XXVII.

E. FRACTIONAL 2^n FACTORIAL

The only complication introduced into the analysis of fractional 2^n over complete 2^n factorial designs, as accomplished by Yates's method, occurs in arriving at the "standard order" of listing, inasmuch as certain treatments are missing. The procedure in a given case requires that we ignore, as far as the standard order listing is concerned, certain factor letter designations and list the treatments called for under the design in standard order with regard to the remaining letters. The number of factor letters by which the treatments are to be listed will be the number required if the experiment were a complete factorial in only those letters. Thus, if the experiment is a $\frac{1}{4}$ replicate of a 2^5, the total number of treatments required under the design will be eight, which corresponds in number to a complete factorial in three factors. Consequently, three factor letters are chosen to implement the listing, whereas the other two are ignored. Unfortunately, an arbitrary choice of factors will not always result in the treatments required for the standard order being a part of the set called for under the design. To determine which letters are appropriate, we must refer to the defining contrast subgroup by which the treatments for the design were determined. With regard to the generators of this subgroup, of which there are p for a $1/2^p$ replicate (see Section IV.C.1.b.), and the device for equating each generator to the identity I, one can arrive at an appropriate set. This set is chosen in such a manner that, through the relationships resulting from equating the generators to I and under the rule of multiplication of effects defined in Section IV.C.1.a.(2), all main effects symbols not chosen in the set to be used for determining the standard order can be expressed in terms of the effects so chosen.

For an example, let us refer to the design of Table 106.VIII. The generators of the defining contrast subgroup are $ABCD$, $CDEF$, and ACE. The resulting equations then are

$$I = ABCD, \quad I = CDEF, \quad \text{and} \quad I = ACE$$

Since there are only eight treatments in the design, this would correspond in number to a complete 2^n factorial in three factors. Thus, from the six factor effect symbols $(A, B, C, D, E,$ and $F)$, we must choose three to be the basis of the standard order listing. Suppose that we try A, B, and C as a tentative set and determine whether the remaining symbols, D, E, and F, can be obtained through the system of three equations above as expressions in A and/or B and/or C.

From the third equation, that is, $I = ACE$, it is apparent that by multiplying both sides by E we have, under the rule of multiplication, that $E = AC$, thus satisfying our requirement for E. Similarly, from the equation $I = ABCD$, on multiplying both sides by D, we have $D = ABC$. Now, since both E and D have been expressed in terms of A and/or B and/or C, and the second equation above gives $I = CDEF$, it is apparent that, by multiplying both sides of this equation by F and making the indicated substitution for D and E, F can also be expressed in terms of combinations of A, B, and C and is in fact the relationship $F = BC$. Thus symbols A, B, and C are acceptable as the base set and result in the standard order listing according to these symbols and the treatments of the design of (1), ade, bdf, abef, cdef, acf, bce, abcd. We noted, however, A, B, and C do not constitute the only set having the desired property. A choice of the symbols D, E, and F for the basic set would also be admissible, the standard order of listing by these symbols being (1), abcd, bce, ade, acf, bdf, abef, cdef.

Once the treatments have been listed in standard order with regard to a particular basic set of effect letters (lower case), the Yates's method is applied to the results in the manner described above for a complete 2^n factorial, the number of cycles or steps required being equivalent to the number of factor effects symbols chosen for the basic set. The linear sums as listed in the last column under this procedure correspond

TABLE 106.XXVIII
Standard Order of Listing and Identification of Effects
in Yates's Method of Analysis for Example of Table 106.VIII

Standard order according to factors A, B, and C	Identification of linear sums by effect from aliased set	Standard order according to factors D, E, and F	Identification of linear sums by effect from aliased set
(1)	T	(1)	T
ade	A	abcd	D
bdf	B	bce	E
abef	AB	ade	DE
cdef	C	acf	F
acf	AC	bdf	DF
bce	BC	abef	EF
abcd	ABC	cdef	DEF

to the combined effect of the aliased subgroups described in Section IV.C.1.b, identified by the symbols corresponding to the effects as listed in standard order, using the letters of the basic set. Table 106.XXVIII gives the standard-order listing for the two different basic sets of symbols considered above, as well as the proper identification by an effect from the corresponding aliased set corresponding to the resulting linear sums, which would appear in column 3 under the Yates's method of analysis. Reference to Table 106.IX shows the equivalence between the two listings, row 1 of the final column under the Yates's method being the same for both listings, item 2 under the A, B, C listing corresponding to 6 under D, E, F, and similarly 3 corresponding to 4, 4 to 7, 5 to 8, 6 to 3, 7 to 5, and 8 to 2.

F. LATIN SQUARE

From Section IV.C.2.a we note that the model for the Latin square design is $X_{hij} = m + a_h + b_i + c_j + e_{j(hi)}$. The statistical analysis of this design results in the resolution of the total variation in the t^2 observations obtained under the design into four identifiable components:

1. Between the a_h's.
2. Between the b_i's.
3. Between the c_j's.
4. Between the $e_{j(hi)}$'s (residual or experimental error).

This resolution is accomplished by computing, under the symbolism of Section VI.A,

$$\frac{\sum_h T_{h..}^2}{} = \text{sum of squares of the totals for each level of } A \text{ factor divided}$$

by t

$$\frac{\sum_i T_{.i.}^2}{t} = \text{sum of squares of totals for each level of } B \text{ factor divided by } t$$

$$\frac{\sum_j T_{..j}^2}{t} = \text{sum of squares of totals for each level of } C \text{ factor divided by } t$$

$$\frac{T_{...}^2}{t^2} = \text{sum of the total of all } t^2 \text{ observations divided by } t^2$$

and

$$\sum_{h,i,j} X_{hij}^2 = \text{sum of squares of all } t^2 \text{ observations}$$

It is to be noted that the subscript j is not independent of subscripts h and i in this design. In other words, by specifying h and i for a given experiment, subscript j (the particular level of factor C) is automatically determined.

Since all factor–level combinations are not involved, the general method of Section VI.A cannot be applied; however, the calculations defined below are very similar to those outlined in Section VI.A. In particular, in this case one computes

$$S_a = \frac{\sum_h T_{h..}^2}{t} - \frac{T_{...}^2}{t^2}$$

$$S_b = \frac{\sum_i T_{.i.}^2}{t} - \frac{T_{...}^2}{t^2}$$

$$S_c = \frac{\sum_j T_{..j}^2}{t} - \frac{T_{...}^2}{t^2}$$

$$S = \sum_{h,i,j} X_{hij}^2 - \frac{T_{...}^2}{t^2}$$

and obtains the error component by differences, that is,

$$S_e = S - S_a - S_b - S_c = \sum_{h,i,j} X_{hij}^2 - \frac{\sum_h T_{h..}^2}{t} - \frac{\sum_i T_{.i.}^2}{t} - \frac{\sum_j T_{..j}^2}{t} + \frac{2T_{...}^2}{t^2}$$

The degrees of freedom to be associated with S_a, S_b, and S_c can be shown to be $t - 1$ in each case. Since the total sum of squares, S, has $t^2 - 1$ degrees of freedom, S_e must necessarily have

$$t^2 - 1 - 3(t - 1) = t^2 - 3t + 2 = (t - 1)(t - 2) \text{ degrees of freedom.}$$

These results are summarized in Table 106.XXIX.

A more rapid method of analysis than the classical one outlined above for Latin square designs has been devised by Tukey (20). The basic mathematical model and fundamental assumptions are the same; however, the technique is slightly less efficient than the classical one. Inherent to the method is a technique by which the means of the treatments can be partitioned into like groups in the event that the effects of the treatments are inferred to be significantly different. With reference to

TABLE 106.XXIX
Analysis of Variance for a $t \times t$ Latin Square

Source of variation	Degrees of freedom	Sum of squares	Mean square	Expected mean square[a]
Between levels of A	$t - 1$	S_a	$S_a/(t - 1)$	$V(e) + tS^2(a)$
Between levels of B	$t - 1$	S_b	$S_b/(t - 1)$	$V(e) + tS^2(b)$
Between levels of C	$t - 1$	S_c	$S_c/(t - 1)$	$V(e) + tS^2(c)$
Residual error	$(t - 1)(t - 2)$	S_e	$S_e/(t - 1)(t - 2)$	$V(e)$

[a] The quantities $S^2(a)$, $S^2(b)$, and $S^2(c)$ are variances associated with the a_h's, b_i's, and c_j's, respectively, if these deviations can be assumed to constitute random samples from hypothetical populations. If such is not the case, then the following definitions hold:

$$S^2(a) = \frac{\sum_h a_h^2}{t - 1}, \quad S^2(b) = \frac{\sum_i b_i^2}{t - 1}, \quad S^2(c) = \frac{\sum_j c_j^2}{t - 1}$$

In any event, under the hypothesis of no difference in effects the $S^2(\ )$'s will be zero.

the results of a given Latin square design being displayed in the characteristic square array corresponding to the design, the analysis is accomplished in the following manner:

1. First, obtain the average of each column of results, and by columns subtract the average of the column from each entry in the column. This is called the array of singly adjusted values.

2. Next, obtain the average of each row of singly adjusted values, and by rows subtract the average of the row from each entry in the row. This is called the array of doubly adjusted values.

3. Finally, arrange the doubly adjusted values by columns according to the levels of the third factor (the row and column factors being the first two) involved in the design, and compute the totals and the range of each column. Compute also the sum of the ranges so obtained.

With reference to Table 106.XXX, with the side of square equal to *t* and the chosen significance level, multiply the corresponding entries by the sum of the doubly adjusted ranges to obtain a new set of critical factors to be interpreted as follows. The upper line in each block corresponds to a significance test at 5% risk ($\alpha = 0.05$), and the lower line to 1% risk. The left-hand column in the left-hand block gives the critical values for comparing at the given significance level the range of the totals resulting from the doubly adjusted values. The right-hand block gives critical factors at the indicated significance levels for comparing the effects of the row and column factor levels. In application, to test for the significance of each factor and to partition the effects of the factor levels in like groups, one in turn arranges in order of magnitude the following:

  a. the *totals* by column of the doubly adjusted values as found under step *3* above,
  b. the column averages of the unadjusted values as found in step *1* above,
  c. the row averages of the singly adjusted values as found in step *2* above,

and compares the entries of the left-hand block with the corresponding values for the doubly adjusted values, and the entries in the right-hand block with the unadjusted and singly adjusted averages. An observed range exceeding the tabulated value indicates a significant difference between associated factor levels, and any difference between algebraically adjacent values which exceeds the critical value for the corresponding gap indicates a real partitioning of factor levels.

TABLE 106.XXX

Critical Factors for Latin Squares[a]

The eight entries are for, respectively:

| Range for 5% risk | Gap for 5% risk | Range for 5% risk | Gap for 5% risk |
| Range for 1% risk | Gap for 1% risk | Range for 1% risk | Gap for 1% risk |

Side of square		Critical factors		
3	3.78	2.76	1.26	0.92
	8.70	6.37	2.90	2.12
4	1.54	1.08	0.38	0.27
	2.23	1.50	0.56	0.38
5	1.07	0.73	0.21	0.15
	1.40	1.03	0.28	0.21
6	0.85	0.56	0.14	0.094
	1.07	0.70	0.18	0.12
7	0.73	0.47	0.10	0.067
	0.88	0.64	0.13	0.091
8	0.64	0.40	0.080	0.050
	0.76	0.54	0.095	0.068
9	0.58	0.36	0.064	0.040
	0.68	0.48	0.076	0.053
10	0.53	0.32	0.053	0.032
	0.62	0.42	0.062	0.042

[a] The values in the table were given by J. W. Tukey in his paper, "Quick and Dirty Methods in Statistics—Part II: Simple Analyses for Standard Designs," delivered to the American Society for Quality Control at the Cleveland Meeting, 1951, and are contained on p. 197 of the published Quality Control Conference papers for that meeting.

G. GRAECO-LATIN SQUARE

The analysis of variance as applied to a Graeco–Latin square of side t corresponding to the model.

$$X_{hijk} = m + a_h + b_i + c_j + d_k + e_{k(hij)}$$

is a simple extension of that required under a Latin square design. In addition to computing S_a, S_b, and S_c in the manner of Section VI.F, we compute S_d by the analogous formula:

$$S_d = \frac{\sum_k T^2_{...k}}{t} - \frac{T^2_{....}}{t^2}$$

TABLE 106.XXXI
Analysis of variance for $t \times t$ Graeco–Latin Square

Source of variation	Degrees of freedom	Sum of squares	Mean square	Expected mean square[a]
Between levels of A	$t - 1$	S_a	$S_a/(t - 1)$	$V(e) + tS^2(a)$
Between levels of B	$t - 1$	S_b	$S_b/(t - 1)$	$V(e) + tS^2(b)$
Between levels of C	$t - 1$	S_c	$S_c/(t - 1)$	$V(e) + tS^2(c)$
Between levels of D	$t - 1$	S_d	$S_d/(t - 1)$	$V(e) + tS^2(d)$
Residual error	$(t - 1)(t - 3)$	S_e	$S_e/(t - 1)(t - 3)$	$V(e)$

[a] The $S^2(\ )$'s reflect the variations within the various sets of effects for the corresponding factor, and under the hypothesis of no difference in effects will be zero.

Again the residual error component is best obtained by subtracting the sum of the component sums of squares from the total sums of squares so that the final analysis is as summarized in Table 106.XXXI.

To test for the significance of each factor, one in turn forms the ratio of the factor mean square to the error mean square and compares each result with the values of Table 106.XX for the chosen significance level and the appropriate degrees of freedom.

H. ORTHOGONAL SQUARES

The analysis of variance for a $t \times t$ orthogonal square involving p factors resolves the total variation in the t^2 observations into $p + 3$ identifiable components (including the row and column factors) and facilitates a test as to the significance of each component. The actual sums of squares to be calculated include those required for the Graeco–Latin square plus those obtained for the additional factors in a straightforward generalization of the manner in which the components of factors C and D were determined under the Latin and Graeco–Latin square designs. Tests of significance follow the same pattern as for the Graeco–Latin square.

I. BALANCED INCOMPLETE BLOCKS

The balanced incomplete block design is characterized as a two-factor experiment (one factor being blocks); and, if we refer to the levels of the other factor as being treatments, every pair of treatments occurs together in the same block the same number of times. In order to describe the analysis we shall adopt the following notation of Davies (13):

N = total number of observations,
t = number of treatments,
r = number of replications (over the entire design) of each treatment,
b = number of blocks
k = number of experimental units per block,
$\lambda = r(k-1)/(t-1) = N(k-1)/t(t-1)$, since $N = tr = bk$.

The observations in any balanced incomplete block design can be arranged in a form analogous to Table 106.XXXII, and the statistical analysis is facilitated by reference to this table under the following definitions:

G = sum of all observations,
$\bar{X} = G/N$,
T_i = total of all observations obtained under treatment i,
B_j = total of all observations in block j,
$Q_i = kT_i -$ (sum of all the totals for all blocks containing the ith treatment),
$P_i = Q_i/t\lambda$.

Table 106.XXXII
Incomplete Block Design: Presentation and Analysis

Treatment	Block					Total	Q	P	Corrected treatment mean
	1	2	3	$\cdots$	b				
1	$\times$	$-$	$\times$	$\cdots$	$\times$	T_1	Q_1	P_1	$P_1 + \bar{X}$
2	$\times$	$\times$	$-$	$\cdots$	$-$	T_2	Q_2	P_2	$P_2 + \bar{X}$
3	$-$	$\times$	$\times$	$\cdots$	$\times$	T_3	Q_3	P_3	$P_3 + \bar{X}$
.									
.									
.									
t	$\times$	$\times$	$-$	$\cdots$	$\times$	T_t	Q_t	P_t	$P_t + \bar{X}$
Total	B_1	B_2	B_3	$\cdots$	B_b	G	0	0	tG/N

The nature of the computation is such that both the Q_i's and P_j's should sum to zero.

Next we compute

1. $\displaystyle\sum_{i,j} X_{ij}^2 - G^2/N$.

2. $\left(\sum_{j} B_j{}^2/k \right) - G^2/N.$

3. $\sum_{i} Q_i{}^2/kt\lambda.$

The analysis of variance may then be summarized as in Table 106.XXXIII.

TABLE 106.XXXIII
Analysis of Variance of Incomplete Block Design

Source of variation	Sum of squares	Degrees of freedom
Between treatments	$\sum_{i} Q_i{}^2/kt\lambda$	$t - 1$
Between blocks, ignoring treatments	$\sum_{j} B_j{}^2/k - G^2/N$	$b - 1$
Residual error	$\sum_{i,j} X_{ij}{}^2 - \sum_{j} B_j{}^2/k - \sum_{i} Q_i{}^2/kt\lambda$	$N - t - b + 1$
Total	$\sum_{i,j} X_{ij}{}^2 - G^2/N$	$N - 1$

The test of significance as to treatment effects is accomplished by forming the ratio of the mean square identified with the between-treatment variation to the mean square identified with the residual error and comparing the result obtained to corresponding values of the F distribution (Table 106.XX) for the chosen significance level and appropriate degrees of freedom. In this analysis the between-block variation has confounded with it certain treatment variation and consequently is uninterpretable. It is computed solely as a mechanism to reduce the error mean square and thus make the test for treatments more discriminating.

J. SYMMETRICALLY BALANCED INCOMPLETE BLOCKS

The analysis of variance to be applied to data obtained under a symmetrically balanced incomplete block design is an extension of that outlined for balanced incomplete blocks. Individual block effects free of treatment differences are obtained by the formula

$Q'_j = rB_j -$ (sum of totals for all treatments appearing in the given

TABLE 106.XXXIV

Analysis of Variance of Symmetrically Balanced Incomplete Block Designs

Source of variation	Sum of squares	Degrees of freedom
Between treatments, adjusting for blocks	$\sum\limits_{i} Q_i{}^2/kt\lambda$	$t-1$
Between blocks, adjusting for treatments	$\sum\limits_{j} Q_j'{}^2/kt\lambda$	$b-1$
Residual error	$\sum\limits_{i,j} X_{ij}{}^2 - \sum\limits_{j} B_j{}^2/k - \sum\limits_{i} Q_i{}^2/kt\lambda$	$N-t-b+1$

block) from which the component sum of squares due to corrected block totals is $\sum\limits_{j} Q_j'{}^2/kt\lambda$, having $b-1$ degrees of freedom.

Thus, in addition to the test of significance for treatment differences, an analogous test on block differences can be constructed by forming the ratio of $\sum\limits_{j} Q_j'{}^2/kt\lambda(b-1)$ to the residual error mean square and comparing the results to the tabulated distribution of F in the manner described for the test on treatment effects.

The analysis of symmetrically balanced incomplete block designs is summarized in Table 106.XXXIV.

REFERENCES

1. Bennett, C. A., and N. L. Franklin, *Statistical Analysis in Chemistry and the Chemical Industry*, Wiley, New York, 1954, p. 40.
2. *Ibid.*, p. 414.
3. Bowker, A. H., and G. J. Lieberman, *Engineering Statistics*, Prentice–Hall, Englewood Cliffs, N.J., 1959, p. 299.
4. Box, G. E. P., *Biometrika*, **39**, 49 (1952).
5. Box, G. E. P., and K. B. Wilson, *J. Roy. Stat. Soc.*, **B13**, 1 (1951).
6. Chew, V., *Experimental Designs in Industry*, Wiley, New York, 1958.
7. Cochran, W. G., and G. M. Cox, *Experimental Design*, 2nd ed., Wiley, New York, 1957.
8. *Ibid.*, p. 376.
9. Davies, O. L., *The Design and Analysis of Industrial Experiments*, Hafner, New York, 1954.
10. *Ibid.*, p. 396.
11. *Ibid.*, p. 495.
12. *Ibid.*, p. 199.

13. *Ibid.,* p. 204.
14. Fisher, R. A., and F. Yates, *Statistical Tables for Biological, Agricultural and Medical Research,* 4th ed., Hafner, New York, 1953, p. 20.
15. *Ibid.,* p. 72.
16. Hald, A., *Statistical Theory with Engineering Application,* Wiley, New York, 1952, p. 374.
17. *Ibid.,* p. 419.
18. *Ibid.,* p. 106.
19. Satterthwaite, F. E., *Biometrics Bull.* **2,** 110 (1946).
20. Tukey, J. W., *Quality Control Conference Papers,* American Society for Quality Control, Cleveland Meeting, 1951, pp. 189–197.
21. Welch, B. L., *J. Roy. Stat. Soc.* (Supplement), **3,** 29 (1936).
22. Wilcoxon, F., *Some Rapid Approximate Statistical Procedures,* American Cyanamid Co., New York, July 1949.
23. Yates, F., *Design and Analysis of Factorial Experiments,* Imperial Bureau of Soil Science, London, 1937.
24. *Ibid.*

APPENDIX

I. GENERAL FACTORIAL

A. Description of Experiment and Resulting Data

An experiment was to be performed relative to the determination of TBP by the acid absorption method in a uranyl nitrate solution. The concentration of uranium in the solution, as well as the subjection or lack of subjection of the solution to Na_2CO_3 scrub, was believed to affect the method. In order to determine whether such was the case, and also to investigate the effects and interactions over a range of TBP concentrations from 3% to 10%, the experiment was designed as a $3 \times 3 \times 2$ factorial. The following table gives the individual experiments and the results obtained.

TBP by Acid Absorption

Milliliters of 0.20N NaOH required to titrate acid in 5 ml aliquot

g/liter uranium	Per Cent TBP		
	3	5	10
	Without Na_2CO_3 Scrub		
0	2.74	4.53	9.05
2.5	2.70	4.53	9.10
5.0	2.69	4.59	9.13
	With Na_2CO_3 Scrub		
0	2.68	4.48	9.05
2.5	2.71	4.53	9.05
5.0	2.72	4.54	9.08

B. Computations

We make the following identification of letters to factors:

A: TBP concentration, B: Uranium concentration, C: Na_2CO_3 scrub

In the manner of Section VI.A we compute:

$T_{11.} = 5.42$	$T_{3.1} = 27.28$	$T_{.2.} = 32.62$
$T_{12.} = 5.41$	$T_{3.2} = 27.18$	$T_{.3.} = 32.75$
$T_{13.} = 5.41$	$T_{.11} = 16.32$	$T_{..1} = 49.06$
$T_{21.} = 9.01$	$T_{.12} = 16.21$	$T_{..2} = 48.84$
$T_{22.} = 9.06$	$T_{.21} = 16.33$	$T_{...} = 97.90$
$T_{23.} = 9.13$	$T_{.22} = 16.29$	$S^* = S^*_{abc}\dagger = 129.1250$
$T_{31.} = 18.10$	$T_{.31} = 16.41$	$S^*_{ab} = 129.1177$
$T_{32.} = 18.15$	$T_{.32} = 16.34$	$S^*_{ac} = 129.1144$
$T_{33.} = 18.21$	$T_{1..} = 16.24$	$S^*_{bc} = 0.0072$
$T_{1.1} = 8.13$	$T_{2..} = 27.20$	$S^*_a = 129.1110$
$T_{1.2} = 8.11$	$T_{3..} = 54.46$	$S^*_b = 0.0041$
$T_{2.1} = 13.65$	$T_{.1.} = 32.53$	$S^*_c = 0.0027$
$T_{2.2} = 13.55$		

† Since only a single observation is obtained for each combination of factor levels, $S^* = S^*_{abc}$, and as a consequence $S(e)$ will be found to be identically zero by the computational scheme. This means that S_{abc} when divided by its degrees of freedom in fact estimates $V(e) + V(abc)$, a separation of these components not being possible under the single observation per combination design. As a consequence, these two variances will always appear in combination in any expected mean square, as will the corresponding effect terms in the model for the observations. Thus it is convenient to represent the combined sum of variances, $V(e) + V(abc)$, as $V(e)$ and the corresponding sum of deviations, $e_{j(hi)} + (abc)_{hij}$, as simply $e_{j(hi)}$ in such cases.

From which we compute:

$S_a = 129.1110$	$S_{ab} = 0.0026$	$S_{bc} = 0.0004$
$S_b = 0.0041$	$S_{ac} = 0.0007$	$S_{abc} = 0.0035$
$S_c = 0.0027$		

If the model for the observations is

$$X_{hij} = a_h + b_i + c_j + (ab)_{hi} + (ac)_{hj} + (bc)_{ij} + e_{j(hi)}$$

and the inferences are assumed to apply only to the levels of the factors used in the experiment, a summary in the form of Table 106.XXIII is as follows.

Complete Model and Reduced Model

$$X_{hij} = m + a_h + b_i + c_j + (ab)_{hi} + (ac)_{hi} + (bc)_{ij} + e_{j(hi)}$$

Sum of squares	Degrees of freedom	Expected mean square (Infinite populations)	Effect	Index h	Index i	Index j	Expected mean square for reduced model
$S_a = 129.1110$	2	$V(e) + 2V(ab) + 3V(ac) + 6V(a)$	a_h	0	3	2	$V(e) + 6V(a)$
$S_b = 0.0041$	2	$V(e) + 3V(bc) + 2V(ab) + 6V(b)$	b_i	3	0	2	$V(e) + 6V(b)$
$S_c = 0.0027$	1	$V(e) + 3V(ac) + 3V(bc) + 9V(c)$	c_j	3	3	0	$V(e) + 9V(c)$
$S_{ab} = 0.0026$	4	$V(e) + 2V(ab)$	$(ab)_{hi}$	0	0	2	$V(e) + 2V(ab)$
$S_{ac} = 0.0007$	2	$V(e) + 3V(ac)$	$(ac)_{hi}$	0	3	0	$V(e) + 3V(ac)$
$S_{bc} = 0.0004$	2	$V(e) + 3V(bc)$	$(bc)_{ij}$	3	0	0	$V(e) + 3V(bc)$
$S_{abc} = 0.0035$	4	$V(e)$	$e_{j(hi)}$	1	1	1	$V(e)$

6473

C. Analysis of Variance

Source	Mean square	Degrees of freedom	Expected mean square	F ratio
Between TPB concentrations	64.5550	2	$V(e) + 6V(a)$	[a]
Between uranium concentrations	0.0020	2	$V(e) + 6V(b)$	3.33
With and without scrub	0.0027	1	$V(e) + 9V(c)$	4.50
TPB × uranium interaction	0.0006	4	$V(e) + 2V(ab)$	[b]
TPB × scrub interaction	0.0004	2	$V(e) + 3V(ac)$	[b]
Uranium × scrub interaction	0.0002	2	$V(e) + 3V(bc)$	[b]
Error	0.0009	4	$V(e)$	
Pooled error[b]	0.0006	12	$V(e)$	

[a] This factor is the subject of test only inasmuch as it interacts with the other factors of the experiment. The nature of the experiment was such as to assure the significance of this factor.

[b] It is apparent that forming the ratio of each interaction mean square to the error mean square results in a number less than 1 in each case. Thus no significance can be inferred to the interactions. Consequently the interaction components can be pooled with the error to obtain a more discriminating test with regard to the main effects.

D. Inferences

At a 10% significance level ($\alpha = 0.10$) the following inferences appear to be valid:

1. Interaction effects between any pair of factors are not significantly different from zero.

2. The use of a Na_2CO_3 scrub significantly *reduced* the absorption. This is determined by observing that the calculated F ratio of 4.50 exceeds the tabled value of the F distribution with 1 and 12 degrees of freedom for $\alpha = .10$, (Fv_1,v_2,α); this value, $F_{1,12,.10}$, is found from Table 106.XX to be 3.18.

3. The uranium concentration, at least at the levels used in this experiment, significantly affected the absorption, since $3.33 > F_{2,12,.10} = 2.81$.

Further analyses are possible from the basic data with regard to comparison of particular treatments, or estimation of the precision of analysis corresponding to a particular set of conditions; however, the methods involved are beyond the scope of this chapter.

II. RANDOMIZED BLOCK

A. Description of Experiment and Resulting Data

It was proposed to extend the colorimetric method of uranium determination to higher concentration levels than had been used previously.

In so doing the effect of the size of aliquot and of day-to-day variations in the performance of equipment was of interest. Since approximately six observations were all that could be run in a day's time, the experiment was designed as a randomized-block type with time of analysis (days) constituting the blocks, and combinations of concentration levels and aliquots the experiments within the blocks. The concentrations chosen for study were 5.0 and 50.0 mg U/ml. The aliquot sizes were 0.1, 0.2, and 0.5 ml. The individual experiments and corresponding absorptions are given in the following table.

Colorimeter Data (Absorption)

Concentration, mg U/ml	Day of analysis[a]					
	1		2		3	
	5.0	50.0	5.0	50.0	5.0	50.0
Aliquot size, ml						
0.1	.116	.122	.128	.122	.128	.123
0.2	.242	.246	.232	.248	.243	.243
0.5	.602	.610	.616	.621	.607	.6I9

[a] Experiments were run in random order within any one day.

B. Computations

We proceed with the analysis in the framework of Section VI.A, in which "day of analysis" is considered as the third factor and no interaction between this and the other two factors is assumed. The following identification of letters to factors is made:

A: Concentration of uranium, B: Aliquot size, C: Day of analysis

We compute:

$T_{1..} = 2.914$ $T_{..1} = 1.938$ $T_{13.} = 1.825$
$T_{2..} = 2.954$ $T_{..2} = 1.967$ $T_{21.} = 0.367$
$T_{.1.} = 0.739$ $T_{..3} = 1.963$ $T_{22.} = 0.737$
$T_{.2.} = 1.454$ $T_{11.} = 0.372$ $T_{23.} = 1.850$
$T_{.3.} = 3.675$ $T_{12.} = 0.717$ $T_{1.1} = 0.960$

$$T_{1.2} = 0.976 \qquad T_{.23} = 0.486 \qquad S_b^* = 0.781342$$
$$T_{1.3} = 0.978 \qquad T_{.31} = 1.212 \qquad S_c^* = 0.000082$$
$$T_{2.1} = 0.978 \qquad T_{.32} = 1.237 \qquad S_a = 0.000089$$
$$T_{2.2} = 0.991 \qquad T_{.33} = 1.226 \qquad S_b = 0.781342$$
$$T_{2.3} = 0.985 \qquad T_{...} = 5.868 \qquad S_c = 0.000082$$
$$T_{.11} = 0.238 \qquad S^* = S_{abc}^* = 0.781870\dagger \qquad S_{ab} = 0.000086$$
$$T_{.12} = 0.250 \qquad S_{ab}^* = 0.781517 \qquad S_{ac} = 0.000011$$
$$T_{.13} = 0.251 \qquad S_{ac}^* = 0.000182 \qquad S_{bc} = 0.000144$$
$$T_{.21} = 0.488 \qquad S_{bc}^* = 0.781569 \qquad S_{abc} = 0.000115$$
$$T_{.22} = 0.480 \qquad S_a^* = 0.000089$$

† See note on middle of p. 6472.

The model which best describes the observations in this case is

$$X_{hij} = m + a_h + b_i + c_j + (ab)_{hi} + e_{j(hi)}$$
$$n_a = N_a = 2, \qquad n_b = N_b = 3, \qquad n_c = 3, \qquad N_c = \infty$$

A summary in the form of Table 106.XXIII is as follows.

Sum of squares	Degrees of freedom	Complete Model[a] $X_{hij} = m + a_h + b_i + c_j + (ab)_{hi}$ $+ (ac)_{hj} + (bc)_{ij} + e_{j(hi)}$ Expected mean square (infinite populations)	Reduced Model[a] $X_{hij} = m + a_h + b_i + c_j$ $+ (ab)_{hi} + e_{j(hi)}$ Expected mean square (infinite populations)
$S_a = 0.000089$	1	$V(e) + 3V(ab) + 3V(ac) + 9V(a)$	$V(e) + 3V(ab) + 9V(a)$
$S_b = 0.781342$	2	$V(e) + 2V(bc) + 3V(ab) + 6V(b)$	$V(e) + 3V(ab) + 6V(b)$
$S_c = 0.000082$	2	$V(e) + 3V(ac) + 2V(bc) + 6V(c)$	$V(e) + 6V(c)$
$S_{ab} = 0.000086$	2	$V(e) + 3V(ab)$	$V(e) + 3V(ab)$
$S_{ac} = 0.000011$	2	$V(e) + 3V(ac)$	$V(e)$
$S_{bc} = 0.000144$	4	$V(e) + 2V(bc)$	$V(e)$
$S_{abc} = 0.000115$	4	$V(e)$	$V(e)$

Pooled sum of squares	Degrees of freedom	Mean square	Effect	Index h	Index i	Index j	Expected mean square for reduced model
$\tilde{S}_a = 0.000089$	1 .	0.000089	a_h	0	3	3	$V(e) + 9V(a)$
$\tilde{S}_b = 0.781342$	2	0.390671	b_i	2	0	3	$V(e) + 6V(b)$
$\tilde{S}_c = 0.000082$	2	0.000041	c_j	2	3	1	$V(e) + 6V(c)$
$\tilde{S}_{ab} = 0.000086$	2	0.000043	$(ab)_{hi}$	0	0	3	$V(e) + 3V(ab)$
$\tilde{S}_e = 0.000270$	10	0.000027	$e_{j(hi)}$	1	1	1	$V(e)$

[a] The error as noted here includes the three-factor interaction, $(abc)_{hij}$.

C. Analysis

To test the hypothesis of no interaction between concentration and quantity of aliquot we form the ratio

$$F = \frac{0.000043}{0.000027} = 1.59$$

and compare this to $F_{2,10,\alpha}$ at the chosen significance level (α). For $\alpha = 0.10$, we find from Table 106.XX that

$$F_{2,10,.10} = 2.92$$

As a consequence we conclude that there is no interaction effect. We can obtain a new error estimate, therefore, by pooling $\tilde{S}_{ab}$ and $\tilde{S}_e$. This results in 0.000030. To test the significance of differences in uranium concentration we form the ratio

$$\frac{0.000089}{0.000030} = 2.97$$

Since $F_{1,12,.10} = 3.18$ is greater than 2.97, we cannot, strictly speaking, infer a uranium concentration effect. In practice, however, our actions are sometimes influenced by the closeness of the calculated value to the critical value, and as a consequence uranium concentration is suspect.

The significance of aliquot size is obvious without test.

III. CONFOUNDED 2^n FACTORIAL

A. Description of Experiment and Data Obtained

One of the by-products of atomic fission is radioactive iodine. In the processing, of irradiated fuel elements, therefore, care is taken that this constituent is removed from the dissolver off-gas before venting to the atmosphere. A question was raised as to the efficiency of the dissolver off-gas scrub solution in removing the iodine. Before this investigation, however, it was believed advisable to investigate the sources of sampling error associated with the iodine analysis in the scrub solution. The most likely source appeared to be that identified with plating out on the sampling equipment. The type of sampling container, the contact area between the solution and the container, the elapsed time between sampling and analysis, and the acidity of the solution were all believed to be possible factors in introducing systematic error into the subsequent analysis. To investigate which if any of these were significant, an experiment of the 2^4 factorial type was conducted. Since only eight analyses could be run in any one day, the entire experiment was confounded in two blocks of eight, in such a manner that the four-factor interaction was confounded with whatever effect the day of analysis might have on the results. The individual experiments and the results obtained are given in the table.

Experiment[a]	Day of analysis	Type of container	Total volume, ml	Elapsed time between sampling and analysis, days	Acidity of solution	Net result,[b] (^{131}I d/m/ml) × 10^{-5}
(1)	1	Glass	2	3	Acid	−0.413
ab	1	Dri-film	4	3	Acid	−0.037
ad	1	Dri-film	2	3	Base	−0.090
bd	1	Glass	4	3	Base	−0.020
ac	1	Dri-film	2	6	Acid	0.097
bc	1	Glass	4	6	Acid	0.127
cd	1	Glass	2	6	Base	0.290
abcd	1	Dri-film	4	6	Base	0.430
a	2	Dri-film	2	3	Acid	−0.203
b	2	Glass	4	3	Acid	−0.043
d	2	Glass	2	3	Base	−0.030
abd	2	Dri-film	4	3	Base	−0.300
c	2	Glass	2	6	Acid	−0.113
abc	2	Dri-film	4	6	Acid	0.087
acd	2	Dri-film	2	6	Base	0.310
bcd	2	Glass	4	6	Base	0.430

Factor	Levels low	high
[a] A: Sampling container	Glass	Dri-film
B: Contact area (volume)	2 ml	4 ml
C: Elapsed time between sampling and analysis	3 days	6 days
D: Acidity of solution	Acid	Base

[b] Net result is the analytical result minus the average of the makeup solution.

B. Computations

Yates's method of computation, as presented in the manner of Table 106.XXVI, is as follows on top of page 6479 when applied to this example.

It is reasonable, from the nature of the factors and the levels chosen for experimentation, to assume all factors crossed and all populations totally sampled in the experiment. Under these assumptions, the expected mean squares corresponding to the computed mean squares can be ob-

Effect	Treat-ment	Result	(1)	(2)	(3)	(4)	Mean square $= \dfrac{(4)^2}{16}$
Total	(1)	−0.413	−0.616	−0.696	−0.498	0.522	
A	a	−0.203	−0.080	0.198	1.020	0.066	0.000272
B	b	−0.043	−0.016	−0.440	0.386	0.826	0.042642
AB	ab	−0.037	0.214	1.460	−0.320	−0.694	0.030102
C	c	−0.113	−0.120	0.216	0.766	2.794	0.487902
AC	ac	0.097	−0.320	0.170	0.060	0.314	0.006162
BC	bc	0.127	0.600	−0.340	−0.454	0.154	0.001482
ABC	abc	0.087	0.860	0.020	−0.240	0.154	0.001482
D	d	−0.030	0.210	0.536	0.894	1.518	0.144020
AD	ad	−0.090	0.006	0.230	1.900	−0.706	0.031152
BD	bd	−0.020	0.210	−0.200	−0.046	−0.706	0.031152
ABD	abd	−0.300	−0.040	0.260	0.360	0.214	0.002862
CD	cd	0.290	−0.060	−0.204	−0.306	1.006	0.063252
ACD	acd	0.310	−0.280	−0.250	0.460	0.406	0.010302
BCD	bcd	0.430	0.020	−0.220	−0.046	0.766	0.036672
ABCD	abcd	0.430	0.000	−0.020	0.200	0.246	0.003782

tained in the manner of Section VI.A, remembering that the four-factor interaction $ABCD$ is confounded with blocks (day of analysis). Thus in the form of Table 106.XXII we have:

Mean square	Degrees of freedom	Effect	Index				Expected mean square
			h	i	j	k	
0.000272	1	a_h	0	2	2	2	$V(e)^a + 8V(a)$
0.042642	1	b_i	2	0	2	2	$V(e) + 8V(b)$
0.030102	1	$(ab)_{hi}$	0	0	2	2	$V(e) + 4V(ab)$
0.487902	1	$(c)_j$	2	2	0	2	$V(e) + 8V(c)$
0.006162	1	$(ac)_{hj}$	0	2	0	2	$V(e) + 4V(ac)$
0.001482	1	$(bc)_{ij}$	2	0	0	2	$V(e) + 4V(bc)$
0.001482	1	$(abc)_{hij}$	0	0	0	2	$V(e) + 2V(abc)$
0.144020	1	$(d)_k$	2	2	2	0	$V(e) + 8V(d)$
0.031152	1	$(ad)_{hk}$	0	2	2	0	$V(e) + 4V(ad)$
0.031152	1	$(bd)_{ik}$	2	0	2	0	$V(e) + 4V(bd)$
0.002862	1	$(abd)_{hik}$	0	0	2	0	$V(e) + 2V(acd)$
0.063252	1	$(cd)_{jk}$	2	2	0	0	$V(e) + 4V(cd)$
0.010302	1	$(acd)_{hjk}$	0	2	0	0	$V(e) + 2V(acd)$
0.036672	1	$(bcd)_{ijk}$	2	0	0	0	$V(e) + 2V(bcd)$
0.003782[b]	1	$(abcd)_{k(hijk)}$	1	1	1	1	$V(e) + V$ (blocks)

[a] See note, middle of page 6472.

[b] This component is discarded since it is confounded with blocks.

In order to carry out tests of significance with regard to the various effects assumed under the model we make a further assumption that three-factor interaction variances are small compared to the error variance. This allows us to pool the mean squares associated with the three-factor interaction effects in the preceding table and consequently to reduce the analysis to the following.

Effect	Mean square	Degrees of freedom	Expected mean square	Ratio of mean square
A	0.000272	1	$V(e) + 8V(a)$	[a]
B	0.042642	1	$V(e) + 8V(b)$	3.32
C	0.487902	1	$V(e) + 8V(c)$	38.03
D	0.144020	1	$V(e) + 8V(d)$	11.23
AB	0.030102	1	$V(e) + 4V(ab)$	2.35
AC	0.006162	1	$V(e) + 4V(ac)$	[a]
AD	0.031152	1	$V(e) + 4V(ad)$	2.43
BC	0.001482	1	$V(e) + 4V(bc)$	[a]
BD	0.031152	1	$V(e) + 4V(bd)$	2.43
CD	0.063252	1	$V(e) + 4V(cd)$	4.93
Error	0.012830	4	$V(e)$	

[a] Less than 1, hence obviously not significant.

We note from Table 106.XX that

$$F_{1,4,.10} = 4.54, \quad F_{1,4,.05} = 7.71, \quad F_{1,4,.01} = 21.2$$

Consequently the following inferences appear to be valid, at least with regard to the levels of the factors used in this experiment:

1. Both the elapsed time and the acidity of solution had a marked effect on the net result, whereas no difference between types of container or contact areas could be detected.

2. The only interaction of significance appears to be that identified with the two factors which were found to be significant, namely, time and acidity. An examination of the data indicates that this is primarily a result of the high analytical readings consistently observed when the solution was basic and was analyzed after 6 days' elapsed time. This combination with its associated high readings no doubt contributed toward making the corresponding main effects significant.

IV. ½ REPLICATE OF A 2⁵ FACTORIAL

A. Description of Experiment and Resulting Data

The extraction of lead, as determined by measuring the absorption on a standard containing 30 μg of this element, into a carbon tetrachloride dithizone media was believed to be dependent on five different variables proposed for study:

A. Quantity of 1.0M̲ hydroxylamine by hydrochloride.
B. Quantity of 0.5M̲ ammonium citrate.
C. Quantity of 0.5M̲ ammonium hydroxide–sodium cyanide.
D. Acidity of solution.
E. Per cent dithizone (25 ml quantity).

Since three-factor interactions were believed to be of little significance and economy of effort was of prime importance, the experiment was

Sample No.	Hydroxylamine hydrochloride[a]	Ammonium citrate[b]	Ammonium hydroxide–sodium cyanide[c]	pH[d]	Dithizone in carbon-tetrachloride[e]	Absorption 30 μg lead
1	L	L	L	L	L	0.352
2	L	L	L	H	H	0.361
3	L	L	H	L	H	0.338
4	L	L	H	H	L	0.320
5	L	H	L	L	H	0.344
6	L	H	L	H	L	0.288
7	L	H	H	L	L	0.287
8	L	H	H	H	H	0.301
9	H	L	L	L	H	0.340
10	H	L	L	H	L	0.362
11	H	L	H	L	L	0.348
12	H	L	H	H	H	0.355
13	H	H	L	L	L	0.304
14	H	H	L	H	H	0.299
15	H	H	H	L	H	0.317
16	H	H	H	H	L	0.287

[a] L = 0.5 ml, H = 1.0 ml.
[b] L = 5 ml, H = 10 ml.
[c] L = 5 ml, H = 10 ml.
[d] L = 10.5, H = 10.6.
[e] L = 10⁻³, H = 1.25 × 10⁻³

designed as a $\frac{1}{2}$ replicate of a 2^5 factorial (the defining subgroup was $I, ABCDE$) in accordance with the following scheme. The result of each experiment is given in the last column of the table shown at the bottom of page 6481.

B. Computations

Using Yates's method, and arranging the experiments in standard order with respect to factors $A, B, C,$ and $D,$ we can summarize in the following table the computation and the nature of the aliasing.

Yates Method of Analysis

Aliased set	Exp.	0	1	2	3	4	Mean square
$T, ABCDE$	(1)	0.352	0.692	1.340	2.630	5.203	1.69195056
$A, BCDE$	ae	0.340	0.648	1.290	2.573	0.021	0.00002756
$B, ACDE$	be	0.344	0.686	1.310	-0.012	-0.349	0.00761256
AB, CDE	ab	0.304	0.604	1.263	0.033	-0.047	0.00013806
$C, ABDE$	ce	0.338	0.723	-0.052	-0.126	-0.097	0.00058806
AC, BDE	ac	0.348	0.587	0.040	-0.223	0.101	0.00063756
BC, ADE	bc	0.287	0.675	0.012	-0.008	0.011	0.00000756
ABC, DE	$abce$	0.317	0.588	0.021	-0.039	-0.011	0.00000756
$D, ABCE$	de	0.361	-0.012	-0.044	-0.050	-0.057	0.00020306
AD, BCE	ad	0.362	-0.040	-0.082	-0.047	0.045	0.00012656
BD, ACE	bd	0.288	0.010	-0.136	0.092	-0.097	0.00058806
ABD, CE	$abde$	0.299	0.030	-0.087	0.009	-0.031	0.00006006
CD, ABE	cd	0.320	0.001	-0.028	-0.038	0.003	0.00000056
ACD, BE	$acde$	0.355	0.011	0.020	0.049	-0.083	0.00043056
BCD, AE	$bcde$	0.301	0.035	0.010	0.048	0.087	0.00047306
$ABCD, E$	$abcd$	0.287	-0.014	-0.049	-0.059	-0.107	0.00071556

C. Analysis of Variance

The experiment does not provide an estimate of the error variance free of interactions. In this particular case, however, there was reason to believe that the interaction of factor A with any of the other factors could be considered negligible. Consequently, interactions involving A were pooled to obtain an error variance estimate. The result of this pooling and a listing of the principal effects in all of the aliased sets, under the assumption that three-factor interactions were negligible, are summarized in the following analysis of variance table.

Analysis of Variance

Effect	Mean square	Degrees of freedom	Ratio of mean square[a] to error mean square
A	0.00002756	1	[b]
B	0.00761256	1	22.14
C	0.00058806	1	1.71
D	0.00020306	1	[b]
E	0.00071556	1	2.08
BC	0.00000756	1	[b]
BD	0.00058806	1	1.71
BE	0.00043056	1	1.25
CD	0.00000056	1	[b]
CE	0.00006006	1	[b]
DE	0.00000756	1	[b]
Error	0.00034381	4	

[a] The determination of the expected values of the mean squares under a fractional design is beyond the scope of this chapter. The usual test conducted for each effect in fractional designs is to compare the mean square for each effect to the error mean square. This provides a practical solution to the problem; however, in so doing one should realize that the apparent significance of certain main effects or low-order interactions by this procedure may be attributable to significant higher-order interactions involving the same factors.

[b] The ratios are less than 1 and hence are obviously not significant. These components could be pooled with the error mean square if so desired.

D. Inferences

From Table 106.XX we note that

$$F_{1,4,.10} = 4.54, \quad F_{1,4,.05} = 7.71, \quad F_{1,4,.01} = 21.2$$

Hence we conclude that the only effect of significance is the B effect, that is, quantity of $0.5\underline{M}$ ammonium citrate used. We can observe from the data that, when a 10-ml quantity was used, as contrasted to 5 ml, the absorption on the average was lower. The test of significance indicates that this observed difference is real.

V. STEEPEST ASCENT

A. Description of Experiment and Resulting Data

In the preparation of examples of radioactively "hot" uranyl nitrate solution for chemical analysis, a high decontamination factor is desirable.

This factor is influenced by, among other things, the aluminum and the tetrapropylammonium nitrate concentration of the sample solution. An experiment was proposed to determine the concentration of these two constituents so as to result in a maximum decontamination factor. The method of steepest ascent was selected as the optimizing procedure. The initial design chosen was a 2^2 factorial, twice replicated. The particular experiments and results are given in the following table.

Decontamination Experiment

Run No.	Treatment	Al, M	TPAN, %	Decontamination factor
1	(1)	2.4	15	199
2	(1)	2.4	15	195
3	b	2.4	20	110
4	b	2.4	20	110
5	a	2.6	15	133
6	a	2.6	15	125
7	ab	2.6	20	66
8	ab	2.6	20	74

B. Computations

The determination of the slopes of the tangent plane is facilitated by the following table, similar in form to Table 106.X.

Factor Levels

Run no.	Treatment	Al, M x_1	TPAN, % x_2	Decontamination factor, y
1	(1)	−1	−1	199
2	(1)	−1	−1	195
3	b	−1	+1	110
4	b	−1	+1	110
5	a	+1	−1	133
6	a	+1	−1	125
7	ab	+1	+1	66
8	ab	+1	+1	74

$$b_1 = \frac{\Sigma y x_1}{8} = -27.0, \qquad b_2 = \frac{\Sigma y x_2}{8} = -36.5$$

C. Analysis

The analysis is best summarized in a table of the form of Table 106.XI.

Calculation of Path of Steepest Ascent
and Subsequent Trials on the Path

		Al, $\underline{M}$ x_1	TPAN, % x_2
1	Base level	2.5	17.5
2	Unit	0.1	2.5
3	Estimated slope b (change in decontamination factor per unit)	−27.0	−36.5
4	Unit × b	−2.7	−91.25
5	Change in level per −0.1$\underline{M}$ change in x_1	−0.1	−3.4
6	Possible trial on the path of steepest ascent	2.5	17.5
		2.4	14.1
		2.3	10.7
		2.2	7.3
		2.1	3.9
		2.0	0.5

Possible trials on the path of steepest ascent are given at the bottom of the table. One would conduct the experiments along this path, that is, at the factor–level combinations indicated, and compare the results obtained to previous data. Variations from the anticipated values should result in a recalculation of the path and subsequent experiments along the path, using all data accumulated to date.

VI. LATIN SQUARE

A. Description of Experiment and Resulting Data

In the recovery of uranium from spent reactor fuel it is necessary to effect a high level of decontamination on a sample of the dissolver effluvium before proceeding with the chemical analysis for uranium. Liquid–liquid solvent extraction is frequently used for this purpose as well as for divesting the sample from diverse ions. Factors believed to be of importance in this sample preparation step were (1) per cent

by volume of extractant (tributylphosphate), (2) time of contact, and
(3) acidity of the sample solution. Although interactions between factors
were not completely discounted, they were believed to be small compared
to the main effects. Consequently the experiment was designed as a
3 × 3 Latin square. The individual experiments and the results expressed
as percentage of uranium extracted were as given in the following table.

Per Cent Uranium Extracted[a]

Time of con-tact, min	Per cent TBP		
	10	20	40
5	2M 90%	8M 94%	4M 95%
10	8M 93%	4M 92%	2M 92%
20	4M 92%	2M 92%	8M 97%

[a] The levels of acidity of solution are given in the body of the table.

B. Computations

Make the following identification of letters to factors:
A: Per cent TBP, B: Time of contact, C: Acid molarity

In the manner of Section VI.F we compute:

$T_{1..} = 275$	$T_{.3.} = 281$	$S_a = 14.00$
$T_{2..} = 278$	$T_{..1} = 274$	$S_b = 2.67$
$T_{3..} = 284$	$T_{..2} = 279$	$S_c = 16.67$
$T_{.1.} = 279$	$T_{..3} = 284$	$S_e = 0.67$
$T_{.2.} = 277$	$T_{...} = 837$	$S = 34.00$

Computations by the Tukey method are as follows:

Array of Singly Adjusted Values

Average by runs

−1.67	1.33	0.33	0.00
1.33	−0.67	−2.67	−0.67
0.33	−0.67	2.33	0.67

Array of Doubly Adjusted Values

−1.67	1.33	0.33
2.00	0.00	−2.00
−0.34	−1.34	1.67

Arrangement of Doubly Adjusted Values
by Level of the Third Factor
Acidity

	2M	4M	8M
	−1.67	0.33	1.33
	−2.00	0.00	2.00
	−1.34	−0.34	1.67
Total	−5.01	−0.01	5.00
Range	0.66	0.67	0.67

Sum of ranges = 2.00

C. Analysis of Data

1. Corresponding to Table 106.XXIX, we have the following summary of the classical method of computation:

Analysis of Variance of Uranium Extraction Data

Source	Degrees of freedom	Sum of squares	Mean square	Expected mean square
Between levels of TBP per cent	2	14.00	7.00	$V(e) + 3V(a)$
Between contact times	2	2.67	1.33	$V(e) + 3V(b)$
Between acidities	2	16.67	8.33	$V(e) + 3V(c)$
Error	2	0.67	0.33	$V(e)$

We note from Table 106.XX the following:

$F_{2,2,.10} = 9.0, \quad F_{2,2,.05} = 19.0, \quad F_{2,2,.01} = 99.0$

2. By the Tukey method, with reference to Table 106.XXX, we obtain for $t = 3$ critical values of

| 3.78 | 2.76 | 1.26 | 0.92 |
| 8.70 | 6.37 | 2.90 | 2.12 |

which, when multiplied by 2, the sum of the doubly adjusted ranges, gives a new set of critical factors:

| 7.56 | 5.52 | 2.52 | 1.84 |
| 17.40 | 12.74 | 5.80 | 4.24 |

D. Inferences

1. BY ANALYSIS OF VARIANCE

By forming the ratio of the error mean square to the effect mean square and comparing with the tabulated values of the F distribution corresponding to the appropriate degrees of freedom, we observe that this ratio is 21.0 for the TBP factor, 4.0 for contact time, and 25.0 for acidity. We infer, therefore, that with regard to the levels of the factors used in this particular experiment, both the percentage of TBP and the acidity exercised a significant effect on the percentage of uranium extracted, whereas the contact time effect, if there was any, could not be detected.

2. BY THE TUKEY METHOD

Testing at the 5% significance level, that is, $\alpha = 0.05$, we note from the adjusted set of critical values, as computed in Section C, that a range of totals, found by arranging the doubly adjusted values by the level of the third factor, in excess of 7.56 should be considered as significant and any gap in excess of 5.52 between totals should be regarded as a real partitioning of the factor levels. The appropriate critical values to be applied to the column (per cent TBP) averages of the unadjusted values, as found in step *1* of the Tukey Method, and the row (time of contact) averages of the singly adjusted values, as found in step *2*, are 2.52 and 1.84.

The data, arranged in algebraic order, to which the critical values apply are as follows:

Acidity:	−5.01 (2M̲)	−0.01 (4M̲)	5.00 (8M̲)
Per cent TBP:	91.67 (10%)	92.67 (20%)	94.67 (40%)
Time of contact:	−0.67 (10 min)	0.00 (5 min)	0.67 (20 min)

Since the range of totals identified with the acidity factors is 10.01, which exceeds 7.56, we infer that the acidity exercised a significant effect. No difference (gap) between adjacent totals, when arranged algebraically, exceeds the critical value 5.52, and hence no partitioning into like groups is inferred. With regard to the per cent TBP, the range of the averages is 3.0, which exceeds the critical value of 2.52; hence we infer a TBP effect. We note that the gap between average extractions corresponding to 20% level and 40% level of TBP is 2.00. This exceeds the gap critical value of 1.84, and therefore we conclude that there is a real difference between the 40% TBP level and the other two levels used in the experiment. Similarly, when we apply the same criteria to the averages (as identified by "time of contact") of the singly adjusted values, we fail to detect any overall difference between the levels of this factor.

We note that the results of the analysis of this example by this method are in complete agreement with those obtained by the classical analysis of variance technique.

VII. GRAECO-LATIN SQUARE

A. Description of Experiment and Resulting Data

In order to compare the cyclical corrosion rates of four different materials proposed for fabrication of a spent reactor fuel dissolver vessel, an experiment was to be performed in which specimens of each material would be subjected to various decontamination media, applied in sequence in the manner anticipated for the vessel. In simulating the conditions under which the media effects would be exercised on the vessel it was necessary in one part of the cycle to bring the decontamination media to a predetermined high temperature level. It was proposed to do this by means of a four-position hot plate. Other media used in the cycle also required controlled temperatures, but since these were somewhat lower it was proposed that a controlled-temperature water bath be used in these cases. Small differences between hot-plate-position temperatures for a given setting were believed to exist, as well as thermal gradients through the water bath. In order to eliminate these effects from the comparisons to be made between cycles and between specimen types the experiment was designed as a 4×4 Graeco–Latin square.

Since the removal and preparation for measurement of a specimen after each cycle could have an effect on the subsequent corrosion rate, eight specimens of each type were initially prepared and placed in each of four different flasks. Two specimens were removed and measured (weighed) after each cycle; the measured specimens were not returned to the flask. These results were converted to inches per month of uniform penetration. The individual experiments and the corresponding results obtained are given in the following table,

	Corrosion Rates (inches per month)			
Material type	1	2	3	4
Cycle No. I	A	B	C	D
	P	Q	R	S
	0.0492	0.0355	0.0370	0.0291
Cycle No. II	B	A	D	C
	R	S	P	Q
	0.0450	0.0345	0.0261	0.0270
Cycle No. III	C	D	A	B
	S	R	Q	P
	0.0332	0.0301	0.0396	0.0331
Cycle No. IV	D	C	B	A
	Q	P	S	R
	0.0340	0.0329	0.0370	0.0357

where the hot plates are labeled

$$\begin{array}{cc} C & B \\ D & A \end{array}$$

and the water-bath positions are designated as

$$P \quad Q \quad R \quad S$$

B. Computations

We make the following identification of letters to factors:

A: Material type, B: Cycle, C: Hot plate, D: Water-bath position

In the manner of Section VI.G we compute

$T_{1...} = 0.1614$ $\quad T_{.3..} = 0.1360$ $\quad T_{..4.} = 0.1193$
$T_{2...} = 0.1330$ $\quad T_{.4..} = 0.1396$ $\quad T_{...1} = 0.1413$
$T_{3...} = 0.1397$ $\quad T_{..1.} = 0.1590$ $\quad T_{...2} = 0.1361$
$T_{4...} = 0.1249$ $\quad T_{..2.} = 0.1506$ $\quad T_{...3} = 0.1478$
$T_{.1..} = 0.1508$ $\quad T_{..3.} = 0.1301$ $\quad T_{...4} = 0.1338$
$T_{.2..} = 0.1326$

$$S_a = 0.0001837$$
$$S_b = 0.00004683$$
$$S_c = 0.00024990$$
$$S_d = 0.00002898$$
$$S = 0.00054602$$
$$S_e = 0.00003661$$

C. Analysis of Variance

In the form of Table 106.XXXI, the computations are summarized in an analysis of variance table as follows:

Analysis of Variance of Corrosion Data

Source of variation	Degrees of freedom	Sum of squares	Mean square	Expected mean square
Between material types	3	0.00018370	0.00006123	$V(e) + 4S^2(a)$
Between cycles	3	0.00004683	0.00001561	$V(e) + 4S^2(b)$
Between hot-plate positions	3	0.00024990	0.00008330	$V(e) + 4S^2(c)$
Between water-bath positions	3	0.00002898	0.00000966	$V(e) + 4S^2(d)$
Error	3	0.00003661	0.00001220	$V(e)$

D. Inferences

We note from Table 106.XX that

$$F_{3,3,.10} = 5.39, \quad F_{3,3,.05} = 9.28, \quad F_{3,3,.01} = 29.5$$

From the analysis of variance table we determine the ratios of the effect mean square to error mean square to be as follows:

Source of variation (effect)	F ratio
Between material types	5.02
Between cycles	1.28
Between hot-plate positions	6.83
Between water-bath positions	0.79

Comparison of these ratios with the critical values above indicates that the materials differ with respect to their abilities to resist the various decontaminating media when applied in cycles. Also we note the indicated significance (at the 0.10 level) of hot-plate position, justifying our concern with regard to this factor and the allowances made for it in the original design. Contrary to what we might expect, however, the corrosion rate did not significantly differ with time (cycles), at least with regard to the four cycles of this experiment.

VIII. SYMMETRICALLY BALANCED INCOMPLETE BLOCKS

A. Description of Experiment and Resulting Data

In the isotope dilution method of analysis for uranium concentration in uranyl nitrate solutions, a question was raised as to the effect, if any, of the ratio of sample volume to spike volume on the measured concentration by this method. A wide range of ratios was of interest. Furthermore, it was believed that periodic variations in instrument performance existed and must be taken into consideration in the experiment. Instrument stability could be assured at best only over the period required for three analyses. As a consequence an experiment was conducted involving spike to sample ratios, in seven different sets (blocks) of three determinations. The design corresponded to a symmetrically balanced incomplete block of seven blocks and seven treatments. The individual experiments and the corresponding results were as follows.

Concentration of Uranium by Isotopic Dilution (mg/ml)

Sample to spike volume ratio	Period of assay						
	1	2	3	4	5	6	7
4:1	1.98	2.01	1.93				
3:1			1.90	2.21	1.87		
2:1	1.93			2.14		2.03	
1:1			1.85			2.02	1.98
1:2	1.86				1.90		1.90
1:3		1.98			1.87	2.00	
1:4		1.96		2.23			1.99

Sample ratios were run in random order for any one period of assay.

B. Computations

In the notation of Sections VI.I and VI.J the following computations are made:

Index	T	Q	B	Q'
1	5.92	0.36	5.77	−0.37
2	5.98	0.04	5.95	−0.10
3	6.10	−0.10	5.68	−0.71
4	5.85	−0.05	6.58	1.48
5	5.66	−0.30	5.64	−0.57
6	5.85	−0.09	6.05	0.35
7	6.18	0.14	5.87	−0.08
Total	41.54	0.0	41.54	0.02

$$\Sigma X_{ij}^2 = 82.3966, \quad \Sigma Q_i^2 = 0.2614, \quad \Sigma B_j^2 = 247.1232, \quad \Sigma Q'^2 = 3.2952$$

$$N = 21 \qquad b = 7$$
$$t = 7 \qquad k = 3$$
$$r = 3 \qquad \lambda = \frac{3(2)}{6} = 1$$
$$kt\lambda = 21$$

C. Analysis of Variance

In the form of Table 106.XXXIV the above computations are summarized in an analysis of variance table for the purpose of effecting tests of significance.

Analysis of Variance of Test Data

Source of variation	Sum of squares	Degrees of freedom	Mean square
Between volume ratios adjusting for periods of assay	$\dfrac{0.2614}{21} = 0.012448$	6	0.002075
Between periods of assay adjusting for volume ratios	$\dfrac{3.2952}{21} = 0.156914$	6	0.026152
Residual error	$82.3966 - \dfrac{247.1232}{3} - \dfrac{.2614}{21}$ $= 0.009752$	8	0.001219

D. Inferences

We note from Table 106.XX the following:

$$F_{6,8,.10} = 2.67, \quad F_{6,8,.05} = 3.58, \quad F_{6,8,.01} = 6.37$$

We calculate from the analysis of variance table the following ratios to be associated with the two effects:

Source of variation (effect)	F ratio
Between volume ratios	1.70
Between periods of assay	21.45

From this table we infer that the sample to spike volume ratio over the range used in the experiment could not be detected as exercising any significant effect on the results. We note, however, that the blocking with regard to period of assay was of great importance, as shown by the significance of this effect. Thus through proper design we were able to remove from the residual error component a considerable amount of variation which otherwise would have been present.

Chapter 107

THE LITERATURE OF ANALYTICAL CHEMISTRY

By M. G. Mellon, *Department of Chemistry, Purdue University, Lafayette, Indiana*

Contents

I. INTRODUCTION

Although the Treatise of which this chapter is a part is a vast storehouse of information on analytical chemistry, it constitutes a mere sampling of the total amount now available. Furthermore, the annual accumulation of information is growing at an ever-increasing rate. The objective of this section is threefold: (*1*) to outline the nature of the different kinds of primary sources from which the treatise was compiled and to which one must turn for further information as progress continues; (*2*) to survey the secondary sources designed to facilitate searching for desired information; and (*3*) to consider briefly problems of using the literature.

In order to understand the nature of the literature of the field, it is necessary first to define what is meant by analytical chemistry and chemical analysis. The periodic table now lists 104 chemical elements. By general convention, the chemistry of element number 6, carbon, and of nearly all of its compounds comprises organic chemistry. The chemistry of the other 103 elements and of their compounds comprises inorganic chemistry. The organometallic and metal-organic compounds are hybrids of these two. Obviously, these definitions include all basic chemistry.

This chemistry consists of endless facts about the physical and chemical properties of elements, compounds, and other entities, such as radicals (free and otherwise), ions, and the so-called fundamental particles. Two examples from inorganic chemistry are sodium chloride, which is white, has a face-centered cubic crystal structure, and melts at 801°C, and silver chloride, which dissolves in aqueous ammonia, has a solubility in water of 1.5 mg/liter at 18°C, and melts at 451°C.

Interpretation of all such facts, and generalizations about them, constitute the role of physical chemistry. Thus it is basic to all other chemistry in providing, as far as presently feasible, answers to the questions of "how" and "why" in regard to the facts.

Where, then, does this leave what has long been known as analytical chemistry? Strictly speaking, there is no such subdivision, if one means thereby a kind of chemistry other than organic and inorganic. This view, of course, likewise excludes various other well-known areas of chemistry commonly cited, such as biological, ceramic, metallurgical, and pharmaceutical. What is really meant in common usage by all such terms is that, for some more or less specific purpose, one selects the inorganic and/or organic chemistry relevant to his purpose.

Chemical analysis, according to Robert Boyle, is the determination of the composition of substances. His definition, made some 300 years ago, implies nothing about how the information is determined, that is, what kind of measurement is used. Some like to extend this definition to include the determination of properties which are important in evaluating the performance of the substances, such as the crystal structure of a steel, or the color of a dye (2,6). This viewpoint is exemplified in the combined analytical and testing laboratories so common in industry.

Analytical chemistry may be considered, then, as a kind of applied chemistry. In determining many single components, such as iron in an ore, or polycomponents, like those in an alloy steel, one usually has to subject the sample to one or more chemical processes in order to render the desired constituent(s) measurable. Primarily these reactions are involved in preliminary treatment(s) and/or separations, and many varied examples are familiar to every analyst of practical experience.

In addition, much qualitative, and nearly all quantitative, chemical analysis includes physics. Every quantitative determination ends with measurement of some kind. Included are both the means for making the measurement (the instrument), and the operations of calibrating and using it. The most familiar example of an analytical instrument is the balance, weighing being the operation. There is no chemistry in any kind of measurement. This physical operation (2) consists in finding the number of times that a particular unit (gram, milliliter, volt) goes into the unknown. Very often some kind of calculation is involved.

II. LITERATURE

One may devise an analytical method merely to ascertain whether it is feasible. Thus a standard solution of hydrochloric acid may be prepared by generating and purifying hydrogen chloride gas and then absorbing it in a known mass of water. From the increase in mass, the concentration of the solution may be calculated. The method is fundamental, but, in actual practice, no one would prepare a titrant in this way. An appreciable percentage of current analytical literature may not prove much more useful.

In contrast, much of the annual production of the publications designated later as primary sources is devoted to improvements in old methods and equipment, to adapting them to new situations, or to proposing readily applicable new methods and/or equipment.

In all of this literature the author has long found it convenient and helpful to differentiate between what is essentially chemistry and what is physics. In general, chemistry is used here to designate the literature which relates to the processes used to accomplish various kinds of chemical transformations. There are many such processes, familiar examples being absorption, complexation, condensation, decomposition, dehydration, dissolution, electrodeposition, extraction, fusion, hydration, oxidation, precipitation, reduction, and volatilization. Along with all the chemical changes which take place, with the production of different substances, there is included the description of the physical properties, examples of which have been mentioned.

Physics is used here to designate the literature which relates to the instruments used in detecting the entities in qualitative chemical analysis, and to measuring their amounts in quantitative chemical analysis. Any method yielding information on chemical composition is a method of *chemical* analysis.

Very generally the instruments used by analysts have been designed and made by physicists and engineers. The analytical chemist is concerned chiefly with the testing, calibration, and use of the instruments for specific purposes. Thus his interests are highly practical, rather than theoretical. Consequently, for example, he must know the merits and limitations of sources, dispersing systems, and detectors in spectrophotometers, together with the merits of different designs of such instruments. He might well appreciate, but is not handicapped by not knowing, why irradiation by a given source produces such different effects in photoconductive, photoemissive, and photovoltaic cells.

Various processes are involved here, too, common terms in the literature being absorption, conduction, irradiation, radiation, reflection, and transmission. As chemists use these terms, chemical transformations are not involved in the processes, although such transformations may result from the processes.

III. PRIMARY LITERATURE SOURCES

In general, the primary sources of both the chemical and physical literature consist of the publications which are devoted essentially to the dissemination of new information. The usual kinds of these sources may be distinguished best by the nature of their contents, along with their method of publication, especially the regularity of appearance.

A. PERIODICALS

Publications appearing at regular intervals of time, varying from a week to a year, are known as periodicals. The word serial is also used for them. Probably periodicals are the most important primary sources, chiefly because of their large number and the consequent volume of their annual contributions. More than 10,000 such publications, appearing in 55 different languages, are abstracted by Chemical Abstracts Service.

Periodicals appear under many titles, such as *Annals, Archives, Bulletins, Communications, Journals, Proceedings, Transactions*, and their equivalents in other languages. Such a word is usually associated with the name of some scientific or technical society, of which the periodical is the official publication. An example is the *Journal of the American Chemical Society*. However, *Analytical Chemistry*, which is also an official publication of the same society, does not show the relationship by the title. Many journals are not sponsored by societies.

A large number of these periodicals contain methods of chemical analysis, or carry some analytically useful information, such as new compounds or reactions or new ideas in instrumentation. In fact, what is analytically useful or interesting is so widespread that about half of all abstractors of *Chemical Abstracts* prepare at least an occasional analytical abstract. In addition, many facts, such as instability constants in some nonanalytical paper, might well be of future analytical value.

Only the small number of periodicals generally recognized as analytical can be mentioned specifically here. Table 107.I contains probably the best-known examples. The bold-face portion of the title is the abbreviation used in *Chemical Abstracts* (and approved by the International Union of Pure and Applied Chemistry). The founding date is included.

A much larger number of other journals is cited, at least occasionally, in Sections **79** and **80** of *Chemical Abstracts*. Two uncommon examples are *Milchwissenschaft* ("Choosing a Gas Chromatograph") and *Science and Culture* ("Photometric Methods for the Determination of Manganese in Manganese Bronze").

Even the wide range of journals cited in Sections **79** and **80** (*C.A.*) accounts for little more than one-half of all analytical abstracts. Further reference is made to this problem in the discussion of *Chemical Abstracts*.

Anyone concerned with analytical methods in a specific area, particularly applied chemistry of various kinds, should watch the periodicals of that area. In practically all cases the analytical papers will be incidental contributions scattered among the more general articles.

The analytical periodicals cited, along with many others, may contain articles on physics, that is, instruments and measurement, as such. More probably, however, such information will appear in periodicals on physics and engineering. Some periodicals from which *Chemical Abstracts* abstracts articles on instruments or their testing include the following:

TABLE 107.I
Important Analytical Chemical Periodicals

American Journal of Clinical Pathology, 1931
American Journal of Medical Technology, 1934
Analyst, The, 1876
Analytical Biochemistry, 1960
Analytical Chemistry, 1929
Analytica Chimica Acta,[a] 1947
Analytical Letters, 1967
Annales des Falsifications et de l'Expertise Chimique, 1887
Applied Spectroscopy, 1946
Bunseki Kagaku (Japan Analyst), 1952
Chemia Analityczna, 1960
Chemical Instrumentation, 1968
Chimie Analytique, 1919
Chromatographia, 1968
Clinical Chemistry, 1955
Clinica Chimica Acta, 1956
Informacion de Quimica Analitica, 1947
Journal of the Association of Official Analytical Chemists, 1915
Journal of Chromatographic Science, 1963
Journal of Chromatography, 1958
Journal of Electroanalytical Chemistry and Interfacial Electrochemistry, 1959
Journal of the Polarographic Society, 1948
Journal of Radioanalytical Chemistry, 1968
Microchemical Journal, 1957
Mikrochimica Acta, 1937
Separation Science, 1966
Spectrochimica Acta, 1939
Spectroscopy Letters, 1968
Talanta, 1958
Zavodskaya Laboratoria, 1935
Zeitschrift für Analytische Chemie, 1862
Zhurnal Analiticheskoi Khimii (J. Anal. Chem. USSR), 1946
Zhurnal Prikladnoi Spektroskopii (J. Appl. Spectr.), 1965

---

[a] Official journal of the Analytical Section of the International Union of Pure and Applied Chemistry. The general official journal of this body is *Pure and Applied Chemistry*.

Instrument Engineering, Instrumentation Technology, Journal of Research of the National Bureau of Standards, Section C, *Journal of Scientific Instruments, Review of Scientific Instruments*, and *Chemical Instrumentation*.

Occasionally the papers presented at a symposium are published in a separate volume rather than in a periodical.

B. GOVERNMENTAL PUBLICATIONS

Many nonperiodical publications are issued by municipal, state, and federal organizations. Probably the contributions of the research laboratories are most important to analysts. Only those from federal laboratories in the United States are considered here.* Of the federal research laboratories it seems likely that the following are chemically most important: National Bureau of Standards, U.S. Bureau of Mines, U.S. Public Health Service (National Institutes of Health), U.S. Atomic Energy Commission, U.S. Geological Survey, and the Agricultural Research Service.

The work of the National Bureau of Standards is devoted chiefly to standardization activities. In the Chemistry Section this involves provision of analyzed samples, meticulous study of certain methods of analysis, and development and refinement of certain techniques, such as emission spectroscopy. In several physics sections standards and instruments are checked, and methods of testing and calibration recommended.

From other laboratories the chief contributions are carefully checked methods of analysis, the kind being dependent on the nature of the materials of interest. For example, the U.S. Bureau of Mines has long been concerned with coal, coke, natural gas, and petroleum, and the Agricultural Research Service with a variety of standards related to agriculture and agricultural products.

Some of the publications of all of these laboratories appear as journal articles. Many of them, however, are issued in the form of bulletins, papers, or similarly named publications.

One can keep in touch with what is going on by examination of the annual reports of the directors of the bureaus, the list of publications of each bureau, and the price list of publications available from each one.† Since 1895 all such publications have been reported in the *Monthly Catalogue of United States Public Documents*. Other listings preceded this one.

For some years the National Technical Information Service has been issuing *U.S. Government Research and Development Reports,* with a *Fast Announcement Service* for selected documents.

* There seems to be no coordinated, general information on the governmental publications of most other countries. Those of Great Britain are described by Bottle (1), Chapter 15. The first issue of each volume of *Chemical Abstracts* lists other sources of government reports for some twenty countries.

† The following price lists have probably the greatest analytical interest: **15**, Geology; **46**, Soils and Fertilizers; **51**, Health and Hygiene; **58**, Mines; **62**, Commerce; **64**, Scientific Tests, Standards; **84**, Atomic Energy.

C. PATENTS

Until recently patents have not been a very important literature source for analytical chemists. From current trends in *Chemical Abstracts*, however, it appears that this type of publication is increasing in importance.

To appreciate the possibilities, one must keep in mind the kinds of patents of possible interest. In the words of the United States Patent Office, these are (*1*) a machine, (*2*) a process, and (*3*) a composition of matter. Analytical examples illustrating these three types are, respectively, a balance, a titration process, and a complexant. In every case, of course, the item must be new, useful, and adequately described in the patent to enable one skilled in the art to make and/or use it.

As more than 3,500,000 United States patents have been issued since 1836, and as 41,000 patents of possible chemical interest were abstracted in 1968 in *Chemical Abstracts*, the total patent literature, including that of all countries granting patents, is enormous. No estimate is available of the number having possible analytical interest. However, the alert analyst can not afford to ignore patents today.

The *Official Gazette* of the U.S. Patent Office,* appearing each Tuesday, lists all patents granted the previous week, together with one claim for each patent. The indexes include names of patentees and subjects of inventions, the latter arranged by class and subclass. *Chemical Abstracts* has a patent number index, classified by countries, for all patents abstracted.

The "Manual of Classification" of the U.S. Patent Office and the "United States Patent Office Index to Classification" merit study by the searcher to locate classes containing the kinds of items which concern him. It is then feasible to obtain copies of all patents in such classes issued in the United States.

Beginning in 1961, an annual "Uniterm Index to Chemical Patents" has been issued by Information for Industry, Inc. Part I contains a key-word index to United States patents, and Part II consists of copies of the entries in the *Official Gazette* for the patents indexed.

Because the description and claims contained in a patent are usually highly detailed, an abstract is rarely adequate to enable one to make or use the invention. The patent itself must be consulted for details. Copies of United States patents may be obtained from the Commissioner of Patents, Patent Office, Washington, D.C., for 50 cents each. The first

---

* The patent offices of a number of other countries issue similar publications for their patents.

issue of each volume of *Chemical Abstracts* contains directions for obtaining copies of patents issued in more than 20 other countries.

D. MISCELLANEOUS PUBLICATIONS

There are at least two kinds of primary sources which do not belong in the classes already considered. Often they are very important in specific cases.

1. Dissertations

Theses contain complete reports of the research work done for advanced academic degrees. By far the most important are doctoral dissertations. Few of them are published as such and in full; usually they appear in abbreviated form as journal articles. With increasing editorial insistence on brevity, many details frequently remain unpublished; to see them one must consult the dissertation.

In case no article is published, a general idea of the nature of the material in dissertations submitted in the United States may be found in *Dissertation Abstracts,* which began publication in 1938. Usually this summary will indicate whether the original should be consulted.

2. Manufacturer's Technical Publications

Many pamphlets, descriptive circulars, and serials (house organs) of interest to analysts are issued by various manufacturers. Occasionally these publications deal with chemicals, but most often they concern equipment, particularly electrical, optical, and other types of instruments. As a current example, many are descriptions of apparatus for making separations by gas chromatography. For some years many have dealt with various kinds of electrical and optical devices applicable to analytical problems.

Such publications vary greatly in quality and detail. Some are designed chiefly for sales promotion, but many are sound scientific presentations. Many contain instructions for testing and using specific instruments. Often no other published information is available about the items concerned.

To keep in touch with these possibilities one should watch current advertising in periodicals. Also it is usually possible to keep one's name on the mailing lists of manufacturers and/or supply houses for specific kinds of such publications. Libraries are unlikely to have files of these pamphlets.

IV. SECONDARY SOURCES

Although the primary sources contain almost all of the new information published, it has become practically impossible, through them alone, to learn what is already known about innumerable subjects or to follow current developments in the field. For these purposes one is largely dependent on sources which collect, organize, and evaluate the facts from the primary sources.

These organizing sources are secondary only in the sense that they usually do not contain new information. In fact, several kinds are so important that they are indispensable to one engaging in research or trying to keep up with the passing chemical show.

The general nature of the principal secondary sources is described in the following sections. For the most part these sources, like the majority of primary sources, are not limited to analytical chemistry.

A. PERIODICALS

In sequence of publication, secondary periodicals precede other kinds of collecting and organizing publications. The three distinct types are considered in the order in which they deal with the information first appearing in the four primary sources.

1. Index Serials

A few periodicals merely list alphabetically by title the separate items, such as articles and patents. Probably the oldest one of possible current analytical interest is biological in nature, *Index Medicus* (1879+). In the engineering area, one nearly as old is *Engineering Index* (1884+), which now is an abstracting journal. *Industrial Arts Index* (1903–1957) is less likely to be of value; in 1958 it was divided into *Applied Science and Technology Index* and *Business Periodicals Index*. *Agricultural Index* (1916+) is largely nonanalytical. *British Technology Index* (1962+) is the newest periodical of this type.

Indexing publications for chemistry of more recent date, and no doubt of more analytical value, include the biweekly American *Chemical Titles* (1960+), which covers some 650 journals. The weekly *Current Contents* (1958+) lists the tables of contents of over 700 primary science journals; in 1967 a *Chemical Sciences* section was started to cover some 170 journals.

The major difficulty with indexing serials is the inadequate titles so often used by authors. Thus a new analytical method, published under the unrevealing title "Metallurgical Analysis," actually described a simultaneous spectrophotometric method for the determination of chromium and manganese in steel. This same brief title was used also for a book devoted to methods for the absorptimetric determination of some half-dozen minor components in certain nonferrous alloys. A different example of an unrevealing title is "B5C-Noal Direct Reading Device of High Quality and Moderate Price." Such deficiencies could be eliminated by making the titles one-sentence abstracts.

2. Abstracting Journals

Ideally the function of abstracting periodicals is to prepare, classify, and publish abstracts of all new material of chemical interest appearing in primary sources. Such publications are among the most nearly indispensable ones that we possess.

Probably the best-known example, *Chemical Abstracts*, aims to cover all that is new in all areas of chemistry. Beginning in 1907, it had expanded its coverage to over 10,000 periodicals by 1970. The abstracts are now arranged in 80 sections, Section 79 being Inorganic Analytical Chemistry and Section 80 being Organic Analytical Chemistry.

There is some organization within the sections, the order being (*1*) reviews, (*2*) general qualitative methods, (*3*) general separations and determinations, and (*4*) alphabetical arrangement for specific constituents determined. This has not been satisfactory for all purposes, but no generally acceptable scheme seems possible for all users. Notices of new books and of patents appear at the ends of the two sections.

Most disturbing to analysts is the failure to include in Sections 79 and 80 all abstracts of analytical interest. Foremost among the missing are many methods of analysis reported in various sections dealing with applied chemistry of different kinds. Examples are electrochemistry, biochemical methods, foods, pharmaceuticals, water, air pollution, and toxicology. However, many cross references, at the ends of the two sections, refer to other sections. Since 1967, general cross references have been carried to Section 6, Biochemical Methods; Section 17, Foods; Section 74, Pharmaceutical Analysis; also, in Section 79, to Section 80, Organic Analytical Chemistry; and, in Section 80, to Section 79, Inorganic Analytical Chemistry.

To indicate the current scattering of analytical abstracts in *Chemical Abstracts* a count was made for the thirteen issues of Volume 69 (1968). The number of abstracts was the actual count of those in Sections 79

TABLE 107.II

Distribution of Analytical Abstracts by *C.A.* Section Numbers

Number of abstracts	Part of *C.A.*				
	I	II	III	IV	V
0	2,3,5,8, 9,10,11, 12,13,15	32			72
1–10	1,4,7, 14,16,18, 20	21,22,23, 24,25,26, 27,28,29, 30,31,33, 34	37,38,40, 41,42	48,49,50, 54,56,63	65,67,69, 70,74,75
11–25	19		35,39,43, 44,45,46	52,53,55, 57,58,62	71
26–100			36	47,51,59, 60	66,68,73, 76,77,78
101–1000	6,17			61,64	80
>1000					79

and 80, and those cited by the cross references therein to other sections, except Sections 6, 17, and 64. The numbers for these three represent the author's judgment of the abstracts of analytical interest. Tables 107.II–107.IV summarize the data.

Even the cross referencing has limitations beyond the scattering of abstracts. One is always at the mercy of the abstractors in the other areas who decide whether a given abstract has analytical interest and so should be cross referenced. Because of the wide range of modern analytical chemistry, these abstractors may easily fail to realize what is (or might be) analytically significant. Only a broadly trained and experienced analyst can sense all that might be of value.

All abstracts must be considered primarily as guides to complete publications. For any significant work reliance should not be placed entirely on an abstract if the original publication can be obtained. This, of course, may mean time and an expensive translation.

The noteworthy volume and cumulative indexes of *Chemical Abstracts*

TABLE 107.III

Distribution of Analytical Abstracts by Numbers of $C.A.$ Sections

Number of abstracts	Part of $C.A.$					Total abstracts
	I (Secs. 1–20)	II (Secs. 21–34)	III (Secs. 35–46)	IV (Secs. 47–64)	V (Secs. 65–80)	
0	10	1			1	0
1–10	7	13	5	6	6	147
11–25	1		6	6	1	242
26–100			1	4	6	508
101–1000	2			2	1	3110
>1000					1	2693
Total abstracts						6700

include authors, empirical formulas, patent numbers, rings, and subjects. Since 1963, a key-word subject index is included in each issue.

Two other current abstracting journals are of interest for checking. *Referativnyi Zhurnal, Khimiya* (1953+), is little used in America except by those with adequate knowledge of the Russian language. *Current Abstracts of Chemistry and Index Chemicus* (1960+), published weekly by the Institute for Scientific Information, covers over 200 periodicals. The abstracts deal with the synthesis, isolation, and identification of new compounds, and with new reactions and syntheses of old compounds

TABLE 107.IV

Relative Standing for Six Largest Sections

Section number	Name of section	Number of abstracts	Percentage of abstracts
79	Inorganic Analytical Chemistry	2693	40.19
6	Biochemical Methods	990	14.78
80	Organic Analytical Chemistry	782	11.68
17	Foods	770	11.47
64	Pharmaceutical Analysis	446	6.67
61	Water	121	1.81
Others		898	13.40
Total		6700	100.00

(chiefly organic). The index issues, *Index Chemicus*, are published monthly and cumulated semiannually and annually. They include journal, author, subject, and molecular formula indexes. The cumulated indexes include a rotaform index of molecular formulas.

For retrospective searches the famous *Chemisches Zentralblatt* (1830–1969) is still useful, as well as *British Abstracts* (1871–1953).

The British *Analytical Abstracts,* begun in 1954 but appearing as Part C of *British Abstracts* from 1944 to 1954, is the only such periodical devoted entirely to analytical chemistry. The abstracts are classified in the following sections: 1. General Analytical Chemistry; 2. Inorganic Chemistry; 3. Organic chemistry; 4. Biochemistry; 5. Pharmaceutical Chemistry; 6. Foods; 7. Agriculture; 8. Air, Water, and effluents; 9. Techniques; Apparatus. Separate publication of all analytical abstracts appearing in *Chemical Abstracts* would constitute a similar journal.

Electroanalytical Abstracts (1964+) covers the application of various electrical phenomena in analytical chemistry.

Many of the numerous abstracting services* are maintained in connection with the publications of societies devoted to specialized areas of chemistry. Many analytical abstracts for such areas may be found there. Examples are *Metals Abstracts* for metallurgy, *Physics Abstracts* for physics (especially instruments), and *Nuclear Science Abstracts.*

Also important is the semimonthly *U.S. Government Research Reports,* which covers research and development released by the Army, Navy, Air Force, Atomic Energy Commission, and certain other agencies of the federal government. Unless otherwise indicated, the reports are available from the National Technical Information Service. The subject divisions include chemistry, physics, and several related areas.

3. Review Serials

A review of the developments in a specific area during some given period, usually a year, is frequently of great value to a searcher. This is especially true when the reviewer has the knowledge and experience to evaluate the relative merits of the various contributions.

Such reviews appear as separate publications if the area covered is sufficiently extensive to warrant this form. Various titles are used, such

* See Report 102, "A Guide to the World's Abstracting and Indexing Services in Science and Technology," by the National Federation of Science Abstracting and Indexing Services, Washington, D.C., 1963. This compilation lists 1855 titles, originating in 40 countries.

as *Advances, Annual Reports, Progress, Reviews, Yearbook,* and the foreign equivalents of these words. Some of these publications, particularly several of the *Advances* and *Progress* annual series, do not give a balanced perspective of overall developments in the field. Instead, they are merely collected papers presented at a meeting, often a symposium on some specified topic. Usually they are limited to a specific area and are more or less related to each other. They may or may not be accounts of new developments. *Advances in Spectroscopy, Developments in Applied Spectroscopy, and Progress in Infrared Spectroscopy,* for example, include much material on analytical spectrometry.

Thus far, analytical chemistry has not been accorded an overall separate treatment of this kind. The general practice has been to include certain areas (e.g., some method of separation or measurement) in a general review serial, such as the *Annual Reports of the Chemical Society* (London). Hopefully, the serial, *Advances in Analytical Chemistry and Instrumentation* (1960+), will develop comprehensive coverage. The new *Critical Reviews in Analytical Chemistry* (1970+) is also a promising publication.

Perhaps the best-known current reviews appear as an extra part of the April issue of *Analytical Chemistry.* In even years they cover some 30 areas of so-called fundamental methods of separation and measurement, although they are not always thus entitled. In odd years various areas of applied analysis are covered. There is some duplication. The popularity of these reviews apparently indicates the utility of this kind of publication.

What amounts to a review serial is Glick's *Methods of Biochemical Analysis,* started in 1954. It covers recently developed or improved methods for biological chemistry.

B. REFERENCE WORKS*

Reference works contain, in some systematic form, more or less complete factual information on all kinds of chemical systems. As the last step in the organizing process, they are designed to provide efficient answers to a great variety of questions. Without them, operation today

---

* Most of the examples of reference works cited are in English. They are probably the most widely used compilations, but in many cases there is a paucity of information about competing works in other languages. The books cited in references 9, 10, and 12 list some works in Czechoslovakian, Polish, and German, respectively (as well as in some other languages). Certain works, such as the great treatises of Beilstein and Gmelin, are unchallenged, of course.

with only the primary and the periodical secondary sources would often be impossible.

With some exceptions, examples of which are noted later, reference works usually are very general in nature. Thus, for example, the analytically useful facts about the chemistry of heteropoly systems, which are inorganic, are given in works on inorganic chemistry. However, such systems have important reactions with certain kinds of organic compounds, and these reactions probably will be treated in organic works in the discussion of the appropriate organic compounds. Moreover, the applications of the heteropoly systems to specific analytical purposes, such as the determination of arsenic, germanium, phosphorus, and silicon, probably will appear in appropriate analytical works under discussions of methods for these elements; it is likely, however, that no specific mention will be made that heteropoly chemistry is involved.

As chemical literature continues to expand, there is some trend toward isolating and compiling from the general works all that is of value for specific areas. No doubt this trend will continue. A specific example of a desirable compilation would be one based on the role of heteropoly systems in analytical chemistry. It would cover the period back to 1826, when Berzelius described ammonium molybdophosphate.

One should note the date of publication of all such works. They are supposed to be up to date nearly to the time of publication, but often there is a large gap between their date of issue and the present.

In the following sections several types of reference works and the kinds of information contained therein are considered briefly.

1. Formula Indexes

Lists of chemical compounds, arranged systematically by empirical formulas, are known as formula indexes. Their primary use is to check quickly whether information is available on a compound containing given elements in given proportions. For example, is perfluorodimethyl glyoxime known? As its elemental composition is $C_4H_2O_2N_2F_6$, one would look under this entry for the answer. It must be kept in mind, however, that there might be isomeric or other related compounds having the same empirical formula.

Several different indexing arrangements have been used for inorganic and organic compounds. The best known today is the combined inorganic–organic system used by *Chemical Abstracts* and adopted for the second supplement (1930) of the Beilstein treatise (see Section IV.B.5).

There are, of course, no analytical formula indexes. This might be expected from the earlier statement that fundamentally chemistry is inorganic and organic.

An analyst looking for fluorine-containing organic reactants would like to have the formulas arranged in this case to have the symbol F come first. To meet this need *Chemical Abstracts* now includes a Hetero-Atom-in-Context index. Details on its construction and use may be found in the Introduction to the index for Volume 66.

When searching for organic compounds having ring structures, *The Ring Index* by A. M. Patterson, L. T. Capell, and D. F. Walker is very useful. *Chemical Abstracts* uses this system (see introductions to the annual and cumulative indexes).

2. Formularies

Formularies are compilations of recipes for making and doing a thousand and one things. A composition for etching glass, an electroplating bath, a fading ink, and a color-changing temperature indicator are examples.

Such works cover a very broad range of chemical arts. Included are various items of interest in applied analysis, but there are no formularies limited to analytical chemistry.

A number of single-volume formularies are published. A comprehensive one of twelve volumes is *The Chemical Formulary*, edited by H. Bennett, Chemical Publishing Co., New York.

3. Dictionaries and Encyclopedias

In general, a dictionary is an alphabetical list of *words*, with their definitions. In contrast, an encyclopedia is an alphabetical compilation of *subjects*, usually with quite brief discussions of each subject. As some chemical writers have not adhered to this distinction, titles do not always indicate the difference between these two kinds of publications.

Practically all of them are general in coverage, but they include many analytical terms and subjects. In fact, many analysts might be surprised to find a fairly extensive discussion of manometry in the *Encyclopaedia Britannica*, which is not even a chemical work. The capsule coverage of a topic in an encyclopedia may well serve to orient a searcher before he begins a more detailed investigation.

Examples of chemical dictionaries are the following:

Bennett, H., ed., *Concise Chemical and Technical Dictionary*, Chemical Publishing Co., New York.

Grant, J., ed., *Hackh's Chemical Dictionary*, McGraw–Hill, New York.
Michels, W. C., ed., *International Dictionary of Physics and Electronics*, Van Nostrand, Princeton, N.J.
. . . . *Van Nostrand's Chemists' Dictionary*, Van Nostrand, Princeton, N.J.

Encyclopedias are more numerous and varied, as the following examples show:

Clark, G. L., ed., *Encyclopedia of Microscopy*, Reinhold Publishing Corp., New York.
Clark, G. L., ed., *Encyclopedia of Spectroscopy*, Reinhold Publishing Corp., New York.
Clark, G. L., and G. G. Hawley, eds., *Encyclopedia of Chemistry*, Reinhold Publishing Corp., New York.
Crouse, W. H., et al., eds., *Encyclopedia of Science and Technology*, 19 vols., McGraw–Hill, New York.
Foerst, W., ed., *Ullmann's Enzyklopaedie der technischen Chemie*, 13 vols., Urban Schwarzenberg, Berlin.
Heilbron, I., et al., eds. *Thorpe's Dictionary of Applied Chemistry*, 12 vols., Longmans, Green, New York.
Hey, D. H., ed., *Kingzett's Chemical Encyclopedia*, Van Nostrand, Princeton, N.J.
Miall, L. M., ed., *A New Dictionary of Chemistry*, Wiley, New York.
Römpp, H., *Chemie-Lexikon*, Franck'sche Verlagshandlung, Stuttgart.
Rose, A., and E. Rose, eds., *Condensed Chemical Dictionary*, Reinhold Publishing Corp., New York.
Standen, A., ed., *Kirk–Othmer Encyclopedia of Chemical Technology*, 21 vols. Interscience–Wiley New York.
Stecher, P. G., ed. *The Merck Index of Chemicals and Drugs*, Merck and Co., Rahway, N.J.
Thewlis, J., ed., *Encyclopedic Dictionary of Physics* (7 vols.), Pergamon, New York.

A new analytical encyclopedia, projected as a multivolume set, is the *Encyclopedia of Industrial Chemical Analysis*, edited by F. D. Snell and C. L. Hilton, Interscience–Wiley, New York.

4. Tabular Compilations

The table of atomic weights (masses) is the best example of a chemical tabular compilation. Vast numbers of a considerable variety of other constants have been collected and published for ready reference. As illustrated by atomic masses, most of these constants do not belong to any particular area or kind of chemistry. They are simply available for whatever interest they may have or for whatever use may be made of them.

Depending on the coverage and the nature of the data included, distinction may be made between comprehensive and selective compilations.

Ideally, the comprehensive work would cover all of the reliable numerical data for all kinds of chemical systems. The best-known example of this type in English is *International Critical Tables*. Completed in 1930 it is now seriously out of date.

The famous German set, *Physikalisch–chemischen Tabellen*, was originated by H. H. Landolt and R. Börnstein and first appeared in 1923. By 1936 three supplements had been issued. The great new edition, edited by A. Eucken et al., bears the title *Landolt–Börnstein's Zahlenwerke und Funktionen aus Physik, Chemie, Astronomie, Geophysik, und Technik*, indicating the broad coverage of physical constants. Twenty-four parts of the projected four volumes, begun in 1950, had appeared by the end of 1970. Several volumes of a completely new series have already been issued.

Analytical chemists use such comprehensive publications chiefly in research. One finds in them not only the constant of interest but also information as to who determined it and where the original publication appeared. Thus the value and the method of determination may be checked.

The shorter, selective compilations take one of two forms. Some are merely condensations of comprehensive sets, the data included being those thought likely to be of general interest to most users. Original references usually are not given. The following examples illustrate such works:

Condon, E. U., and H. Odishaw, eds., *Handbook of Physics,* McGraw–Hill, New York.
Feather, N., et al., *Kaye and Laby's Tables of Physical and Chemical Constants,* Longmans, Green, New York.
Gray, D. E., ed., *American Institute of Physics Handbook,* McGraw–Hill, New York.
Lange, N. A., *Handbook of Chemistry,* McGraw–Hill, New York.
Meites, L., ed., *Handbook of Analytical Chemistry,* McGraw–Hill, New York.
Rauscher, K., et al., *Chemische Tabellen und Rechentaffeln für die analytische Praxis,* VEB Deutscher Verlag für Grundstoffindustrie, Leipzig.
Weast, R. C., ed., *Handbook of Chemistry and Physics,* Chemical Rubber Co., Cleveland, Ohio.
Zyka, J., *Analyticka Prirucka,* SNTL, Prague.

The books by Meites and by Zyka are the first such works compiled particularly for analytical chemists. They differ from the other common, general works largely in containing, in addition to tabular data, much descriptive information. In this respect they resemble the handbooks in physics (Condon and Odishaw), biochemistry (Long), and chemical engineering (Perry, Chilton, and Kirkpatrick).

In contrast to the general, small works, some selective works are limited to a particular kind of data, the selection being made for certain kinds of users. The following compilations are examples:

Albert, A., and E. P. Serjeant, *Ionization Constants of Acids and Bases,* Wiley, New York.

Harrison, G. R., *Wavelength Tables,* Wiley, New York.

Koch, R. C., ed., *Activation Analysis Handbook,* Academic Press, New York.

Linke, W. F., ed., *Solubilities,* 3 vols., Van Nostrand, Princeton, N.J.

Long, C., ed., *Biochemists' Handbook,* Van Nostrand, Princeton, N.J.

Lyman, T., ed., *Metals Handbook,* American Society for Metals, Metals Park, Ohio.

Sadtler Standard Spectra, Sadtler Research Laboratories, Philadelphia, Pa.

Spencer, G. L., and G. P. Meade, *Cane Sugar Handbook,* Chapman and Hall, London.

Syzmanski, H. A., *Infrared Band Handbook,* Plenum Press, New York.

5. Treatises

As the word is used here, a treatise is a comprehensive presentation of the chemistry and/or physics of a broad area, such as inorganic or organic chemistry or physics. For the two chemical areas mentioned, the dominant treatises are, respectively, *Gmelin's Handbuch der anorganischen Chemie,* edited now by R. J. Meyer, and *Beilstein's Handbuch der organischen Chemie,* edited most recently by F. Richter. The great *Handbuch der Physik,* edited by Flügge, occupies a similar position in physics.

In the Beilstein treatise, for example, one may find the essential chemistry of every known organic compound, as far as the set has been completed. Thus, for a given compound, such as 2,2'-bipyridine, one would look here for a summary of the analytically useful properties, or for possibilities of determining it through some kind of chemical reaction. The greatest deficiency of all such sets is their time lag. International cooperation, as yet unachieved, is needed to bring them up to date.

Ideally, a treatise for analytical chemistry would include discussions of the chemical processes and physical operations involved in the detection, separation, and determination of any chemical entity in any and every kind of substance, in any determinable amount. Very often, although not always, in industry this means (*1*) selection, measurement, and preliminary treatment of the sample; (*2*) separation of one or more interfering or desired constituents; and (*3*) measurement of the desired constituent(s). These three steps would cover samples on a macro, micro, and ultramicro scale, and substances in the gas, liquid, and solid states.

Compilation of such a work would be a tremendous undertaking. The analytically applicable chemistry for inorganic elements, radicals, isotopes, and compounds is enormous; but much greater is this kind of chemistry for the millions of known organic compounds, including all of the isomers, polymers, and other forms. There is also the vast number of naturally occurring materials. To all of this must be added the physics of all of the analytically usable means of measurement for these systems. In the last direction, for example, a quarter-century ago J. A. Van den Akker outlined 72 different kinds of spectrophotometers, based on the possible geometry, spectrometry, and photometry of this one kind of instrument.

Of the few attempted comprehensive analytical treatises, perhaps the best early one, as far as it was published, was A. Rüdisüle's *Nachweis, Bestimmung, und Trennung der chemischen Elemente*, M. Drechsel, Bern. The volume for sulfur totaled over 700 pages. The set was limited to the elements and some of their compounds, and contained nothing on any of the kinds of measurement or on their applications.

The treatise by W. Fresenius and G. Jander, *Handbuch der analytischen Chemie*, Springer, Berlin, appeared much later and presumably is nearing completion. It covers the elements according to families of the periodic table, with separate treatment of detection, separation, and determination. The contents of the volumes are shown herewith, the incomplete parts being indicated by italicized symbols.

Part I. General Methods

  I. Not announced
 II. Not announced

Part II. Qualitative Analysis

   I. H, Li, Na, K, NH₄, Rb, Cs, Cu, Ag, Au (1944–55)
  II. Be, Mg, Ca, Sr, Ba, Zn, Cd, Hg (1955)
 III. B, Al, Ga, In, Tl, Sc, Y, rare earths, As (1944)
 IV. C, Si, Ge, Sn, *Pb*, Ti, Zr, Hf, Th (1956–   )
  V. N, P, As, Sb, Bi, V, Nb, Ta, Pa (1956)
 VI. O, S, Se, Te, Cr, Mo, W, U (1948)
 VII. F, Cl, Br, I, Mn, Tc, Re (1953)
VIII. Fe, Co, Ni, Pt, Pd, Ir, Rh, Ru, Os (1951–56)
 IX. Preliminary tests and separations of cations and anions (1956)

Part III. Quantitative Methods of Separation and Determination

  I. Li, Na, K, NH₄, Rb, Cs, *Cu*, Ag, Au (1940–   )
 II. Be, Mg, Ca, Sr, Ba, Ra, Zn, Cd, Hg (1940–45)
 III. B, Al, Ga, In, Tl, Sc, Y, rare earths, Ac (1942–56)
 IV. C, Si, Ge, Sn, Pb, Ti, Zr, Hf, Th (1950–67)
 V. N, P, As, Sb, Bi, V, Nb, Ta, Pa (1951–57)

VI. O, O₃, H₂O₂, *S, Se, Te, Po*, Cr, *Mo*, W, U (1953–   )
VII. H, H₂O, F, Cl, Br, I, *Mn, Tc, Re* (1950–   )
VIII. He, Ne, Ar, Kr, Xe, Rn, *Fe, Co, Ni*, Pt, Pd, Rh, Ir, Os, Ru (1949–   )

The Treatise of which this chapter is a part is the latest comprehensive effort. Comparison of its table of contents with those of the two earlier general sets shows their similarities and differences. The present set is by far the most ambitious in its coverage.

Two shorter treatises are important in their designated special areas. One is the five-volume (in seven parts) *Handbuch der Physiologisch– und Pathologisch–chemischen Analyse*, edited by K. Lang and E. Lehnartz, Springer–Verlag, Berlin, 1953–60. The other is the four-volume (in five parts) *Handbuch der mikorchemischen Methoden*, edited by F. Hecht and M. K. Zacherl, Springer–Verlag, Vienna, 1954–61.

A number of other treatises, differing in viewpoint, coverage, and merit, deserve mention. In general, the presentations emphasize kinds of measurement or the analysis of materials. A few carry titles which do not accurately reflect the coverage. Examples of such sets follow.

Berl, W. G., ed., *Physical Methods in Chemical Analysis,* 4 vols., Academic Press, New York.
Erdey, L., *Theorie und Praxis der gravimetrischen Analyse,* 3 vols., Akadémiai Kiadó, Budapest.
Furman, N. H., and F. J. Welcher, eds., *Standard Methods of Chemical Analysis,* 3 vols., Van Nostrand, Princeton, N.J.
Kline, G. M., ed., *Analytical Chemistry of Polymers,* 3 vols., Interscience, New York.
Kolthoff, I. M., and V. A. Stenger, *Volumetric Analysis,* 3 vols., Interscience, New York.
Snell, F. D., and C. T. Snell, *Colorimetric Methods of Analysis,* 8 vols., Van Nostrand, Princeton, N.J.
Strouts, C. R. N., H. N. Wilson, and R. T. Parry–Jones, *Chemical Analysis: The Working Tools,* 3 vols., Clarendon Press, Oxford, England.
Welcher, F. J., *Organic Analytical Reagents,* 5 vols., Van Nostrand, Princeton, N.J.
Wilson, C. L., and D. W. Wilson, *Comprehensive Analytical Chemistry,* 5 vols., American Elsevier, New York.
Zweig, G., ed., *Analytical Methods for Pesticides, Plant Growth Regulators, and Food Additives,* 4 vols., Academic Press, New York.

There are many equally valuable single-volume reference books in this area. An outstanding example is W. F. Hillebrand and G. E. F. Lundell's *Applied Inorganic Analysis*, as revised by H. A. Bright and J. I. Hoffman. When such books are rather specialized, they are hardly distinguishable from monographs (see later).

Closely related to treatises on the analysis of materials are the official methods of analysis of a number of scientific societies. Usually the compilations are single-volume works; but their great importance to practicing analysts, especially those in industry and in public health and welfare work, merits special consideration.

In order to have standards for the enactment and enforcement of various laws, and for the arbitration of disputes in industry, standardized specifications and methods of testing are needed. They deal with the testing and analysis of many classes of materials. To provide for new compositions in some kinds of materials, to change the methods by incorporating improvements, or to substitute new and better procedures, the official methods are under continual review by committees and their task forces assigned to specific problems. Usually revision occurs every five years, although the period may be shorter or longer. The following works are well known examples of this kind:

ASTM Standards, 33 vols., American Society for Testing and Materials, Philadelphia, Pa.

Farber, L., ed., *Standard Methods for the Examination of Water and Wastewater,* American Public Health Association, New York.

Horwitz, W., ed., *Official Methods of Analysis,* Association of Official Analytical Chemists, Washington, D.C.

MacMasters, M. M., et al., eds., *Cereal Laboratory Methods,* American Association of Cereal Chemists, St. Paul, Minn.

Pharmacopeia of the United States of America, American Pharmaceutical Association, Washington, D.C.

Sallee, E. M., et al., eds., *Official and Tentative Methods of the American Oil Chemists' Society,* American Oil Chemists' Society, Chicago.

Seligson, D., et al., eds., *Standard Methods of Clinical Chemistry,* Academic Press, New York.

Stenger, V. A., et al., eds., *Reagent Chemicals; American Chemical Society Specifications,* American Chemical Society, Washington, D.C.

An unofficial compilation, similar to that edited by V. A. Stenger, et al. is *Reagent Chemicals and Standards,* edited by J. Rosin, Van Nostrand, Princeton, N.J.

V. MONOGRAPHS AND TEXTBOOKS

Two other kinds of publications will be considered briefly. One is primarily of value for reference; the other is not.

Monographs are contemporary surveys (to date of publication) of present knowledge for specific, limited areas. Although resembling the discussions of subjects in encyclopedias, they are much more detailed.

When written competently, by one who knows the subject from personal experience, monographs are among the most valuable reference works. They enable one to obtain fairly quickly a perspective of the current status, to near the date of publication, of the field covered. The references usually included provide direction to the original literature sources, if further details are desired.

Probably the majority of the usual books (not bound periodicals) in a chemical library are monographs. Many examples are classified under analytical chemistry. Because the analytical research chemist or teacher is often concerned so broadly with inorganic and organic chemistry, many monographs in these areas may be valuable. A fine example of a monograph is *Determination of pH; Theory and Practice*, by R. G. Bates.

Some monographs are issued as a series under a general editorship. An important American series has been appearing for some years under the general title *Chemical Analysis*, issued by Interscience Publishers (John Wiley), New York. A wide variety of subjects is included in the more than 30 volumes published.

Under the editorship of A. P. Vinogradov, a new series is being published by the Vernadskii Institute of Geochemistry and Analytical Chemistry of the USSR Academy of Sciences. The English translations (Ann Arbor–Humphrey Science Publishers, Ann Arbor, Mich.) have the title *Analytical Chemistry of the Elements*. Thus far, separate volumes have appeared for the following elements; Be; B; Co; Ga; Mo; Ni; Nb and Ta; Pu; K; Pa; Ru; Te, Pm, At, and Fr; Tl; Th; U; Yt and rare earths; Zr and Hf. Except for the lack of some kind of classification, such as families of the periodic table, these volumes might well have been called a treatise. Some 50 volumes are projected. As stated in the Forword, "the monographs contain general information on the properties of the elements and their compounds, followed by a discussion of the chemical reactions which are the basis of the analytical work. The physical, physicochemical, and chemical methods for the quantitative determination of the element are given in the following order: first, the analysis of raw materials; next, the analysis of typical semimanufactured products; and last, that of finished products—metals or alloys, oxides, salts, and other compounds and materials."

Textbooks, as compared to monographs, are primarily manuals of instruction for classes, rather than reference works. Occasionally, however, the more comprehensive ones may be of some value for reference. Examples are Kolthoff, Meehan, Sandell, and Bruckenstein's *Textbook of Quantitative Chemical Analysis*, and Vogel's *Textbook of Quantitative Inorganic Analysis*.

In the present writer's opinion, many textbooks do not compare in reliability with the best monographs. The latter cover a very limited area, which, presumably, the author knows from personal experience and knowledge. In contrast, the textbook often presumes to cover a fairly broad area, such as general quantitative chemical analysis. No author these days can know from experience about all the methods of separation and measurement, applied to all of the kinds of substances which he selects to discuss. The result may be glaring errors of fact. For example, one does not dry, weigh, and dissolve a water sample for the titrimetric determination of the chloride ion therein, although this is implied in a well-known textbook. Also, the general treatment is often unbalanced.

VI. BIBLIOGRAPHIES

In the most limited sense, bibliographies are simply lists of references pertaining to specific subjects. They may refer to any kind of publication, but most often the citations are to periodicals, government documents, patents, and books. Perhaps the most familiar examples are the literature citations included in most articles describing the results of experimental researches.

Some bibliographies are sufficiently extensive and important that they are published separately in book form. A fine example, of some analytical interest, is Deitz's *Solid Adsorbents*, which covers the period 1900–1942. It contains over 6000 references, and a supplement for 1943–1953 adds 13,763 more.

A bibliography should be arranged in some systematic way for efficient use. In addition, each entry should be annotated to indicate the specific items of interest in the original publications cited. Unfortunately, many writers do not approach this ideal.

There are few analytical bibliographies in book form. Nearly all review articles carry extensive citations of the publications reviewed, but they seldom include annotations and generally omit titles. Of course, any list, competently compiled, saves at least some time if a search must be made.

VII. USING THE LITERATURE

Except as a record of facts, and of theories involving these facts, none of the kinds of publications discussed are of any value to the practicing analyst or analytical chemist unless he can find these facts in the various kinds of sources. This involves searching. Efficient operation means knowing where to turn, that is, in which kind(s) of source(s) he is most likely to find the information desired. Also involved is knowing how to use each source and how to evaluate what is found.

Space precludes discussing in any detail the methods of searching for the great variety of specific problems likely to arise in a modern analytical laboratory. Only very general suggestions seem appropriate. Many papers have been written on the art of searching, both for general information and for very specific items. Reference may be made to the present author's textbook (11), and to several others (1, 4, 9, 10, 12, 13, 14) for general suggestions. In addition, *Advances in Chemistry, Series 30*, entitled *Searching the Chemical Literature*, contains 31 papers on diverse aspects of searching (8).

The secondary sources are generally of importance first, either to provide specific facts at once, or to indicate where details may be found in the primary sources. Examples of what may be found in specific secondary sources follow.

Abstracts: brief summaries, chiefly of articles and patents in primary sources. They may provide desired information directly.

Bibliographies: lists of references on specific subjects.

Dictionaries: lists of chemical terms, with definitions, but thus far only a limited selection of names of compounds.

Encyclopedias: short, general discussions of subjects, such as separation by electrodeposition.

Formula and ring indexes: empirical formulas of known compounds; ring structures of organic compounds.

Formularies: directions for doing things, such as making a bath for electroplating silver.

Index serials: lists of titles of publications in primary sources.

Monographs: reasonably comprehensive discussions of contemporary knowledge in limited fields, such as the glass electrode.

Review serials: annual or biennial (usually) reviews of progress in specific areas.

Tabular compilations: collections of physical constants.

Treatises: more or less comprehensive, systematic collections of facts about elements and their compounds; about chemical transformations to make elements, radicals, compounds, and other entities detectable and/or measurable; about instruments and methods of using them to detect, separate, and determine these entities.

If one wants details of processes, operations, equipment, and other items, original sources must usually be consulted. Even an article based on a doctoral dissertation is not likely to include all experimental details. It is practically impossible to abstract from patents the (usually) numerous specifications and claims.

Efficient use of sources, primary and secondary, comes only from knowledge gained by extensive experience. Intelligent practice may not make perfect, but it is of enormous value in cultivating a sense of where to look for what and in developing a skill in using individual sources, such as the subject indexes of *Chemical Abstracts*. Of prime importance is a clear concept of what is wanted.

Listed below is a classification of the kinds of questions analytical chemists take to the library, together with representative examples of questions. Suggested sources are included only for the first question of each type.

I. Specific questions. Those in which the information desired relates to only one subject.

1. *Bibliography.* Partial or complete lists of references, with or without annotations.
   a. Use of the mercury cathode for quantitative electrodeposition. (See: ACS *Bibliography of Chemical Reviews;* monographs or treatises on analytical separations.)
   b. Early densimetric methods.
   c. Chemical resistance of platinum alloys.
2. *History and biography.* Events in the life of, and contributions by, an individual; developments and/or applications of a method or technique.
   a. Analytical contributions of Frenchmen before 1800. (See: histories of chemistry; periodical literature of the time.)
   b. Activation analysis.
   c. Organic applications of polarography.
3. *Source, occurrence, existence.* The availability of materials and/or instruments.
   a. Primary standards for titrimetry. (See: publications of national standardizing agencies; abstracting journals for late recommendations.)
   b. Pure solvents for absorptimetry.
   c. Quartz fiber balances.
4. *Composition.* Naturally occurring materials and artificial products; specifications and standards; formulas for workshop recipes.
   a. Specifications (ACS) for reagent-grade sodium carbonate. (See: V. A. Stenger, et al., *Reagent Chemicals; American Chemical Society Specifications.*)
   b. Formula for platinizing platinum electrodes.
   c. Composition of alloyed steels containing niobium.

5. *Methods of production, preparation, or manipulation.* Laboratory or industrial processes; details of procedures; materials required; equipment employed.
   a. Instructions for testing and operating a General Electric spectrophotometer. (See: manufacturer's technical publication accompanying the instrument.)
   b. Preparation of a calomel electrode.
   c. Construction of a working (calibration) curve in photographic emission spectrometry.
6. *Properties.* Physical and/or chemical; general and specific reactions.
   a. Instability constant of FeF_6^{3-} ion. (See: compilations of tabular data.)
   b. Complexing action of thioglycolic acid.
   c. Reactions for volatilizing chromium from steels.
7. *Uses and applications.* Laboratory and industrial; general and specific applications.
   a. Fluorine-containing organic reactants. (See: permuted formula index, if available; review articles; abstracting journals.)
   b. Substitutes for platinum ware in analytical work.
   c. Applications of gas chromatography in metallurgical analysis.
8. *Identification, testing, analysis.* Methods available; interpretation of results.
   a. Quality-control statistics and experimental design in analytical work. (See: books and articles dealing with chemical applications of statistics.)
   b. Use of absorption spectra in qualitative chemical analysis.
   c. Fluorescimetric detection of uranium.
9. *Patents.* Date of expiration; details of specifications; objects of this kind previously patented.
   a. Patented absorptimetric methods. (See: abstracting journals; recent review articles.)
   b. Specifications for U.S. Patent 3,078,647 by L. C. Mosier on continuous gas chromatography.
   c. Patents on analytical balances.
10. *Statistical data.* Production, consumption, price, suppliers, uses.
   a. Price of 2-propanol. (See: market reports in periodicals; catalogs of supply houses.)
   b. Source of 2,2'-bipyrimidine.
   c. Uses of 2,3-butanedionedioxime.

II. General questions. Those in which the information desired relates to more than one of the classes mentioned above. Two variations occur in the questions: (*1*) those indicating the particular classes involved, and (*2*) those not expressing or implying any limitations on the kind of information desired.

The examples cited show that such questions may be extremely variable. Many involve only a single fact which usually can be found easily by one familiar with the literature. Others, such as the one calling for a complete survey of all methods ever used for determining the platinum elements, are very comprehensive. Persistence would probably suffice in this case, even though the documented chemistry for these metals goes back at least to 1750.

In recent years there has been a trend toward distinguishing between what have come to be called current awareness and retrospective searches. To illustrate the viewpoints involved, and the adaptation of searching techniques thereto, two specific problems are suggested.

Current awareness searches are made in order to keep as nearly up to date as possible with the latest developments in specific directions. Atomic absorption spectrometry is an example of a relatively new technique of measurement, at least for analytical purposes. Examples of relevant questions are the following: What instruments are available? What do they cost? Are they patented? Which one is the latest? What are the new analytical applications? For information on these and related questions one may turn first to several kinds of primary sources, that is, periodicals, government documents, patents, and manufacturers' technical bulletins.

Next come the periodicals in the secondary sources. First of all, one should watch the chemical and physical periodicals most likely to carry articles on these instruments and their uses. They include the analytical examples already cited, along with such others as *Instrument Engineering* and the *Review of Scientific Instruments*. Next come index serials, such as *Chemical Titles*. A little later *Chemical Abstracts* provides abstracts of all of the primary publications. Each issue now contains the new key-word index.

Although *Chemical Abstracts* covers items from government documents, the information may be slow in arriving. The best way is to know the people who are working in this area, if any, and to keep in touch with them about the announcement of any new publications. Patents are abstracted, too, but again time is required for the process to operate. If one knows the class covering the instrument of interest, a standing order may be maintained for all such U.S. patents granted. They will then be sent as soon as printed.

Finally, one may watch advertisements in periodicals for late announcements about instruments. Keeping one's name on the mailing lists of companies making and/or dealing in the instruments should result in the receipt of pertinent bulletins.

Rather than being concerned with the latest information available, retrospective searches concern the past. The questions cover a vast range, as illustrated by these examples. Who reported the first analytically useful heteropoly compound and when? Which element was first determined by atomic absorption spectrometry? Who has determined the solubility of palladium dimethylglyoxime? Which derivatives of 1,10-phenanthroline have been found analytically useful? What analytical methods have been used to determine the platinum metals in their ores?

Space can be taken for only a brief suggestion of procedure for the last question. Before we start, however, the problem should be better defined. Must provision be made for all six metals in the same material or only for particular ones in certain ores? What is the nature of the ores—especially, what other elements may be present which would interfere in given methods? Will the platinum elements be present in major, minor, or trace amounts? Is the amount of sample material available limited?

As J. L. Howe's *Bibliography of the Metals of the Platinum Group* covers the period 1748–1896, advantage may be taken of it for the older literature. One must keep in mind, however, that the titles in it may not indicate the analytical content of the publications cited. The present writer prefers to begin with the latest publications cited in the annual indexes of *Chemical Abstracts* and work backwards, as someone may already have published an adequate bibliography. In working backward the cumulative indexes should be used as soon as one reaches them. Unless a satisfactory bibliography is found, the search is continued backward until all that is available is located. In our example this would include at least the following information: methods of obtaining the various ores (or any particular one) and of preparing them for analysis; methods of preliminary treatment and separation, to the extent that such are necessary for particular methods of measurement; and possible methods of measuring each of the six metals, singly or in any of the possible combinations.

Somewhere along the way, before going too far, it is desirable, in a search such as this one, to examine the great treatises, such as the present work, and those by Fresenius and Jander and by Rüdisüle. They contain recommended methods for the elements and bibliographies relating to them.

There is uncertainty when one should begin to read the various publications located to this point. Most of the authors probably cite previous references which they found. These citations provide a check on one's own searching.

Ultimately, searching will stop. The report may be simply a compilation, in suitable form, of what has been found; any use of the information is left to others. If a recommended specific procedure is desired, there is the problem of selecting what seems best for the situation at hand. Only the pragmatic test of workability remains.

An example of a different kind of comprehensive search is the compilation of a bibliography on the analytical uses of ethanol. It serves to emphasize the fact that any analytical applications of a compound depend on taking advantage of the chemical and physical properties of the system.

In the Beilstein classification ethanol is a Class 2, hydroxy compound, Subclass A, monohydroxy compound, Rubric I, monohydroxy alcohol $(C_nH_{2n+2}O)$, and Series B, a compound containing two carbon atoms. One may find the properties of the compound thus located, and note those which might find application in an analytical procedure. Being a long-known compound, ethanol can also be located by using the Beilstein formula indexes, now completed as far as 1930.

Unless someone has published a paper on the analytical uses of ethanol, searching subject indexes for such information is very unlikely to yield results. The subject is so specialized that any such uses will probably appear quite incidentally in a publication primarily on some other topic. The abstractor will not consider this information worth mentioning in his abstract, and, consequently no index entry will be made by the indexer. This is a fine example of hidden facts.

In the author's opinion, about the only practical method of searching, even though it is very laborious, is to compile a list of the properties of ethanol which seem likely to offer analytical utility, and then to search, book by book, or article by article, through analytical procedures which might conceivably apply such properties. A search made in this way revealed a surprising variety of such uses of ethanol (1).

A. AUTOMATION

The analyst should be alert to take advantage of any developments in automating the use of the literature of chemistry and physics which are relevant to his problems. Literature scientists and engineers have made encouraging progress in facilitating the recovery of recorded information, and further developments are certain. For the present, at least, the machines available seem satisfactory. The weakness is lack of adequate coding for them.

To illustrate some of the problems, consider a special steel containing Al, Co, Cr, Cu, Mo, Nb, Ni, Pb, Sn, V, and W, in addition to the usual elements: Fe, C, Mn, P, S, and Si. What methods are available for determining any one of these elements in the presence of the other sixteen? What methods can be used to determine all of the elements in this material? What is the applicability of particular techniques, such as mass spectrometry or X-ray fluorescimetry, to such a sample?

It should be obvious that, no matter how sophisticated the machine, its output of coded information on such a problem cannot be better than the input. If there has been no input, there will be no output. Consequently, the user should always be aware of what the machine can and cannot do.

One of the most recent publications in this area is *Selected Mechanized Scientific and Technical Systems,* Herner and Company, Washington, D.C. A product of the Committee on Scientific and Technical Information of the Federal Council for Science and Technology, it is a description of thirteen computer-based, operational systems designed primarily for the announcement, storage, retrieval, and secondary distribution of scientific and technical reports. Among others, the systems included are for the U.S. Department of Defense, the National Aeronautical and Space Agency, the National Library of Medicine, the U.S. Geological Survey, and the U.S. Patent Office.

B. NOMENCLATURE

Important in cultivating proficiency in searching analytical sources is an understanding of the difficulties arising with chemical nomenclature. Although the use of subject indexes is of chief concern, there are other items which involve analytical chemistry.

The problem has two aspects. The first, a rather general situation throughout chemistry, is encountered in at least the following items: names of chemical compounds, especially organic compounds; definitions, such as those for light, absorbance index, and volumetric; and ambiguous or unrevealing subjects and titles.

In regard to the naming of compounds, the practice of Chemical Abstracts Service is summarized in the introductory sections of the various subject indexes. Especially to be noted is the "Combined Introductions to Indexes to Volume 66." This treatise contains general suggestions for inorganic systems by W. C. Fernelius (Part II, Vol. I, pp. 1–33), and for organic systems by L. T. Capell and K. L. Loening (Part II, Vol. II, pp. 1–44).

The second aspect is primarily analytical in nature. It concerns unexplained abbreviations or trade names for reagents and equipment, and the names of analytical processes, methods, and instruments. With increasing frequency one encounters in analytical titles and abstracts unfamiliar trade names and abbreviations of long names, especially of organic compounds. Examples are Tofranil, Trilon B, Sephadex G-100, and Complexion III for reagents; EDTA, EGTA, PAN, and TDT for reactants; EDE-10, FN, and KU-2 for resins; and M5-8450 for a drug. Even if one had an up-to-date dictionary for all such trade names and acronyms, frequent use of it would become irksome.

Analytical chemists have shown little interest in adopting and using some kind of systematic nomenclature, especially for methods of separation and measurement, the two principal operations encountered in the analysis of many types of polycomponent materials. Free enterprise has been rampant, with the result that methods of determination have been designated in at least the following terms:

1. The originator(s) of the method
   a. The individual(s)—for example, Kjeldahl method
   b. The institution or laboratory—for example, Harvard, or Bureau of Mines method
2. The organization approving the method
   a. Name only—for example, USP (U.S. Pharmacopoeia) method
   b. Name and status of adoption—for example, ASTM (American Society for Testing and Materials) official or tentative method
   c. Name and applicability—for example, AOAC (Association of Official Analytical Chemists) general or specific method
   d. Name and number—for example, APHA (American Public Health Association) Method I
3. The kind of material analyzed—for example, steel-analysis method
4. The physical state of the sample material—for example, gas-analysis method
5. The scale of operations used—for example, micro or macro method
6. The kind of general physical method used—for example, electrical method
7. The effect or action of the desired constituent—for example, emanation method
8. The instrument used

a. In preliminary treatment of the sample—for example, Parr bomb method

b. In separation of the desired constituent—for example, vacuum-oven method

c. In measurement of the desired constituent—for example, refractometer method

Sometimes current usage is absurd, as in designating as "wet" the determination of carbon in an alloy steel by burning out the carbon as carbon dioxide at 1200°C. To the present author the term instrumental is meaningless, for the coordinate term is noninstrumental; as all measurement is instrumental, there are no noninstrumental quantitative methods.

All of these practices are important to the analytical chemist searching subject indexes, such as those of *Chemical Abstracts*.* Ordinarily the indexer must depend on the usage of authors and abstractors. Then, unless the index user can think of the entries employed by the indexer, the indexed item will not be found (3).

REFERENCES

1. Bottle, R. T., ed., *Use of the Chemical Literature,* Butterworths, London, 1962.
2. Chirnside, R. C., *Analyst,* **86,** 314 (1961).
3. Crane, E. J., *Ind. Eng. Chem.,* **14,** 901 (1922).
4. Crane, E. J., A. M. Patterson, and E. B. Marr, *A Guide to the Literature of Chemistry,* Wiley, New York, 1957.
5. Dyson, G. M., *A Short Guide to Chemical Literature,* Longmans, Green, London, 1958.
6. Elving, P. J., *Anal. Chem.,* **22,** 962 (1950).
7. Ferner, G. W., and M. G. Mellon, *J. Chem. Educ.,* **10,** 243 (1933).
8. Gould, R. F., ed., *Advances in Chemistry,* Ser. 30: *Searching the Chemical Literature,* 1961.
9. Hanč, O., B. Hlarica, V. Hummel, and J. Jelinek, *Chemická Literatura,* Statni Nakladatelstrí Technické Literatury, Praha, 1961.
10. Janiszewska-Drabarek, S., J. Surowinski, and A. Szuchnik, *Literatura Chemiczna,* Technika poslugiwania sie, Warsaw, 1960.
11. Mellon, M. G., *Chemical Publications,* McGraw-Hill, New York, 1965; also *Searching the Chemical Literature,* American Chemical Society, Washington, D.C., 1964.
12. Novak, A., *Fachliteratur des Chemikers,* VEB Deutscher Verlag der Wissenschaften, Berlin, 1962.
13. Reid, E. E., *Invitation to Research,* Franklin Publishing Co., Palisade, N.Y., 1961, pp. 225–329.
14. Whitford, R. H., *Physics Literature,* Scarecrow Press, Metuchen, N.J., 1968.

* For a resume of some of the difficulties encountered by the author, see *Advances in Chemistry, Series* 30, pp. 67–74 (1961).

SUBJECT INDEX

6529